TABLE 3 *(continued)*

z	.00	.01	.02	.03	.04	.05	.06	.07	.08	.09
0.0	.5000	.5040	.5080	.5120	.5160	.5199	.5239	.5279	.5319	.5359
0.1	.5398	.5438	.5478	.5517	.5557	.5596	.5636	.5675	.5714	.5753
0.2	.5793	.5832	.5871	.5910	.5948	.5987	.6026	.6064	.6103	.6141
0.3	.6179	.6217	.6255	.6293	.6331	.6368	.6406	.6443	.6480	.6517
0.4	.6554	.6591	.6628	.6664	.6700	.6736	.6772	.6808	.6844	.6879
0.5	.6915	.6950	.6985	.7019	.7054	.7088	.7123	.7157	.7190	.7224
0.6	.7257	.7291	.7324	.7357	.7389	.7422	.7454	.7486	.7517	.7549
0.7	.7580	.7611	.7642	.7673	.7704	.7734	.7764	.7794	.7823	.7852
0.8	.7881	.7910	.7939	.7967	.7995	.8023	.8051	.8078	.8106	.8133
0.9	.8159	.8186	.8212	.8238	.8264	.8289	.8315	.8340	.8365	.8389
1.0	.8413	.8438	.8461	.8485	.8508	.8531	.8554	.8577	.8599	.8621
1.1	.8643	.8665	.8686	.8708	.8729	.8749	.8770	.8790	.8810	.8830
1.2	.8849	.8869	.8888	.8907	.8925	.8944	.8962	.8980	.8997	.9015
1.3	.9032	.9049	.9066	.9082	.9099	.9115	.9131	.9147	.9162	.9177
1.4	.9192	.9207	.9222	.9236	.9251	.9265	.9279	.9292	.9306	.9319
1.5	.9332	.9345	.9357	.9370	.9382	.9394	.9406	.9418	.9429	.9441
1.6	.9452	.9463	.9474	.9484	.9495	.9505	.9515	.9525	.9535	.9545
1.7	.9554	.9564	.9573	.9582	.9591	.9599	.9608	.9616	.9625	.9633
1.8	.9641	.9649	.9656	.9664	.9671	.9678	.9686	.9693	.9699	.9706
1.9	.9713	.9719	.9726	.9732	.9738	.9744	.9750	.9756	.9761	.9767
2.0	.9772	.9778	.9783	.9788	.9793	.9798	.9803	.9808	.9812	.9817
2.1	.9821	.9826	.9830	.9834	.9838	.9842	.9846	.9850	.9854	.9857
2.2	.9861	.9864	.9868	.9871	.9875	.9878	.9881	.9884	.9887	.9890
2.3	.9893	.9896	.9898	.9901	.9904	.9906	.9909	.9911	.9913	.9916
2.4	.9918	.9920	.9922	.9925	.9927	.9929	.9931	.9932	.9934	.9936
2.5	.9938	.9940	.9941	.9943	.9945	.9946	.9948	.9949	.9951	.9952
2.6	.9953	.9955	.9956	.9957	.9959	.9960	.9961	.9962	.9963	.9964
2.7	.9965	.9966	.9967	.9968	.9969	.9970	.9971	.9972	.9973	.9974
2.8	.9974	.9975	.9976	.9977	.9977	.9978	.9979	.9979	.9980	.9981
2.9	.9981	.9982	.9982	.9983	.9984	.9984	.9985	.9985	.9986	.9986
3.0	.9987	.9987	.9987	.9988	.9988	.9989	.9989	.9989	.9990	.9990
3.1	.9990	.9991	.9991	.9991	.9992	.9992	.9992	.9992	.9993	.9993
3.2	.9993	.9993	.9994	.9994	.9994	.9994	.9994	.9995	.9995	.9995
3.3	.9995	.9995	.9995	.9996	.9996	.9996	.9996	.9996	.9996	.9997
3.4	.9997	.9997	.9997	.9997	.9997	.9997	.9997	.9997	.9997	.9998

Index of Applets on the CourseMate Web site

Introduction to Probability and Statistics

EDITION

William Mendenhall, III

Robert J. Beaver
University of California, Riverside, Emeritus

Barbara M. Beaver
University of California, Riverside, Emeritus

BROOKS/COLE
CENGAGE Learning™

Australia • Brazil • Japan • Korea • Mexico • Singapore • Spain • United Kingdom • United States

BROOKS/COLE
CENGAGE Learning™

Introduction to Probability and Statistics, Fourteenth Edition
Mendenhall/Beaver/Beaver

Editor in Chief: Michelle Julet

Publisher: Richard Stratton

Senior Sponsoring Editor: Molly Taylor

Assistant Editor: Shaylin Walsh

Editorial Assistant: Alexander Gontar

Associate Media Editor: Andrew Coppola

Marketing Director: Mandee Eckersley

Senior Marketing Manager: Barb Bartoszek

Marketing Coordinator: Michael Ledesma

Marketing Communications Manager:
Mary Anne Payumo

Content Project Manager: Jill Quinn

Art Director: Linda Helcher

Senior Manufacturing Print Buyer: Diane
Gibbons

Rights Acquisition Specialist: Shalice
Shah-Caldwell

Production Service: MPS Limited, a
Macmillan Company

Cover Designer: Rokusek Design

Cover Image: Vera Volkova/©
Shutterstock

Compositor: MPS Limited, a Macmillan
Company

For product information and technology assistance, contact us at
Cengage Learning Customer & Sales Support, 1-800-354-9706

For permission to use material from this text or product, submit all requests online at **www.cengage.com/permissions**. Further permissions questions can be emailed to **permissionrequest@cengage.com**.

Library of Congress Control Number: 2011933688

Student Edition
ISBN-13: 978-1-133-10375-2
ISBN-10: 1-133-10375-8

Brooks/Cole
20 Channel Center Street
Boston, MA 02210
USA

Cengage Learning is a leading provider of customized learning solutions with office locations around the globe, including Singapore, the United Kingdom, Australia, Mexico, Brazil and Japan. Locate your local office at
international.cengage.com/region

Cengage Learning products are represented in Canada by Nelson Education, Ltd.

For your course and learning solutions, visit
www.cengage.com.

Purchase any of our products at your local college store or at our preferred online store **www.cengagebrain.com**.
Instructors: Please visit **login.cengage.com** and log in to access instructor-specific resources.

Printed in United States of America
4 5 6 7 8 9 10   22 21 20 19 18

Preface

Every time you pick up a newspaper or a magazine, watch TV, or surf the Internet, you encounter statistics. Every time you fill out a questionnaire, register at an online website, or pass your grocery rewards card through an electronic scanner, your personal information becomes part of a database containing your personal statistical information. You cannot avoid the fact that in this information age, data collection and analysis are an integral part of our day-to-day activities. In order to be an educated consumer and citizen, you need to understand how statistics are used and misused in our daily lives.

THE SECRET TO OUR SUCCESS

The first college course in introductory statistics that we ever took used *Introduction to Probability and Statistics* by William Mendenhall. Since that time, this text—currently in the fourteenth edition—has helped several generations of students understand what statistics is all about and how it can be used as a tool in their particular area of application. The secret to the success of *Introduction to Probability and Statistics* is its ability to blend the old with the new. With each revision we try to build on the strong points of previous editions, while always looking for new ways to motivate, encourage, and interest students using new technological tools.

HALLMARK FEATURES OF THE FOURTEENTH EDITION

The fourteenth edition retains the traditional outline for the coverage of descriptive and inferential statistics. This revision maintains the straightforward presentation of the thirteenth edition. In this spirit, we have continued to simplify and clarify the language and to make the language and style more readable and "user friendly"—without sacrificing the statistical integrity of the presentation. Great effort has been taken to explain not only how to apply statistical procedures, but also to explain

- how to meaningfully describe real sets of data
- what the results of statistical tests mean in terms of their practical applications
- how to evaluate the validity of the assumptions behind statistical tests
- what to do when statistical assumptions have been violated

Exercises

In the tradition of all previous editions, the variety and number of real applications in the exercise sets is a major strength of this edition. We have revised the exercise sets to provide new and interesting real-world situations and real data sets, many of which are drawn from current periodicals and journals. The fourteenth edition contains over 1300 problems, many of which are new to this edition. A set of classic exercises compiled from previous editions is available on the website (http://www.cengage. com/statistics/mendenhall). Exercises are graduated in level of difficulty; some, involving only basic techniques, can be solved by almost all students, while others, involving practical applications and interpretation of results, will challenge students to use more sophisticated statistical reasoning and understanding.

Organization and Coverage

We believe that Chapters 1 through 10—with the possible exception of Chapter 3—should be covered in the order presented. The remaining chapters can be covered in any order. The analysis of variance chapter precedes the regression chapter, so that the instructor can present the analysis of variance as part of a regression analysis. Thus, the most effective presentation would order these three chapters as well.

Chapters 1–3 present descriptive data analysis for both one and two variables, using both *MINITAB* and Microsoft Excel® graphics. Chapter 4 includes a full presentation of probability and probability distributions. Three optional sections—Counting Rules, the Total Law of Probability, and Bayes' Rule—are placed into the general flow of text, and instructors will have the option of complete or partial coverage. The sections that present event relations, independence, conditional probability, and the Multiplication Rule have been rewritten in an attempt to clarify concepts that often are difficult for students to grasp. As in the thirteenth edition, the chapters on analysis of variance and linear regression include both calculational formulas and computer printouts in the basic text presentation. These chapters can be used with equal ease by instructors who wish to use the "hands-on" computational approach to linear regression and ANOVA and by those who choose to focus on the interpretation of computer-generated statistical printouts.

One important feature in the hypothesis testing chapters involves the emphasis on *p*-values and their use in judging statistical significance. With the advent of computer-generated *p*-values, these probabilities have become essential components in reporting the results of a statistical analysis. As such, the observed value of the test statistic and its *p*-value are presented together at the outset of our discussion of statistical hypothesis testing as equivalent tools for decision-making. Statistical significance is defined in terms of preassigned values of α, and the *p-value approach* is presented as an alternative to the *critical value approach* for testing a statistical hypothesis. Examples are presented using both the *p-value* and *critical value* approaches to hypothesis testing. Discussion of the practical interpretation of statistical results, along with the difference between statistical significance and practical significance, is emphasized in the practical examples in the text.

Special Features of the Fourteenth Edition

- NEED TO KNOW. . .: A special feature of this edition are highlighted sections called "NEED TO KNOW. . ." and identified by this icon. 🔲 **NEED TO KNOW...** These sections provide information consisting of definitions, procedures or step-by-step hints on problem solving for specific questions such as "NEED TO KNOW... How to Construct a Relative Frequency Histogram?" or "NEED TO KNOW... How to Decide Which Test to Use?"

- Applets: Easy access to the Internet has made it possible for students to visualize statistical concepts using an interactive webtool called an applet. Applets written by Gary McClelland, author of *Seeing Statistics*™, are found on the CourseMate ⃗CourseMate Website that accompanies the text. Following each applet, appropriate exercises are available that provide visual reinforcement of the concepts presented in the text. Applets allow the user to perform a statistical experiment, to interact with a statistical graph, to change its form, or to access an interactive "statistical table."

- Graphical and numerical data description includes both traditional and EDA methods, using computer graphics generated by *MINITAB 16* for Windows and MS Excel.

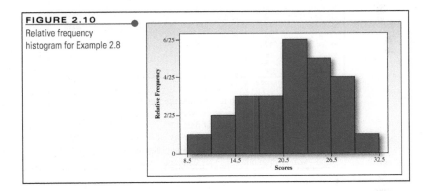

FIGURE 2.10

Relative frequency histogram for Example 2.8

FIGURE 2.18

(b)

	E	F	G	H
	Front Leg Room		*Rear Leg Room*	
Mean		41.333	Mean	28.944
Standard Error		0.344	Standard Error	0.543
Median		41	Median	29.5
Mode		41	Mode	30
Standard Deviation		1.031	Standard Deviation	1.629
Sample Variance		1.063	Sample Variance	2.653
Kurtosis		-0.745	Kurtosis	1.735
Skewness		0.245	Skewness	-1.212
Range		3	Range	5.5
Minimum		40	Minimum	25.5
Maximum		43	Maximum	31
Sum		372	Sum	260.5
Count		9	Count	9

- All examples and exercises in the text contain printouts based on *MINITAB 16* and consistent with earlier versions of *MINITAB* or MS Excel. Printouts are provided for some exercises, while other exercises require the student to obtain solutions without using a computer.

1.47 Presidential Vetoes Here is a list of the 44 presidents of the United States along with the number of regular vetoes used by each:[5]

Washington	2	B. Harrison	19
J. Adams	0	Cleveland	42
Jefferson	0	McKinley	6
Madison	5	T. Roosevelt	42
Monroe	1	Taft	30
J. Q. Adams	0	Wilson	33
Jackson	5	Harding	5
Van Buren	0	Coolidge	20
W. H. Harrison	0	Hoover	21
Tyler	6	F. D. Roosevelt	372
Polk	2	Truman	180
Taylor	0	Eisenhower	73
Fillmore	0	Kennedy	12
Pierce	9	L. Johnson	16
Buchanan	4	Nixon	26
Lincoln	2	Ford	48
A. Johnson	21	Carter	13
Grant	45	Reagan	39
Hayes	12	G. H. W. Bush	29
Garfield	0	Clinton	36
Arthur	4	G. W. Bush	11
Cleveland	304	Obama	1

Source: The World Almanac and Book of Facts 2011

Use an appropriate graph to describe the number of vetoes cast by the 44 presidents. Write a summary paragraph describing this set of data.

1.48 Windy Cities Are some cities more windy than others? Does Chicago deserve to be

(1950)	121.3	122.3	121.3	122.0	123.0	121.4	123.2	122.1	125.0	122.1
(1960)	122.2	124.0	120.2	121.4	120.0	121.1	122.0	120.3	122.1	121.4
(1970)	123.2	123.1	121.4	119.2[†]	124.0	122.0	121.3	122.1	121.1	122.2
(1980)	122.0	122.0	122.2	122.1	122.2	120.1	122.4	123.2	122.2	125.0
(1990)	122.0	123.0	123.0	122.2	123.3	121.1	121.0	122.4	122.2	123.2
(2000)	121.0	119.97	121.13	121.19	124.06	122.75	121.36	122.17	121.86	122.66
(2010)	124.4									

[†]Record time set by Secretariat in 1973.
Source: www.kentuckyderby.com

a. Do you think there will be a trend in the winning times over the years? Draw a line chart to verify your answer.

b. Describe the distribution of winning times using an appropriate graph. Comment on the shape of the distribution and look for any unusual observations.

1.50 Gulf Oil Spill Cleanup On April 20, 2010, the United States experienced a major environmental disaster when a Deepwater Horizon drilling rig exploded in the Gulf of Mexico. The number of personnel and equipment used in the Gulf oil spill cleanup, beginning May 2, 2010 (Day 13) through June 9, 2010 (Day 51) is given in the following table.[13]

	Day 13	Day 26	Day 39	Day 51
Number of personnel (1000s)	3.0	17.5	20.0	24.0
Federal Gulf fishing areas closed	3%	8%	25%	32%
Booms laid (miles)	46	315	644	909
Dispersants used (1000 gallons)	156	500	870	1143

TECHNOLOGY TODAY

The Role of Computers in the Fourteenth Edition—TECHNOLOGY TODAY

Computers are now a common tool for college students in all disciplines. Most students are accomplished users of word processors, spreadsheets, and databases, and they have no trouble navigating through software packages in the Windows environment. We believe, however, that advances in computer technology should not turn statistical analyses into a "black box." Rather, we choose to use the computational shortcuts and interactive visual tools that modern technology provides to give us more time to emphasize statistical reasoning as well as the understanding and interpretation of statistical results.

In this edition, students will be able to use computers for both standard statistical analyses and as a tool for reinforcing and visualizing statistical concepts. Both MS Excel and *MINITAB* 16 (consistent with earlier versions of *MINITAB*) are used exclusively as the computer packages for statistical analysis. However, we have chosen to isolate the instructions for generating computer output into individual sections called Technology Today at the end of each chapter. Each discussion uses numerical examples to guide the student through the MS Excel commands and option necessary for the procedures presented in that chapter, and then present the equivalent steps and commands needed to produce the same or similar results using *MINITAB*. We have included screen captures from both MS Excel and *MINITAB* 16, so that the student can actually work through these sections as "mini-labs."

If you do not need "hands-on" knowledge of *MINITAB* or MS Excel, or if you are using another software package, you may choose to skip these sections and simply use the printouts as guides for the basic understanding of computer printouts.

TECHNOLOGY TODAY

Numerical Descriptive Measures in *Excel*

MS Excel provides most of the basic descriptive statistics presented in Chapter a single command on the **Data** tab. Other descriptive statistics can be calculate the **Function** command on the **Formulas** tab.

EXAMPLE 2.15

The following data are the front and rear leg rooms (in inches) for nine differer utility vehicles:[14]

Make & Model	Front Leg Room	Rear Leg Room
Acura MDX	41.0	28.5
Buick Enclave	41.5	30.0
Chevy TrailBlazer	40.0	25.5
Chevy Tahoe Hybrid V8 CVT	41.0	27.5
GMC Terrain 1LT 4-cyl	43.0	31.0

Numerical Descriptive Measures in *MINITAB*

MINITAB provides most of the basic descriptive statistics presented in Chapter 2 using a single command in the drop-down menus.

The following data are the front and rear leg rooms (in inches) for nine different sports utility vehicles:[14]

Make and Model	Front Leg Room	Rear Leg Room
Acura MDX	41.0	28.5
Buick Enclave	41.5	30.0
Chevy TrailBlazer	40.0	25.5
Chevy Tahoe Hybrid V8 CVT	41.0	27.5
GMC Terrain 1LT 4-cyl	43.0	31.0
Honda CR-V	41.0	29.5

   Any student who has Internet access can use the applets found on the CourseMate Website to visualize a variety of statistical concepts (access instructions for the CourseMate Website are listed on the Printed Access Card that is an optional bundle with this text). In addition, some of the applets can be used instead of computer software to perform simple statistical analyses. Exercises written specifically for use with these applets also appear on the CourseMate Website. Students can use the applets at home or in a computer lab. They can use them as they read through the text material, once they have finished reading the entire chapter, or as a tool for exam review. Instructors can use the applets as a tool in a lab setting, or use them for visual demonstrations during lectures. We believe that these applets will be a powerful tool that will increase student enthusiasm for, and understanding of, statistical concepts and procedures.

STUDY AIDS

The many and varied exercises in the text provide the best learning tool for students embarking on a first course in statistics. The answers to all odd-numbered exercises are given in the back of the text, and a detailed solution appears in the *Student Solutions Manual,* which is available as a supplement for students. Each application exercise has a title, making it easier for students and instructors to immediately identify both the context of the problem and the area of application.

Use Table 2 to find the following probabilities:

a. $P(x \le 3)$ **b.** $P(x > 3)$

c. $P(x = 3)$ **d.** $P(3 \le x \le 5)$

5.38 Consider a Poisson random variable with $\mu = 0.8$. Use Table 2 to find the following probabilities:

a. $P(x = 0)$ **b.** $P(x \le 2)$

c. $P(x > 2)$ **d.** $P(2 \le x \le 4)$

5.39 Let x be a Poisson random variable with mean $\mu = 2$. Calculate these probabilities:

a. $P(x = 0)$ **b.** $P(x = 1)$

c. $P(x > 1)$ **d.** $P(x = 5)$

APPLICATIONS

5.43 Airport Safety The increased number of small commuter planes in major airports has heightened concern over air safety. An eastern airport has recorded a monthly average of five near misses on landings and takeoffs in the past 5 years.

a. Find the probability that during a given month there are no near misses on landings and takeoffs at the airport.

b. Find the probability that during a given month there are five near misses.

c. Find the probability that there are at least five near-

Students should be encouraged to use the "**NEED TO KNOW. . .**" sections as they occur in the text. The placement of these sections is intended to answer questions as they would normally arise in discussions. In addition, there are numerous hints called "**NEED A TIP?**" that appear in the margins of the text. The tips are short and concise.

 NEED A TIP

Empirical Rule $\Leftrightarrow$ mound-shaped data

Tchebysheff $\Leftrightarrow$ any shaped data

Is Tchebysheff's Theorem applicable? Yes, because it can be used for any set of data. According to Tchebysheff's Theorem,

- at least 3/4 of the measurements will fall between 10.6 and 32.6.
- at least 8/9 of the measurements will fall between 5.1 and 38.1.

Finally, sections called **Key Concepts and Formulas** appear in each chapter as a review in outline form of the material covered in that chapter.

CHAPTER REVIEW

Key Concepts and Formulas

I. Measures of the Center of a Data Distribution

1. Arithmetic mean (mean) or average
   a. Population: μ
   b. Sample of n measurements: $\bar{x} = \dfrac{\Sigma x_i}{n}$

2. Median; **position** of the median $= .5(n + 1)$
3. Mode
4. The median may be preferred to the mean if the data are highly skewed.

II. Measures of Variability

1. Range: $R = \text{largest} - \text{smallest}$
2. Variance
   a. Population of N measurements:
   $$\sigma^2 = \frac{\Sigma(x_i - \mu)^2}{N}$$
   b. Sample of n measurements:
   $$s^2 = \frac{\Sigma(x_i - \bar{x})^2}{n - 1} = \frac{\Sigma x_i^2 - \dfrac{(\Sigma x_i)^2}{n}}{n - 1}$$

68%, 95%, and 99.7% of the measurements are within one, two, and three standard deviations of the mean, respectively.

IV. Measures of Relative Standing

1. Sample z-score: $z = \dfrac{x - \bar{x}}{s}$
2. pth percentile; $p\%$ of the measurements are smaller, and $(100 - p)\%$ are larger.
3. Lower quartile, Q_1; **position** of $Q_1 = .25(n + 1)$
4. Upper quartile, Q_3; **position** of $Q_3 = .75(n + 1)$
5. Interquartile range: $\text{IQR} = Q_3 - Q_1$

V. The Five-Number Summary and Box Plots

1. The **five-number summary**:

   Min Q_1 Median Q_3 Max

   One-fourth of the measurements in the data set lie between each of the four adjacent pairs of numbers.

2. Box plots are used for detecting outliers and

The CourseMate Website, a password-protected resource that can be accessed with a Printed Access Card (optional bundle item), provides students with an array of study resources, including the complete set of Java applets, the TI Calculator Tech Guide that includes instructions for performing many of the techniques in the text using the popular TI 83/84/89 graphing calculator, an interactive eBook, online Quizzes, flashcards, and more. The data sets (saved in a variety of formats) can be found on the book's website (www.CengageBrain.com) as well as the CourseMate Website. ⁊CourseMate

INSTRUCTOR RESOURCES

The **Instructor's Website** (http://www.cengage.com/statistics/mendenhall), available to adopters of the fourteenth edition, provides a variety of teaching aids, including

- All the material from the CourseMate website including exercises using the Large Data Sets, which is accompanied by three large data sets that can be used throughout the course. A file named "Fortune" contains the revenues (in millions) for the *Fortune* 500 largest U.S. industrial corporations in a recent year; a file named "Batting" contains the batting averages for the National and American baseball league batting champions from 1976 to 2010; and a file named "Blood Pressure" contains the age and diastolic and systolic blood pressures for 965 men and 945 women compiled by the National Institutes of Health.
- Classic exercises with data sets and solutions
- PowerPoint lecture slides
- Applets by Gary McClelland (the complete set of Java applets used for the MyApps exercises on the website)
- TI Calculator Tech Guide, which includes instructions for performing many of the techniques in the text using the Tl-83/84/89 graphing calculators.

Also available for instructors:

Aplia
Aplia is a web-based learning solution that increases student effort and engagement. It helps make statistics relevant and engaging to students by connecting real-world examples to course concepts. When combined with the textual material of *Introduction to Probability and Statistics* (IPS) 14,

- Students receive immediate, detailed explanations for every answer.

- Math and graphing tutorials help students ovecome deficiencies in these crucial areas.

- Grades are automatically recorded in the instructor's Aplia gradebook.

Solution Builder
This online instructor database offers complete worked-out solutions to all exercises in the text, allowing you to create customized, secure solutions printouts (in PDF format) matched exactly to the problems you assign in class. Sign up for access at www.cengage.com/solutionbuilder.

PowerLecture™

PowerLecture with ExamView® for *Introduction to Probability and Statistics* contains the Instructor's Solutions Manual, PowerPoint lectures, ExamView Computerized Testing, Classic Exercises, and TI-83/84/89 calculator Tech Guide which includes instructions for performing many of the techniques in the text using the Tl-83/84/89 graphing calculators.

ACKNOWLEDGMENTS

The authors are grateful to Molly Taylor and the editorial staff of Cengage Learning for their patience, assistance, and cooperation in the preparation of this edition. A special thanks to Gary McClelland for the Java applets used in the text.

Thanks are also due to fourteenth edition reviewers Ronald C. Degges, Bob C. Denton, Dr. Dorothy M. French, Jungwon Mun, Kazuhiko Shinki, Florence P. Shu and thirteenth edition reviewers Bob Denton, Timothy Husband, Rob LaBorde, Craig McBride, Marc Sylvester, Kanapathi Thiru, and Vitaly Voloshin. We wish to thank authors and organizations for allowing us to reprint selected material; acknowledgments are made wherever such material appears in the text.

Robert J. Beaver
Barbara M. Beaver

Brief Contents

Contents

Introduction
What is Statistics?

What is statistics? Have you ever met a statistician? Do you know what a statistician does? Perhaps you are thinking of the person who sits in the broadcast booth at the Rose Bowl, recording the number of pass completions, yards rushing, or interceptions thrown on New Year's Day. Or perhaps the mere mention of the word *statistics* sends a shiver of fear through you. You may think you know nothing about statistics; however, it is almost inevitable that you encounter statistics in one form or another every time you pick up a daily newspaper. Here are some examples concerning the California 2010 elections:

- **Rowdy crowd jeers Whitman.** GOP candidate criticizes unions; earlier stop draws friendlier audience.
  GLENDALE— . . . Whitman, a billionaire, has spent $142 million from her personal fortune in the race so far.
  A Field Poll released Thursday showed her trailing Jerry Brown 49 percent to 39 percent among likely voters.[1]

- **Fiorina calls herself similar to Feinstein, who supports Boxer.**
  MENLO PARK—Republican Carly Fiorina said Friday she would be a like-minded colleague of Democratic Sen. Dianne Feinstein if she unseats Barbara Boxer next week, drawing sharp responses from both Democratic senators.
  . . . Fiorina, the former CEO of Hewlett-Packard Co., disputed a Field Poll released Friday showing Boxer leading her among likely voters, 49 percent to 41 percent.[2]

- **Race for attorney general tight. Field Poll:** Nearly a quarter of those surveyed are undecided. Newsom holds a slim lead over Maldonado for lieutenant governor.

SACRAMENTO—Tuesday's election for attorney general is a tossup, with Democrat Kamala Harris and Republican Steve Cooley virtually tied as Harris gains ground in voter-rich Los Angeles County and among women according to the latest Field Poll.

. . . Today's poll shows Cooley with 39 percent and Harris with 38 percent among likely voters. Almost a quarter of likely voters remain undecided.

. . . Newsom, the mayor of San Francisco, leads Maldonado, who was appointed lieutenant governor this year, 42 percent to 37 percent. A fifth of voters are undecided.

Today's poll was conducted for *The Press-Enterprise* and other California media subscribers. It was conducted October 14 through October 26 and included 1092 voters. It has a margin of error of plus or minus 3.2 percent.[3]

—*The Press-Enterprise,* Riverside, CA

Articles similar to these are commonplace in our newspapers and magazines, and in the period just prior to a presidential or congressional election, a new poll is reported almost every day. The language of these articles are very familiar to us; however, they leave the inquisitive reader with some unanswered questions. How were the people in the poll selected? Will these people give the same response tomorrow? Will they give the same response on election day? Will they even vote? Are these people representative of all those who will vote on election day? It is the job of a statistician to ask these questions and to find answers for them in the language of the poll.

Most Believe "Cover-Up" of JFK Assassination Facts
A majority of the public believes the assassination of President John F. Kennedy was part of a larger conspiracy, not the act of one individual. In addition, most Americans think there was a cover-up of facts about the 1963 shooting. Almost 50 years after JFK's assassination, a FOX news poll shows many Americans disagree with the government's conclusions about the killing. The **Warren Commission** found that **Lee Harvey Oswald** acted alone when he shot Kennedy, but 66 percent of the public today think the assassination was "part of a larger conspiracy" while only 25 percent think it was the "act of one individual."

"For older Americans, the Kennedy assassination was a traumatic experience that began a loss of confidence in government," commented Opinion Dynamics President John Gorman. "Younger people have grown up with movies and documentaries that have pretty much pushed the 'conspiracy' line. Therefore, it isn't surprising there is a fairly solid national consensus that we still don't know the truth."

(The poll asked): "Do you think that we know all the facts about the assassination of President John F. Kennedy or do you think there was a cover-up?"

	We Know All the Facts (%)	There Was a Cover-Up	(Not Sure)
All	14	74	12
Democrats	11	81	8
Republicans	18	69	13
Independents	12	71	17

—www.foxnews.com[4]

When you see an article like this one in a magazine, do you simply read the title and the first paragraph, or do you read further and try to understand the meaning of the numbers? How did the authors get these numbers? Did they really interview every American with each political affiliation? It is the job of the statistician to interpret the language of this study.

Hot News: 98.6 Not Normal
After believing for more than a century that 98.6 was the normal body temperature for humans, researchers now say normal is not normal anymore.

For some people at some hours of the day, 99.9 degrees could be fine. And readings as low as 96 turn out to be highly human.

The 98.6 standard was derived by a German doctor in 1868. Some physicians have always been suspicious of the good doctor's research. His claim: 1 million readings—in an epoch without computers.

So Mackowiak & Co. took temperature readings from 148 healthy people over a three-day period and found that the mean temperature was 98.2 degrees. Only 8 percent of the readings were 98.6.

—*The Press-Enterprise*[5]

What questions come to your mind when you read this article? How did the researcher select the 148 people, and how can we be sure that the results based on these 148 people are accurate when applied to the general population? How did the researcher arrive at the normal "high" and "low" temperatures given in the article? How did the German doctor record 1 million temperatures in 1868? Again, we encounter a statistical problem with an application to everyday life.

Statistics is a branch of mathematics that has applications in almost every facet of our daily life. It is a new and unfamiliar language for most people, however, and, like any new language, statistics can seem overwhelming at first glance. But once the language of statistics is learned and understood, it provides a powerful tool for data analysis in many different fields of application.

THE POPULATION AND THE SAMPLE

In the language of statistics, one of the most basic concepts is **sampling.** In most statistical problems, a specified number of measurements or data—a **sample**—is drawn from a much larger body of measurements, called the **population.**

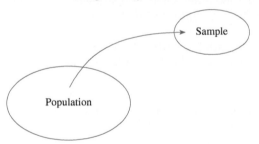

For the body-temperature experiment, the sample is the set of body-temperature measurements for the 148 healthy people chosen by the experimenter. We hope that the sample is representative of a much larger body of measurements—the population—the body temperatures of all healthy people in the world!

Which is of primary interest, the sample or the population? In most cases, we are interested primarily in the population, but the population may be difficult or impossible to enumerate. Imagine trying to record the body temperature of every healthy person on earth or the presidential preference of every registered voter in the United States! Instead, **we try to describe or predict the behavior of the population on the basis of information obtained from a representative sample from that population.**

The words *sample* and *population* have two meanings for most people. For example, you read in the newspapers that a Gallup poll conducted in the United States was based on a sample of 1823 people. Presumably, each person interviewed is asked a particular question, and that person's response represents a single measurement in the sample. Is the sample the set of 1823 people, or is it the 1823 responses that they give?

In statistics, we distinguish between the set of objects on which the measurements are taken and the measurements themselves. To experimenters, the objects on which measurements are taken are called **experimental units.** The sample survey statistician calls them **elements of the sample.**

DESCRIPTIVE AND INFERENTIAL STATISTICS

When first presented with a set of measurements—whether a sample or a population—you need to find a way to organize and summarize it. The branch of statistics that presents techniques for describing sets of measurements is called **descriptive statistics.** You have seen descriptive statistics in many forms: bar charts, pie charts, and line charts presented by a political candidate; numerical tables in the newspaper; or the average rainfall amounts reported by the local television weather forecaster. Computer-generated graphics and numerical summaries are commonplace in our everyday communication.

Definition **Descriptive statistics** consists of procedures used to summarize and describe the important characteristics of a set of measurements.

If the set of measurements is the entire population, you need only to draw conclusions based on the descriptive statistics. However, it might be too expensive or too time consuming to enumerate the entire population. Perhaps enumerating the population would destroy it, as in the case of "time to failure" testing. For these or other reasons, you may have only a sample from the population. By looking at the sample, you want to answer questions about the population as a whole. The branch of statistics that deals with this problem is called **inferential statistics.**

Definition **Inferential statistics** consists of procedures used to make inferences about population characteristics from information contained in a sample drawn from this population.

The **objective of inferential statistics** is to make inferences (that is, draw conclusions, make predictions, make decisions) about the characteristics of a population from information contained in a sample.

ACHIEVING THE OBJECTIVE OF INFERENTIAL STATISTICS: THE NECESSARY STEPS

How can you make inferences about a population using information contained in a sample? The task becomes simpler if you organize the problem into a series of logical steps.

1. **Specify the questions to be answered and identify the population of interest.** In the California election poll, the objective is to determine who will get the most votes on election day. Hence, the population of interest is the set of all votes in the California election. When you select a sample, it is important that

the sample be representative of *this* population, not the population of voter preferences on October 30 or on some other day prior to the election.

2. **Decide how to select the sample.** This is called the *design of the experiment* or the *sampling procedure*. Is the sample representative of the population of interest? For example, if a sample of registered voters is selected from the city of San Francisco, will this sample be representative of all voters in California? Will it be the same as a sample of "likely voters"—those who are likely to actually vote in the election? Is the sample large enough to answer the questions posed in step 1 without wasting time and money on additional information? A good sampling design will answer the questions posed with minimal cost to the experimenter.

3. **Select the sample and analyze the sample information.** No matter how much information the sample contains, you must use an appropriate method of analysis to extract it. Many of these methods, which depend on the sampling procedure in step 2, are explained in the text.

4. **Use the information from step 3 to make an inference about the population.** Many different procedures can be used to make this inference, and some are better than others. For example, 10 different methods might be available to estimate human response to an experimental drug, but one procedure might be more accurate than others. You should use the best inference-making procedure available (many of these are explained in the text).

5. **Determine the reliability of the inference.** Since you are using only a fraction of the population in drawing the conclusions described in step 4, you might be wrong! How can this be? If an agency conducts a statistical survey for you and estimates that your company's product will gain 34% of the market this year, how much confidence can you place in this estimate? Is this estimate accurate to within 1, 5, or 20 percentage points? Is it reliable enough to be used in setting production goals? Every statistical inference should include a measure of reliability that tells you how much confidence you have in the inference.

Now that you have learned a few basic terms and concepts, we again pose the question asked at the beginning of this discussion: Do you know what a statistician does? The statistician's job is to implement all of the preceding steps.

KEYS FOR SUCCESSFUL LEARNING

As you begin to study statistics, you will find that there are many new terms and concepts to be mastered. Since statistics is an applied branch of mathematics, many of these basic concepts are mathematical—developed and based on results from calculus or higher mathematics. However, you do not have to be able to derive results in order to apply them in a logical way. In this text, we use numerical examples and common-sense arguments to explain statistical concepts, rather than more complicated mathematical arguments.

In recent years, computers have become readily available to many students and provide them with an invaluable tool. In the study of statistics, even the beginning student can use packaged programs to perform statistical analyses with a high degree of speed and accuracy. Some of the more common statistical packages available at computer facilities are *MINITAB*™, SAS (Statistical Analysis System), and SPSS

(Statistical Package for the Social Sciences); personal computers will support packages such as *MINITAB, MS Excel,* and others. There are even online statistical programs and interactive "applets" on the Internet.

These programs, called **statistical software,** differ in the types of analyses available, the options within the programs, and the forms of printed results (called **output**). However, they are all similar. In this book, we use both *MINITAB* and *Microsoft Excel* as statistical tools. Understanding the basic output of these packages will help you interpret the output from other software systems.

At the end of most chapters, you will find a section called *"Technology Today."* These sections present numerical examples to guide you through the *MINITAB* and *MS Excel* commands and options that are used for the procedures in that chapter. If you are using *MINITAB* or *MS Excel* in a lab or home setting, you may want to work through this section at your own computer so that you become familiar with the hands-on methods in computer analysis. If you do not need hands-on knowledge of *MINITAB* or *MS Excel,* you may choose to skip this section and simply use the computer printouts for analysis as they appear in the text.

Another learning tool called statistical **applets** can be found on the CourseMate Web site. Also found on this Web site are explanatory sections called "Using the Applets," which will help you understand how the applets can be used to visualize many of the chapter concepts. An accompanying section called "Applet APPs" provides some exercises (with solutions) that can be solved using the statistical applets. Whenever there is an applet available for a particular concept or application, you will find an icon in the left margin of the text, together with the name of the appropriate applet.

Most important, using statistics successfully requires common sense and logical thinking. For example, if we want to find the average height of all students at a particular university, would we select our entire sample from the members of the basketball team? In the body-temperature example, the logical thinker would question an 1868 average based on 1 million measurements—when computers had not yet been invented.

As you learn new statistical terms, concepts, and techniques, remember to view every problem with a critical eye and be sure that the rule of common sense applies. Throughout the text, we will remind you of the pitfalls and dangers in the use or misuse of statistics. Benjamin Disraeli once said that there are three kinds of lies: *lies, damn lies,* and *statistics*! Our purpose is to dispel this claim—to show you how to make statistics *work* for you and not *lie* for you!

As you continue through the book, refer back to this introduction periodically. Each chapter will increase your knowledge of statistics and should, in some way, help you achieve one of the steps described here. Each of these steps is essential in attaining the overall objective of inferential statistics: to make inferences about a population using information contained in a sample drawn from that population.

Describing Data with Graphs

© Ocean/Corbis

NEED TO KNOW...

How to Construct a Stem and Leaf Plot
How to Construct a Relative Frequency Histogram

How Is Your Blood Pressure?

Is your blood pressure normal, or is it too high or too low? The case study at the end of this chapter examines a large set of blood pressure data. You will use graphs to describe these data and compare your blood pressure with that of others of your same age and gender.

1.1 VARIABLES AND DATA

In Chapters 1 and 2, we will present some basic techniques in *descriptive statistics*—the branch of statistics concerned with describing sets of measurements, both *samples* and *populations*. Once you have collected a set of measurements, how can you display this set in a clear, understandable, and readable form? First, you must be able to define what is meant by measurements or "data" and to categorize the types of data that you are likely to encounter in real life. We begin by introducing some definitions.

Definition   A **variable** is a characteristic that changes or varies over time and/or for different individuals or objects under consideration.

For example, body temperature is a variable that changes over time within a single individual; it also varies from person to person. Religious affiliation, ethnic origin, income, height, age, and number of offspring are all variables—characteristics that vary depending on the individual chosen.

In the Introduction, we defined an *experimental unit* or an *element of the sample* as the object on which a measurement is taken. Equivalently, we could define an experimental unit as the object on which a variable is measured. When a variable is actually measured on a set of experimental units, a set of measurements or **data** result.

Definition   An **experimental unit** is the individual or object on which a variable is measured. A single **measurement** or data value results when a variable is actually measured on an experimental unit.

If a measurement is generated for every experimental unit in the entire collection, the resulting data set constitutes the *population* of interest. Any smaller subset of measurements is a *sample*.

Definition   A **population** is the set of all measurements of interest to the investigator.

Definition   A **sample** is a subset of measurements selected from the population of interest.

EXAMPLE 1.1

A set of five students is selected from all undergraduates at a large university, and measurements are entered into a spreadsheet as shown in Figure 1.1. Identify the various elements involved in generating this set of measurements.

Solution   There are several *variables* in this example. The *experimental unit* on which the variables are measured is a particular undergraduate student on the campus, identified in column A. Five variables are measured for each student: grade point average (GPA), gender, year in college, major, and current number of units enrolled. Each of these characteristics varies from student to student. If we consider the GPAs of all students at this university to be the population of interest, the five GPAs in column B represent a *sample* from this population. If the GPA of each undergraduate student at the university had been measured, we would have generated the entire *population* of measurements for this variable.

FIGURE 1.1

Measurements on five
undergraduate students

	A	B	C	D	E	F
1	Student	GPA	Gender	Year	Major	Number of Units
2	1	2	F	Fr	Psychology	16
3	2	2.3	F	So	Mathematics	15
4	3	2.9	M	So	English	17
5	4	2.7	M	Fr	English	15
6	5	2.6	F	Jr	Business	14

Sheet1 / Sheet2 / Sheet3

The second variable measured on the students is gender, in column C. This variable is somewhat different from GPA, since it can take only one of two values—male (M) or female (F). The population, if it could be enumerated, would consist of a set of Ms and Fs, one for each student at the university. Similarly, the third and fourth variables, year and major, generate nonnumerical data. Year has four categories (Fr, So, Jr, Sr), and major has one category for each undergraduate major on campus. The last variable, current number of units enrolled, is numerically valued, generating a set of numbers rather than a set of qualities or characteristics.

Although we have discussed each variable individually, remember that we have measured each of these five variables on a single experimental unit: the student. Therefore, in this example, a "measurement" really consists of five observations, one for each of the five measured variables. For example, the measurement taken on student 2 produces this observation:

(2.3, F, So, Mathematics, 15)

You can see that there is a difference between a *single* variable measured on a single experimental unit and *multiple* variables measured on a single experimental unit as in Example 1.1.

Definition   **Univariate data** result when a single variable is measured on a single experimental unit.

Definition   **Bivariate data** result when two variables are measured on a single experimental unit. **Multivariate data** result when more than two variables are measured.

If you measure the body temperatures of 148 people, the resulting data are *univariate*. In Example 1.1, five variables were measured on each student, resulting in *multivariate* data.

TYPES OF VARIABLES

1.2

Variables can be classified into one of two types: **qualitative** or **quantitative.**

Definition   **Qualitative variables** measure a quality or characteristic on each experimental unit. **Quantitative variables** measure a numerical quantity or amount on each experimental unit.

NEED A TIP?

Qualitative ⇔ "quality" or characteristic

Quantitative ⇔ "quantity" or number

Qualitative variables produce data that can be categorized according to similarities or differences in kind; hence, they are often called **categorical data.** The variables gender, year, and major in Example 1.1 are qualitative variables that produce categorical data. Here are some other examples:

- Political affiliation: Republican, Democrat, Independent
- Taste ranking: excellent, good, fair, poor
- Color of an M&M'S® candy: brown, yellow, red, orange, green, blue

Quantitative variables, often represented by the letter x, produce numerical data, such as those listed here:

- x = Prime interest rate
- x = Number of passengers on a flight from Los Angeles to New York City
- x = Weight of a package ready to be shipped
- x = Volume of orange juice in a glass

Notice that there is a difference in the types of numerical values that these quantitative variables can assume. The number of passengers, for example, can take on only the values $x = 0, 1, 2, \ldots,$ whereas the weight of a package can take on any value greater than zero, or $0 < x < \infty$. To describe this difference, we define two types of quantitative variables: **discrete** and **continuous.**

Definition A **discrete variable** can assume only a finite or countable number of values. A **continuous variable** can assume the infinitely many values corresponding to the points on a line interval.

NEED A TIP?

Discrete ⇔ "listable"

Continuous ⇔ "unlistable"

The name *discrete* relates to the discrete gaps between the possible values that the variable can assume. Variables such as number of family members, number of new car sales, and number of defective tires returned for replacement are all examples of discrete variables. On the other hand, variables such as height, weight, time, distance, and volume are *continuous* because they can assume values at any point along a line interval. For any two values you pick, a third value can always be found between them!

EXAMPLE (1.2) Identify each of the following variables as qualitative or quantitative:

1. The most frequent use of your microwave oven (reheating, defrosting, warming, other)
2. The number of consumers who refuse to answer a telephone survey
3. The door chosen by a mouse in a maze experiment (A, B, or C)
4. The winning time for a horse running in the Kentucky Derby
5. The number of children in a fifth-grade class who are reading at or above grade level

 NEED A TIP?

Discrete variables often involve the "number of" items in a set.

Solution Variables 1 and 3 are both *qualitative* because only a quality or characteristic is measured for each individual. The categories for these two variables are shown in parentheses. The other three variables are *quantitative*. Variables 2 and 5 are *discrete* variables that can take on any of the values $x = 0, 1, 2, \ldots,$ with a maximum value depending on the number of consumers called or the number of children in the class, respectively. Variable 4, the winning time for a Kentucky Derby horse, is the only *continuous* variable in the list. The winning time, if it could be measured with sufficient accuracy, could be 121 seconds, 121.5 seconds, 121.25 seconds, or any values between any two times we have listed.

Why should you be concerned about different kinds of variables (shown in Figure 1.2) and the data that they generate? The reason is that different types of data require you to use different methods for description, so that the data can be presented clearly and understandably to your audience!

FIGURE 1.2

Types of data

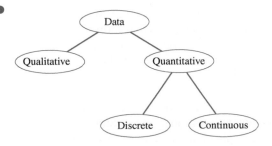

GRAPHS FOR CATEGORICAL DATA

1.3

After the data have been collected, they can be consolidated and summarized to show the following information:

- What values of the variable have been measured
- How often each value has occurred

For this purpose, you can construct a *statistical table* that can be used to display the data graphically as a **data distribution.** The type of graph you choose depends on the type of variable you have measured.

When the variable of interest is *qualitative* or *categorical,* the statistical table is a list of the categories being considered along with a measure of how often each value occurred. You can measure "how often" in three different ways:

- The **frequency,** or number of measurements in each category
- The **relative frequency,** or proportion of measurements in each category
- The **percentage** of measurements in each category

If you let n be the total number of measurements in the set, you can find the relative frequency and percentage using these relationships:

$$\text{Relative frequency} = \frac{\text{Frequency}}{n}$$

$$\text{Percent} = 100 \times \text{Relative frequency}$$

You will find that the sum of the frequencies is always *n,* the sum of the relative frequencies is 1, and the sum of the percentages is 100%.

When the variable is qualitative, the categories should be chosen so that

- a measurement will belong to one and only one category
- each measurement has a category to which it can be assigned

For example, if you categorize meat products according to the type of meat used, you might use these categories: beef, chicken, seafood, pork, turkey, other. To categorize ranks of college faculty, you might use these categories: professor, associate professor, assistant professor, instructor, lecturer, other. The "other" category is included in both cases to allow for the possibility that a measurement cannot be assigned to one of the earlier categories.

Once the measurements have been categorized and summarized in a *statistical table,* you can use either a pie chart or a bar chart to display the distribution of the data. A **pie chart** is the familiar circular graph that shows how the measurements are distributed among the categories. A **bar chart** shows the same distribution of measurements among the categories, with the height of the bar measuring how often a particular category was observed.

EXAMPLE 1.3

In a survey concerning public education, 400 school administrators were asked to rate the quality of education in the United States. Their responses are summarized in Table 1.1. Construct a pie chart and a bar chart for this set of data.

Solution To construct a pie chart, assign one sector of a circle to each category. The angle of each sector should be proportional to the proportion of measurements (or *relative frequency*) in that category. Since a circle contains 360°, you can use this equation to find the angle:

$$\text{Angle} = \text{Relative frequency} \times 360°$$

TABLE 1.1 • **U.S. Education Rating by 400 Educators**

Rating	Frequency
A	35
B	260
C	93
D	12
Total	400

Table 1.2 shows the ratings along with the frequencies, relative frequencies, percentages, and sector angles necessary to construct the pie chart. Figure 1.3 shows the pie chart constructed from the values in the table. While pie charts use percentages to determine the relative sizes of the "pie slices," bar charts usually plot frequency against the categories. A bar chart for these data is shown in Figure 1.4.

TABLE 1.2 ● **Calculations for the Pie Chart in Example 1.3**

Rating	Frequency	Relative Frequency	Percent	Angle
A	35	35/400 = .09	9%	.09 × 360 = 32.4°
B	260	260/400 = .65	65%	234.0°
C	93	93/400 = .23	23%	82.8°
D	12	12/400 = .03	3%	10.8°
Total	400	1.00	100%	360°

The visual impact of these two graphs is somewhat different. The pie chart is used to display the relationship of the parts to the whole; the bar chart is used to emphasize the actual quantity or frequency for each category. Since the categories in this example are ordered "grades" (A, B, C, D), we would not want to rearrange the bars in the chart to change its *shape*. In a pie chart, the order of presentation is irrelevant.

FIGURE 1.3 ●

Pie chart for Example 1.3

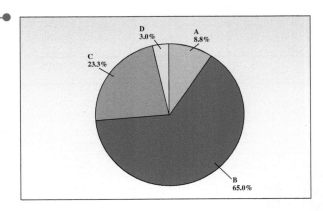

FIGURE 1.4 ●

Bar chart for Example 1.3

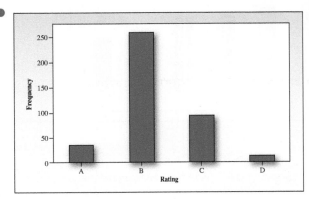

EXAMPLE (1.4) A snack size bag of peanut M&M'S candies contains 21 candies with the colors listed in Table 1.3. The variable "color" is *qualitative,* so Table 1.4 lists the six categories along with a tally of the number of candies of each color. The last three columns of Table 1.4 show how often each category occurred. Since the categories are colors and have no particular order, you could construct bar charts with many different *shapes* just by reordering the bars. To emphasize that brown is the most frequent color, followed by blue, green, and orange, we order the bars from largest to smallest and create the bar chart in Figure 1.5. A bar chart in which the bars are ordered from largest to smallest is called a **Pareto chart.**

TABLE 1.3 ● **Raw Data: Colors of 21 Candies**

Brown	Green	Brown	Blue
Red	Red	Green	Brown
Yellow	Orange	Green	Blue
Brown	Blue	Blue	Brown
Orange	Blue	Brown	Orange
Yellow			

TABLE 1.4 ● **Statistical Table: M&M'S Data for Example 1.4**

Category	Tally	Frequency	Relative Frequency	Percent
Brown	JHT I	6	6/21	28%
Green	III	3	3/21	14
Orange	III	3	3/21	14
Yellow	II	2	2/21	10
Red	II	2	2/21	10
Blue	JHT	5	5/21	24
Total		21	1	100%

FIGURE 1.5

Pareto chart for Example 1.4

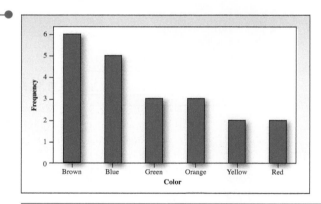

EXERCISES

UNDERSTANDING THE CONCEPTS

1.1 Experimental Units Identify the experimental units on which the following variables are measured:

a. Gender of a student

b. Number of errors on a midterm exam

c. Age of a cancer patient

d. Number of flowers on an azalea plant

e. Color of a car entering a parking lot

1.2 Qualitative or Quantitative? Identify each variable as quantitative or qualitative:

a. Amount of time it takes to assemble a simple puzzle

b. Number of students in a first-grade classroom

c. Rating of a newly elected politician (excellent, good, fair, poor)

d. State in which a person lives

1.3 Discrete or Continuous? Identify the following quantitative variables as discrete or continuous:

a. Population in a particular area of the United States

b. Weight of newspapers recovered for recycling on a single day

c. Time to complete a sociology exam

d. Number of consumers in a poll of 1000 who consider nutritional labeling on food products to be important

1.4 Discrete or Continuous? Identify each quantitative variable as discrete or continuous.

a. Number of boating accidents along a 50-mile stretch of the Colorado River

b. Time required to complete a questionnaire

c. Cost of a head of lettuce

d. Number of brothers and sisters you have

e. Yield in kilograms of wheat from a 1-hectare plot in a wheat field

1.5 Parking on Campus Six vehicles are selected from the vehicles that are issued campus parking permits, and the following data are recorded:

Vehicle	Type	Make	Carpool?	One-way Commute Distance (miles)	Age of Vehicle (years)
1	Car	Honda	No	23.6	6
2	Car	Toyota	No	17.2	3
3	Truck	Toyota	No	10.1	4
4	Van	Dodge	Yes	31.7	2
5	Motor-cycle	Harley-Davidson	No	25.5	1
6	Car	Chevrolet	No	5.4	9

a. What are the experimental units?

b. What are the variables being measured? What types of variables are they?

c. Is this univariate, bivariate, or multivariate data?

1.6 Past U.S. Presidents A data set consists of the ages at death for each of the 38 past presidents of the United States now deceased.

a. Is this set of measurements a population or a sample?

b. What is the variable being measured?

c. Is the variable in part b quantitative or qualitative?

1.7 Voter Attitudes You are a candidate for your state legislature, and you want to survey voter attitudes regarding your chances of winning. Identify the population that is of interest to you and from which you would like to select your sample. How is this population dependent on time?

1.8 Cancer Survival Times A medical researcher wants to estimate the survival time of a patient after the onset of a particular type of cancer and after a particular regimen of radiotherapy.

a. What is the variable of interest to the medical researcher?

b. Is the variable in part a qualitative, quantitative discrete, or quantitative continuous?

c. Identify the population of interest to the medical researcher.

d. Describe how the researcher could select a sample from the population.

e. What problems might arise in sampling from this population?

1.9 New Teaching Methods An educational researcher wants to evaluate the effectiveness of a new method for teaching reading to deaf students. Achievement at the end of a period of teaching is measured by a student's score on a reading test.

a. What is the variable to be measured? What type of variable is it?

b. What is the experimental unit?

c. Identify the population of interest to the experimenter.

BASIC TECHNIQUES

1.10 Fifty people are grouped into four categories— A, B, C, and D—and the number of people who fall into each category is shown in the table:

Category	Frequency
A	11
B	14
C	20
D	5

a. What is the experimental unit?

b. What is the variable being measured? Is it qualitative or quantitative?

c. Construct a pie chart to describe the data.

d. Construct a bar chart to describe the data.

e. Does the shape of the bar chart in part d change depending on the order of presentation of the four categories? Is the order of presentation important?

f. What *proportion* of the people are in category B, C, or D?

g. What *percentage* of the people are *not* in category B?

1.11 Jeans A manufacturer of jeans has plants in California, Arizona, and Texas. A group of 25 pairs of jeans is randomly selected from the computerized database, and the state in which each is produced is recorded:

CA	AZ	AZ	TX	CA
CA	CA	TX	TX	TX
AZ	AZ	CA	AZ	TX
CA	AZ	TX	TX	TX
CA	AZ	AZ	CA	CA

a. What is the experimental unit?

b. What is the variable being measured? Is it qualitative or quantitative?

c. Construct a pie chart to describe the data.

d. Construct a bar chart to describe the data.

e. What *proportion* of the jeans are made in Texas?

f. What state produced the most jeans in the group?

g. If you want to find out whether the three plants produced equal numbers of jeans, or whether one produced more jeans than the others, how can you use the charts from parts c and d to help you? What conclusions can *you* draw from these data?

APPLICATIONS

1.12 Election 2012 During the spring of 2010, the news media were already conducting opinion polls that tracked the fortunes of the major candidates hoping to become the president of the United States. One such poll conducted by *CNN/Opinion Research Corporation Poll* showed the following results:[1]

"If Barack Obama were the Democratic Party's candidate and [see below] were the Republican Party's candidate, who would you be more likely to vote for: Obama, the Democrat, or [see below], the Republican?" If unsure: "As of today, who do you lean more toward?"

	Barack Obama (D) %	Mitt Romney (R) %	Neither (vol.) %
4/9–11/10	53	45	1

	Barack Obama (D) %	Mike Huckabee (R) %	Neither (vol.) %
4/9–11/10	54	45	1

	Barack Obama (D) %	Sarah Palin (R) %	Neither (vol.) %
4/9–11/10	55	42	3

	Barack Obama (D) %	Newt Gingrich (R) %	Neither (vol.) %
4/9–11/10	55	43	1

Source: www.pollingreport.com

The results were based on a sample taken April 9–11, 2010, of 907 registered voters nationwide.

a. If the pollsters were planning to use these results to predict the outcome of the 2012 presidential election, describe the population of interest to them.

b. Describe the actual population from which the sample was drawn.

c. Some pollsters prefer to select a sample of "likely" voters. What is the difference between "registered voters" and "likely voters"? Why is this important?

d. Is the sample selected by the pollsters representative of the population described in part a? Explain.

1.13 Want to Be President? Would you want to be the president of the United States? Although many teenagers think that they could grow up to be the president, most don't want the job. In an opinion poll conducted by *ABC News,* nearly 80% of the teens were not interested in the job.[2] When asked "What's the main reason you would not want to be president?" they gave these responses:

Other career plans/no interest	40%
Too much pressure	20%
Too much work	15%
Wouldn't be good at it	14%
Too much arguing	5%

a. Are all of the reasons accounted for in this table? Add another category if necessary.

b. Would you use a pie chart or a bar chart to graphically describe the data? Why?

c. Draw the chart you chose in part b.

d. If you were the person conducting the opinion poll, what other types of questions might you want to investigate?

1.14 Facebook Fanatics The social networking site called *Facebook* has grown quickly since its inception in 2004. In fact, *Facebook's* United States user base grew from 42 million users to 103 million users between 2009 and 2010. The table below shows the age distribution of *Facebook* users (in thousands) as it changed from January 2009 to January 2010.[3]

Age	As of 1/04/2009	As of 1/04/2010
13–17	5675	10,680
18–24	17,192	26,076
25–34	11,255	25,580
35–54	6989	29,918
55+	955	9764
Unknown	23	1068
Total	42,089	103,086

a. Define the variable that has been measured in this table.

b. Is the variable quantitative or qualitative?

c. What do the numbers represent?

d. Construct a pie chart to describe the age distribution of *Facebook* users as of January 4, 2009.

e. Construct a pie chart to describe the age distribution of *Facebook* users as of January 4, 2010.

f. Refer to parts d and e. How would you describe the changes in the age distributions of *Facebook* users during this 1-year period?

1.15 Back to Work How long does it take you to adjust to your normal work routine after coming back from vacation? A bar graph with data from the Snapshots section of *USA Today* is shown below:[4]

a. Are all of the opinions accounted for in the table? Add another category if necessary.

b. Is the bar chart drawn accurately? That is, are the three bars in the correct proportion to each other?

c. Use a pie chart to describe the opinions. Which graph is more interesting to look at?

GRAPHS FOR QUANTITATIVE DATA

Quantitative variables measure an amount or quantity on each experimental unit. If the variable can take only a finite or countable number of values, it is a *discrete* variable. A variable that can assume an infinite number of values corresponding to points on a line interval is called *continuous*.

Pie Charts and Bar Charts

Sometimes information is collected for a quantitative variable measured on different segments of the population, or for different categories of classification. For example, you might measure the average incomes for people of different age groups, different genders, or living in different geographic areas of the country. In such cases, you can use pie charts or bar charts to describe the data, using the amount measured in each category rather than the frequency of occurrence of each category. The *pie chart* displays how the total quantity is distributed among the categories, and the *bar chart* uses the height of the bar to display the amount in a particular category.

EXAMPLE 1.5   The amount of money expended in fiscal year 2009 by the U.S. Department of Defense in various categories is shown in Table 1.5.[5] Construct both a pie chart and a bar chart to describe the data. Compare the two forms of presentation.

TABLE 1.5 ● **Expenses by Category**

Category	Amount ($ billions)
Military personnel	147.3
Operation and maintenance	259.3
Procurement	129.2
Research and development	79.0
Military construction	17.6
Other	4.3
Total	636.7

Source: The World Almanac and Book of Facts 2011

Solution   Two variables are being measured: the category of expenditure (*qualitative*) and the amount of the expenditure (*quantitative*). The bar chart in Figure 1.6 displays the categories on the horizontal axis and the amounts on the vertical axis.

FIGURE 1.6 ●

Bar chart for Example 1.5

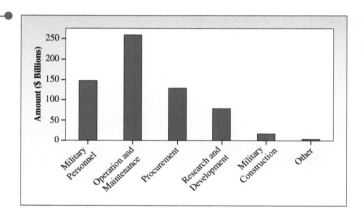

For the pie chart in Figure 1.7, each "pie slice" represents the proportion of the total expenditures ($636.7 billion) corresponding to its particular category. For example, for the research and development category, the angle of the sector is

$$\frac{79.0}{636.7} \times 360° = 44.7°$$

FIGURE 1.7 ●

Pie chart for Example 1.5

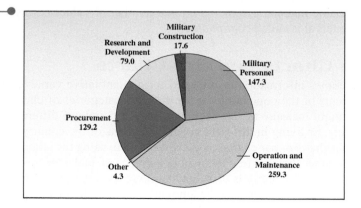

Both graphs show that the largest amounts of money were spent on personnel and operations. Since there is no inherent order to the categories, you are free to rearrange the bars or sectors of the graphs in any way you like. The *shape* of the bar chart has no bearing on its interpretation.

Line Charts

When a quantitative variable is recorded over time at equally spaced intervals (such as daily, weekly, monthly, quarterly, or yearly), the data set forms a **time series.** Time series data are most effectively presented on a **line chart** with time as the horizontal axis. The idea is to try to discern a pattern or **trend** that will likely continue into the future, and then to use that pattern to make accurate predictions for the immediate future.

EXAMPLE **1.6**

In the year 2025, the oldest "baby boomers" (born in 1946) will be 79 years old, and the oldest "Gen Xers" (born in 1965) will be 2 years from Social Security eligibility. How will this affect the consumer trends in the next 25 years? Will there be sufficient funds for "baby boomers" to collect Social Security benefits? The United States *Bureau of the Census* gives projections for the portion of the U.S. population that will be 85 and over in the coming years, as shown in Table 1.6.[5] Construct a line chart to illustrate the data. What is the effect of stretching and shrinking the vertical axis on the line chart?

TABLE 1.6

Population Growth Projections

Year	2020	2030	2040	2050
85 and over (millions)	6.6	8.7	14.2	19.0

Source: The World Almanac and Book of Facts 2011

NEED A TIP?

Beware of stretching or shrinking axes when you look at a graph!

Solution   The quantitative variable "85 and over" is measured over four time intervals, creating a *time series* that you can graph with a line chart. The time intervals are marked on the horizontal axis and the projections on the vertical axis. The data points are then connected by line segments to form the line charts in Figure 1.8. Notice the marked difference in the vertical scales of the two graphs. *Shrinking* the scale on the vertical axis causes large changes to appear small, and vice versa. To avoid misleading conclusions, you must look carefully at the scales of the vertical and horizontal axes. However, from both graphs you get a clear picture of the steadily increasing number of those 85 and older in the early years of the new millennium.

FIGURE 1.8
Line charts for Example 1.6

Dotplots

Many sets of quantitative data consist of numbers that cannot easily be separated into categories or intervals of time. You need a different way to graph this type of data!

The simplest graph for quantitative data is the **dotplot.** For a small set of measurements—for example, the set 2, 6, 9, 3, 7, 6—you can simply plot the measurements as points on a horizontal axis, as shown in Figure 1.9(a). For a large data set, however, such as the one in Figure 1.9(b), the dotplot can be uninformative and tedious to interpret.

FIGURE 1.9

Dotplots for small and large data sets

(a)

Small Set

ONLINE APPLET

Building a Dotplot

(b)

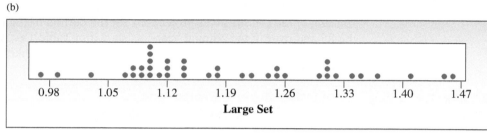

Large Set

Stem and Leaf Plots

Another simple way to display the distribution of a quantitative data set is the **stem and leaf plot.** This plot presents a graphical display of the data using the actual numerical values of each data point.

? NEED TO KNOW...

How to Construct a Stem and Leaf Plot

1. Divide each measurement into two parts: the **stem** and the **leaf.**
2. List the stems in a column, with a vertical line to their right.
3. For each measurement, record the leaf portion in the same row as its corresponding stem.
4. Order the leaves from lowest to highest in each stem.
5. Provide a key to your stem and leaf coding so that the reader can re-create the actual measurements if necessary.

EXAMPLE 1.7

Table 1.7 lists the prices (in dollars) of 19 different brands of walking shoes. Construct a stem and leaf plot to display the distribution of the data.

TABLE 1.7 **Prices of Walking Shoes**

90	70	70	70	75	70
65	68	60	74	70	95
75	70	68	65	40	65
70					

Solution   To create the stem and leaf, you could divide each observation between the ones and the tens place. The number to the left is the stem; the number to the right is the leaf. Thus, for the shoes that cost $65, the stem is 6 and the leaf is 5. The stems, ranging from 4 to 9, are listed in Figure 1.10, along with the leaves for each of the 19 measurements. If you indicate that the leaf unit is 1, the reader will realize that the stem and leaf 6 and 8, for example, represent the number 68, recorded to the nearest dollar.

FIGURE 1.10

Stem and leaf plot for the data in Table 1.7

4	0		Leaf unit = 1	4	0	
5				5		
6	5 8 0 8 5 5		Reordering ⟶	6	0 5 5 5 8 8	
7	0 0 0 5 0 4 0 5 0 0			7	0 0 0 0 0 0 0 4 5 5	
8				8		
9	0 5			9	0 5	

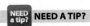 **NEED A TIP?**

Stem | Leaf

Sometimes the available stem choices result in a plot that contains too few stems and a large number of leaves within each stem. In this situation, you can stretch the stems by dividing each one into several lines, depending on the leaf values assigned to them. Stems are usually divided in one of two ways:

- Into two lines, with leaves 0–4 in the first line and leaves 5–9 in the second line
- Into five lines, with leaves 0–1, 2–3, 4–5, 6–7, and 8–9 in the five lines, respectively

EXAMPLE   **1.8**

The data in Table 1.8 are the weights at birth of 30 full-term babies, born at a metropolitan hospital and recorded to the nearest tenth of a pound.[6] Construct a stem and leaf plot to display the distribution of the data.

TABLE 1.8 **Birth Weights of 30 Full-Term Newborn Babies**

7.2	7.8	6.8	6.2	8.2
8.0	8.2	5.6	8.6	7.1
8.2	7.7	7.5	7.2	7.7
5.8	6.8	6.8	8.5	7.5
6.1	7.9	9.4	9.0	7.8
8.5	9.0	7.7	6.7	7.7

Solution   The data, though recorded to an accuracy of only one decimal place, are measurements of the continuous variable x = weight, which can take on any positive value. By examining Table 1.8, you can quickly see that the highest and lowest weights are 9.4 and 5.6, respectively. But how are the remaining weights distributed?

If you use the decimal point as the dividing line between the stem and the leaf, you have only five stems, which does not produce a very good picture. When you divide each stem into two lines, there are eight stems, since the first line of stem 5 and the second line of stem 9 are empty! This produces a more descriptive plot, as shown in Figure 1.11. For these data, the leaf unit is .1, and the reader can infer that the stem and leaf 8 and 2, for example, represent the measurement $x = 8.2$.

FIGURE 1.11

Stem and leaf plot for the data in Table 1.8

5	8 6
6	1 2
6	8 8 8 7
7	2 2 1
7	8 7 9 5 7 7 5 8 7
8	0 2 2 2
8	5 6 5
9	0 4 0

Reordering →   Leaf unit = .1

5	6 8
6	1 2
6	7 8 8 8
7	1 2 2
7	5 5 7 7 7 7 8 8 9
8	0 2 2 2
8	5 5 6
9	0 0 4

If you turn the stem and leaf plot sideways, so that the vertical line is now a horizontal axis, you can see that the data have "piled up" or been "distributed" along the axis in a pattern that can be described as "mound-shaped"—much like a pile of sand on the beach. This plot again shows that the weights of these 30 newborns range between 5.6 and 9.4; many weights are between 7.5 and 8.0 pounds.

Interpreting Graphs with a Critical Eye

Once you have created a graph or graphs for a set of data, what should you look for as you attempt to describe the data?

- First, check the horizontal and vertical **scales,** so that you are clear about what is being measured.
- Examine the **location** of the data distribution. Where on the horizontal axis is the center of the distribution? If you are comparing two distributions, are they both centered in the same place?
- Examine the **shape** of the distribution. Does the distribution have one "peak," a point that is higher than any other? If so, this is the most frequently occurring measurement or category. Is there more than one peak? Are there an approximately equal number of measurements to the left and right of the peak?
- Look for any unusual measurements or **outliers.** That is, are any measurements much bigger or smaller than all of the others? These outliers may not be representative of the other values in the set.

Distributions are often described according to their shapes.

Definition   A distribution is **symmetric** if the left and right sides of the distribution, when divided at the middle value, form mirror images.

A distribution is **skewed to the right** if a greater proportion of the measurements lie to the right of the peak value. Distributions that are **skewed right** contain a few unusually large measurements.

A distribution is **skewed to the left** if a greater proportion of the measurements lie to the left of the peak value. Distributions that are **skewed left** contain a few unusually small measurements.

A distribution is **unimodal** if it has one peak; a **bimodal** distribution has two peaks. Bimodal distributions often represent a mixture of two different populations in the data set.

EXAMPLE 1.9

Examine the three dotplots shown in Figure 1.12. Describe these distributions in terms of their locations and shapes.

FIGURE 1.12

Shapes of data distributions for Example 1.9

NEED a tip? **NEED A TIP?**

Symmetric ⇔ mirror images
Skewed right ⇔ long right tail
Skewed left ⇔ long left tail

Solution The first dotplot shows a *relatively symmetric* distribution with a single peak located at $x = 4$. If you were to fold the page at this peak, the left and right halves would *almost* be mirror images. The second dotplot, however, is far from symmetric. It has a long "right tail," meaning that there are a few unusually large observations. If you were to fold the page at the peak, a larger proportion of measurements would be on the right side than on the left. This distribution is *skewed to the right*. Similarly, the third dotplot with the long "left tail" is *skewed to the left*.

EXAMPLE 1.10

An administrative assistant for the athletics department at a local university is monitoring the GPAs for eight members of the women's volleyball team. He enters the GPAs into the database but accidentally misplaces the decimal point in the last entry.

2.8   3.0   3.0   3.3   2.4   3.4   3.0   .21

Use a dotplot to describe the data and uncover the assistant's mistake.

Solution   The dotplot of this small data set is shown in Figure 1.13(a). You can clearly see the *outlier* or unusual observation caused by the assistant's data entry error. Once the error has been corrected, as in Figure 1.13(b), you can see the correct distribution of the data set. Since this is a very small set, it is difficult to describe the shape of the distribution, although it seems to have a peak value around 3.0 and it appears to be relatively symmetric.

FIGURE 1.13

Distributions of GPAs for Example 1.10

(a)

(b)

NEED A TIP?

Outliers lie out, away from the main body of data.

When comparing graphs created for two data sets, you should compare their *scales of measurement, locations,* and *shapes,* and look for unusual measurements or outliers. Remember that outliers are not always caused by errors or incorrect data entry. Sometimes they provide very valuable information that should not be ignored. You may need additional information to decide whether an outlier is a valid measurement that is simply unusually large or small, or whether there has been some sort of mistake in the data collection. If the scales differ widely, be careful about making comparisons or drawing conclusions that might be inaccurate!

1.5   RELATIVE FREQUENCY HISTOGRAMS

A relative frequency histogram resembles a bar chart, but it is used to graph quantitative rather than qualitative data. The data in Table 1.9 are the birth weights of 30 full-term newborn babies, reproduced from Example 1.8 and shown as a dotplot in Figure 1.14(a). First, divide the interval from the smallest to the largest measurements into subintervals or *classes of equal length*. If you stack up the dots in each subinterval (Figure 1.14(b)), and draw a bar over each stack, you will have created a **frequency histogram** or a **relative frequency histogram,** depending on the scale of the vertical axis.

TABLE 1.9 ● **Birth Weights of 30 Full-Term Newborn Babies**

7.2	7.8	6.8	6.2	8.2
8.0	8.2	5.6	8.6	7.1
8.2	7.7	7.5	7.2	7.7
5.8	6.8	6.8	8.5	7.5
6.1	7.9	9.4	9.0	7.8
8.5	9.0	7.7	6.7	7.7

FIGURE 1.14

How to construct a histogram

(a)

(b)

Definition A **relative frequency histogram** for a quantitative data set is a bar graph in which the height of the bar shows "how often" (measured as a proportion or relative frequency) measurements fall in a particular class or subinterval. The classes or subintervals are plotted along the horizontal axis.

As a rule of thumb, the number of classes should range from 5 to 12; the more data available, the more classes you need.[†] The classes must be chosen so that each measurement falls into one and only one class. For the birth weights in Table 1.9, we decided to use eight intervals of equal length. Since the total span of the birth weights is

$$9.4 - 5.6 = 3.8$$

the minimum class width necessary to cover the range of the data is (3.8 ÷ 8) = .475. For convenience, we round this approximate width up to .5. Beginning the first interval at the lowest value, 5.6, we form subintervals from 5.6 up to *but not including* 6.1, 6.1 up to *but not including* 6.6, and so on. By using the **method of left inclusion,** and including the left class boundary point but not the right boundary point in the class, we eliminate any confusion about where to place a measurement that happens to fall on a class boundary point.

Table 1.10 shows the eight classes, labeled from 1 to 8 for identification. The boundaries for the eight classes, along with a tally of the number of measurements that fall in each class, are also listed in the table. As with the charts in Section 1.3, you can now measure *how often* each class occurs using *frequency* or *relative frequency.*

[†]You can use this table as a guide for selecting an appropriate number of classes. Remember that this is only a guide; you may use more or fewer classes than the table recommends if it makes the graph more descriptive.

Sample Size	25	50	100	200	500
Number of Classes	6	7	8	9	10

To construct the relative frequency histogram, plot the class boundaries along the horizontal axis. Draw a bar over each class interval, with height equal to the relative frequency for that class. The relative frequency histogram for the birth weight data, Figure 1.15, shows at a glance how birth weights are distributed over the interval 5.6 to 9.4.

TABLE 1.10 ● **Relative Frequencies for the Data of Table 1.9**

NEED A TIP?

Relative frequencies add to 1; frequencies add to n.

Class	Class Boundaries	Tally	Class Frequency	Class Relative Frequency
1	5.6 to <6.1	II	2	2/30
2	6.1 to <6.6	II	2	2/30
3	6.6 to <7.1	IIII	4	4/30
4	7.1 to <7.6	ℍℍ	5	5/30
5	7.6 to <8.1	ℍℍ III	8	8/30
6	8.1 to <8.6	ℍℍ	5	5/30
7	8.6 to <9.1	III	3	3/30
8	9.1 to <9.6	I	1	1/30

FIGURE 1.15 ●

Relative frequency histogram

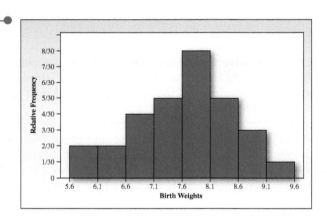

EXAMPLE 1.11 Twenty-five Starbucks® customers are polled in a marketing survey and asked, "How often do you visit Starbucks in a typical week?" Table 1.11 lists the responses for these 25 customers. Construct a relative frequency histogram to describe the data.

TABLE 1.11 ● **Number of Visits in a Typical Week for 25 Customers**

6	7	1	5	6
4	6	4	6	8
6	5	6	3	4
5	5	5	7	6
3	5	7	5	5

Solution The variable being measured is "number of visits to Starbucks," which is a discrete variable that takes on only integer values. In this case, it is simplest to choose the classes or subintervals as the integer values over the range of observed values: 1, 2, 3, 4, 5, 6, and 7. Table 1.12 shows the classes and their corresponding frequencies and relative frequencies. The relative frequency histogram is shown in Figure 1.16.

TABLE 1.12 • **Frequency Table for Example 1.11**

Number of Visits to Starbucks	Frequency	Relative Frequency
1	1	.04
2	—	—
3	2	.08
4	3	.12
5	8	.32
6	7	.28
7	3	.12
8	1	.04

FIGURE 1.16 •

Relative frequency histogram for Example 1.11

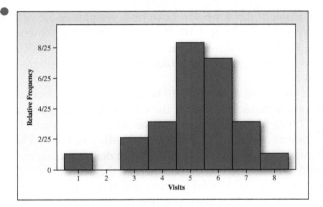

Notice that the distribution is skewed to the left and that there is a gap between 1 and 3.

 NEED TO KNOW...

How to Construct a Relative Frequency Histogram

1. Choose the number of classes, usually between 5 and 12. The more data you have, the more classes you should use.

2. Calculate the approximate class width by dividing the difference between the largest and smallest values by the number of classes.

3. Round the approximate class width up to a convenient number.

4. If the data are discrete, you might assign one class for each integer value taken on by the data. For a large number of integer values, you may need to group them into classes.

5. Locate the class boundaries. The lowest class must include the smallest measurement. Then add the remaining classes using the left inclusion method.

6. Construct a statistical table containing the classes, their frequencies, and their relative frequencies.

7. Construct the histogram like a bar graph, plotting class intervals on the horizontal axis and relative frequencies as the heights of the bars.

A relative frequency histogram can be used to describe the distribution of a set of data in terms of its *location* and *shape,* and to check for *outliers* as you did with other graphs. For example, the birth weight data were relatively symmetric, with no unusual measurements, while the Starbucks data were skewed left. Since the bar constructed above each class represents the *relative frequency* or proportion of the measurements in that class, these heights can be used to give us further information:

- The proportion of the measurements that fall in a particular class or group of classes
- The probability that a measurement drawn at random from the set will fall in a particular class or group of classes

Consider the relative frequency histogram for the birth weight data in Figure 1.15. What proportion of the newborns have birth weights of 7.6 or higher? This involves all classes beyond 7.6 in Table 1.10. Because there are 17 newborns in those classes, the proportion who have birth weights of 7.6 or higher is 17/30, or approximately 57%. This is also the percentage of the total area under the histogram in Figure 1.15 that lies to the right of 7.6.

Suppose you wrote each of the 30 birth weights on a piece of paper, put them in a hat, and drew one at random. What is the chance that this piece of paper contains a birth weight of 7.6 or higher? Since 17 of the 30 pieces of paper fall in this category, you have 17 chances out of 30; that is, the probability is 17/30. The word *probability* is not unfamiliar to you; we will discuss it in more detail in Chapter 4.

Although we are interested in describing the set of $n = 30$ birth weights, we might also be interested in the population from which the sample was drawn, which is the set of birth weights of *all* babies born at this hospital. Or, if we are interested in the weights of newborns in general, we might consider our sample as representative of the population of birth weights for newborns at similar metropolitan hospitals. A sample histogram provides valuable information about the population histogram—the graph that describes the distribution of the entire population. Remember, though, that different samples from the same population will produce *different* histograms, even if you use the same class boundaries. However, you can expect that the sample and population histograms will be similar. As you add more and more data to the sample, the two histograms become more and more alike. If you enlarge the sample to include the entire population, the two histograms will be identical!

1.5 EXERCISES

BASIC TECHNIQUES

 1.16 Construct a stem and leaf plot for these
EX0116 50 measurements:

3.1	4.9	2.8	3.6	2.5	4.5	3.5	3.7	4.1	4.9
2.9	2.1	3.5	4.0	3.7	2.7	4.0	4.4	3.7	4.2
3.8	6.2	2.5	2.9	2.8	5.1	1.8	5.6	2.2	3.4
2.5	3.6	5.1	4.8	1.6	3.6	6.1	4.7	3.9	3.9
4.3	5.7	3.7	4.6	4.0	5.6	4.9	4.2	3.1	3.9

a. Describe the shape of the data distribution. Do you see any outliers?

b. Use the stem and leaf plot to find the smallest observation.

c. Find the eighth and ninth largest observations.

1.17 Refer to Exercise 1.16. Construct a relative frequency histogram for the data.

a. Approximately how many class intervals should you use?

b. Suppose you decide to use classes starting at 1.6 with a class width of .5 (i.e., 1.6 to <2.1, 2.1 to <2.6). Construct the relative frequency histogram for the data.

c. What fraction of the measurements are less than 5.1?

d. What fraction of the measurements are larger than 3.6?

e. Compare the relative frequency histogram with the stem and leaf plot in Exercise 1.16. Are the shapes similar?

1.18 Consider this set of data:

EX0118

4.5	3.2	3.5	3.9	3.5	3.9
4.3	4.8	3.6	3.3	4.3	4.2
3.9	3.7	4.3	4.4	3.4	4.2
4.4	4.0	3.6	3.5	3.9	4.0

a. Construct a stem and leaf plot by using the leading digit as the stem.

b. Construct a stem and leaf plot by using each leading digit twice. Does this technique improve the presentation of the data? Explain.

1.19 A discrete variable can take on only the values 0, 1, or 2. A set of 20 measurements on this variable is shown here:

1	2	1	0	2
2	1	1	0	0
2	2	1	1	0
0	1	2	1	1

a. Construct a relative frequency histogram for the data.

b. What proportion of the measurements are greater than 1?

c. What proportion of the measurements are less than 2?

d. If a measurement is selected at random from the 20 measurements shown, what is the probability that it is a 2?

e. Describe the shape of the distribution. Do you see any outliers?

1.20 Refer to Exercise 1.19.

a. Draw a dotplot to describe the data.

b. How could you define the stem and the leaf for this data set?

c. Draw the stem and leaf plot using your decision from part b.

d. Compare the dotplot, the stem and leaf plot, and the relative frequency histogram (Exercise 1.19). Do they all convey roughly the same information?

1.21 Navigating a Maze An experimental psychologist measured the length of time it took for a rat to successfully navigate a maze on each of 5 days. The results are shown in the table. Create a line chart to describe the data. Do you think that any learning is taking place?

Day	1	2	3	4	5
Time (seconds)	45	43	46	32	25

1.22 Measuring over Time The value of a quantitative variable is measured once a year for a 10-year period. Here are the data:

EX0122

Year	Measurement	Year	Measurement
1	61.5	6	58.2
2	62.3	7	57.5
3	60.7	8	57.5
4	59.8	9	56.1
5	58.0	10	56.0

a. Create a line chart to describe the variable as it changes over time.

b. Describe the measurements using the chart constructed in part a.

1.23 Cheeseburgers Create a dotplot for the number of cheeseburgers consumed in a given week by 10 college students.

4	5	4	2	1
3	3	4	2	7

a. How would you describe the shape of the distribution?

b. What proportion of the students ate more than 4 cheeseburgers that week?

1.24 Test Scores The test scores on a 100-point test were recorded for 20 students:

EX0124

61	93	91	86	55	63	86	82	76	57
94	89	67	62	72	87	68	65	75	84

a. Use an appropriate graph to describe the data.

b. Describe the shape and location of the scores.

c. Is the shape of the distribution unusual? Can you think of any reason the distribution of the scores would have such a shape?

APPLICATIONS

1.25 Survival Times Altman and Bland report the survival times for patients with active hepatitis, half treated with prednisone and half receiving no treatment.[7] The data that follow are adapted from their data for those treated with prednisone. The survival times are recorded to the nearest month:

8	87	127	147
11	93	133	148
52	97	139	157
57	109	142	162
65	120	144	165

a. Look at the data. Can you guess the approximate shape of the data distribution?

b. Construct a relative frequency histogram for the data. What is the shape of the distribution?

c. Are there any outliers in the set? If so, which survival times are unusually short?

Data set EX0126 **1.26 A Recurring Illness** The length of time (in months) between the onset of a particular illness and its recurrence was recorded for $n = 50$ patients:

2.1	4.4	2.7	32.3	9.9	9.0	2.0	6.6	3.9	1.6
14.7	9.6	16.7	7.4	8.2	19.2	6.9	4.3	3.3	1.2
4.1	18.4	.2	6.1	13.5	7.4	.2	8.3	.3	1.3
14.1	1.0	2.4	2.4	18.0	8.7	24.0	1.4	8.2	5.8
1.6	3.5	11.4	18.0	26.7	3.7	12.6	23.1	5.6	.4

a. Construct a relative frequency histogram for the data.

b. Would you describe the shape as roughly symmetric, skewed right, or skewed left?

c. Give the fraction of recurrence times less than or equal to 10 months.

Data set EX0127 **1.27 Education Pays Off!** Education pays off, according to a snapshot provided by the *Bureau of Labor Statistics*.[8] The median weekly earnings for six different levels of education are shown in the table:

Educational Level	Median Weekly Earnings ($)
Less than a high school diploma	454
High school graduate	626
Some college, no degree	699
Associate degree	761
Bachelor's degree	1025
Master's degree	1257
Professional degree	1529
Doctoral degree	1532

Source: Bureau of Labor Statistics, Current Population Survey

a. What graphical methods could you use to describe the data?

b. Select the method from part a that you think best describes the data and create the appropriate graph.

c. How would you summarize the information that you see in the graph regarding educational levels and salary?

Data set EX0128 **1.28 Preschool** The ages (in months) at which 50 children were first enrolled in a preschool are listed below.

38	40	30	35	39	40	48	36	31	36
47	35	34	43	41	36	41	43	48	40
32	34	41	30	46	35	40	30	46	37
55	39	33	32	32	45	42	41	36	50
42	50	37	39	33	45	38	46	36	31

a. Construct a stem and leaf display for the data.

b. Construct a relative frequency histogram for these data. Start the lower boundary of the first class at 30 and use a class width of 5 months.

c. Compare the graphs in parts a and b. Are there any significant differences that would cause you to choose one as the better method for displaying the data?

d. What proportion of the children were 35 months (2 years, 11 months) or older, but less than 45 months (3 years, 9 months) of age when first enrolled in preschool?

e. If one child were selected at random from this group of children, what is the probability that the child was less than 50 months old (4 years, 2 months) when first enrolled in preschool?

Data set EX0129 **1.29 Organized Religion** Statistics of the world's religions are only very rough approximations, since many religions do not keep track of their membership numbers. An estimate of these numbers (in millions) is shown in the table.[9]

Religion	Members (millions)	Religion	Members (millions)
Buddhism	376	Judaism	14
Christianity	2100	Sikhism	23
Hinduism	900	Chinese traditional	394
Islam	1500	Other	61
Primal indigenous and African traditional	400		

a. Construct a pie chart to describe the total membership in the world's organized religions.

b. Construct a bar chart to describe the total membership in the world's organized religions.

c. Order the religious groups from the smallest to the largest number of members. Construct a Pareto chart to describe the data. Which of the three displays is most effective?

Data set
EX0130
1.30 How Long Is the Line? To decide on the number of service counters needed for stores to be built in the future, a supermarket chain wanted to obtain information on the length of time (in minutes) required to service customers. To find the distribution of customer service times, a sample of 60 customers' service times was recorded and are shown here:

3.6	1.9	2.1	.3	.8	.2	1.0	1.4	1.8	1.6
1.1	1.8	.3	1.1	.5	1.2	.6	1.1	.8	1.7
1.4	.2	1.3	3.1	.4	2.3	1.8	4.5	.9	.7
.6	2.8	2.5	1.1	.4	1.2	.4	1.3	.8	1.3
1.1	1.2	.8	1.0	.9	.7	3.1	1.7	1.1	2.2
1.6	1.9	5.2	.5	1.8	.3	1.1	.6	.7	.6

a. Construct a stem and leaf plot for the data.

b. What fraction of the service times are less than or equal to 1 minute?

c. What is the smallest of the 60 measurements?

1.31 Service Times, continued Refer to Exercise 1.30. Construct a relative frequency histogram for the supermarket service times.

a. Describe the shape of the distribution. Do you see any outliers?

b. Assuming that the outliers in this data set are valid observations, how would you explain them to the management of the supermarket chain?

c. Compare the relative frequency histogram with the stem and leaf plot in Exercise 1.30. Do the two graphs convey the same information?

Data set
EX0132
1.32 Calcium Content The calcium (Ca) content of a powdered mineral substance was analyzed 10 times with the following percent compositions recorded:

.0271	.0282	.0279	.0281	.0268
.0271	.0281	.0269	.0275	.0276

a. Draw a dotplot to describe the data. (HINT: The scale of the horizontal axis should range from .0260 to .0290.)

b. Draw a stem and leaf plot for the data. Use the numbers in the hundredths and thousandths places as the stem.

c. Are any of the measurements inconsistent with the other measurements, indicating that the technician may have made an error in the analysis?

Data set
EX0133
1.33 American Presidents The following table lists the ages at the time of death for the 38 deceased American presidents from George Washington to Ronald Reagan:[5]

Washington	67	Arthur	56
J. Adams	90	Cleveland	71
Jefferson	83	B. Harrison	67
Madison	85	McKinley	58
Monroe	73	T. Roosevelt	60
J. Q. Adams	80	Taft	72
Jackson	78	Wilson	67
Van Buren	79	Harding	57
W. H. Harrison	68	Coolidge	60
Tyler	71	Hoover	90
Polk	53	F. D. Roosevelt	63
Taylor	65	Truman	88
Fillmore	74	Eisenhower	78
Pierce	64	Kennedy	46
Buchanan	77	L. Johnson	64
Lincoln	56	Nixon	81
A. Johnson	66	Ford	93
Grant	63	Carter	93
Hayes	70	Reagan	93
Garfield	49		

a. Before you graph the data, try to visualize the distribution of the ages at death for the presidents. What shape do you think it will have?

b. Construct a stem and leaf plot for the data. Describe the shape. Does it surprise you?

c. The five youngest presidents at the time of death appear in the lower "tail" of the distribution. Three of the five youngest have one common trait. Identify the five youngest presidents at death. What common trait explains these mea-surements?

Data set
EX0134
1.34 RBC Counts The red blood cell count of a healthy person was measured on each of 15 days. The number recorded is measured in 10^6 cells per microliter (μL).

5.4	5.2	5.0	5.2	5.5
5.3	5.4	5.2	5.1	5.3
5.3	4.9	5.4	5.2	5.2

a. Use an appropriate graph to describe the data.

b. Describe the shape and location of the red blood cell counts.

c. If the person's red blood cell count is measured today as $5.7 \times 10^6/\mu L$, would you consider this unusual? What conclusions might you draw?

Data set
EX0135
1.35 Batting Champions The officials of major league baseball have crowned a batting champion in the National League each year since 1876. A sample of winning batting averages is listed in the table:[5]

Year	Name	Average
2000	Todd Helton	.372
1915	Larry Doyle	.320
1917	Edd Roush	.341
1934	Paul Waner	.362
1911	Honus Wagner	.334
1898	Willie Keeler	.379
1924	Roger Hornsby	.424
1963	Tommy Davis	.326
1992	Gary Sheffield	.330
1954	Willie Mays	.345
1975	Bill Madlock	.354
1958	Richie Ashburn	.350
1942	Ernie Lombardi	.330
1948	Stan Musial	.376
1971	Joe Torre	.363
1996	Tony Gwynn	.353
1961	Roberto Clemente	.351
1968	Pete Rose	.335
1885	Roger Connor	.371
2009	Hanley Ramirez	.342

a. Construct a relative frequency histogram to describe the batting averages for these 20 champions.

b. If you were to randomly choose one of the 20 names, what is the chance that you would choose a player whose average was above .400 for his championship year?

Data set EX0136 **1.36 Top 20 Movies** The table that follows shows the weekend gross ticket sales for the top 20 movies for the weekend of June 25, 2010:[10]

Movie	Weekend Gross ($ millions)
1. Toy Story 3	59.3
2. Grown Ups	40.5
3. Knight and Day	20.1
4. The Karate Kid	15.5
5. The A-Team	6.2
6. Get Him to the Greek	3.1
7. Shrek Forever After	3.1
8. Prince of Persia	2.8
9. Killers	1.9
10. Jonah Hex	1.6
11. Iron Man 2	1.4
12. Sex and the City 2	1.2
13. Marmaduke	1.0
14. Robin Hood	0.6
15. Solitary Man	0.5
16. How to Train Your Dragon	0.5
17. Winter's Bone	0.4
18. Letters to Juliet	0.4
19. Joan Rivers: A Piece of Work	0.4
20. Cyrus	0.3

Source: www.radiofree.com/mov-tops.shtml

a. Draw a stem and leaf plot for the data. Describe the shape of the distribution. Are there any outliers?

b. Construct a dotplot for the data. Which of the two graphs is more informative? Explain.

Data set EX0137 **1.37 Hazardous Waste** How safe is your neighborhood? Are there any hazardous waste sites nearby? The table shows the number of hazardous waste sites in each of the 50 states and the District of Columbia in the year 2009:[5]

AL	15	HI	3	MA	32	NM	14	SD	2
AK	6	ID	9	MI	69	NY	90	TN	15
AZ	9	IL	48	MN	25	NC	36	TX	50
AR	9	IN	32	MS	6	ND	0	UT	19
CA	96	IA	12	MO	31	OH	41	VT	11
CO	20	KS	12	MT	17	OK	9	VA	31
CT	15	KY	14	NE	13	OR	13	WA	48
DE	15	LA	12	NV	1	PA	97	WV	9
DC	1	ME	12	NH	21	RI	12	WI	39
FL	55	MD	19	NJ	114	SC	26	WY	2
GA	16								

a. What variable is being measured? Is the variable discrete or continuous?

b. Describe the shape of the data distribution using the stem and leaf plot shown here. Identify the unusually large measurements marked "HI" by state.

Stem and Leaf Display: Hazardous Waste

```
Stem-and-leaf of Sites  N = 51
Leaf Unit = 1.0
   6   0   011223
  13   0   6699999
  23   1   1222223344
  (8)  1   55556799
  20   2   01
  18   2   56
  16   3   1122
  12   3   69
  10   4   1
   9   4   88
   7   5   0
   6   5   5

       HI 69, 90, 96, 97, 114
```

c. Can you think of any reason these five states would have a large number of hazardous waste sites? What other variable might you measure to help explain why the data behave as they do?

As you continue to work through the exercises in this chapter, you will become more experienced in recognizing different types of data and in determining the most appropriate graphical method to use. Remember that the type of graphic you use is not as important as the interpretation that accompanies the picture. Look for these important characteristics:

- Location of the center of the data
- Shape of the distribution of data
- Unusual observations in the data set

Using these characteristics as a guide, you can interpret and compare sets of data using graphical methods, which are only the first of many statistical tools that you will soon have at your disposal.

CHAPTER REVIEW

Key Concepts

I. How Data Are Generated
1. Experimental units, variables, measurements
2. Samples and populations
3. Univariate, bivariate, and multivariate data

II. Types of Variables
1. Qualitative or categorical
2. Quantitative
   a. Discrete
   b. Continuous

III. Graphs for Univariate Data Distributions
1. Qualitative or categorical data
   a. Pie charts
   b. Bar charts
2. Quantitative data
   a. Pie and bar charts
   b. Line charts
   c. Dotplots
   d. Stem and leaf plots
   e. Relative frequency histograms
3. Describing data distributions
   a. Shapes—symmetric, skewed left, skewed right, unimodal, bimodal
   b. Proportion of measurements in certain intervals
   c. Outliers

 TECHNOLOGY TODAY

Introduction to *Microsoft Excel*

MS Excel is a spreadsheet program in the Microsoft Office system. It is designed for a variety of analytical applications, including statistical applications. We will assume that you are familiar with Windows, and that you know the basic techniques necessary for executing commands from the tabs, groups, and drop-down menus at the top of the screen. If not, perhaps a lab or teaching assistant can help you to master the basics. The current version of *MS Excel* at the time of this printing is *Excel 2010*, used in the Windows 7 environment. When the program opens, a **spreadsheet** appears (see Figure 1.17), containing rows and columns into which you can enter data. Tabs at the bottom of the screen identify the three spreadsheets available for use; when saved as a collection, these spreadsheets are called a **workbook.**

FIGURE 1.17

Graphing with *Excel*

Pie charts, bar charts, and **line charts** can all be created in *MS Excel*. Data is entered into an *Excel* spreadsheet, including labels if needed. Highlight the data to be graphed, and then click the chart type that you want on the **Insert** tab in the **Charts** group. Once the chart has been created, it can be edited in a variety of ways to change its appearance.

EXAMPLE 1.12

(Pie and Bar Charts) The class status of 105 students in an introductory statistics class are listed in Table 1.13. The qualitative variable "class status" has been recorded for each student, and the frequencies have already been recorded.

Status of Students in Statistics Class

Status	Freshman	Sophomore	Junior	Senior	Grad Student
Frequency	5	23	32	35	10

1. Enter the *categories* into column A of the first spreadsheet and the *frequencies* into column B. You should have two columns of data, including the labels.

2. Highlight the data, using your left mouse to *click-and-drag* from cell A1 to cell B6 (sometimes written as **A1:B6**). Click the **Insert** tab and select **Pie** in the **Charts** group. In the drop-down list, you will see a variety of styles to choose from. Select the first option to produce the pie chart. Double click on the title "Frequency" and change the title to "Student Status."

3. ***Editing the pie chart:*** Once the chart has been created, use your mouse to make sure that the chart is selected. You should see a green area above the tabs marked "Chart Tools." Click the **Design** tab, and look at the drop-down lists in the **Chart Layout** and **Chart Styles** groups. These lists allow you to alter the appearance of your chart. In Figure 1.18(a), the pie chart has been changed so that the percentages are shown in the appropriate sectors. By clicking on the legend, we have dragged it so that it is closer to the pie chart.

FIGURE 1.18

(a)

(b)

4. Click on various parts of the pie chart (legend, chart area, sector) and a box with round and/or square handles will appear. Double-click, and a dialog box will appear. You can adjust the appearance of the selected object or region in this box and click **OK.** Click **Cancel** to exit the dialog box without change!

5. Still in the **Design** section, but in the **Type** group, click on **Change Chart Type** and choose the simplest **Column** type. Click **OK** to create a bar chart for the same data set, shown in Figure 1.18(b).

6. *Editing the bar chart:* Again, you can experiment with the various options in the **Chart Layout** and **Chart Styles** groups to change the look of the chart. You can click the entire bar chart ("chart area") or the interior "Plot area" to stretch the chart. You can change colors by double-clicking on the appropriate region. We have chosen a design that allows axis titles and have deleted the "frequency legend entry." We have also chosen to delete the minor gridlines, by clicking the **Layout** tab in the **Chart Tools,** using the **Gridlines** drop-down list, and selecting **Primary Horizontal Gridlines ▶ Major Gridlines.** We have decreased the gaps between the bars by right-clicking on one of the bars, selecting **Format Data Series,** and changing the **Gap Width to 50%.**

EXAMPLE 1.13 (Line Charts) The Dow Jones Industrial Average was monitored at the close of trading for 10 days in a recent year, with the results shown in Table 1.14.

TABLE 1.14 **Dow Jones Industrial Average**

Day	1	2	3	4	5	6	7	8	9	10
DJIA	10,636	10,680	10,674	10,653	10,698	10,644	10,378	10,319	10,303	10,302

1. Click the tab at the bottom of the screen marked "Sheet 2." Enter the *Days* into column A of this second spreadsheet and the *DJIA* into column B. You should have two columns of data, including the labels.

2. Highlight the DJIA data in column B, using your left mouse to *click-and-drag* from cell B1 to cell B11 (sometimes written as **B1:B11).** Click the **Insert** tab and select **Line** in the **Charts** group. In the drop-down list, you will see a variety of styles to choose from. Select the first option to produce the line chart.

3. *Editing the line chart:* Again, you can experiment with the various options in the **Chart Layout** and **Chart Styles** groups to change the look of the chart. We have chosen a design that allows titles on both axes, which we have changed to "Day" and "DJIA," and we have deleted the title and the "frequency legend entry." The line chart is shown in Figure 1.19.

FIGURE 1.19 ●

4. *Note:* If your time series involves time periods that are *not equally spaced,* it is better to use a **scatterplot** with points connected to form a line chart. This procedure is described in the *Technology Today* section in Chapter 3 of the text.

EXAMPLE (1.14) **(Frequency Histograms)** The top 40 stocks on the over-the-counter (OTC) market, ranked by percentage of outstanding shares traded on a particular day, are listed in Table 1.15.

TABLE 1.15 ● **Percentage of OTC Stocks Traded**

11.88	6.27	5.49	4.81	4.40	3.78	3.44	3.11	2.88	2.68
7.99	6.07	5.26	4.79	4.05	3.69	3.36	3.03	2.74	2.63
7.15	5.98	5.07	4.55	3.94	3.62	3.26	2.99	2.74	2.62
7.13	5.91	4.94	4.43	3.93	3.48	3.20	2.89	2.69	2.61

1. Many of the statistical procedures that we will use in this textbook require the installation of the **Analysis ToolPak** add-in. To load this add-in, click **File ▶ Options ▶ Add-ins.** Select **Analysis ToolPak** and click **OK.**

2. Click the tab at the bottom of the screen marked "Sheet 3." Enter the data into the first column of this spreadsheet and include the label "Stocks" in the first cell.

3. *Excel* refers to the maximum value for each class interval as a **bin.** This means that *Excel* is using a **method of right inclusion,** which is slightly different from the method presented in Section 1.5. For this example, we choose to use the class intervals >2.5–3.5, >3.5–4.5, >4.5–5.5, etc. Enter the *bin values* (3.5, 4.5, 5.5, ..., 12.5) into the second column of the spreadsheet, labeling them as "Percent Traded" in cell **B1.**

4. Select **Data ▶ Data Analysis ▶ Histogram** and click **OK.** The Histogram dialog box will appear, as shown in Figure 1.20.

FIGURE 1.20 ●

5. Highlight or type in the appropriate Input Range and Bin Range for the data. Notice that you can click the minimize button on the right of the box before you *click-and-drag* to highlight. Click the minimize button again to see the entire dialog box. The Input Range will appear as $A$1:$A$41, with the dollar sign indicating an *absolute cell range*. Make sure to click the "Labels" and "Chart Output" check boxes. Pick a convenient cell location for the output (we picked **D1**) and click **OK.** The frequency table and histogram will appear in the spreadsheet. The histogram (Figure 1.21 (a)) doesn't appear quite like we wanted.

FIGURE 1.21

(a)

(b)

6. ***Editing the histogram:*** Click on the frequency legend entry and press the Delete key. Then select the Data Series by double-clicking on a bar. In the **Series Options** box that appears, change the **Gap Width** to 0% (no gap) and click **Close.** Stretch the graph by dragging the lower right corner, and edit the colors, title, and labels if necessary to finish your histogram, as shown in Figure 1.21 (b). Remember that the numbers shown along the horizontal axis are the **bins,** the upper limit of the class interval, *not the midpoint of the interval*.

7. You can save your *Excel* workbook for use at a later time using **File ▶ Save** or **File ▶ Save As** and naming it "Chapter 1."

TECHNOLOGY TODAY

Introduction to *MINITAB*™

MINITAB computer software is a Windows-based program designed specifically for statistical applications. We will assume that you are familiar with Windows, and that you know the basic techniques necessary for executing commands from the tabs and drop-down menus at the top of the screen. If not, perhaps a lab or teaching assistant can help you to master the basics. The current version of *MINITAB* at the time of this printing is *MINITAB 16,* used in the Windows 7 environment. When the program opens, the main screen (see Figure 1.22) is displayed, containing two windows: the Data window, similar to an *Excel* spreadsheet, and the Session window, in which your results will appear. Just as with *MS Excel, MINITAB* allows you to save worksheets (similar to *Excel* spreadsheets), projects (collections of worksheets), or graphs.

FIGURE 1.22

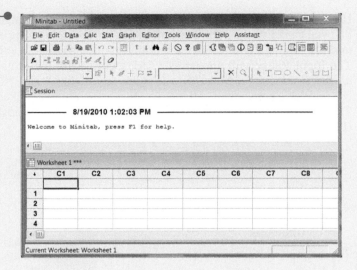

Graphing with *MINITAB*

All of the graphical methods that we have discussed in this chapter can be created in *MINITAB*. Data is entered into a *MINITAB* worksheet, with labels entered in the gray cells just below the column name (C1, C2, etc.) in the Data window.

EXAMPLE 1.15

(Pie and Bar Charts) The class status of 105 students in an introductory statistics class are listed in Table 1.16. The qualitative variable "class status" has been recorded for each student, and the frequencies have already been recorded.

TABLE 1.16

Status of Students in Statistics Class

Status	Freshman	Sophomore	Junior	Senior	Grad Student
Frequency	5	23	32	35	10

1. Enter the *categories* into column C1, with your own descriptive name, perhaps "Status" in the gray cell. Notice that the name **C1** has changed to **C1-T** because you are entering text rather than numbers. Continue by naming column 2 (C2) "Frequency," and enter the five numerical frequencies into C2.

2. To construct a pie chart for these data, click on **Graph ▶ Pie Chart,** and a Dialog box will appear (see Figure 1.23). Click the radio button marked **Chart values from a table.** Then place your cursor in the box marked "Categorical variable." Either (1) highlight C1 in the list at the left and choose **Select,** (2) double-click on C1 in the list at the left, or (3) type C1 in the "Categorical variable" box. Similarly, place the cursor in the box marked "Summary variables" and select C2. Click **Labels** and select the tab marked **Slice Labels.** Check the boxes marked "Category names" and "Percent." When you click **OK** twice, *MINITAB* will create the pie chart in Figure 1.24(a). We have removed the legend by selecting and deleting it.

FIGURE 1.23) ●

3. As you become more proficient at using the pie chart command, you may want to take advantage of some of the options available. Once the chart is created, *right-click* on the pie chart and select **Edit Pie.** You can change the colors and format of the chart, "explode" important sectors of the pie, and change the order of the categories. If you *right-click* on the pie chart and select **Update Graph Automatically,** the pie chart will automatically update when you change the data in columns C1 and C2 of the *MINITAB* worksheet.

4. If you would rather construct a bar chart, use the command **Graph ▶ Bar Chart.** In the Dialog box that appears, choose **Simple.** Choose an option in the "Bars represent" drop-down list, depending on the way that the data has been entered into the worksheet. For the data in Table 1.13, we choose "Values from a table" and click **OK.** When the Dialog box appears, place your cursor in the "Graph variables" box and select **C2.** Place your cursor in the "Categorical variable" box, and select **C1.** Click **OK** to finish the bar chart, shown in Figure 1.24(b). Once the chart is created, *right-click* on various parts of the bar chart and choose **Edit** to change the look of the chart.

FIGURE 1.24) ●

(a)

(b)

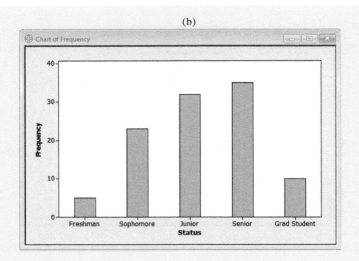

EXAMPLE 1.16

(Line Charts) The Dow Jones Industrial Average was monitored at the close of trading for 10 days in a recent year with the results shown in Table 1.17.

TABLE 1.17

Dow Jones Industrial Average

Day	1	2	3	4	5	6	7	8	9	10
DJIA	10,636	10,680	10,674	10,653	10,698	10,644	10,378	10,319	10,303	10,302

1. Although we could simply enter this data into third and fourth columns of the current worksheet, let's create a new worksheet using **File ▶ New ▶ Minitab Worksheet.** Enter the *Days* into column C1 of this second spreadsheet and the *DJIA* into column C2. You should have two columns of data, including the labels.

2. To create the line chart, use **Graph ▶ Time Series Plot ▶ Simple.** In the Dialog box that appears, place your cursor in the "Series" box and select "DJIA" from the list to the left. Under **Time/Scale,** choose "Stamp" and select column **C1** ("Day") in the box labeled "Stamp Columns." Click **OK** twice. You can select the numbered days shown above the line and delete them to obtain the line chart shown in Figure 1.25.

FIGURE 1.25

EXAMPLE 1.17 (**Dotplots, Stem and Leaf Plots, Histograms**) The top 40 stocks on the over-the-counter (OTC) market, ranked by percentage of outstanding shares traded on a particular day, are listed in Table 1.18. Create a new worksheet (**File ▶ New ▶ Minitab Worksheet**). Enter the data into column C1 and name it "Stocks" in the gray cell just below the C1.

TABLE 1.18 ● **Percentage of OTC Stocks Traded**

11.88	6.27	5.49	4.81	4.40	3.78	3.44	3.11	2.88	2.68
7.99	6.07	5.26	4.79	4.05	3.69	3.36	3.03	2.74	2.63
7.15	5.98	5.07	4.55	3.94	3.62	3.26	2.99	2.74	2.62
7.13	5.91	4.94	4.43	3.93	3.48	3.20	2.89	2.69	2.61

1. To create a dotplot, use **Graph ▶ Dotplot.** In the Dialog box that appears, choose **One Y ▶ Simple** and click **OK.** To create a stem and leaf plot, use **Graph ▶ Stem-and-Leaf.** For either graph, place your cursor in the "Graph variables" box, and select "Stocks" from the list to the left (see Figure 1.26).

FIGURE 1.26

2. You can choose from a variety of formatting options before clicking **OK.** The dotplot appears as a graph, while the stem and leaf plot appears in the Session window. To print either a Graph window or the Session window, click on the window to make it active and use **File ▶ Print Graph** (or **Print Session Window**).

3. To create a histogram, use **Graph ▶ Histogram.** In the Dialog box that appears, choose **Simple** and click **OK,** selecting "Stocks" for the "Graph variables" box. Select **Scale ▶ Y-Scale Type** and click the radio button marked "Frequency." (You can edit the histogram later to show relative frequencies.) Click **OK** twice. Once the histogram has been created, *right-click* on the Y-axis and choose **Edit Y-Scale.** Under the tab marked "Scale," you can click the radio button marked "Position of ticks" and type in **0 5 10 15.** Then click the tab marked "Labels," the radio button marked "Specified" and type **0 5/40 10/40 15/40.** Click **OK.** This will reduce the number of ticks on the y-axis and change them to relative frequencies. Finally, double-click on the word "Frequency" along the y-axis. Change the box marked "Text" to read "Relative frequency" and click **OK.**

4. To adjust the type of boundaries for the histogram, *right-click* on the bars of the histogram and choose **Edit Bars.** Use the tab marked "Binning" to choose either "Cutpoints" or "Midpoints" for the histogram; you can specify the cutpoint or midpoint positions if you want. In this same **Edit** box, you can change the colors, fill type, and font style of the histogram. If you *right-click* on the bars and select **Update Graph Automatically,** the histogram will automatically update when you change the data in the "Stocks" column.

As you become more familiar with *MINITAB* for Windows, you can explore the various options available for each type of graph. It is possible to plot more than one variable at a time, to change the axes, to choose the colors, and to modify graphs in many ways. However, even with the basic default commands, it is clear that the distribution of OTC stocks in Figure 1.27 is highly skewed to the right.

FIGURE 1.27

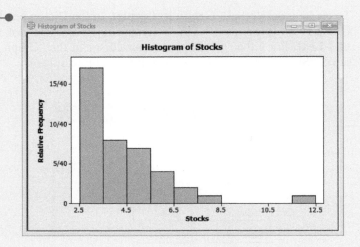

Supplementary Exercises

1.38 Quantitative or Qualitative? Identify each variable as quantitative or qualitative:

a. Ethnic origin of a candidate for public office

b. Score (0–100) on a placement examination

c. Fast-food establishment preferred by a student (McDonald's, Burger King, or Carl's Jr.)

d. Mercury concentration in a sample of tuna

1.39 Symmetric or Skewed? Do you expect the distributions of the following variables to be symmetric or skewed? Explain.

a. Size in dollars of nonsecured loans

b. Size in dollars of secured loans

c. Price of an 8-ounce can of peas

d. Height in inches of freshman women at your university

e. Number of broken taco shells in a package of 100 shells

f. Number of ticks found on each of 50 trapped cottontail rabbits

1.40 Continuous or Discrete? Identify each variable as continuous or discrete:

a. Number of homicides in Detroit during a 1-month period

b. Length of time between arrivals at an outpatient clinic

c. Number of typing errors on a page of manuscript

d. Number of defective lightbulbs in a package containing four bulbs

e. Time required to finish an examination

1.41 Continuous or Discrete, again Identify each variable as continuous or discrete:

a. Weight of two dozen shrimp

b. A person's body temperature

c. Number of people waiting for treatment at a hospital emergency room

d. Number of properties for sale by a real estate agency

e. Number of claims received by an insurance company during one day

1.42 Continuous or Discrete, again Identify each variable as continuous or discrete:

a. Number of people in line at a supermarket checkout counter

b. Depth of a snowfall

c. Length of time for a driver to respond when faced with an impending collision

d. Number of aircraft arriving at the Atlanta airport in a given hour

1.43 Aqua Running Aqua running has been suggested as a method of cardiovascular conditioning for injured athletes and others who want a low-impact aerobics program. A study reported in the *Journal of Sports Medicine* investigated the relationship between exercise cadence and heart rate by measuring the heart rates of 20 healthy volunteers at a cadence of 96 steps per minute.[11] The data are listed here:

87	109	79	80	96	95	90	92	96	98
101	91	78	112	94	98	94	107	81	96

Construct a stem and leaf plot to describe the data. Discuss the characteristics of the data distribution.

1.44 Major World Lakes A lake is a body of water surrounded by land. Hence, some bodies of water named "seas," like the Caspian Sea, are actual salt lakes. In the table that follows, the length in miles is listed for the major natural lakes of the world, excluding the Caspian Sea, which has a length of 760 miles.[5]

Name	Length (mi)	Name	Length (mi)
Superior	350	Titicaca	122
Victoria	250	Nicaragua	102
Huron	206	Athabasca	208
Michigan	307	Reindeer	143
Aral Sea	260	Tonle Sap	70
Tanganyika	420	Turkana	154

Baykal	395	Issyk Kul	115
Great Bear	192	Torrens	130
Nyasa	360	Vänern	91
Great Slave	298	Nettilling	67
Erie	241	Winnipegosis	141
Winnipeg	266	Albert	100
Ontario	193	Nipigon	72
Balkhash	376	Gairdner	90
Ladoga	124	Urmia	90
Maracaibo	133	Manitoba	140
Onega	145	Chad	175
Eyre	90		

Source: *The World Almanac and Book of Facts 2011*

a. Use a stem and leaf plot to describe the lengths of the world's major lakes.

b. Use a histogram to display these same data. How does this compare to the stem and leaf plot in part a?

c. Are these data symmetric or skewed? If skewed, what is the direction of the skewing?

1.45 Ages of Pennies We collected 50 pennies and recorded their ages, by calculating AGE = CURRENT YEAR − YEAR ON PENNY.

5	1	9	1	2	20	0	25	0	17
1	4	4	3	0	25	3	3	8	28
5	21	19	9	0	5	0	2	1	0
0	1	19	0	2	0	20	16	22	10
19	36	23	0	1	17	6	0	5	0

a. Before drawing any graphs, try to visualize what the distribution of penny ages will look like. Will it be mound-shaped, symmetric, skewed right, or skewed left?

b. Draw a relative frequency histogram to describe the distribution of penny ages. How would you describe the shape of the distribution?

1.46 Ages of Pennies, continued The data below represent the ages of a different set of 50 pennies, again calculated using AGE = CURRENT YEAR − YEAR ON PENNY.

41	9	0	4	3	0	3	8	21	3
2	10	4	0	14	0	25	12	24	19
3	1	14	7	2	4	4	5	1	20
14	9	3	5	3	0	8	17	16	0
0	7	3	5	23	7	28	17	9	2

a. Draw a relative frequency histogram to describe the distribution of penny ages. Is the shape similar to the shape of the relative frequency histogram in Exercise 1.45?

b. Draw a stem and leaf plot to describe the penny ages. Are there any unusually large or small measurements in the set?

Data set EX0147   **1.47 Presidential Vetoes** Here is a list of the 44 presidents of the United States along with the number of regular vetoes used by each:[5]

Washington	2	B. Harrison	19
J. Adams	0	Cleveland	42
Jefferson	0	McKinley	6
Madison	5	T. Roosevelt	42
Monroe	1	Taft	30
J. Q. Adams	0	Wilson	33
Jackson	5	Harding	5
Van Buren	0	Coolidge	20
W. H. Harrison	0	Hoover	21
Tyler	6	F. D. Roosevelt	372
Polk	2	Truman	180
Taylor	0	Eisenhower	73
Fillmore	0	Kennedy	12
Pierce	9	L. Johnson	16
Buchanan	4	Nixon	26
Lincoln	2	Ford	48
A. Johnson	21	Carter	13
Grant	45	Reagan	39
Hayes	12	G. H. W. Bush	29
Garfield	0	Clinton	36
Arthur	4	G. W. Bush	11
Cleveland	304	Obama	1

Source: The World Almanac and Book of Facts 2011

Use an appropriate graph to describe the number of vetoes cast by the 44 presidents. Write a summary paragraph describing this set of data.

Data set EX0148   **1.48 Windy Cities** Are some cities more windy than others? Does Chicago deserve to be nicknamed "The Windy City"? These data are the average wind speeds (in miles per hour) for 54 selected cities in the United States:[5]

8.9	12.3	10.7	8.4	7.8	11.5	8.2	9.0	8.8
7.1	11.8	10.3	7.7	9.0	10.5	9.1	8.7	8.7
9.1	9.0	10.5	11.2	7.7	8.8	12.2	7.9	8.8
8.7	7.1	8.7	7.6	5.1	35.1	10.5	10.4	11.0
10.2	8.6	10.7	9.6	8.3	8.0	9.5	7.7	9.4
8.7	7.8	10.2	6.9	9.2	10.2	6.2	9.6	12.2

Source: The World Almanac and Book of Facts 2011

a. Construct a relative frequency histogram for the data. (HINT: Choose the class boundaries without including the value $x = 35.1$ in the range of values.)

b. The value $x = 35.1$ was recorded at Mt. Washington, New Hampshire. Does the geography of that city explain the observation?

c. The average wind speed in Chicago is recorded as 10.3 miles per hour. Do you consider this unusually windy?

Data set EX0149   **1.49 Kentucky Derby** The following data set shows the winning times (in seconds) for the Kentucky Derby races from 1950 to 2010.[12]

(1950)	121.3	122.3	121.3	122.0	123.0	121.4	123.2	122.1	125.0	122.1
(1960)	122.2	124.0	120.2	121.4	120.0	121.1	122.0	120.3	122.1	121.4
(1970)	123.2	123.1	121.4	119.2†	124.0	122.0	121.3	122.1	121.1	122.2
(1980)	122.0	122.0	122.2	122.1	122.2	120.1	122.4	123.2	122.2	125.0
(1990)	122.0	123.0	123.0	122.2	123.3	121.1	121.0	122.4	122.2	123.2
(2000)	121.0	119.97	121.13	121.19	124.06	122.75	121.36	122.17	121.86	122.66
(2010)	124.4									

†Record time set by Secretariat in 1973.
Source: www.kentuckyderby.com

a. Do you think there will be a trend in the winning times over the years? Draw a line chart to verify your answer.

b. Describe the distribution of winning times using an appropriate graph. Comment on the shape of the distribution and look for any unusual observations.

Data set EX0150   **1.50 Gulf Oil Spill Cleanup** On April 20, 2010, the United States experienced a major environmental disaster when a Deepwater Horizon drilling rig exploded in the Gulf of Mexico. The number of personnel and equipment used in the Gulf oil spill cleanup, beginning May 2, 2010 (Day 13) through June 9, 2010 (Day 51) is given in the following table.[13]

	Day 13	Day 26	Day 39	Day 51
Number of personnel (1000s)	3.0	17.5	20.0	24.0
Federal Gulf fishing areas closed	3%	8%	25%	32%
Booms laid (miles)	46	315	644	909
Dispersants used (1000 gallons)	156	500	870	1143
Vessels deployed (100s)	1.0	6.0	14.0	35.0

a. What graphical methods could you use to display these data?

b. Before you draw your graphs, what trends do you see in each of the variables displayed?

c. Use a line chart to display the number of personnel deployed over this 51-day period.

d. Use a bar graph to display the percentage of federal Gulf fishing areas closed.

e. Use a line chart to display the amounts of dispersants used. Is there any underlying straight line relationship over time?

Data set EX0151   **1.51 Election Results** The 2008 election was a race in which Barack Obama defeated John McCain and other candidates, receiving 53% of the popular vote. The popular vote (in thousands) for Barack Obama in each of the 50 states is listed below:[14]

AL	813	HI	326	MA	1904	NM	472	SD	171
AK	124	ID	236	MI	2873	NY	4805	TN	1087
AZ	1035	IL	3420	MN	1573	NC	2143	TX	3529
AR	422	IN	1374	MS	555	ND	141	UT	328
CA	8274	IA	829	MO	1442	OH	2933	VT	219

CO	1289	KS	515	MT	232	OK	502	VA	1960
CT	998	KY	752	NE	333	OR	1037	WA	1751
DE	255	LA	783	NV	534	PA	3276	WV	304
FL	4282	ME	422	NH	385	RI	297	WI	1677
GA	1844	MD	1629	NJ	2215	SC	862	WY	83

a. By just looking at the table, what shape do you think the data distribution for the popular vote by state will have?

b. Draw a relative frequency histogram to describe the distribution of the popular vote for President Obama in the 50 states.

c. Did the histogram in part b confirm your guess in part a? Are there any outliers? How can you explain them?

1.52 **Election Results, continued** Refer to Exercise 1.51. Listed here is the *percentage* of the popular vote received by President Obama in each of the 50 states:[14]

AL	39	HI	72	MA	62	NM	57	SD	45
AK	38	ID	36	MI	57	NY	63	TN	42
AZ	45	IL	62	MN	54	NC	50	TX	44
AR	39	IN	50	MS	43	ND	45	UT	34
CA	61	IA	54	MO	49	OH	52	VT	68
CO	54	KS	42	MT	47	OK	34	VA	53
CT	61	KY	41	NE	42	OR	57	WA	58
DE	62	LA	40	NV	55	PA	55	WV	43
FL	51	ME	58	NH	54	RI	63	WI	56
GA	47	MD	62	NJ	57	SC	45	WY	33

a. By just looking at the table, what shape do you think the data distribution for the *percentage* of the popular vote by state will have?

b. Draw a relative frequency histogram to describe the distribution. Describe the shape of the distribution and look for outliers. Did the graph confirm your answer to part a?

1.53 **Election Results, continued** Refer to Exercises 1.51 and 1.52. The accompanying stem and leaf plots were generated using *MINITAB* for the variables named "Popular Vote" and "Percent Vote."

Stem and Leaf Display: Popular Vote

```
Stem-and-leaf of
Popular Vote N = 50
Leaf Unit = 100
17    0 01112222233333444
(10)  0 5555778889
23    1 000234
17    1 5667899
10    2 12
8     2 89
6     3 24
4     3 5
```

Stem and Leaf Display: Percent Vote

```
Stem-and-leaf of
Percent Vote N = 50
Leaf Unit = 1.0
3     3 344
7     3 6899
15    4 01222334
22    4 5555779
(9)   5 001234444
19    5 556777788
10    6 11222233
2     6 8
1     7 2
```

HI 42, 48, 82

a. Describe the shapes of the two distributions. Are there any outliers?

b. Do the stem and leaf plots resemble the relative frequency histograms constructed in Exercises 1.51 and 1.52?

c. Explain why the distribution of the popular vote for President Obama by state is skewed while the percentage of popular votes by state is mound-shaped.

1.54 **Student Heights** The self-reported heights of 105 students in a biostatistics class are described in the relative frequency histogram below.

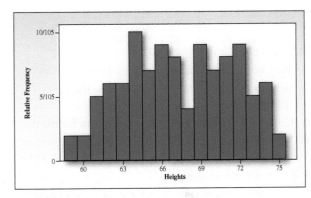

a. Describe the shape of the distribution.

b. Do you see any unusual feature in this histogram?

c. Can you think of an explanation for the two peaks in the histogram? Is there some other factor that is causing the heights to mound up in two separate peaks? What is it?

1.55 **Diamonds are Forever!** Much of the world's diamond industry is located in Africa, with Russia and Canada also showing large revenues from their diamond mining industry. A visual representation of the various shares of the world's diamond revenues, adapted from *Time Magazine*,[15] is shown below:

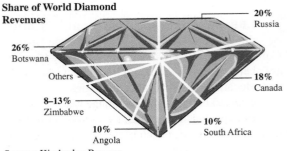

Source: Kimberley Process

a. Draw a pie chart to describe the various shares of the world's diamond revenues.

b. Draw a bar chart to describe the various shares of the world's diamond revenues.

c. Draw a Pareto chart to describe the various shares of the world's diamond revenues.

d. Which of the charts is the most effective in describing the data?

1.56 Pulse Rates A group of 50 biomedical students recorded their pulse rates by counting the number of beats for 30 seconds and multiplying by 2.

80	70	88	70	84	66	84	82	66	42
52	72	90	70	96	84	96	86	62	78
60	82	88	54	66	66	80	88	56	104
84	84	60	84	88	58	72	84	68	74
84	72	62	90	72	84	72	110	100	58

a. Why are all of the measurements even numbers?

b. Draw a stem and leaf plot to describe the data, splitting each stem into two lines.

c. Construct a relative frequency histogram for the data.

d. Write a short paragraph describing the distribution of the student pulse rates.

1.57 Starbucks Students at the University of California, Riverside (UCR), along with many other Californians love their Starbucks! The distances in miles from campus for the 41 Starbucks stores within 10 miles of UCR are shown below:[16]

0.4	0.5	4.3	5.2	9.3	4.2	8.3	9.4	7.3	7.1	8.7
6.1	3.7	8.0	4.8	3.9	5.3	9.8	1.2	6.4	9.8	
8.0	7.8	9.4	8.4	5.9	9.8	7.3	2.5	9.6	0.7	
6.7	7.6	9.7	3.5	9.5	8.8	9.0	7.9	5.2	6.8	

Construct a relative frequency to describe the distances from the UCR campus, using 10 classes of width 1, starting at 0.0.

a. What is the shape of the histogram? Do you see any unusual features?

b. Can you explain why the histogram looks the way it does?

1.58 Stressful Times In the spring of 2010, almost all of the 50 U.S. states plus the District of Columbia were facing drastic financial crises, with many planning across-the-board budget cuts, layoffs, higher education fees, and other strategies to cut next year's expected budget gap. The image that follows shows the expected shortfall next year as a percentage of this year's budget for each of the 50 U.S. states and the District of Columbia.[17]

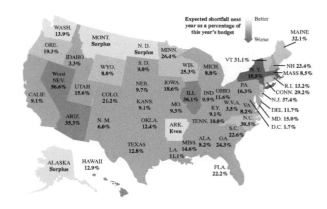

a. Construct a relative frequency histogram to describe the percentages for the 48 states that are expecting to face shortfalls next year.

b. What is the shape of the histogram? Do you see any unusual features in the histogram? If there are any outliers, can you explain them?

c. There are three states, Alaska, Montana, and North Dakota, that are expecting a surplus next year. Can you think of a reason why this might be?

1.59 An Archeological Find An article in *Archaeometry* involved an analysis of 26 samples of Romano-British pottery, found at four different kiln sites in the United Kingdom.[18] The samples were analyzed to determine their chemical composition, and the percentage of aluminum oxide in each of the 26 samples is shown in the following table.

Llanederyn		Caldicot	Island Thorns	Ashley Rails
14.4	11.6	11.8	18.3	17.7
13.8	11.1	11.6	15.8	18.3
14.6	13.4		18.0	16.7
11.5	12.4		18.0	14.8
13.8	13.1		20.8	19.1
10.9	12.7			
10.1	12.5			

a. Construct a relative frequency histogram to describe the aluminum oxide content in the 26 pottery samples.

b. What unusual feature do you see in this graph? Can you think of an explanation for this feature?

c. Draw a dotplot for the data, using a letter (L, C, I, or A) to locate the data point on the horizontal scale. Does this help explain the unusual feature in part b?

1.60 The Great Calorie Debate Want to lose weight? You can do it by cutting calories, as long as you get enough nutritional value from the foods that you do eat! Below you will see a visual representation of the number of calories in some of America's favorite foods adapted from an article in *The Press-Enterprise*.[19]

Number of Calories

26	53	140	145	330	800
Hershey's Kiss	Oreo cookie	12-ounce can of Coke	12-ounce bottle of Budweiser beer	Slice of a large Papa John's pepperoni pizza	Burger King Whopper with cheese

a. Comment on the accuracy of the graph shown above. Do the sizes, heights, and volumes of the six items accurately represent the number of calories in the item?

b. Draw an actual bar chart to describe the number of calories in these six food favorites.

1.61 Laptops and Learning An informal experiment was conducted at McNair Academic High School in Jersey City, New Jersey, to investigate the use of laptop computers as a learning tool in the study of algebra.[20] A freshman class of 20 students was given laptops to use at school and at home, while another freshman class of 27 students was not given laptops; however, many of these students were able to use computers at home. The final exam scores for the two classes are shown below.

Laptops		No Laptops		
98	84	63	83	97
97	93	93	52	74
88	57	83	63	88
100	84	86	81	84
100	81	99	91	49
78	83	80	81	89
68	84	78	29	64
47	93	74	72	89
90	57	67	89	70
94	83			

The histograms that follow show the distribution of final exam scores for the two groups.

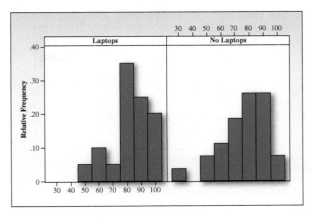

Write a summary paragraph describing and comparing the distribution of final exam scores for the two groups of students.

1.62 Old Faithful The data below are the waiting times between eruptions of the Old Faithful geyser in Yellowstone National Park.[21] Use one of the graphical methods from this chapter to describe the distribution of waiting times. If there are any unusual features in your graph, see if you can think of any practical explanation for them.

56	89	51	79	58	82	52	88	52	78
69	75	77	53	80	54	79	74	65	78
55	87	53	85	61	93	54	76	80	81
59	86	78	71	77	89	45	93	72	71
76	94	75	50	83	82	72	77	75	65
79	72	78	77	79	72	82	74	80	49
75	78	64	80	49	49	88	51	78	85
65	75	77	69	92	91	53	86	49	79
68	87	61	81	55	93	53	84	70	73
93	50	87	77	74	89	87	76	59	80

1.63 Gasoline Tax The following are the 2010 state gasoline tax in cents per gallon for the 50 U.S. states and the District of Columbia.[5]

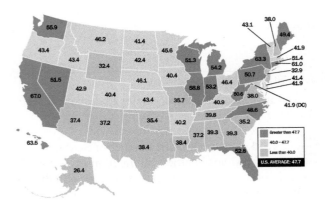

AK	26.4	HI	63.5	MA	41.9	NM	37.2	SD	42.4
AL	39.3	ID	43.4	MI	54.2	NY	63.3	TN	39.8
AR	40.2	IL	58.8	MN	45.6	NC	48.6	TX	38.4
AZ	37.4	IN	53.2	MS	37.2	ND	41.4	UT	42.9
CA	67.0	IA	40.4	MO	35.7	OH	46.4	VT	43.1
CO	40.4	KS	43.4	MT	46.2	OK	35.4	VA	38.0
CT	61.0	KY	40.9	NE	46.1	OR	43.4	WA	55.9
DE	41.4	LA	38.4	NV	51.5	PA	50.7	WV	50.6
DC	41.9	ME	49.4	NH	38.0	RI	51.4	WI	51.3
FL	52.8	MD	41.9	NJ	32.9	SC	35.2	WY	32.4
GA	39.3								

Source: http://www.api.org/statistics/fueltaxes/upload/GASOLINE_ TAX_MAP_ APRIL2010.pdf, July 6, 2010

a. Construct a stem and leaf display for the data.

b. How would you describe the shape of this distribution?

c. Are there states with unusually high or low gasoline taxes? If so, which states are they?

Data set EX0164 **1.64 Hydroelectric Plants** The following data represent the planned rated capacities in megawatts (millions of watts) for the world's 20 largest hydroelectric plants.[5]

18,200	4,500	3,000
12,600	4,200	2,940
10,000	4,200	2,715
8,370	3,840	2,700
6,400	3,444	2,541
6,300	3,300	2,512
6,000	3,100	

Source: The World Almanac and Book of Facts, 2011

a. Construct a stem and leaf display for the data.

b. How would you describe the shape of this distribution?

Data set EX0165 **1.65 Car Colors** The most popular colors for compact and sports cars in a recent year are given in the table.[5]

Color	Percentage	Color	Percentage
Silver	19	White/white pearl	12
Black/black effect	17	Beige/brown	3
Gray	17	Yellow/gold	2
Blue	15	Green	2
Red	12	Other	1

Source: The World Almanac and Book of Facts 2011

Use an appropriate graphical display to describe these data.

Data set EX0166 **1.66 Starbucks** The number of Starbucks coffee shops in cities within 20 miles of the University of California, Riverside is shown in the following table.[16]

City	Starbucks	City	Starbucks
Riverside	18	Ontario	12
Grand Terrace	1	Norco	4
Rialto	6	Fontana	7
Colton	2	Mira Loma	2
San Bernardino	6	Perris	3
Redlands	8	Highland	1
Corona	10	Rancho Cucamonga	10
Yucaipa	3	Lake Elsinore	2
Chino	11	Moreno Valley	5
Upland	2	Montclair	1
Bloomington	1		

a. Draw a dotplot to describe the data.

b. Describe the shape of the distribution.

c. Is there another variable that you could measure that might help to explain why some cities have more Starbucks than others? Explain.

Data set EX0167 **1.67 What's Normal?** The 98.6 degree standard for human body temperature was derived by a German doctor in 1868. In an attempt to verify his claim, Mackowiak, Wasserman, and Levine[22] took temperatures from 148 healthy people over a 3-day period. A data set closely matching the one in Mackowiak's article was derived by Allen Shoemaker, and appears in the *Journal of Statistics Education*.[23] The body temperatures for these 130 individuals are shown in the relative frequency histogram that follows.

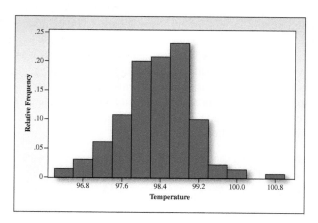

a. Describe the shape of the distribution of temperatures.

b. Are there any unusual observations? Can you think of any explanation for these?

c. Locate the 98.6-degree standard on the horizontal axis of the graph. Does it appear to be near the center of the distribution?

CASE STUDY

 Blood
Pressure

How Is Your Blood Pressure?

Blood pressure is the pressure that the blood exerts against the walls of the arteries. When physicians or nurses measure your blood pressure, they take two readings. The systolic blood pressure is the pressure when the heart is contracting and therefore pumping. The diastolic blood pressure is the pressure in the arteries when the heart is relaxing. The diastolic blood pressure is always the lower of the two readings. Blood pressure varies from one person to another. It will also vary for a single individual from day to day and even within a given day.

If your blood pressure is too high, it can lead to a stroke or a heart attack. If it is too low, blood will not get to your extremities and you may feel dizzy. Low blood pressure is usually not serious.

So, what should *your* blood pressure be? A systolic blood pressure of 120 would be considered normal. One of 150 would be high. But since blood pressure varies with gender and increases with age, a better gauge of the relative standing of your blood pressure would be obtained by comparing it with the population of blood pressures of all persons of your gender and age in the United States. Of course, we cannot supply you with that data set, but we can show you a very large sample selected from it. The blood pressure data on 1910 persons, 965 men and 945 women between the ages of 15 and 20, are found at the CourseMate Web site. The data are part of a health survey conducted by the National Institutes of Health (NIH). Entries for each person include that person's age and systolic and diastolic blood pressures at the time the blood pressure was recorded.

1. Describe the variables that have been measured in this survey. Are the variables quantitative or qualitative? Discrete or continuous? Are the data univariate, bivariate, or multivariate?

2. What types of graphical methods are available for describing this data set? What types of questions could be answered using various types of graphical techniques?

3. Using the systolic blood pressure data set, construct a relative frequency histogram for the 965 men and another for the 945 women. Use a statistical software package if you have access to one. Compare the two histograms.

4. Consider the 965 men and 945 women as the entire population of interest. Choose a sample of $n = 50$ men and $n = 50$ women, recording their systolic blood pressures and their ages. Draw two relative frequency histograms to graphically display the systolic blood pressures for your two samples. Do the shapes of the histograms resemble the population histograms from part 3?

5. How does your blood pressure compare with that of others of your same gender? Check your systolic blood pressure against the appropriate histogram in part 3 or 4 to determine whether your blood pressure is "normal" or whether it is unusually high or low.

2

Describing Data with Numerical Measures

© Joe Sohm-VisionsofAmerica/Photodisc/Getty

NEED TO KNOW...

How to Calculate Sample Quartiles

The Boys of Summer

Are the baseball champions of today better than those of "yesteryear"? Do players in the National League hit better than players in the American League? The case study at the end of this chapter involves the batting averages of major league batting champions. Numerical descriptive measures can be used to answer these and similar questions.

DESCRIBING A SET OF DATA WITH NUMERICAL MEASURES

2.1

Graphs can help you describe the basic shape of a data distribution; "a picture is worth a thousand words." There are limitations, however, to the use of graphs. Suppose you need to display your data to a group of people and the bulb on the data projector blows out! Or you might need to describe your data over the telephone—no way to display the graphs! You need to find another way to convey a mental picture of the data to your audience.

A second limitation is that graphs are somewhat imprecise for use in statistical inference. For example, suppose you want to use a sample histogram to make inferences about a population histogram. How can you measure the similarities and differences between the two histograms in some concrete way? If they were identical, you could say "They are the same!" But, if they are different, it is difficult to describe the "degree of difference."

One way to overcome these problems is to use **numerical measures,** which can be calculated for either a sample or a population of measurements. You can use the data to calculate a set of *numbers* that will convey a good mental picture of the frequency distribution. These measures are called **parameters** when associated with the population, and they are called **statistics** when calculated from sample measurements.

Definition   Numerical descriptive measures associated with a population of measurements are called **parameters;** those computed from sample measurements are called **statistics.**

MEASURES OF CENTER

2.2

In Chapter 1, we introduced dotplots, stem and leaf plots, and histograms to describe the distribution of a set of measurements on a quantitative variable x. The horizontal axis displays the values of x, and the data are "distributed" along this horizontal line. One of the first important numerical measures is a **measure of center**—a measure along the horizontal axis that locates the center of the distribution.

The birth weight data presented in Table 1.9 ranged from a low of 5.6 to a high of 9.4, with the center of the histogram located in the vicinity of 7.5 (see Figure 2.1). Let's consider some rules for locating the center of a distribution of measurements.

FIGURE 2.1

Center of the birth weight data

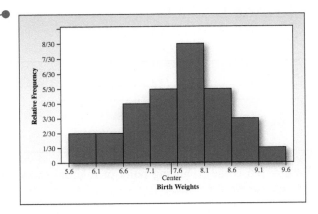

The arithmetic average of a set of measurements is a very common and useful measure of center. This measure is often referred to as the **arithmetic mean,** or simply the **mean,** of a set of measurements. To distinguish between the mean for the sample and the mean for the population, we will use the symbol $\bar{x}$ (x-bar) for a sample mean and the symbol μ (Greek lowercase mu) for the mean of a population.

Definition The **arithmetic mean** or **average** of a set of n measurements is equal to the sum of the measurements divided by n.

Since statistical formulas often involve adding or "summing" numbers, we use a shorthand symbol to indicate the process of summing. Suppose there are n measurements on the variable x—call them $x_1, x_2, \ldots, x_n$. To add the n measurements together, we use this shorthand notation:

$$\sum_{i=1}^{n} x_i \quad \text{which means } x_1 + x_2 + x_3 + \cdots + x_n$$

The Greek capital sigma (Σ) tells you to add the items that appear to its right, beginning with the number below the sigma ($i = 1$) and ending with the number above ($i = n$). However, since the typical sums in statistical calculations are almost always made on the total set of n measurements, you can use a simpler notation:

Σx_i which means "the sum of all the x measurements"

Using this notation, we write the formula for the sample mean:

NOTATION

Sample mean: $\quad \bar{x} = \dfrac{\Sigma x_i}{n}$

Population mean: $\quad \mu$

EXAMPLE **2.1**

Draw a dotplot for the $n = 5$ measurements 2, 9, 11, 5, 6. Find the sample mean and compare its value with what you might consider the "center" of these observations on the dotplot.

Solution The dotplot in Figure 2.2 seems to be centered between 6 and 8. To find the sample mean, calculate

$$\bar{x} = \frac{\Sigma x_i}{n} = \frac{2 + 9 + 11 + 5 + 6}{5} = 6.6$$

FIGURE 2.2

Dotplot for Example 2.1

The statistic $\bar{x} = 6.6$ is the balancing point or fulcrum shown on the dotplot. It does seem to mark the center of the data.

NEED a tip? NEED A TIP?

mean = balancing point
or fulcrum

Remember that samples are measurements drawn from a larger population that is usually unknown. An important use of the sample mean $\bar{x}$ is as an estimator of the unknown population mean μ. The birth weight data in Table 1.9 are a sample from a larger population of birth weights, and the distribution is shown in Figure 2.1. The mean of the 30 birth weights is

$$\bar{x} = \frac{\Sigma x_i}{30} = \frac{227.2}{30} = 7.57$$

shown in Figure 2.1; it marks the balancing point of the distribution. The mean of the entire population of newborn birth weights is unknown, but if you had to guess its value, your best estimate would be 7.57. Although the sample mean $\bar{x}$ changes from sample to sample, the population mean μ stays the same.

A second measure of central tendency is the **median,** which is the value in the middle position in the set of measurements ordered from smallest to largest.

Definition The **median** m of a set of n measurements is the value of x that falls in the middle position when the measurements are ordered from smallest to largest.

EXAMPLE 2.2

Find the median for the set of measurements 2, 9, 11, 5, 6.

Solution Rank the $n = 5$ measurements from smallest to largest:

2   5   6   9   11
↑

The middle observation, marked with an arrow, is in the center of the set, or $m = 6$.

EXAMPLE 2.3

Find the median for the set of measurements 2, 9, 11, 5, 6, 27.

Solution Rank the measurements from smallest to largest:

2   5 $\boxed{6 \quad 9}$ 11   27
↑

NEED a tip? NEED A TIP?

Roughly 50% of the measurements are smaller, 50% are larger than the median.

Now there are two "middle" observations, shown in the box. To find the median, choose a value halfway between the two middle observations:

$$m = \frac{6 + 9}{2} = 7.5$$

The value **.5(n + 1)** indicates the **position of the median** in the ordered data set. If the position of the median is a number that ends in the value **.5,** you need to average the two adjacent values.

EXAMPLE 2.4

For the $n = 5$ ordered measurements from Example 2.2, the position of the median is .5(n + 1) =.5(6) = 3, and the median is the *3rd ordered observation,* or $m = 6$. For the $n = 6$ ordered measurements from Example 2.3, the position of the median is .5(n + 1) = .5(7) = 3.5, and the median is the *average of the 3rd and 4th ordered observations,* or $m = (6 + 9)/2 = 7.5$.

NEED A TIP?

Symmetric:
mean = median

Skewed right:
mean > median

Skewed left:
mean < median

Although both the mean and the median are good measures of the center of a distribution, the median is less sensitive to extreme values or *outliers*. For example, the value $x = 27$ in Example 2.3 is much larger than the other five measurements. The median, $m = 7.5$, is not affected by the outlier, whereas the sample average,

$$\bar{x} = \frac{\Sigma x_i}{n} = \frac{60}{6} = 10$$

is affected; its value is not representative of the remaining five observations.

When a data set has extremely small or extremely large observations, the sample mean is drawn toward the direction of the extreme measurements (see Figure 2.3).

FIGURE 2.3 ●

Relative frequency distributions showing the effect of extreme values on the mean and median

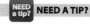 **ONLINE APPLET**

How Extreme Values Affect the Mean and Median

If a distribution is skewed to the right, the mean shifts to the right; if a distribution is skewed to the left, the mean shifts to the left. The median is not affected by these extreme values because the numerical values of the measurements are not used in its calculation. When a distribution is symmetric, the mean and the median are equal. If a distribution is strongly skewed by one or more extreme values, you should use the median rather than the mean as a measure of center.

Another way to locate the center of a distribution is to look for the value of x that occurs with the highest frequency. This measure of the center is called the **mode.**

Definition   The **mode** is the category that occurs most frequently, or the most frequently occurring value of x. When measurements on a continuous variable have been grouped as a frequency or relative frequency histogram, the class with the highest peak or frequency is called the **modal class,** and the midpoint of that class is taken to be the mode.

NEED A TIP?

Remember that there can be several modes or no mode (if each observation occurs only once).

The mode is generally used to describe large data sets, whereas the mean and median are used for both large and small data sets. From the data in Example 1.11, reproduced in Table 2.1**(a),** the mode of the distribution of the number of reported weekly visits to Starbucks for 30 Starbucks customers is 5. The modal class and the value of x occurring with the highest frequency are the same, as shown in Figure 2.4(a).

For the birth weight data in Table 2.**1(b),** a birth weight of 7.7 occurs four times, and therefore the mode for the distribution of birth weights is 7.7. Using the histogram to find the modal class, you find that the class with the highest peak is the fifth class, from 7.6 to 8.1. Our choice for the mode would be the midpoint of this class, or 7.85. See Figure 2.4(b).

It is possible for a distribution of measurements to have more than one mode. These modes would appear as "local peaks" in the relative frequency distribution. For example, if we were to tabulate the length of fish taken from a lake during one season, we might get a *bimodal distribution,* possibly reflecting a mixture of young and old fish in the population. Sometimes bimodal distributions of sizes or weights reflect a mixture of measurements taken on males and females. In any case, a set or distribution of measurements may have more than one mode.

TABLE 2.1

Starbucks and birth weight data

(a) Starbucks data					(b) Birth weight data				
6	7	1	5	6	7.2	7.8	6.8	6.2	8.2
4	6	4	6	8	8.0	8.2	5.6	8.6	7.1
6	5	6	3	4	8.2	7.7	7.5	7.2	7.7
5	5	5	7	6	5.8	6.8	6.8	8.5	7.5
3	5	7	5	5	6.1	7.9	9.4	9.0	7.8
					8.5	9.0	7.7	6.7	7.7

FIGURE 2.4

Relative frequency histograms for the Starbucks and birth weight data

2.2 EXERCISES

BASIC TECHNIQUES

2.1 You are given $n = 5$ measurements: 0, 5, 1, 1, 3.

a. Draw a dotplot for the data. (HINT: If two measurements are the same, place one dot above the other.) Guess the approximate "center."

b. Find the mean, median, and mode.

c. Locate the three measures of center on the dotplot in part a. Based on the relative positions of the mean and median, are the measurements symmetric or skewed?

2.2 You are given $n = 8$ measurements: 3, 2, 5, 6, 4, 4, 3, 5.

a. Find $\bar{x}$.

b. Find m.

c. Based on the results of parts a and b, are the measurements symmetric or skewed? Draw a dotplot to confirm your answer.

2.3 You are given $n = 10$ measurements: 3, 5, 4, 6, 10, 5, 6, 9, 2, 8.

a. Calculate $\bar{x}$.

b. Find m.

c. Find the mode.

APPLICATIONS

2.4 Auto Insurance The cost of automobile insurance has become a sore subject in California because insurance rates are dependent on so many different variables, such as the city in which you live, the number of cars you insure, and the company with which you are insured. The website www.insurance.ca.gov reports the annual 2010 premium for a male, licensed for 6–8 years, who drives a Honda Accord 12,600–15,000 miles per year and has no violations or accidents.[1]

City	GEICO ($)	21st Century ($)
Long Beach	2780	2352
Pomona	2411	2462
San Bernardino	2261	2284
Moreno Valley	2263	2520

Source: www.insurance.ca.gov

a. What is the average premium for GEICO Insurance?

b. What is the average premium for 21st Century Insurance?

c. If you were a consumer, would you be interested in the average premium cost? If not, what would you be interested in?

2.5 DVRs The digital video recorder (DVR) is a common fixture in most American households. In fact, most American households have DVRs, and many have more than one. A sample of 25 households produced the following measurements on x, the number of DVRs in the household:

1	0	2	1	1
1	0	2	1	0
0	1	2	3	2
1	1	1	0	1
3	1	0	1	1

a. Is the distribution of x, the number of DVRs in a household, symmetric or skewed? Explain.

b. Guess the value of the mode, the value of x that occurs most frequently.

c. Calculate the mean, median, and mode for these measurements.

d. Draw a relative frequency histogram for the data set. Locate the mean, median, and mode along the horizontal axis. Are your answers to parts a and b correct?

2.6 Fortune 500 Revenues Ten of the 50 largest businesses in the United States, randomly selected from the *Fortune 500,* are listed below along with their revenues (in millions of dollars):[2]

Company	Revenues ($)	Company	Revenues ($)
General Motors	104,589	Target	65,357
IBM	95,758	Morgan Stanley	31,515
Bank of America	150,450	Johnson & Johnson	61,867
Home Depot	66,176	Apple	36,537
Boeing	68,281	Exxon Mobil	284,650

a. Draw a stem and leaf plot for the data. Are the data skewed?

b. Calculate the mean revenue for these 10 businesses. Calculate the median revenue.

c. Which of the two measures in part b best describes the center of the data? Explain.

2.7 Birth Order and Personality Does birth order have any effect on a person's personality? A report on a study by an MIT researcher indicates that later-born children are more likely to challenge the establishment, more open to new ideas, and more accepting of change.[3] In fact, the number of later-born children is increasing. During the Depression years of the 1930s, families averaged 2.5 children (59% later born), whereas the parents of baby boomers averaged 3 to 4 children (68% later born). What does the author mean by an average of 2.5 children?

2.8 Tuna Fish An article in *Consumer Reports* gives the price—an estimated average for a 6-ounce can or a 7.06-ounce pouch—for 14 different brands of water-packed light tuna, based on prices paid nationally in supermarkets:[4]

.99	1.92	1.23	.85	.65	.53	1.41
1.12	.63	.67	.69	.60	.60	.66

a. Find the average price for the 14 different brands of tuna.

b. Find the median price for the 14 different brands of tuna.

c. Based on your findings in parts a and b, do you think that the distribution of prices is skewed? Explain.

2.9 Sports Salaries As professional sports teams become a more and more lucrative business for their owners, the salaries paid to the players have also increased. In fact, sports superstars are paid astronomical salaries for their talents. If you were asked by a sports management firm to describe the distribution of players' salaries in several different categories of professional sports, what measure of center would you choose? Why?

2.10 Time on Task In a psychological experiment, the time on task was recorded for 10 subjects under a 5-minute time constraint. These measurements are in seconds:

175	190	250	230	240
200	185	190	225	265

a. Find the average time on task.

b. Find the median time on task.

c. If you were writing a report to describe these data, which measure of central tendency would you use? Explain.

2.11 Starbucks The number of Starbucks coffee shops in 21 cities within 20 miles of the University of California, Riverside is shown in the following table.[5]

18	1	6	2	6	8
10	3	11	2	1	12
4	7	2	3	1	10
2	5	1			

a. Find the mean, the median, and the mode.

b. Compare the median and the mean. What can you say about the shape of this distribution?

c. Draw a dotplot for the data. Does this confirm your conclusion about the shape of the distribution from part b?

2.12 Nintendo's Wii The "Wii" is an interactive gaming console popular among many gamers. Its cost can vary dramatically, depending on where it is purchased. The website www.pricegrabber.com lists 14 online sellers with various prices, including shipping and taxes:[6]

Seller	Price ($)	Seller	Price ($)
Buy.com	216.49	Dell	184.86
Sears	222.84	Kmart	222.84
Sam's Club	180.17	EagleDirectUSA	231.04
USA Sales	279.90	Wii4family	262.95
PalaceToys	280.98	QuickShip USA	299.48
Simbaoo7	289.97	BUY-IT-NOW	384.99
jandk425	433.00	SW Evolution	1024.24

a. What is the average price of the Wii for these 14 sellers?

b. What is the median price of the Wii for these 14 sellers?

c. As a consumer, would you be interested in the average price of the Wii? What other variables would be important to you?

MEASURES OF VARIABILITY

Data sets may have the same center but look different because of the way the numbers *spread out* from the center. Consider the two distributions shown in Figure 2.5. Both distributions are centered at $x = 4$, but there is a big difference in the way the measurements spread out, or *vary*. The measurements in Figure 2.5(a) vary from 3 to 5; in Figure 2.5(b) the measurements vary from 0 to 8.

FIGURE 2.5

Variability or dispersion
of data

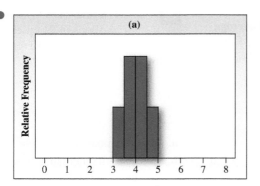

Variability or **dispersion** is a very important characteristic of data. For example, if you were manufacturing bolts, extreme variation in the bolt diameters would cause a high percentage of defective products. On the other hand, if you were trying to discriminate between good and poor accountants, you would have trouble if the examination always produced test grades with little variation, making discrimination very difficult.

Measures of variability can help you create a mental picture of the spread of the data. We will present some of the more important ones. The simplest measure of variation is the **range.**

Definition   The **range, *R*,** of a set of *n* measurements is defined as the difference between the largest and smallest measurements.

For example, the measurements 5, 7, 1, 2, 4 vary from 1 to 7. Hence, the range is $7 - 1 = 6$. The range is easy to calculate, easy to interpret, and is an adequate measure of variation for small sets of data. But, for large data sets, the range is not an adequate measure of variability. For example, the two relative frequency distributions in Figure 2.6 have the same range but very different shapes and variability.

FIGURE 2.6

Distributions with equal
range and unequal
variability

Is there a measure of variability that is more sensitive than the range? Consider again the sample measurements 5, 7, 1, 2, 4, displayed as a dotplot in Figure 2.7. The mean of these five measurements is

$$\bar{x} = \frac{\Sigma x_i}{n} = \frac{19}{5} = 3.8$$

FIGURE 2.7

Dotplot showing the deviations of points from the mean

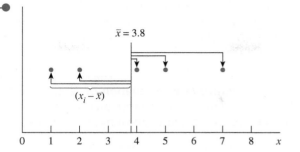

$\bar{x} = 3.8$

$(x_i - \bar{x})$

as indicated on the dotplot. The horizontal distances between each dot (measurement) and the mean $\bar{x}$ will help you to measure the variability. If the distances are large, the data are more spread out or more *variable* than if the distances are small. If x_i is a particular dot (measurement), then the **deviation** of that measurement from the mean is $(x_i - \bar{x})$. Measurements to the right of the mean produce positive deviations, and those to the left produce negative deviations. The values of x and the deviations for our example are listed in the first and second columns of Table 2.2.

TABLE 2.2 ● **Computation of $\Sigma(x_i - \bar{x})^2$**

x_i	$(x_i - \bar{x})$	$(x_i - \bar{x})^2$
5	1.2	1.44
7	3.2	10.24
1	−2.8	7.84
2	−1.8	3.24
4	.2	.04
19	0.0	22.80

Because the deviations in the second column of the table contain information on variability, one way to combine the five deviations into one numerical measure is to average them. Unfortunately, the average will not work because some of the deviations are positive, some are negative, and the sum is always zero (unless round-off errors have been introduced into the calculations). Note that the deviations in the second column of Table 2.2 sum to zero.

Another possibility might be to disregard the signs of the deviations and calculate the average of their absolute values.[†] This method has been used as a measure of variability in exploratory data analysis and in the analysis of time series data. We prefer, however, to overcome the difficulty caused by the signs of the deviations by working with their sum of squares. From the sum of squared deviations, a single measure called the **variance** is calculated. To distinguish between the variance of a

[†]The absolute value of a number is its magnitude, ignoring its sign. For example, the absolute value of −2, represented by the symbol $|-2|$, is 2. The absolute value of 2—that is, $|2|$—is 2.

sample and the variance of a *population,* we use the symbol s^2 for a sample variance and σ^2 (Greek lowercase sigma) for a population variance. *The variance will be relatively large for highly variable data and relatively small for less variable data.*

Definition The **variance of a population** of N measurements is the average of the squares of the deviations of the measurements about their mean μ. The population variance is denoted by σ^2 and is given by the formula

$$\sigma^2 = \frac{\Sigma(x_i - \mu)^2}{N}$$

Most often, you will not have all the population measurements available but will need to calculate the *variance of a sample* of n measurements.

Definition The **variance of a sample** of n measurements is the sum of the squared deviations of the measurements about their mean $\bar{x}$ divided by $(n - 1)$. The sample variance is denoted by s^2 and is given by the formula

$$s^2 = \frac{\Sigma(x_i - \bar{x})^2}{n - 1}$$

NEED A TIP?

The variance and the standard deviation *cannot* be negative numbers.

For the set of $n = 5$ sample measurements presented in Table 2.2, the square of the deviation of each measurement is recorded in the third column. Adding, we obtain

$$\Sigma(x_i - \bar{x})^2 = 22.80$$

and the sample variance is

$$s^2 = \frac{\Sigma(x_i - \bar{x})^2}{n - 1} = \frac{22.80}{4} = 5.70$$

The variance is measured in terms of the square of the original units of measurement. If the original measurements are in inches, the variance is expressed in square inches. Taking the square root of the variance, we obtain the **standard deviation,** which returns the measure of variability to the original units of measurement.

Definition The **standard deviation** of a set of measurements is equal to the positive square root of the variance.

NOTATION

n: number of measurements in the sample

s^2: sample variance

$s = \sqrt{s^2}$: sample standard deviation

N: number of measurements in the population

σ^2: population variance

$\sigma = \sqrt{\sigma^2}$: population standard deviation

NEED A TIP?

If you are using your calculator, make sure to choose the correct key for the sample standard deviation.

For the set of $n = 5$ sample measurements in Table 2.2, the sample variance is $s^2 = 5.70$, so the sample standard deviation is $s = \sqrt{s^2} = \sqrt{5.70} = 2.39$. The more variable the data set is, the larger the value of s.

For the small set of measurements we used, the calculation of the variance is not too difficult. However, for a larger set, the calculations can become very tedious. Most scientific calculators have built-in programs that will calculate $\bar{x}$ and s or μ and σ, so that your computational work will be minimized. The sample or population mean key is usually marked with $\bar{x}$. The sample standard deviation key is usually marked with s, s_x, or σ_{xn-1}, and the population standard deviation key with σ, σ_x, or σ_{xn}. In using any calculator with these built-in function keys, be sure you know which calculation is being carried out by each key!

If you need to calculate s^2 and s by hand, it is much easier to use the alternative computing formula given next. This computational form is sometimes called the **shortcut method for calculating s^2.**

THE COMPUTING FORMULA FOR CALCULATING s^2

$$s^2 = \frac{\sum x_i^2 - \dfrac{(\sum x_i)^2}{n}}{n - 1}$$

The symbols $(\sum x_i)^2$ and $\sum x_i^2$ in the computing formula are shortcut ways to indicate the arithmetic operation you need to perform. You know from the formula for the sample mean that $\sum x_i$ is the sum of all the measurements. To find $\sum x_i^2$, you square each individual measurement and then add them together.

$\sum x_i^2 =$ Sum of the squares of the individual measurements
$(\sum x_i)^2 =$ Square of the sum of the individual measurements

---

The *sample standard deviation, s,* is the positive square root of s^2.

EXAMPLE **2.5**

Calculate the variance and standard deviation for the five measurements in Table 2.3, which are 5, 7, 1, 2, 4. Use the computing formula for s^2 and compare your results with those obtained using the original definition of s^2.

TABLE 2.3 ● **Table for Simplified Calculation of s^2 and s**

x_i	x_i^2
5	25
7	49
1	1
2	4
4	16
19	95

Don't round off partial results as you go along!

Solution The entries in Table 2.2 are the individual measurements, x_i, and their squares, x_i^2, together with their sums. Using the computing formula for s^2, you have

$$s^2 = \frac{\sum x_i^2 - \frac{(\sum x_i)^2}{n}}{n-1} = \frac{95 - \frac{(19)^2}{5}}{4} = \frac{22.80}{4} = 5.70$$

and $s = \sqrt{s^2} = \sqrt{5.70} = 2.39$, as before.

ONLINE APPLET

Why Divide by n-1

You may wonder why you need to divide by $(n-1)$ rather than n when computing the sample variance. Just as we used the sample mean $\bar{x}$ to estimate the population mean μ, you may want to use the sample variance s^2 to estimate the population variance σ^2. It turns out that the sample variance s^2 with $(n-1)$ in the denominator provides better estimates of σ^2 than would an estimator calculated with n in the denominator. **For this reason, we always divide by $(n-1)$ when computing the sample variance s^2 and the sample standard deviation s.**

Now that you have learned how to compute the variance and standard deviation, remember these points:

- The value of s is always greater than or equal to zero.
- The larger the value of s^2 or s, the greater the variability of the data set.
- If s^2 or s is equal to zero, all the measurements must have the same value.
- In order to measure the variability in the same units as the original observations, we compute the standard deviation $s = \sqrt{s^2}$.

This information allows you to compare several sets of data with respect to their locations and their variability. How can you use these measures to say something more specific about a single set of data? The theorem and rule presented in the next section will help answer this question.

2.3 EXERCISES

BASIC TECHNIQUES

2.13 You are given $n = 5$ measurements: 2, 1, 1, 3, 5.

a. Calculate the sample mean, $\bar{x}$.

b. Calculate the sample variance, s^2, using the formula given by the definition.

c. Find the sample standard deviation, s.

d. Find s^2 and s using the computing formula. Compare the results with those found in parts b and c.

2.14 Refer to Exercise 2.13.

a. Use the data entry method in your scientific calculator to enter the five measurements. Recall the

proper memories to find the sample mean and standard deviation.

b. Verify that the calculator provides the same values for $\bar{x}$ and s as in Exercise 2.13, parts a and c.

2.15 You are given $n = 8$ measurements: 4, 1, 3, 1, 3, 1, 2, 2.

a. Find the range.

b. Calculate $\bar{x}$.

c. Calculate s^2 and s using the computing formula.

d. Use the data entry method in your calculator to find $\bar{x}$, s, and s^2. Verify that your answers are the same as those in parts b and c.

2.16 You are given $n = 8$ measurements: 3, 1, 5, 6, 4, 4, 3, 5.

a. Calculate the range.

b. Calculate the sample mean.

c. Calculate the sample variance and standard deviation.

d. Compare the range and the standard deviation. The range is approximately how many standard deviations?

APPLICATIONS

2.17 An Archeological Find, again An article in *Archaeometry* involved an analysis of 26 samples of Romano-British pottery found at four different kiln sites in the United Kingdom.[7] The samples were analyzed to determine their chemical composition. The percentage of iron oxide in each of five samples collected at the Island Thorns site was:

  1.28, 2.39, 1.50, 1.88, 1.51

a. Calculate the range.

b. Calculate the sample variance and the standard deviation using the computing formula.

c. Compare the range and the standard deviation. The range is approximately how many standard deviations?

 2.18 Utility Bills in Southern California

EX0218   The monthly utility bills for a household in Riverside, California, were recorded for 12 consecutive months starting in January 2010:

Month	Amount ($)	Month	Amount ($)
January	288.02	July	311.20
February	230.60	August	370.23
March	216.85	September	368.57
April	243.74	October	301.79
May	236.96	November	271.99
June	288.57	December	298.12

a. Calculate the range of the utility bills for the year 2010.

b. Calculate the average monthly utility bill for the year 2010.

c. Calculate the standard deviation for the 2010 utility bills.

ON THE PRACTICAL SIGNIFICANCE OF THE STANDARD DEVIATION

2.4

We now introduce a useful theorem developed by the Russian mathematician Tchebysheff. Proof of the theorem is not difficult, but we are more interested in its application than its proof.

Tchebysheff's Theorem   Given a number k greater than or equal to 1 and a set of n measurements, at least $[1 - (1/k^2)]$ of the measurements will lie within k standard deviations of their mean.

Tchebysheff's Theorem applies to *any set of measurements* and can be used to describe either a sample or a population. We will use the notation appropriate for populations, but you should realize that we could just as easily use the mean and the standard deviation for the sample.

The idea involved in Tchebysheff's Theorem is illustrated in Figure 2.8. An interval is constructed by measuring a distance $k\sigma$ on either side of the mean μ. The number k can be any number as long as it is greater than or equal to 1. Then Tchebysheff's Theorem states that at least $1 - (1/k^2)$ of the total number n measurements lies in the constructed interval.

FIGURE 2.8

Illustrating Tchebysheff's
Theorem ●

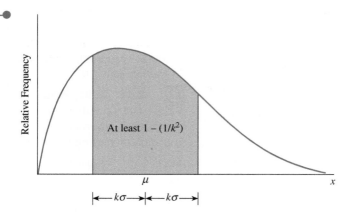

In Table 2.4, we choose a few numerical values for k and compute $[1 - (1/k^2)]$.

TABLE 2.4 ● **Illustrative Values of $[1 - (1/k^2)]$**

k	$1 - (1/k^2)$
1	$1 - 1 = 0$
2	$1 - 1/4 = 3/4$
3	$1 - 1/9 = 8/9$

From the calculations in Table 2.4, the theorem states:

- At least none of the measurements lie in the interval $\mu - \sigma$ to $\mu + \sigma$.
- At least 3/4 of the measurements lie in the interval $\mu - 2\sigma$ to $\mu + 2\sigma$.
- At least 8/9 of the measurements lie in the interval $\mu - 3\sigma$ to $\mu + 3\sigma$.

Although the first statement is not at all helpful, the other two values of k provide valuable information about the proportion of measurements that fall in certain intervals. The values $k = 2$ and $k = 3$ are not the only values of k you can use; for example, the proportion of measurements that fall within $k = 2.5$ standard deviations of the mean is at least $1 - [1/(2.5)^2] = .84$.

EXAMPLE 2.6   The mean and variance of a sample of $n = 25$ measurements are 75 and 100, respectively. Use Tchebysheff's Theorem to describe the distribution of measurements.

Solution   You are given $\bar{x} = 75$ and $s^2 = 100$. The standard deviation is $s = \sqrt{100} = 10$. The distribution of measurements is centered about $\bar{x} = 75$, and Tchebysheff's Theorem states:

- *At least* 3/4 of the 25 measurements lie in the interval $\bar{x} \pm 2s = 75 \pm 2(10)$ —that is, 55 to 95.
- *At least* 8/9 of the measurements lie in the interval $\bar{x} \pm 3s = 75 \pm 3(10)$—that is, 45 to 105.

Since Tchebysheff's Theorem applies to *any* distribution, it is very conservative. This is why we emphasize "at least $1 - (1/k^2)$" in this theorem.

Another rule for describing the variability of a data set does not work for *all* data sets, but it does work very well for data that "pile up" in the familiar mound shape shown in Figure 2.9. The closer your data distribution is to the mound-shaped curve in Figure 2.9, the more accurate the rule will be. Since mound-shaped data distributions occur quite frequently in nature, the rule can often be used in practical applications. For this reason, we call it the **Empirical Rule.**

FIGURE 2.9
Mound-shaped distribution

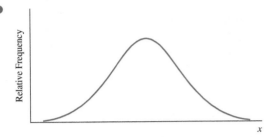

Empirical Rule   Given a distribution of measurements that is approximately mound-shaped:

The interval ($\mu \pm \sigma$) contains approximately 68% of the measurements.

The interval ($\mu \pm 2\sigma$) contains approximately 95% of the measurements.

The interval ($\mu \pm 3\sigma$) contains approximately 99.7% of the measurements.

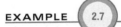 **NEED A TIP?**

Remember these three numbers:

68—95—99.7

The mound-shaped distribution shown in Figure 2.9 is commonly known as the **normal distribution** and will be discussed in detail in Chapter 6.

EXAMPLE 2.7

In a time study conducted at a manufacturing plant, the length of time to complete a specified operation is measured for each of $n = 40$ workers. The mean and standard deviation are found to be 12.8 and 1.7, respectively. Describe the sample data using the Empirical Rule.

Solution   To describe the data, calculate these intervals:

$(\bar{x} \pm s) = 12.8 \pm 1.7$ or   11.1 to 14.5

$(\bar{x} \pm 2s) = 12.8 \pm 2(1.7)$ or   9.4 to 16.2

$(\bar{x} \pm 3s) = 12.8 \pm 3(1.7)$ or   7.7 to 17.9

According to the Empirical Rule, you expect approximately 68% of the measurements to fall into the interval from 11.1 to 14.5, approximately 95% to fall into the interval from 9.4 to 16.2, and approximately 99.7% to fall into the interval from 7.7 to 17.9.

If you doubt that the distribution of measurements is mound-shaped, or if you wish for some other reason to be conservative, you can apply Tchebysheff's Theorem and be absolutely certain of your statements. Tchebysheff's Theorem tells you that at least 3/4 of the measurements fall into the interval from 9.4 to 16.2 and at least 8/9 into the interval from 7.7 to 17.9.

EXAMPLE 2.8 — Student teachers are trained to develop lesson plans, on the assumption that the written plan will help them to perform successfully in the classroom. In a study to assess the relationship between written lesson plans and their implementation in the classroom, 25 lesson plans were scored on a scale of 0 to 34 according to a Lesson Plan Assessment Checklist. The 25 scores are shown in Table 2.5. Use Tchebysheff's Theorem and the Empirical Rule (if applicable) to describe the distribution of these assessment scores.

TABLE 2.5 ● **Lesson Plan Assessment Scores**

26.1	26.0	14.5	29.3	19.7
22.1	21.2	26.6	31.9	25.0
15.9	20.8	20.2	17.8	13.3
25.6	26.5	15.7	22.1	13.8
29.0	21.3	23.5	22.1	10.2

Solution Use your calculator or the computing formulas to verify that $\bar{x} = 21.6$ and $s = 5.5$. The appropriate intervals are calculated and listed in Table 2.6. We have also referred back to the original 25 measurements and counted the actual number of measurements that fall into each of these intervals. These frequencies and relative frequencies are shown in Table 2.6.

TABLE 2.6 ● **Intervals $\bar{x} \pm ks$ for the Data of Table 2.5**

k	Interval $\bar{x} \pm ks$	Frequency in Interval	Relative Frequency
1	16.1–27.1	16	.64
2	10.6–32.6	24	.96
3	5.1–38.1	25	1.00

NEED A TIP?

Empirical Rule ⇔ mound-shaped data
Tchebysheff ⇔ any shaped data

Is Tchebysheff's Theorem applicable? Yes, because it can be used for any set of data. According to Tchebysheff's Theorem,

- at least 3/4 of the measurements will fall between 10.6 and 32.6.
- at least 8/9 of the measurements will fall between 5.1 and 38.1.

You can see in Table 2.6 that Tchebysheff's Theorem is true for these data. In fact, the proportions of measurements that fall into the specified intervals exceed the lower bound given by this theorem.

Is the Empirical Rule applicable? You can check for yourself by drawing a graph—either a stem and leaf plot or a histogram. The relative frequency histogram in Figure 2.10 shows that the distribution is *relatively* mound-shaped, so the Empirical Rule should work *relatively well*. That is,

- approximately 68% of the measurements will fall between 16.1 and 27.1.
- approximately 95% of the measurements will fall between 10.6 and 32.6.
- approximately 99.7% of the measurements will fall between 5.1 and 38.1.

The relative frequencies in Table 2.6 closely approximate those specified by the Empirical Rule.

FIGURE 2.10

Relative frequency histogram for Example 2.8

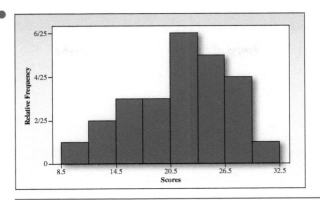

USING TCHEBYSHEFF'S THEOREM AND THE EMPIRICAL RULE

Tchebysheff's Theorem can be proven mathematically. It applies to any set of measurements—sample or population, large or small, mound-shaped or skewed.

Tchebysheff's Theorem gives a *lower bound* to the fraction of measurements to be found in an interval constructed as $\bar{x} \pm ks$. *At least* $1 - (1/k^2)$ of the measurements will fall into this interval, and probably more!

The Empirical Rule is a "rule of thumb" that can be used as a descriptive tool only when the data tend to be roughly mound-shaped (the data tend to pile up near the center of the distribution).

When you use these two tools for describing a set of measurements, Tchebysheff's Theorem will always be satisfied, but it is a very conservative estimate of the fraction of measurements that fall into a particular interval. If it is appropriate to use the Empirical Rule (mound-shaped data), this rule will give you a more accurate estimate of the fraction of measurements that fall into the interval.

2.5 A CHECK ON THE CALCULATION OF s

Tchebysheff's Theorem and the Empirical Rule can be used to detect gross errors in the calculation of s. Roughly speaking, these two tools tell you that *most of the time,* measurements lie within *two* standard deviations of their mean. This interval is marked off in Figure 2.11, and it implies that the total range of the measurements, from smallest to largest, should be somewhere around four standard deviations. This is, of course, a very rough approximation, but it can be very useful in checking for large errors in your calculation of s. If the range, R, is about four standard deviations, or $4s$, you can write

$$R \approx 4s \quad \text{or} \quad s \approx \frac{R}{4}$$

The computed value of s using the shortcut formula should be of roughly the same order as the approximation.

FIGURE 2.11

Range approximation to s

EXAMPLE 2.9

Use the range approximation to check the calculation of s for Table 2.3.

Solution   The range of the five measurements—5, 7, 1, 2, 4—is

$$R = 7 - 1 = 6$$

Then

$$s \approx \frac{R}{4} = \frac{6}{4} = 1.5$$

This is the same order as the calculated value $s = 2.39$.

NEED A TIP?

$s \approx R/4$ gives only an **approximate** value for s.

The range approximation is *not* intended to provide an accurate value for s. Rather, its purpose is to detect gross errors in calculating, such as the failure to divide the sum of squares of deviations by $(n - 1)$ or the failure to take the square root of s^2. If you make one of these mistakes, your answer will be many times larger than the range approximation of s.

EXAMPLE 2.10

Use the range approximation to determine an approximate value for the standard deviation for the data in Table 2.5.

Solution   The range $R = 31.9 - 10.2 = 21.7$. Then

$$s \approx \frac{R}{4} = \frac{21.7}{4} = 5.4$$

Since the exact value of s is 5.5 for the data in Table 2.5, the approximation is very close.

The range for a sample of n measurements will depend on the sample size, n. For larger values of n, a larger range of the x values is expected. The range for large samples (say, $n = 50$ or more observations) may be as large as $6s$, whereas the range for small samples (say, $n = 5$ or less) may be as small as or smaller than $2.5s$.

The range approximation for s can be improved if it is known that the sample is drawn from a mound-shaped distribution of data. Thus, the calculated s should not differ substantially from the range divided by the appropriate ratio given in Table 2.7.

TABLE 2.7 ● **Divisor for the Range Approximation of s**

Number of Measurements	Expected Ratio of Range to s
5	2.5
10	3
25	4

2.5 EXERCISES

BASIC TECHNIQUES

2.19 A set of $n = 10$ measurements consists of the values 5, 2, 3, 6, 1, 2, 4, 5, 1, 3.

a. Use the range approximation to estimate the value of s for this set. (HINT: Use the table at the end of Section 2.5.)

b. Use your calculator to find the actual value of s. Is the actual value close to your estimate in part a?

c. Draw a dotplot of this data set. Are the data mound-shaped?

d. Can you use Tchebysheff's Theorem to describe this data set? Why or why not?

e. Can you use the Empirical Rule to describe this data set? Why or why not?

2.20 Suppose you want to create a mental picture of the relative frequency histogram for a large data set consisting of 1000 observations, and you know that the mean and standard deviation of the data set are 36 and 3, respectively.

a. If you are fairly certain that the relative frequency distribution of the data is mound-shaped, how might you picture the relative frequency distribution? (HINT: Use the Empirical Rule.)

b. If you have no prior information concerning the shape of the relative frequency distribution, what can you say about the relative frequency histogram? (HINT: Construct intervals $\bar{x} \pm ks$ for several choices of k.)

2.21 A distribution of measurements is relatively mound-shaped with mean 50 and standard deviation 10.

a. What proportion of the measurements will fall between 40 and 60?

b. What proportion of the measurements will fall between 30 and 70?

c. What proportion of the measurements will fall between 30 and 60?

d. If a measurement is chosen at random from this distribution, what is the probability that it will be greater than 60?

2.22 A set of data has a mean of 75 and a standard deviation of 5. You know nothing else about the size of the data set or the shape of the data distribution.

a. What can you say about the proportion of measurements that fall between 60 and 90?

b. What can you say about the proportion of measurements that fall between 65 and 85?

c. What can you say about the proportion of measurements that are less than 65?

APPLICATIONS

2.23 Driving Emergencies The length of time required for an automobile driver to respond to a particular emergency situation was recorded for $n = 10$ drivers. The times (in seconds) were .5, .8, 1.1, .7, .6, .9, .7, .8, .7, .8.

a. Scan the data and use the procedure in Section 2.5 to find an approximate value for s. Use this value to check your calculations in part b.

b. Calculate the sample mean $\bar{x}$ and the standard deviation s. Compare with part a.

2.24 Packaging Hamburger Meat The data listed here are the weights (in pounds) of 27 packages of ground beef in a supermarket meat display:

1.08	.99	.97	1.18	1.41	1.28	.83
1.06	1.14	1.38	.75	.96	1.08	.87
.89	.89	.96	1.12	1.12	.93	1.24
.89	.98	1.14	.92	1.18	1.17	

a. Construct a stem and leaf plot or a relative frequency histogram to display the distribution of weights. Is the distribution relatively mound-shaped?

b. Find the mean and standard deviation of the data set.

c. Find the percentage of measurements in the intervals $\bar{x} \pm s$, $\bar{x} \pm 2s$, and $\bar{x} \pm 3s$.

d. How do the percentages obtained in part c compare with those given by the Empirical Rule? Explain.

e. How many of the packages weigh exactly 1 pound? Can you think of any explanation for this?

2.25 Breathing Rates Is your breathing rate normal? Actually, there is no standard breathing rate for humans. It can vary from as low as 4 breaths per minute to as high as 70 or 75 for a person engaged in strenuous exercise. Suppose that the resting breathing rates for college-age students have a relative frequency distribution that is mound-shaped, with a mean equal to 12 and a standard deviation of 2.3 breaths per minute.

What fraction of all students would have breathing rates in the following intervals?

a. 9.7 to 14.3 breaths per minute

b. 7.4 to 16.6 breaths per minute

c. More than 18.9 or less than 5.1 breaths per minute

Data set **2.26 Ore Samples** A geologist collected 20
EX0226 different ore samples, all the same weight, and randomly divided them into two groups. She measured the titanium (Ti) content of the samples using two different methods.

Method 1					Method 2				
.011	.013	.013	.015	.014	.011	.016	.013	.012	.015
.013	.010	.013	.011	.012	.012	.017	.013	.014	.015

a. Construct stem and leaf plots for the two data sets. Visually compare their centers and their ranges.

b. Calculate the sample means and standard deviations for the two sets. Do the calculated values confirm your visual conclusions from part a?

Data set **2.27 Social Security Numbers** A group of
EX0227 70 students were asked to record the last digit of their social security number.

```
1 6 9 1 5 9 0 2 8 4
0 7 3 4 2 3 5 8 4 2
3 2 0 0 2 1 2 7 7 4
0 0 9 9 5 3 8 4 7 4
6 6 9 0 2 6 2 9 5 8
5 1 7 7 7 8 7 5 1 8
3 4 1 9 3 8 6 6 6 6
```

a. Draw a relative frequency histogram using the values 0 through 9 as the class midpoints. What is the shape of the distribution? Based on the shape, what would be your best estimate for the mean of the data set?

b. Use the range approximation to guess the value of s for this set.

c. Use your calculator to find the actual values of $\bar{x}$ and s. Compare with your estimates in parts a and b.

2.28 Social Security Numbers, continued Refer to the data set in Exercise 2.27.

a. Find the percentage of measurements in the intervals $\bar{x} \pm s$, $\bar{x} \pm 2s$, and $\bar{x} \pm 3s$.

b. How do the percentages obtained in part a compare with those given by the Empirical Rule? Should they be approximately the same? Explain.

2.29 Survival Times A group of laboratory animals is infected with a particular form of bacteria, and their survival time is found to average 32 days, with a standard deviation of 36 days.

a. Visualize the distribution of survival times. Do you think that the distribution is relatively mound-shaped, skewed right, or skewed left? Explain.

b. Within what limits would you expect at least 3/4 of the measurements to lie?

2.30 Survival Times, continued Refer to Exercise 2.29. You can use the Empirical Rule to see why the distribution of survival times could not be mound-shaped.

a. Find the value of x that is exactly one standard deviation below the mean.

b. If the distribution is in fact mound-shaped, approximately what percentage of the measurements should be less than the value of x found in part a?

c. Since the variable being measured is time, is it possible to find any measurements that are more than one standard deviation below the mean?

d. Use your answers to parts b and c to explain why the data distribution cannot be mound-shaped.

Data set **2.31 Timber Tracts** To estimate the amount
EX0231 of lumber in a tract of timber, an owner decided to count the number of trees with diameters exceeding 12 inches in randomly selected 50-by-50-foot squares. Seventy 50-by-50-foot squares were chosen, and the selected trees were counted in each tract. The data are listed here:

7	8	7	10	4	8	6	8	9	10
9	6	4	9	10	9	8	8	7	9
3	9	5	9	9	8	7	5	8	8
10	2	7	4	8	5	10	7	7	7
9	6	8	8	8	7	8	9	6	8
6	11	9	11	7	7	11	7	9	13
10	8	8	5	9	9	8	5	9	8

a. Construct a relative frequency histogram to describe the data.

b. Calculate the sample mean $\bar{x}$ as an estimate of μ, the mean number of timber trees for all 50-by-50-foot squares in the tract.

c. Calculate s for the data. Construct the intervals $\bar{x} \pm s$, $\bar{x} \pm 2s$, and $\bar{x} \pm 3s$. Calculate the percentage of squares falling into each of the three intervals, and compare with the corresponding percentages given by the Empirical Rule and Tchebysheff's Theorem.

2.32 Tuna Fish, again Refer to Exercise 2.8 and data set EX0208. The prices of a 6-ounce can or a 7.06-ounce pouch for 14 different brands of water-packed light tuna, based on prices paid nationally in supermarkets are reproduced here.[4]

```
.99  1.92  1.23  .85  .65  .53  1.41
1.12  .63   .67  .69  .60  .60   .66
```

a. Use the range approximation to find an estimate of s.

b. How does it compare to the computed value of s?

Data set EX0233 **2.33 Old Faithful** The data below are 30 waiting times between eruptions of the Old Faithful geyser in Yellowstone National Park.[8]

```
56  89  51  79  58  82  52  88  52  78  69  75  77  72  71
55  87  53  85  61  93  54  76  80  81  59  86  78  71  77
```

a. Calculate the range.

b. Use the range approximation to approximate the standard deviation of these 30 measurements.

c. Calculate the sample standard deviation s.

d. What proportion of the measurements lie within two standard deviations of the mean? Within three standard deviations of the mean? Do these proportions agree with the proportions given in Tchebysheff's Theorem?

Data set EX0234 **2.34 The President's Kids** The table below shows the names of the 43 presidents of the United States along with the number of their children.[9]

Washington	0	Van Buren	4	Buchanan	0
Adams	5	W.H. Harrison	10	Lincoln	4
Jefferson	6	Tyler*	15	A. Johnson	5
Madison	0	Polk	0	Grant	4
Monroe	2	Taylor	6	Hayes	8
J.Q. Adams	4	Fillmore*	2	Garfield	7
Jackson	0	Pierce	3	Arthur	3
Cleveland	5	Coolidge	2	Nixon	2
B. Harrison*	3	Hoover	2	Ford	4
McKinley	2	F.D. Roosevelt	6	Carter	4
T. Roosevelt*	6	Truman	1	Reagan*	4
Taft	3	Eisenhower	2	G.H.W. Bush	6
Wilson*	3	Kennedy	3	Clinton	1
Harding	0	L.B. Johnson	2	G.W. Bush	2
				Obama	2

*Married twice   Source: The World Almanac and Book of Facts 2011

a. Construct a relative frequency histogram to describe the data. How would you describe the shape of this distribution?

b. Calculate the mean and the standard deviation for the data set.

c. Construct the intervals $\bar{x} \pm s$, $\bar{x} \pm 2s$, and $\bar{x} \pm 3s$. Find the percentage of measurements falling into these three intervals and compare with the corresponding percentages given by Tchebysheff's Theorem and the Empirical Rule.

2.35 An Archeological Find, again Refer to Exercise 2.17. The percentage of iron oxide in each of five pottery samples collected at the Island Thorns site was:

```
1.28   2.39   1.50   1.88   1.51
```

a. Use the range approximation to find an estimate of s, using an appropriate divisor from Table 2.7.

b. Calculate the standard deviation s. How close did your estimate come to the actual value of s?

Data set EX0236 **2.36 Aaron Rodgers** The number of passes completed by Aaron Rodgers, quarterback for the Minnesota Vikings, was recorded for each of the 15 regular season games that he played in the fall of 2010 (www.ESPN.com):[10]

```
19   19   34   12   27   18   21   15
27   22   26   21    7   25   19
```

a. Draw a stem and leaf plot to describe the data.

b. Calculate the mean and standard deviation for Aaron Rodgers' per game pass completions.

c. What proportion of the measurements lie within two standard deviations of the mean?

CALCULATING THE MEAN AND STANDARD DEVIATION FOR GROUPED DATA (OPTIONAL)

2.37 Suppose that some measurements occur more than once and that the data $x_1, x_2, \ldots, x_k$ are arranged in a frequency table as shown here:

Observations	Frequency f_i
x_1	f_1
x_2	f_2
.	.
.	.
.	.
x_k	f_k

The formulas for the mean and variance for grouped data are

$$\bar{x} = \frac{\Sigma x_i f_i}{n}, \qquad \text{where } n = \Sigma f_i$$

and

$$s^2 = \frac{\Sigma x_i^2 f_i - \dfrac{(\Sigma x_i f_i)^2}{n}}{n - 1}$$

Notice that if each value occurs once, these formulas reduce to those given in the text. Although these formulas for grouped data are primarily of value when you have a large number of measurements, demonstrate their use for the sample 1, 0, 0, 1, 3, 1, 3, 2, 3, 0, 0, 1, 1, 3, 2.

a. Calculate $\bar{x}$ and s^2 directly, using the formulas for ungrouped data.

b. The frequency table for the $n = 15$ measurements is as follows:

x	f
0	4
1	5
2	2
3	4

Calculate $\bar{x}$ and s^2 using the formulas for grouped data. Compare with your answers to part a.

2.38 International Baccalaureate High school students in an International Baccalaureate (IB) program are placed in accelerated or advanced courses and must take IB examinations in each of six subject areas at the end of their junior or senior year. Students are scored on a scale of 1–7, with 1–2 being poor, 3 mediocre, 4 average, and 5–7 excellent. During its first year of operation at John W. North High School in Riverside, California, 17 juniors attempted the IB economics exam, with these results:

Exam Grade	Number of Students
7	1
6	4
5	4
4	4
3	4

Calculate the mean and standard deviation for these scores.

2.39 A Skewed Distribution To illustrate the utility of the Empirical Rule, consider a distribution that is heavily skewed to the right, as shown in the accompanying figure.

a. Calculate $\bar{x}$ and s for the data shown. (NOTE: There are 10 zeros, 5 ones, and so on.)

b. Construct the intervals $\bar{x} \pm s$, $\bar{x} \pm 2s$, and $\bar{x} \pm 3s$ and locate them on the frequency distribution.

c. Calculate the proportion of the $n = 25$ measurements that fall into each of the three intervals. Compare with Tchebysheff's Theorem and the Empirical Rule. Note that, although the proportion that falls into the interval $\bar{x} \pm s$ does not agree closely with the Empirical Rule, the proportions that fall into the intervals $\bar{x} \pm 2s$ and $\bar{x} \pm 3s$ agree very well. Many times this is true, even for non-mound-shaped distributions of data.

Distribution for Exercise 2.39

MEASURES OF RELATIVE STANDING

2.6

Sometimes you need to know the position of one observation relative to others in a set of data. For example, if you took an examination with a total of 35 points, you might want to know how your score of 30 compared to the scores of the other students in the class. The mean and standard deviation of the scores can be used to calculate a **z-score,** which measures the relative standing of a measurement in a data set.

Definition   The **sample z-score** is a measure of relative standing defined by

$$z\text{-score} = \frac{x - \bar{x}}{s}$$

 NEED A TIP?

Positive *z*-score ⇔ *x* is above the mean.

Negative *z*-score ⇔ *x* is below the mean.

A *z*-**score measures the distance between an observation and the mean, measured in units of standard deviation.** For example, suppose that the mean and standard deviation of the test scores (based on a total of 35 points) are 25 and 4, respectively. The *z*-score for your score of 30 is calculated as follows:

$$z\text{-score} = \frac{x - \bar{x}}{s} = \frac{30 - 25}{4} = 1.25$$

Your score of 30 lies 1.25 standard deviations above the mean ($30 = \bar{x} + 1.25s$).

The *z*-score is a valuable tool for determining whether a particular observation is likely to occur quite frequently or whether it is unlikely and might be considered an **outlier.**

According to Tchebysheff's Theorem and the Empirical Rule,

- at least 75% and more likely 95% of the observations lie within two standard deviations of their mean: their *z*-scores are between −2 and +2. *Observations with z-scores exceeding 2 in absolute value happen about 5% of the time for mound-shaped data and are considered somewhat unlikely.*

- at least 89% and more likely 99.7% of the observations lie within three standard deviations of their mean: their *z*-scores are between −3 and +3. *Observations with z-scores exceeding 3 in absolute value happen less than 1% of the time for mound-shaped data and are considered very unlikely.*

 NEED A TIP?

z-Scores above 3 in absolute value are very unusual.

You should look carefully at any observation that has a *z*-score exceeding 3 in absolute value. Perhaps the measurement was recorded incorrectly or does not belong to the population being sampled. Perhaps it is just a highly unlikely observation, but a valid one nonetheless!

EXAMPLE   2.11

Consider this sample of $n = 10$ measurements:

1, 1, 0, 15, 2, 3, 4, 0, 1, 3

The measurement $x = 15$ appears to be unusually large. Calculate the *z*-score for this observation and state your conclusions.

Solution   Calculate $\bar{x} = 3.0$ and $s = 4.42$ for the $n = 10$ measurements. Then the *z*-score for the suspected outlier, $x = 15$, is calculated as

$$z\text{-score} = \frac{x - \bar{x}}{s} = \frac{15 - 3}{4.42} = 2.71$$

Hence, the measurement $x = 15$ lies 2.71 standard deviations above the sample mean, $\bar{x} = 3.0$. Although the *z*-score does not exceed 3, it is close enough so that you might suspect that $x = 15$ is an outlier. You should examine the sampling procedure to see whether $x = 15$ is a faulty observation.

A **percentile** is another measure of relative standing and is most often used for large data sets. (Percentiles are not very useful for small data sets.)

Definition   A set of n measurements on the variable x has been arranged in order of magnitude. The **pth percentile** is the value of x that is greater than $p\%$ of the measurements and is less than the remaining $(100 - p)\%$.

EXAMPLE (2.12)

Suppose you have been notified that your score of 610 on the Verbal Graduate Record Examination placed you at the 60th percentile in the distribution of scores. Where does your score of 610 stand in relation to the scores of others who took the examination?

Solution   Scoring at the 60th percentile means that 60% of all the examination scores were lower than your score and 40% were higher.

For any data distribution, regardless of its shape, the 60th percentile for the variable x is a point on the *horizontal axis* of the data distribution that is greater than 60% of the measurements and less than the others. That is, 60% of the measurements are less than the 60th percentile and 40% are greater (see Figure 2.12). Since the total area under the distribution is 100%, 60% of the area is to the left and 40% of the area is to the right of the 60th percentile. Remember that the median, m, of a set of data is the middle measurement; that is, 50% of the measurements are smaller and 50% are larger than the median. Thus, the *median is the same as the 50th percentile!*

FIGURE 2.12

The 60th percentile shown on the relative frequency histogram for a data set

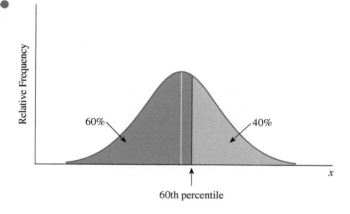

The 25th and 75th percentiles, called the **lower** and **upper quartiles,** along with the median (the 50th percentile), locate points that divide the data into four sets, each containing an equal number of measurements. Twenty-five percent of the measurements will be less than the lower (first) quartile, 50% will be less than the median (the second quartile), and 75% will be less than the upper (third) quartile. Thus, the median and the lower and upper quartiles are located at points on the x-axis so that the area under the relative frequency histogram for the data is partitioned into four equal areas, as shown in Figure 2.13.

FIGURE 2.13

Location of quartiles

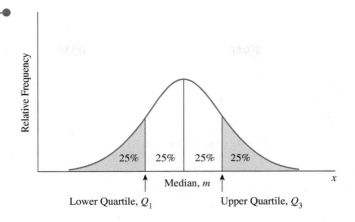

Definition   A set of n measurements on the variable x has been arranged in order of magnitude. The **lower quartile (first quartile), Q_1,** is the value of x that is greater than one-fourth of the measurements and is less than the remaining three-fourths. The **second quartile** is the median. The **upper quartile (third quartile), Q_3,** is the value of x that is greater than three-fourths of the measurements and is less than the remaining one-fourth.

For small data sets, it is often impossible to divide the set into four groups, each of which contains exactly 25% of the measurements. For example, when $n = 10$, you would need to have $2\frac{1}{2}$ measurements in each group! Even when you can perform this task (for example, if $n = 12$), there are many numbers that would satisfy the preceding definition, and could therefore be considered "quartiles." To avoid this ambiguity, we use the following rule to locate sample quartiles.

CALCULATING SAMPLE QUARTILES

- When the measurements are arranged in order of magnitude, the **lower quartile, Q_1,** is the value of x in position $.25(n + 1)$, and the **upper quartile, Q_3,** is the value of x in position $.75(n + 1)$.
- When $.25(n + 1)$ and $.75(n + 1)$ are not integers, the quartiles are found by interpolation, using the values in the two adjacent positions.[†]

EXAMPLE 2.13

Find the lower and upper quartiles for this set of measurements:

16, 25, 4, 18, 11, 13, 20, 8, 11, 9

Solution   Rank the $n = 10$ measurements from smallest to largest:

4, 8, 9, 11, 11, 13, 16, 18, 20, 25

[†]This definition of quartiles is consistent with the one used in *MINITAB 16* and *MS Excel 2010*. Some textbooks use ordinary rounding when finding quartile positions, whereas others compute sample quartiles as the medians of the upper and lower halves of the data set.

Calculate

$$\text{Position of } Q_1 = .25(n + 1) = .25(10 + 1) = 2.75$$
$$\text{Position of } Q_3 = .75(n + 1) = .75(10 + 1) = 8.25$$

Since these positions are not integers, the lower quartile is taken to be the value 3/4 of the distance between the second and third ordered measurements, and the upper quartile is taken to be the value 1/4 of the distance between the eighth and ninth ordered measurements. Therefore,

$$Q_1 = 8 + .75(9 - 8) = 8 + .75 = 8.75$$

and

$$Q_3 = 18 + .25(20 - 18) = 18 + .5 = 18.5$$

Because the median and the quartiles divide the data distribution into four parts, each containing approximately 25% of the measurements, Q_1 and Q_3 are the upper and lower boundaries for the middle 50% of the distribution. We can measure the range of this "middle 50%" of the distribution using a numerical measure called the **interquartile range.**

Definition   The **interquartile range (IQR)** for a set of measurements is the difference between the upper and lower quartiles; that is, IQR $= Q_3 - Q_1$.

For the data in Example 2.13, IQR $= Q_3 - Q_1 = 18.50 - 8.75 = 9.75$. We will use the IQR along with the quartiles and the median in the next section to construct another graph for describing data sets.

NEED TO KNOW...

How to Calculate Sample Quartiles

1. Arrange the data set in order of magnitude from smallest to largest.
2. Calculate the quartile positions:
    - Position of Q_1: $.25(n + 1)$
    - Position of Q_3: $.75(n + 1)$
3. If the positions are integers, then Q_1 and Q_3 are the values in the ordered data set found in those positions.
4. If the positions in step 2 are not integers, find the two measurements in positions just above and just below the calculated position. Calculate the quartile by finding a value either one-fourth, one-half, or three-fourths of the way between these two measurements.

Many of the numerical measures that you have learned are easily found using computer programs or even graphics calculators. The *MINITAB* command **Stat ▶ Basic Statistics ▶ Display Descriptive Statistics** or the *Excel* command **Data ▶ Data Analysis ▶ Descriptive Statistics** (see the "Technology Today" section at the end of this chapter) produce output containing the mean, the standard deviation, the median, the maximum and minimum, as well as the values of other statistics that we have not discussed. The data from Example 2.13 produced the *MINITAB* output shown in Figure 2.14. Notice that the quartiles are identical to the hand-calculated values in that example.

FIGURE 2.14

MINITAB output for the data in Example 2.13

Descriptive Statistics: x

Variable	N	N*	Mean	SE Mean	StDev	Minimum	Q1	Median	Q3	Maximum
X	10	0	13.50	1.98	6.28	4.00	8.75	12.00	18.50	25.00

2.7 THE FIVE-NUMBER SUMMARY AND THE BOX PLOT

The median and the upper and lower quartiles shown in Figure 2.13 divide the data into four sets, each containing an equal number of measurements. If we add the largest number (Max) and the smallest number (Min) in the data set to this group, we will have a set of numbers that provide a quick and rough summary of the data distribution.

> The **five-number summary** consists of the smallest number, the lower quartile, the median, the upper quartile, and the largest number, presented in order from smallest to largest:
>
> **Min** Q_1 **Median** Q_3 **Max**
>
> By definition, one-fourth of the measurements in the data set lie between each of the four adjacent pairs of numbers.

The five-number summary can be used to create a simple graph called a **box plot** to visually describe the data distribution. From the box plot, you can quickly detect any skewness in the shape of the distribution and see whether there are any outliers in the data set. An outlier may result from transposing digits when recording a measurement, from incorrectly reading an instrument dial, from a malfunctioning piece of equipment, or from other problems.

Even when there are no recording or observational errors, a data set may contain one or more valid measurements that, for one reason or another, differ markedly from the others in the set. These outliers can cause a marked distortion in commonly used numerical measures such as $\bar{x}$ and s. In fact, outliers may themselves contain important information not shared with the other measurements in the set. Therefore, isolating outliers, if they are present, is an important step in any preliminary analysis of a data set. The box plot is designed expressly for this purpose.

TO CONSTRUCT A BOX PLOT

- Calculate the median, the upper and lower quartiles, and the IQR for the data set.

• Draw a horizontal line representing the scale of measurement. Form a box just above the horizontal line with the right and left ends at Q_1 and Q_3. Draw a vertical line through the box at the location of the median.

A box plot is shown in Figure 2.15.

FIGURE 2.15

Box plot

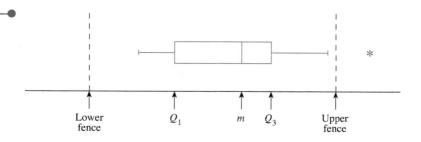

| Lower fence | Q_1 | m | Q_3 | Upper fence |

In Section 2.6, the z-score provided boundaries for finding unusually large or small measurements. You looked for z-scores greater than 2 or 3 in absolute value. The box plot uses the IQR to create imaginary "fences" to separate outliers from the rest of the data set:

DETECTING OUTLIERS—OBSERVATIONS THAT ARE BEYOND:

• Lower fence: $Q_1 - 1.5(\text{IQR})$
• Upper fence: $Q_3 + 1.5(\text{IQR})$

The upper and lower fences are shown with broken lines in Figure 2.15, but they are not usually drawn on the box plot. Any measurement beyond the upper or lower fence is an **outlier;** the rest of the measurements, inside the fences, are not unusual. Finally, the box plot marks the range of the data set using "whiskers" to connect the smallest and largest measurements (*excluding outliers*) to the box.

TO FINISH THE BOX PLOT

• Mark any **outliers** with an asterisk (*) on the graph.
• Extend horizontal lines called "whiskers" from the ends of the box to the smallest and largest observations that are *not* outliers.

EXAMPLE 2.14

As American consumers become more careful about the foods they eat, food processors try to stay competitive by avoiding excessive amounts of fat, cholesterol, and sodium in the foods they sell. The following data are the amounts of sodium per slice (in milligrams) for each of eight brands of regular American cheese. Construct a box plot for the data and look for outliers.

340,   300,   520,   340,   320,   290,   260,   330

Solution   The $n = 8$ measurements are first ranked from smallest to largest:

260,   290,   300,   320,   330,   340,   340,   520

The positions of the median, Q_1, and Q_3 are

$$.5(n + 1) = .5(9) = 4.5$$
$$.25(n + 1) = .25(9) = 2.25$$
$$.75(n + 1) = .75(9) = 6.75$$

so that $m = (320 + 330)/2 = 325$, $Q_1 = 290 + .25(10) = 292.5$, and $Q_3 = 340$. The interquartile range is calculated as

$$\text{IQR} = Q_3 - Q_1 = 340 - 292.5 = 47.5$$

Calculate the upper and lower fences:

Lower fence: $292.5 - 1.5(47.5) = 221.25$
Upper fence: $340 + 1.5(47.5) = 411.25$

Building a Box Plot

The value $x = 520$, a brand of cheese containing 520 milligrams of sodium, is the only *outlier,* lying beyond the upper fence.

The box plot for the data is shown in Figure 2.16. The outlier is marked with an asterisk (*). Once the outlier is excluded, we find (from the ranked data set) that the smallest and largest measurements are $x = 260$ and $x = 340$. These are the two values that form the whiskers. Since the value $x = 340$ is the same as Q_3, there is no whisker on the right side of the box.

FIGURE 2.16

Box plot for Example 2.14

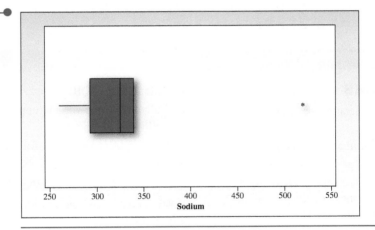

You can use the box plot to describe the shape of a data distribution by looking at the position of the median line compared to Q_1 and Q_3, the left and right ends of the box. If the median is close to the middle of the box, the distribution is fairly symmetric, providing equal-sized intervals to contain the two middle quarters of the data. If the median line is to the left of center, the distribution is skewed to the right; if the median is to the right of center, the distribution is skewed to the left. Also, for most skewed distributions, the whisker on the skewed side of the box tends to be longer than the whisker on the other side.

Figure 2.17 shows two box plots, one for the sodium contents of the eight brands of cheese in Example 2.14, and another for five brands of fat-free cheese with these sodium contents:

300, 300, 320, 290, 180

Look at the long whisker on the left side of both box plots and the position of the median lines. Both distributions are skewed to the left; that is, there are a few unusually small measurements. The regular cheese data, however, also show one brand ($x = 520$) with an unusually large amount of sodium. In general, it appears that the sodium content of the fat-free brands is lower than that of the regular brands, but the variability of the sodium content for regular cheese (excluding the outlier) is less than that of the fat-free brands.

FIGURE 2.17

Box plots for regular and fat-free cheese

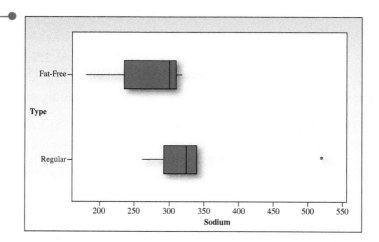

2.7 EXERCISES

BASIC TECHNIQUES

2.40 Given the following data set: 8, 7, 1, 4, 6, 6, 4, 5, 7, 6, 3, 0

a. Find the five-number summary and the IQR.

b. Calculate $\bar{x}$ and s.

c. Calculate the z-score for the smallest and largest observations. Is either of these observations unusually large or unusually small?

2.41 Find the five-number summary and the IQR for these data:

19, 12, 16, 0, 14, 9, 6, 1, 12, 13, 10, 19, 7, 5, 8

2.42 Given the following data set: 2.3, 1.0, 2.1, 6.5, 2.8, 8.8, 1.7, 2.9, 4.4, 5.1, 2.0

a. Find the positions of the lower and upper quartiles.

b. Sort the data from smallest to largest and find the lower and upper quartiles.

c. Calculate the IQR.

2.43 Given the following data set: .23, .30, .35, .41, .56, .58, .76, .80

a. Find the lower and upper quartiles.

b. Calculate the IQR.

c. Calculate the lower and upper fences. Are there any outliers?

2.44 Construct a box plot for these data and identify any outliers:

25, 22, 26, 23, 27, 26, 28, 18, 25, 24, 12

2.45 Construct a box plot for these data and identify any outliers:

3, 9, 10, 2, 6, 7, 5, 8, 6, 6, 4, 9, 22

APPLICATIONS

2.46 If you scored at the 69th percentile on a placement test, how does your score compare with others?

2.47 Mercury Concentration in Dolphins EX0247 Environmental scientists are increasingly concerned with the accumulation of toxic elements in marine mammals and the transfer of such elements to the animals' offspring. The striped dolphin (*Stenella coeruleoalba*), considered to be a top predator in the marine food chain, was the subject of one such study. The mercury concentrations (micrograms/gram) in the livers of 28 male striped dolphins were as follows:

1.70	183.00	221.00	286.00
1.72	168.00	406.00	315.00
8.80	218.00	252.00	241.00
5.90	180.00	329.00	397.00
101.00	264.00	316.00	209.00
85.40	481.00	445.00	314.00
118.00	485.00	278.00	318.00

a. Calculate the five-number summary for the data.

b. Construct a box plot for the data.

c. Are there any outliers?

d. If you knew that the first four dolphins were all less than 3 years old, while all the others were more than 8 years old, would this information help explain the difference in the magnitude of those four observations? Explain.

2.48 Hamburger Meat The weights (in pounds) of the 27 packages of ground beef from Exercise 2.24 (see data set EX0224) are listed here in order from smallest to largest:

.75	.83	.87	.89	.89	.89	.92
.93	.96	.96	.97	.98	.99	1.06
1.08	1.08	1.12	1.12	1.14	1.14	1.17
1.18	1.18	1.24	1.28	1.38	1.41	

a. Confirm the values of the mean and standard deviation, calculated in Exercise 2.24 as $\bar{x} = 1.05$ and $s = .17$.

b. The two largest packages of meat weigh 1.38 and 1.41 pounds. Are these two packages unusually heavy? Explain.

c. Construct a box plot for the package weights. What does the position of the median line and the length of the whiskers tell you about the shape of the distribution?

Data set **2.49 Comparing NFL Quarterbacks** How EX0249 does Aaron Rodgers, quarterback for the 2011 Super Bowl winners, the Green Bay Packers, compare to Drew Brees, quarterback for the 2010 Super Bowl winners, the New Orleans Saints? The table below shows the number of completed passes for each athlete during the 2010 NFL football season:[11]

Aaron Rodgers			Drew Brees		
19	21	7	27	37	25
19	15	25	28	34	29
34	27	19	30	27	35
12	22		33	29	22
27	26		24	23	
18	21		21	24	

a. Calculate five-number summaries for the number of passes completed by both Aaron Rodgers and Drew Brees.

b. Construct box plots for the two sets of data. Are there any outliers? What do the box plots tell you about the shapes of the two distributions?

c. Write a short paragraph comparing the number of pass completions for the two quarterbacks.

2.50 Presidential Vetoes The set of presidential vetoes in Exercise 1.47 and data set EX0147 is listed here, along with a box plot generated by *MINITAB*. Use the box plot to describe the shape of the distribution and identify any outliers.

Washington	2	B. Harrison	19
J. Adams	0	Cleveland	42
Jefferson	0	McKinley	6
Madison	5	T. Roosevelt	42
Monroe	1	Taft	30
J. Q. Adams	0	Wilson	33
Jackson	5	Harding	5
Van Buren	0	Coolidge	20
W. H. Harrison	0	Hoover	21
Tyler	6	F. D. Roosevelt	372
Polk	2	Truman	180
Taylor	0	Eisenhower	73
Fillmore	0	Kennedy	12
Pierce	9	L. Johnson	16
Buchanan	4	Nixon	26
Lincoln	2	Ford	48
A. Johnson	21	Carter	13
Grant	45	Reagan	39
Hayes	12	G. H. W. Bush	29
Garfield	0	Clinton	36
Arthur	4	G. W. Bush	11
Cleveland	304	Obama	1

Source: The World Almanac and Book of Facts 2011

Box plot for Exercise 2.50

2.51 Survival Times Altman and Bland report the survival times for patients with active hepatitis, half treated with prednisone and half receiving no treatment.[12] The survival times (in months) (Exercise 1.25 and EX0125) are adapted from their data for those treated with prednisone.

8	87	127	147
11	93	133	148
52	97	139	157
57	109	142	162
65	120	144	165

a. Can you tell by looking at the data whether it is roughly symmetric? Or is it skewed?

b. Calculate the mean and the median. Use these measures to decide whether or not the data are symmetric or skewed.

c. Draw a box plot to describe the data. Explain why the box plot confirms your conclusions in part b.

Data set **2.52 Utility Bills in Southern California, EX0252 again** The monthly utility bills for a household in Riverside, California, were recorded for 12 consecutive months starting in January 2010:

Month	Amount ($)	Month	Amount ($)
January	288.02	July	311.20
February	230.60	August	370.23
March	216.85	September	368.57
April	243.74	October	301.79
May	236.96	November	271.99
June	288.57	December	298.12

a. Construct a box plot for the monthly utility costs.

b. What does the box plot tell you about the distribution of utility costs for this household?

2.53 What's Normal? again Refer to Exercise 1.67 and data set EX0167. In addition to the normal body temperature in degrees Fahrenheit for the 130 individuals, the data record the gender of the individuals. Box plots for the two groups, male and female, are shown below:[13]

Box plots for Exercise 2.53

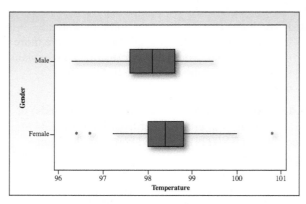

How would you describe the similarities and differences between male and female temperatures in this data set?

CHAPTER REVIEW

Key Concepts and Formulas

I. Measures of the Center of a Data Distribution

1. Arithmetic mean (mean) or average
   a. Population: μ
   b. Sample of n measurements: $\bar{x} = \dfrac{\Sigma x_i}{n}$

2. Median; **position** of the median $= .5(n + 1)$

3. Mode

4. The median may be preferred to the mean if the data are highly skewed.

II. Measures of Variability

1. Range: $R = $ largest $-$ smallest

2. Variance
   a. Population of N measurements:
   $$\sigma^2 = \frac{\Sigma(x_i - \mu)^2}{N}$$

   b. Sample of n measurements:
   $$s^2 = \frac{\Sigma(x_i - \bar{x})^2}{n - 1} = \frac{\Sigma x_i^2 - \dfrac{(\Sigma x_i)^2}{n}}{n - 1}$$

3. Standard deviation
   a. Population: $\sigma = \sqrt{\sigma^2}$
   b. Sample: $s = \sqrt{s^2}$

4. A rough approximation for s can be calculated as $s \approx R/4$. The divisor can be adjusted depending on the sample size.

III. Tchebysheff's Theorem and the Empirical Rule

1. Use Tchebysheff's Theorem for any data set, regardless of its shape or size.
   a. At least $1 - (1/k^2)$ of the measurements lie within k standard deviations of the mean.
   b. This is only a lower bound; there may be more measurements in the interval.

2. The Empirical Rule can be used only for relatively mound-shaped data sets. Approximately 68%, 95%, and 99.7% of the measurements are within one, two, and three standard deviations of the mean, respectively.

IV. Measures of Relative Standing

1. Sample z-score: $z = \dfrac{x - \bar{x}}{s}$

2. pth percentile; $p\%$ of the measurements are smaller, and $(100 - p)\%$ are larger.

3. Lower quartile, Q_1; **position** of $Q_1 = .25 (n + 1)$

4. Upper quartile, Q_3; **position** of $Q_3 = .75 (n + 1)$

5. Interquartile range: $\text{IQR} = Q_3 - Q_1$

V. The Five-Number Summary and Box Plots

1. The **five-number summary:**

   Min Q_1 Median Q_3 Max

   One-fourth of the measurements in the data set lie between each of the four adjacent pairs of numbers.

2. Box plots are used for detecting outliers and shapes of distributions.

3. Q_1 and Q_3 form the ends of the box. The median line is in the interior of the box.

4. Upper and lower fences are used to find outliers, observations that lie outside these fences.
   a. **Lower fence:** $Q_1 - 1.5(\text{IQR})$
   b. **Upper fence:** $Q_3 + 1.5(\text{IQR})$

5. **Outliers** are marked on the box plot with an asterisk (*).

6. **Whiskers** are connected to the box from the smallest and largest observations that are *not* outliers.

7. Skewed distributions usually have a long whisker *in the direction of the skewness,* and the median line is drawn *away from the direction of the skewness.*

TECHNOLOGY TODAY

Numerical Descriptive Measures in *Excel*

MS Excel provides most of the basic descriptive statistics presented in Chapter 2 using a single command on the **Data** tab. Other descriptive statistics can be calculated using the **Function** command on the **Formulas** tab.

EXAMPLE 2.15

The following data are the front and rear leg rooms (in inches) for nine different sports utility vehicles:[14]

Make & Model	Front Leg Room	Rear Leg Room
Acura MDX	41.0	28.5
Buick Enclave	41.5	30.0
Chevy TrailBlazer	40.0	25.5
Chevy Tahoe Hybrid V8 CVT	41.0	27.5
GMC Terrain 1LT 4-cyl	43.0	31.0
Honda CR-V	41.0	29.5
Hyundai Tucson	42.5	29.5
Kia Sportage	40.0	29.0
Lexus GX	42.0	30.0

1. Since the data involve two variables and a third labeling variable, enter the data into the first three columns of an *Excel* spreadsheet, using the labels in the table. Select **Data ▶ Data Analysis ▶ Descriptive Statistics**, and highlight or type the **Input range** (the data in the second and third columns) into the Descriptive Statistics Dialog box (Figure 2.18(a)). Type an Output location, make sure the check boxes for "Labels in First Row" and "Summary Statistics" are both checked, and click **OK**. The summary statistics (Figure 2.18(b)) will appear in the selected location in your spreadsheet.

FIGURE 2.18

(a)

(b)

	E	F	G	H
	Front Leg Room		*Rear Leg Room*	
Mean		41.333	Mean	28.944
Standard Error		0.344	Standard Error	0.543
Median		41	Median	29.5
Mode		41	Mode	30
Standard Deviation		1.031	Standard Deviation	1.629
Sample Variance		1.063	Sample Variance	2.653
Kurtosis		-0.745	Kurtosis	1.735
Skewness		0.245	Skewness	-1.212
Range		3	Range	5.5
Minimum		40	Minimum	25.5
Maximum		43	Maximum	31
Sum		372	Sum	260.5
Count		9	Count	9

2. You may notice that some of the cells in the spreadsheet are overlapping. To adjust this, highlight the affected columns and click the **Home** tab. In the **Cells** group, choose **Format ▶ AutoFit Column Width**. You may want to modify the

appearance of the output by decreasing the decimal accuracy in certain cells. Highlight the appropriate cells and click the **Decrease Decimal** icon $^{+.0}_{.00}$ (**Home** tab, **Number** group) to modify the output. We have displayed the accuracy to three decimal places.

3. Notice that the sample quartiles, Q_1 and Q_3, are not given in the *Excel* output in Figure 2.18(b). You can calculate the quartiles using the function command. Place your cursor into an empty cell and select **Formulas ▶ More Functions ▶ Statistical ▶ QUARTILE.EXC**. Highlight the appropriate cells in the box marked "Array" and type an integer ($0 =$ min, $1 =$ first quartile, $2 =$ median, $3 =$ third quartile, or $4 =$ max) in the box marked "Quart." The quartile (calculated using this textbook's method) will appear in the cell you have chosen. An alternative method for calculating the quartiles will be used if you select **Formulas ▶ More Functions ▶ Statistical ▶ QUARTILE.INC**. (NOTE: This function is called **QUARTILE** in *Excel 2007* and earlier versions.) Using the two quartiles, you can calculate the IQR and construct a box plot by hand.

Numerical Descriptive Measures in *MINITAB*

MINITAB provides most of the basic descriptive statistics presented in Chapter 2 using a single command in the drop-down menus.

EXAMPLE 2.16

The following data are the front and rear leg rooms (in inches) for nine different sports utility vehicles:[14]

Make and Model	Front Leg Room	Rear Leg Room
Acura MDX	41.0	28.5
Buick Enclave	41.5	30.0
Chevy TrailBlazer	40.0	25.5
Chevy Tahoe Hybrid V8 CVT	41.0	27.5
GMC Terrain 1LT 4-cyl	43.0	31.0
Honda CR-V	41.0	29.5
Hyundai Tucson	42.5	29.5
Kia Sportage	40.0	29.0
Lexus GX	42.0	30.0

1. Since the data involve two variables and a third labeling variable, enter the data into the first three columns of a *MINITAB* worksheet, using the labels in the table. Using the drop-down menus, select **Stat ▶ Basic Statistics ▶ Display Descriptive Statistics.** The Dialog box is shown in Figure 2.19(a).

FIGURE 2.19

(a)

(b)

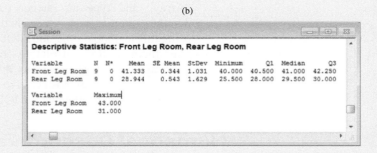

2. Now click on the Variables box and **select** both columns from the list on the left. (You can click on the **Graphs** option and choose one of several graphs if you like. You may also click on the **Statistics** option to select the statistics you would like to see displayed.) Click **OK.** A display of descriptive statistics for both columns will appear in the Session window (see Figure 2.19(b)). You may print this output using **File ▶ Print Session Window** if you choose.

3. To examine the distribution of the two variables and look for outliers, you can create box plots using the command **Graph ▶ Boxplot ▶ One Y ▶ Simple.** Click **OK.** Select the appropriate column of measurements in the Dialog box (see Figure 2.20(a)). You can change the appearance of the box plot in several ways. **Scale ▶ Axes and Ticks** will allow you to transpose the axes and orient the box plot horizontally, when you check the box marked "Transpose value and category scales." **Multiple Graphs** provides printing options for multiple box plots. **Labels** will let you annotate the graph with titles and footnotes. If you have entered data into the worksheet as a frequency distribution (values in one column, frequencies in another), the **Data Options** will allow the data to be read in that format. The box plot for the rear leg rooms is shown in Figure 2.20(b).

4. Save this worksheet in a file called "Leg Room" before exiting *MINITAB.* We will use it again in Chapter 3.

FIGURE 2.20

(a)

(b)

Supplementary Exercises

2.54 Raisins The number of raisins in each of 14 miniboxes (1/2-ounce size) was counted for a generic brand and for Sunmaid brand raisins. The two data sets are shown here:

Generic Brand				Sunmaid			
25	26	25	28	25	29	24	24
26	28	28	27	28	24	28	22
26	27	24	25	25	28	30	27
26	26			28	24		

a. What are the mean and standard deviation for the generic brand?

b. What are the mean and standard deviation for the Sunmaid brand?

c. Compare the centers and variabilities of the two brands using the results of parts a and b.

2.55 Raisins, continued Refer to Exercise 2.54.

a. Find the median, the upper and lower quartiles, and the IQR for each of the two data sets.

b. Construct two box plots on the same horizontal scale to compare the two sets of data.

c. Draw two stem and leaf plots to depict the shapes of the two data sets. Do the box plots in part b verify these results?

d. If we can assume that none of the boxes of raisins are being underfilled (that is, they all weigh approximately 1/2 ounce), what do your results say about the average number of raisins for the two brands?

2.56 TV Viewers The number of television viewing hours per household and the prime viewing times are two factors that affect television advertising income. A random sample of 25 households in a particular viewing area produced the following estimates of viewing hours per household:

3.0	6.0	7.5	15.0	12.0
6.5	8.0	4.0	5.5	6.0
5.0	12.0	1.0	3.5	3.0
7.5	5.0	10.0	8.0	3.5
9.0	2.0	6.5	1.0	5.0

a. Scan the data and use the range to find an approximate value for s. Use this value to check your calculations in part b.

b. Calculate the sample mean $\bar{x}$ and the sample standard deviation s. Compare s with the approximate value obtained in part a.

c. Find the percentage of the viewing hours per household that falls into the interval $\bar{x} \pm 2s$. Compare with the corresponding percentage given by the Empirical Rule.

2.57 A Recurring Illness Refer to Exercise 1.26 and data set EX0126. The lengths of time (in months) between the onset of a particular illness and its recurrence were recorded:

2.1	4.4	2.7	32.3	9.9
9.0	2.0	6.6	3.9	1.6
14.7	9.6	16.7	7.4	8.2
19.2	6.9	4.3	3.3	1.2
4.1	18.4	.2	6.1	13.5
7.4	.2	8.3	.3	1.3
14.1	1.0	2.4	2.4	18.0
8.7	24.0	1.4	8.2	5.8
1.6	3.5	11.4	18.0	26.7
3.7	12.6	23.1	5.6	.4

a. Find the range.

b. Use the range approximation to find an approximate value for s.

c. Compute s for the data and compare it with your approximation from part b.

2.58 A Recurring Illness, continued Refer to Exercise 2.57.

a. Examine the data and count the number of observations that fall into the intervals $\bar{x} \pm s$, $\bar{x} \pm 2s$, and $\bar{x} \pm 3s$.

b. Do the percentages that fall into these intervals agree with Tchebysheff's Theorem? With the Empirical Rule?

c. Why might the Empirical Rule be unsuitable for describing these data?

2.59 A Recurring Illness, again Find the median and the lower and upper quartiles for the data on times until recurrence of an illness in Exercise 2.57. Use these descriptive measures to construct a box plot for the data. Use the box plot to describe the data distribution.

2.60 Tuna Fish, again Refer to Exercise 2.8. The prices of a 6-ounce can or a 7.06-ounce pouch for 14 different brands of water-packed light tuna, based on prices paid nationally in supermarkets, are reproduced here.[4]

.99	1.92	1.23	.85	.65	.53	1.41
1.12	.63	.67	.69	.60	.60	.66

a. Calculate the five-number summary.

b. Construct a box plot for the data. Are there any outliers?

c. The value $x = 1.92$ looks large in comparison to the other prices. Use a z-score to decide whether this is an unusually expensive brand of tuna.

2.61 Electrolysis An analytical chemist wanted to use electrolysis to determine the number of moles of cupric ions in a given volume of solution. The solution was partitioned into $n = 30$ portions of .2 milliliter each, and each of the portions was tested. The average number of moles of cupric ions for the $n = 30$ portions was found to be .17 mole; the standard deviation was .01 mole.

a. Describe the distribution of the measurements for the $n = 30$ portions of the solution using Tchebysheff's Theorem.

b. Describe the distribution of the measurements for the $n = 30$ portions of the solution using the Empirical Rule. (Do you expect the Empirical Rule to be suitable for describing these data?)

c. Suppose the chemist had used only $n = 4$ portions of the solution for the experiment and obtained the readings .15, .19, .17, and .15. Would the Empirical Rule be suitable for describing the $n = 4$ measurements? Why?

2.62 Chloroform According to the EPA, chloroform, which in its gaseous form is suspected of being a cancer-causing agent, is present in small quantities in all of the country's 240,000 public water sources. If the mean and standard deviation of the amounts of chloroform present in the water sources are 34 and 53 micrograms per liter, respectively, describe the distribution for the population of all public water sources.

2.63 Achievement Tests Mathematics achievement test scores for 400 students were found to have a mean and a variance equal to 600 and 4900, respectively. If the distribution of test scores was mound-shaped, approximately how many of the scores would fall into the interval 530 to 670? Approximately how many scores would be expected to fall into the interval 460 to 740?

2.64 Sleep and the College Student How much sleep do you get on a typical school night? A group of 10 college students were asked to report the number of hours that they slept on the previous night with the following results:

$$7, \quad 6, \quad 7.25, \quad 7, \quad 8.5, \quad 5, \quad 8, \quad 7, \quad 6.75, \quad 6$$

a. Find the mean and the standard deviation of the number of hours of sleep for these 10 students.

b. Calculate the z-score for the largest value ($x = 8.5$). Is this an unusually sleepy college student?

c. What is the most frequently reported measurement? What is the name for this measure of center?

d. Construct a box plot for the data. Does the box plot confirm your results in part b? [HINT: Since the z-score and the box plot are two unrelated methods for detecting outliers, and use different types of statistics, they do not necessarily have to (but usually do) produce the same results.]

Data set
EX0265
2.65 Gas Mileage The miles per gallon (mpg) for each of 20 medium-sized cars selected from a production line during the month of March follow.

23.1	21.3	23.6	23.7
20.2	24.4	25.3	27.0
24.7	22.7	26.2	23.2
25.9	24.7	24.4	24.2
24.9	22.2	22.9	24.6

a. What are the maximum and minimum miles per gallon? What is the range?

b. Construct a relative frequency histogram for these data. How would you describe the shape of the distribution?

c. Find the mean and the standard deviation.

d. Arrange the data from smallest to largest. Find the z-scores for the largest and smallest observations. Would you consider them to be outliers? Why or why not?

e. What is the median?

f. Find the lower and upper quartiles.

2.66 Gas Mileage, continued Refer to Exercise 2.65. Construct a box plot for the data. Are there any outliers? Does this conclusion agree with your results in Exercise 2.65?

2.67 Polluted Seawater Petroleum pollution in seas and oceans stimulates the growth of some types of bacteria. A count of petroleumlytic micro-organisms (bacteria per 100 milliliters) in 10 portions of seawater gave these readings:

49, 70, 54, 67, 59, 40, 61, 69, 71, 52

a. Guess the value for s using the range approximation.

b. Calculate $\bar{x}$ and s and compare with the range approximation of part a.

c. Construct a box plot for the data and use it to describe the data distribution.

2.68 Basketball Attendances at a high school's basketball games were recorded and found to have a sample mean and variance of 420 and 25, respectively. Calculate $\bar{x} \pm s$, $\bar{x} \pm 2s$, and $\bar{x} \pm 3s$ and then state the approximate fractions of measurements you would expect to fall into these intervals according to the Empirical Rule.

2.69 SAT Tests The College Board's verbal and mathematics scholastic aptitude tests are scored on a scale of 200 to 800. It seems reasonable to assume that a distribution of all test scores, either verbal or math, is mound-shaped. If σ is the standard deviation of one of these distributions, what is the largest value (approximately) that σ might assume? Explain.

2.70 Summer Camping A favorite summer pastime for many Americans is camping. In fact, camping has become so popular at the California beaches that reservations must sometimes be made months in advance! Data from a *USA Today* Snapshot is shown below.[15]

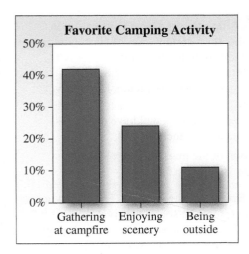

Favorite Camping Activity

The Snapshot also reports that men go camping 2.9 times a year, women go 1.7 times a year; and men are more likely than women to want to camp more often. What does the magazine mean when they talk about 2.9 or 1.7 times a year?

2.71 Long-Stemmed Roses A strain of long-stemmed roses has an approximate normal distribution with a mean stem length of 15 inches and standard deviation of 2.5 inches.

a. If one accepts as "long-stemmed roses" only those roses with a stem length greater than 12.5 inches, what percentage of such roses would be unacceptable?

b. What percentage of these roses would have a stem length between 12.5 and 20 inches?

Data set EX0272 **2.72 Drugs for Hypertension** A pharmaceutical company wishes to know whether an experimental drug being tested in its laboratories has any effect on systolic blood pressure. Fifteen randomly selected subjects were given the drug, and their systolic blood pressures (in millimeters) are recorded.

172	148	123
140	108	152
123	129	133
130	137	128
115	161	142

a. Guess the value of s using the range approximation.

b. Calculate $\bar{x}$ and s for the 15 blood pressures.

c. Find two values, a and b, such that at least 75% of the measurements fall between a and b.

2.73 Lumber Rights A company interested in lumbering rights for a certain tract of slash pine trees is told that the mean diameter of these trees is 14 inches with a standard deviation of 2.8 inches. Assume the distribution of diameters is roughly mound-shaped.

a. What fraction of the trees will have diameters between 8.4 and 22.4 inches?

b. What fraction of the trees will have diameters greater than 16.8 inches?

Data set EX0274 **2.74 Social Ambivalence** The following data represent the social ambivalence scores for 15 people as measured by a psychological test. (The higher the score, the stronger the ambivalence.)

9	13	12
14	15	11
10	4	10
8	19	13
11	17	9

a. Guess the value of s using the range approximation.

b. Calculate $\bar{x}$ and s for the 15 social ambivalence scores.

c. What fraction of the scores actually lie in the interval $\bar{x} \pm 2s$?

2.75 TV Commercials The mean duration of television commercials on a given network is 75 seconds, with a standard deviation of 20 seconds. Assume that durations are approximately normally distributed.

a. What is the approximate probability that a commercial will last less than 35 seconds?

b. What is the approximate probability that a commercial will last longer than 55 seconds?

Data set EX0276 **2.76 Parasites in Foxes** A random sample of 100 foxes was examined by a team of veterinarians to determine the prevalence of a particular type of parasite. Counting the number of parasites per fox, the veterinarians found that 69 foxes had no parasites, 17 had one parasite, and so on. A frequency tabulation of the data is given here:

Number of Parasites, x	0	1	2	3	4	5	6	7	8
Number of Foxes, f	69	17	6	3	1	2	1	0	1

a. Construct a relative frequency histogram for x, the number of parasites per fox.

b. Calculate $\bar{x}$ and s for the sample.

c. What fraction of the parasite counts fall within two standard deviations of the mean? Within three standard deviations? Do these results agree with Tchebysheff's Theorem? With the Empirical Rule?

2.77 College Teachers Consider a population consisting of the number of teachers per college at small 2-year colleges. Suppose that the number of teachers per college has an average $\mu = 175$ and a standard deviation $\sigma = 15$.

a. Use Tchebysheff's Theorem to make a statement about the percentage of colleges that have between 145 and 205 teachers.

b. Assume that the population is normally distributed. What fraction of colleges have more than 190 teachers?

Data set EX0278 **2.78 Is It Accurate?** From the following data, a student calculated s to be .263. On what grounds might we doubt his accuracy? What is the correct value (to the nearest hundredth)?

17.2	17.1	17.0	17.1	16.9	17.0	17.1	17.0	17.3	17.2
17.1	17.0	17.1	16.9	17.0	17.1	17.3	17.2	17.4	17.1

Data set **2.79 Homerun Kings** In the summer of 2001,
EX0279 Barry Bonds began his quest to break Mark McGwire's record of 70 home runs hit in a single season. At the end of the 2003 major league baseball season, the number of home runs hit per season by each of four major league superstars over each player's career were recorded, and are shown in the box plots below:[16]

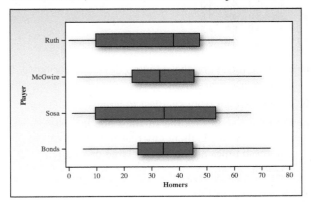

Write a short paragraph comparing the home run hitting patterns of these four players.

Data set **2.80 Barry Bonds** In the seasons that
EX0280 followed his 2001 record-breaking season, Barry Bonds hit 46, 45, 45, 5, 26, and 28 homers, respectively, until he retired from major league baseball in 2007 (www.ESPN.com).[16] Two box plots, one of Bond's homers through 2001, and a second including the years 2002–2007 follow.

The statistics used to construct these box plots are given in the table.

Years	Min	Q_1	Median	Q_3	IQR	Max	n
2001	16	25.00	34.00	41.50	16.5	73	16
2007	5	25.00	34.00	45.00	20.0	73	22

a. Calculate the upper fences for both of these box plots.

b. Can you explain why the record number of homers is an outlier in the 2001 box plot, but not in the 2007 box plot?

2.81 Ages of Pennies Here are the ages of 50 pennies from Exercise 1.45 and data set EX0145. The data have been sorted from smallest to largest.

0	0	0	0	0	0	0	0	0	0
0	0	1	1	1	1	1	1	2	2
2	3	3	3	4	4	5	5	5	5
6	8	9	9	10	16	17	17	19	19
19	20	20	21	22	23	25	25	28	36

a. What is the average age of the pennies?

b. What is the median age of the pennies?

c. Based on the results of parts a and b, how would you describe the age distribution of these 50 pennies?

d. Construct a box plot for the data set. Are there any outliers? Does the box plot confirm your description of the distribution's shape?

2.82 Snapshots Here are a few facts reported as Snapshots in *USA Today*.

• About 12% of America's volunteers spend more than 5 hours per week volunteering.[17]

• Fifty-eight percent of all cars in operation are at least 8 years old.[18]

• Twenty-two percent of all fans are willing to pay $75 or more for a ticket to one of the top 100 concert tours.[19]

Identify the variable x being measured, and any percentiles you can determine from this information.

Data set **2.83 Breathing Patterns** Research
EX0283 psychologists are interested in finding out whether a person's breathing patterns are affected by a particular experimental treatment. To determine the general respiratory patterns of the $n = 30$ people in the study, the researchers collected some baseline measurements—the total ventilation in liters of air per minute adjusted for body size—for each person before the treatment. The data are shown here, along with some descriptive tools generated by *MINITAB* and *MS Excel*.

5.23	4.79	5.83	5.37	4.35	5.54	6.04	5.48	6.58	4.82
5.92	5.38	6.34	5.12	5.14	4.72	5.17	4.99	4.51	5.70
4.67	5.77	5.84	6.19	5.58	5.72	5.16	5.32	4.96	5.63

Descriptive Statistics: Liters

```
Variable    N    N*    Mean   SE Mean   StDev
Liters     30    0   5.3953   0.0997   0.5462


Minimum      Q1  Median     Q3  Variable  Maximum
 4.3500  4.9825  5.3750  5.7850  Liters    6.5800
```

Stem and Leaf Display: Liters

```
Stem-and-leaf of Liters  N = 30
Leaf Unit = 0.10

  1    4  3
  2    4  5
  5    4  677
  8    4  899
 12    5  1111
 (4)   5  2333
 14    5  455
 11    5  6777
  7    5  889
  4    6  01
  2    6  3
  1    6  5
```

MS Excel Descriptive Statistics

Liters	
Mean	5.3953
Standard Error	0.0997
Median	5.3750
Mode	#N/A
Standard Deviation	0.5462
Sample Variance	0.2983
Kurtosis	20.4069
Skewness	0.1301
Range	2.23
Minimum	4.35
Maximum	6.58
Sum	161.86
Count	30

a. Summarize the characteristics of the data distribution using the computer output.

b. Does the Empirical Rule provide a good description of the proportion of measurements that fall within two or three standard deviations of the mean? Explain.

c. How large or small does a ventilation measurement have to be before it is considered unusual?

2.84 Arranging Objects The following data are the response times in seconds for $n = 25$ first graders to arrange three objects by size.

EX0284

5.2	3.8	5.7	3.9	3.7
4.2	4.1	4.3	4.7	4.3
3.1	2.5	3.0	4.4	4.8
3.6	3.9	4.8	5.3	4.2
4.7	3.3	4.2	3.8	5.4

a. Find the mean and the standard deviation for these 25 response times.

b. Order the data from smallest to largest.

c. Find the z-scores for the smallest and largest response times. Is there any reason to believe that these times are unusually large or small? Explain.

2.85 Arranging Objects, continued Refer to Exercise 2.84.

a. Find the five-number summary for this data set.

b. Construct a box plot for the data.

c. Are there any unusually large or small response times identified by the box plot?

d. Construct a stem and leaf display for the response times. How would you describe the shape of the distribution? Does the shape of the box plot confirm this result?

CASE STUDY

 Batting

The Boys of Summer

Which baseball league has had the best hitters? Many of us have heard of baseball greats like Stan Musial, Hank Aaron, Roberto Clemente, and Pete Rose of the National League and Ty Cobb, Babe Ruth, Ted Williams, Rod Carew, and Wade Boggs of the American League. But have you ever heard of Willie Keeler, who batted .432 for the Baltimore Orioles, or Nap Lajoie, who batted .422 for the Philadelphia A's? The batting averages for the batting champions of the National and American Leagues are given on the CourseMate Web site.

The batting averages for the National League begin in 1876 with Roscoe Barnes, whose batting average was .403 when he played with the Chicago Cubs. The last entry for the National League is for the year 2010, when Carlos Gonzalez of the Colorado Rockies averaged .336. The American League records begin in 1901 with Nap Lajoie of the Philadelphia A's, who batted .422, and end in 2010 with Josh Hamilton of the Texas Rangers, who batted .359.[9] How can we summarize the information in this data set?

1. Use *MS Excel, MINITAB*, or another statistical software package to describe the batting averages for the American and National League batting champions. Generate any graphics that may help you in interpreting these data sets.

2. Does one league appear to have a higher percentage of hits than the other? Do the batting averages of one league appear to be more variable than the other?

3. Are there any outliers in either league?

4. Summarize your comparison of the two baseball leagues.

Describing Bivariate Data

© Janis Christie/Photodisc/Getty Images

 NEED TO KNOW...

How to Calculate the Correlation Coefficient
How to Calculate the Regression Line

Do You *Think* Your Dishes Are *Really* Clean?

Does the price of an appliance, such as a dishwasher, convey something about its quality? In the case study at the end of this chapter, we rank 48 different brands of dishwashers according to their prices, and then we rate them on various characteristics, such as how the dishwasher performs, how much noise it makes, its cost for either gas or electricity, its cycle time, and its water use. The techniques presented in this chapter will help to answer our question.

3.1 BIVARIATE DATA

Very often researchers are interested in more than just one variable that can be measured during their investigation. For example, an auto insurance company might be interested in the number of vehicles owned by a policyholder as well as the number of drivers in the household. An economist might need to measure the amount spent per week on groceries in a household and also the number of people in that household. A real estate agent might measure the selling price of a residential property and the square footage of the living area.

NEED A TIP?

"**Bi**" means "two." Bivariate data generate pairs of measurements.

When two variables are measured on a single experimental unit, the resulting data are called **bivariate data.** How should you display these data? Not only are both variables important when studied separately, but you also may want to explore the *relationship between the two variables*. Methods for graphing bivariate data, whether the variables are qualitative or quantitative, allow you to study the two variables together. As with *univariate data,* you use different graphs depending on the type of variables you are measuring.

3.2 GRAPHS FOR CATEGORICAL VARIABLES

When at least one of the two variables is *qualitative* or *categorical*, you can use either simple or more intricate pie charts, line charts, and bar charts to display and describe the data. Sometimes you will have one qualitative and one quantitative variable that have been measured in two different populations or groups. In this case, you can use two **side-by-side pie charts** or a bar chart in which the bars for the two populations are placed side by side. Another option is to use a **stacked bar chart,** in which the bars for each category are stacked on top of each other.

EXAMPLE 3.1

Are professors in private colleges paid more than professors at public colleges? The data in Table 3.1 were collected from a sample of 400 college professors whose rank, type of college, and salary were recorded.[1] The number in each cell is the average salary (in thousands of dollars) for all professors who fell into that category. Use a graph to answer the question posed for this sample.

TABLE 3.1 ● **Salaries of Professors by Rank and Type of College**

	Full Professor	Associate Professor	Assistant Professor
Public	98.1	72.7	61.5
Private	110.5	74.9	61.6

Source: Digest of Educational Statistics

Solution   To display the average salaries of these 400 professors, you can use a side-by-side bar chart, as shown in Figure 3.1. The height of the bars is the average salary, with each pair of bars along the horizontal axis representing a different professorial rank. Salaries are substantially higher for full professors in private colleges, however, there are less striking differences at the lower two ranks.

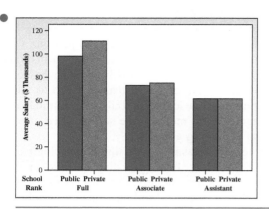

EXAMPLE 3.2

Along with the salaries for the 400 college professors in Example 3.1, the researcher recorded two qualitative variables for each professor: rank and type of college. Table 3.2 shows the number of professors in each of the $2 \times 3 = 6$ categories. Use comparative charts to describe the data. Do the private colleges employ as many high-ranking professors as the public colleges do?

TABLE 3.2 ● **Number of Professors by Rank and Type of College**

	Full Professor	Associate Professor	Assistant Professor	Total
Public	24	57	69	150
Private	60	78	112	250

Solution   The numbers in the table are not quantitative measurements on a single experimental unit (the professor). They are *frequencies,* or counts, of the number of professors who fall into each category. To compare the numbers of professors at public and private colleges, you might draw two pie charts and display them side by side, as in Figure 3.2.

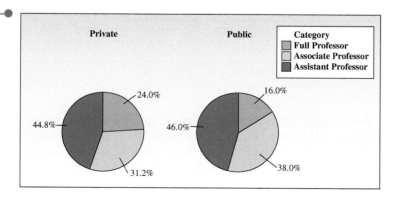

Alternatively, you could draw either a stacked or a side-by-side bar chart. The stacked bar chart is shown in Figure 3.3.

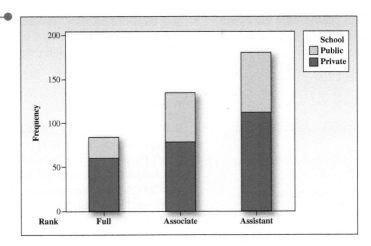

Although the graphs are not strikingly different, you can see that public colleges have fewer full professors and more associate professors than private colleges. The reason for these differences is not clear, but you might speculate that private colleges, with their higher salaries, are able to attract more full professors. Or perhaps public colleges are not as willing to promote professors to the higher-paying ranks. In any case, the graphs provide a means for comparing the two sets of data.

You can also compare the distributions for public versus private colleges by creating *conditional data distributions.* These conditional distributions are shown in Table 3.3. One distribution shows the proportion of professors in each of the three ranks under the *condition* that the college is public, and the other shows the proportions under the *condition* that the college is private. These *relative frequencies* are easier to compare than the *actual frequencies* and lead to the same conclusions:

- The proportion of assistant professors is roughly the same for both public and private colleges.
- Public colleges have a smaller proportion of full professors and a larger proportion of associate professors.

TABLE 3.3   ● **Proportions of Professors by Rank for Public and Private Colleges**

	Full Professor	Associate Professor	Assistant Professor	Total
Public	$\frac{24}{150} = .16$	$\frac{57}{150} = .38$	$\frac{69}{150} = .46$	1.00
Private	$\frac{60}{250} = .24$	$\frac{78}{250} = .31$	$\frac{112}{250} = .45$	1.00

EXERCISES

BASIC TECHNIQUES

3.1 Gender Differences Male and female respondents to a questionnaire about gender differences are categorized into three groups according to their answers to the first question:

	Group 1	Group 2	Group 3
Men	37	49	72
Women	7	50	31

a. Create side-by-side pie charts to describe these data.

b. Create a side-by-side bar chart to describe these data.

c. Draw a stacked bar chart to describe these data.

d. Which of the three charts best depicts the difference or similarity of the responses of men and women?

3.2 State-by-State A group of items are categorized according to a certain attribute—X, Y, Z—and according to the state in which they are produced:

	X	Y	Z
New York	20	5	5
California	10	10	5

a. Create a comparative (side-by-side) bar chart to compare the numbers of items of each type made in California and New York.

b. Create a stacked bar chart to compare the numbers of items of each type made in the two states.

c. Which of the two types of presentation in parts a and b is more easily understood? Explain.

d. What other graphical methods could you use to describe the data?

3.3 Consumer Spending The table below shows the average amounts spent per week by men and women in each of four spending categories:

	A	B	C	D
Men	$54	$27	$105	$22
Women	21	85	100	75

a. What possible graphical methods could you use to compare the spending patterns of women and men?

b. Choose two different methods of graphing and display the data in graphical form.

c. What can you say about the similarities or differences in the spending patterns for men and women?

d. Which of the two methods used in part b provides a better descriptive graph?

APPLICATIONS

3.4 M&M'S The color distributions for two snack-size bags of M&M'S® candies, one plain and one peanut, are displayed in the table. Choose an appropriate graphical method and compare the distributions.

	Brown	Yellow	Red	Orange	Green	Blue
Plain	15	14	12	4	5	6
Peanut	6	2	2	3	3	5

3.5 How Much Free Time? When you were growing up, did you feel that you did not have enough free time? Parents and children have differing opinions on this subject. A research group surveyed 198 parents and 200 children and recorded their responses to the question, "How much free time does your child have?" or "How much free time do you have?" The responses are shown in the table below:[2]

	Just the Right Amount	Not Enough	Too Much	Don't Know
Parents	138	14	40	6
Children	130	48	16	6

a. Define the sample and the population of interest to the researchers.

b. Describe the variables that have been measured in this survey. Are the variables qualitative or quantitative? Are the data univariate or bivariate?

c. What do the entries in the cells represent?

d. Use comparative pie charts to compare the responses for parents and children.

e. What other graphical techniques could be used to describe the data? Would any of these techniques be more informative than the pie charts constructed in part d?

3.6 Consumer Price Index The price of living in the United States has increased dramatically in the past decade, as demonstrated by the consumer price indexes (CPIs) for housing and transportation. These CPIs are listed in the table for the years 1996 through the first half of 2010.[3]

Year	1996	1997	1998	1999	2000	2001	2002	2003
Housing	152.8	156.8	160.4	163.9	169.6	176.4	180.3	184.8
Transportation	143.0	144.3	141.6	144.4	153.3	154.3	152.9	157.6

Year	2004	2005	2006	2007	2008	2009	2010
Housing	189.5	195.7	201.6	209.6	216.3	217.1	215.9
Transportation	163.1	173.9	181.4	184.7	195.5	179.3	192.2

EX0306

a. Create side-by-side comparative bar charts to describe the CPIs over time.

b. Draw two line charts on the same set of axes to describe the CPIs over time.

c. What conclusions can you draw using the two graphs in parts a and b? Which is the most effective?

3.7 How Big Is the Household? A local
EX0307 chamber of commerce surveyed 126 households within its city and recorded the type of residence and the number of family members in each of the households. The data are shown in the table.

	Type of Residence		
Family Members	Apartment	Duplex	Single Residence
1	8	10	2
2	15	4	14
3	9	5	24
4 or more	6	1	28

a. Use a side-by-side bar chart to compare the number of family members living in each of the three types of residences.

b. Use a stacked bar chart to compare the number of family members living in each of the three types of residences.

c. What conclusions can you draw using the graphs in parts a and b?

3.8 Facebook, again Not only is the
EX0308 *Facebook* social networking site growing rapidly in the United States, but the composition of *Facebook* members depends on both age and gender. During a 1-month period in early 2010, *Facebook* reported its growth by both age and gender, as shown in the table.[4]

	Growth (number of users)		
Age category	Female	Male	Total
13–17	270,900	121,280	392,180
18–25	445,920	653,060	1,098,980
26–34	570,920	154,600	725,520
35–44	365,740	305,260	671,000
45–54	90,240	36,680	126,920
55–65	64,960	83,480	148,440
Total	1,808,680	1,354,360	3,163,040

a. Construct a stacked bar chart to display the *Facebook* growth given in the table.

b. Construct two comparative pie charts to display the *Facebook* growth given in the table.

c. Write a short paragraph summarizing the information that can be gained by looking at these graphs. Which of the two types of comparative graphs is more effective?

SCATTERPLOTS FOR TWO QUANTITATIVE VARIABLES

3.3

When both variables to be displayed on a graph are *quantitative,* one variable is plotted along the horizontal axis and the second along the vertical axis. The first variable is often called x and the second is called y, so that the graph takes the form of a plot on the (x, y) axes, which is familiar to most of you. Each pair of data values is plotted as a point on this two-dimensional graph, called a **scatterplot.** It is the two-dimensional extension of the dotplot we used to graph one quantitative variable in Section 1.4.

You can describe the relationship between two variables, x and y, using the patterns shown in the scatterplot.

MY ONLINE APPLET
Building a Scatterplot

- **What type of pattern do you see?** Is there a constant upward or downward trend that follows a straight-line pattern? Is there a curved pattern? Is there no pattern at all, but just a random scattering of points?

- **How strong is the pattern?** Do all of the points follow the pattern exactly, or is the relationship only weakly visible?

- **Are there any unusual observations?** An outlier is a point that is far from the cluster of the remaining points. Do the points cluster into groups? If so, is there an explanation for the observed groupings?

EXAMPLE 3.3 The number of household members, x, and the amount spent on groceries per week, y, are measured for six households in a local area. Draw a scatterplot of these six data points.

x	2	2	3	4	1	5
y	\$95.75	\$110.19	\$118.33	\$150.92	\$85.86	\$180.62

Solution Label the horizontal axis x, the vertical axis y, and plot the points using the coordinates (x, y) for each of the six pairs. The scatterplot in Figure 3.4 shows the six pairs marked as dots. You can see a pattern even with only six data pairs. The cost of weekly groceries increases with the number of household members in an apparent straight-line relationship.

Suppose you found that a seventh household with two members spent \$165 on groceries. This observation is shown as an X in Figure 3.4. It does not fit the linear pattern of the other six observations and is classified as an outlier. Possibly these two people were having a party the week of the survey!

FIGURE 3.4

Scatterplot for
Example 3.3

EXAMPLE 3.4 A distributor of table wines conducted a study of the relationship between price and demand using a type of wine that ordinarily sells for \$10.00 per bottle. He sold this wine in 10 different marketing areas over a 12-month period, using five different price levels—from \$10 to \$14. The data are given in Table 3.4. Construct a scatterplot for the data, and use the graph to describe the relationship between price and demand.

TABLE 3.4 **Cases of Wine Sold at Five Price Levels**

Cases Sold per 10,000 Population	Price per Bottle
23, 21	\$10
19, 18	11
15, 17	12
19, 20	13
25, 24	14

Solution The 10 data points are plotted in Figure 3.5. As the price increases from $10 to $12, the demand decreases. However, as the price continues to increase, from $12 to $14, the demand begins to *increase*. The data show a curved pattern, with the relationship changing as the price changes. How do you explain this relationship? Possibly, the increased price is a signal of increased quality for the consumer, which causes the increase in demand once the cost exceeds $12. You might be able to think of other reasons, or perhaps some other variable, such as the income of people in the marketing areas, that may be causing the change.

FIGURE 3.5

Scatterplot for
Example 3.4

NUMERICAL MEASURES FOR QUANTITATIVE BIVARIATE DATA

A constant rate of increase or decrease is perhaps the most common pattern found in bivariate scatterplots. The scatterplot in Figure 3.4 exhibits this *linear* pattern—that is, a straight line with the data points lying both above and below the line and within a fixed distance from the line. When this is the case, we say that the two variables exhibit a *linear relationship*.

EXAMPLE 3.5 The data in Table 3.5 are the size of the living area (in square feet), *x*, and the selling price, *y*, of 12 residential properties. The scatterplot in Figure 3.6 shows a linear pattern in the data.

TABLE 3.5 **Living Area and Selling Price of 12 Properties**

Residence	x (sq. ft.)	y (in thousands)
1	1360	$278.5
2	1940	375.7
3	1750	339.5
4	1550	329.8
5	1790	295.6
6	1750	310.3
7	2230	460.5
8	1600	305.2
9	1450	288.6
10	1870	365.7
11	2210	425.3
12	1480	268.8

FIGURE 3.6

Scatterplot of *x* versus *y*
for Example 3.5

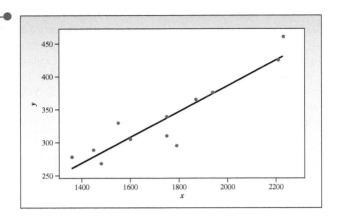

For the data in Example 3.5, you could describe each variable, *x* and *y*, individually using descriptive measures such as the means ($\bar{x}$ and $\bar{y}$) or the standard deviations (s_x and s_y). However, these measures do not describe the relationship between *x* and *y* for a particular residence—that is, how the size of the living space affects the selling price of the home. A simple measure that serves this purpose is called the **correlation coefficient,** denoted by *r*, and is defined as

$$r = \frac{s_{xy}}{s_x s_y}$$

The quantities s_x and s_y are the standard deviations for the variables *x* and *y*, respectively, which can be found by using the statistics function on your calculator or the computing formula in Section 2.3. The new quantity s_{xy} is called the **covariance** between *x* and *y* and is defined as

MY ONLINE APPLET

Exploring Correlation

$$s_{xy} = \frac{\Sigma(x_i - \bar{x})(y_i - \bar{y})}{n - 1}$$

There is also a computing formula for the covariance:

$$s_{xy} = \frac{\Sigma x_i y_i - \dfrac{(\Sigma x_i)(\Sigma y_i)}{n}}{n - 1}$$

where $\Sigma x_i y_i$ is the sum of the products $x_i y_i$ for each of the *n* pairs of measurements. How does this quantity detect and measure a linear pattern in the data?

Look at the signs of the cross products $(x_i - \bar{x})(y_i - \bar{y})$ in the numerator of *r*, or s_{xy}. When a data point (*x*, *y*) is in either area I or III in the scatterplot shown in Figure 3.7,

FIGURE 3.7

The signs of the cross
products $(x_i - \bar{x})(y_i - \bar{y})$
in the covariance formula

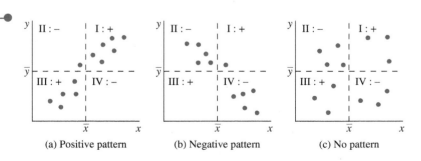

(a) Positive pattern   (b) Negative pattern   (c) No pattern the cross product will be positive; when a data point is in area II or IV, the cross product will be negative. We can draw these conclusions:

- If most of the points are in areas I and III (forming a positive pattern), s_{xy} and r will be positive.

- If most of the points are in areas II and IV (forming a negative pattern), s_{xy} and r will be negative.

- If the points are scattered across all four areas (forming *no* pattern), s_{xy} and r will be close to 0.

Most scientific and graphics calculators can compute the correlation coefficient, r, when the data are entered in the proper way. Check your calculator manual for the proper sequence of entry commands. Computer programs are also programmed to perform these calculations. The output in Figure 3.8 shows the covariance and correlation coefficient for x and y in Example 3.5. In the covariance table, you will find these values:

$$s_{xy} = 15{,}545.20 \qquad s_x^2 = 79{,}233.33 \qquad s_y^2 = 3571.16$$

and in the correlation output, you find $r = .924$.

However you decide to calculate the correlation coefficient, it can be shown that the value of r always lies between -1 and 1. When r is positive, x increases when y increases, and vice versa. When r is negative, x decreases when y increases, or x increases when y decreases. When r takes the value 1 or -1, all the points lie exactly on a straight line. If $r = 0$, then there is no apparent linear relationship between the two variables. The closer the value of r is to 1 or -1, the stronger the linear relationship between the two variables.

NEED A TIP?

$r > 0 \Leftrightarrow$ positive *linear* relationship

$r < 0 \Leftrightarrow$ negative *linear* relationship

$r \approx 0 \Leftrightarrow$ no *linear* relationship

FIGURE 3.8

MINITAB output of covariance and *EXCEL* correlation output for Example 3.5

Covariances: x, y

	x	y
x	79233.33	
y	15545.20	3571.16

Correlations: x, y

	x	y
x	1	
y	0.92414	1

EXAMPLE 3.6

Find the correlation coefficient for the number of square feet of living area and the selling price of a home for the data in Example 3.5.

Solution Three quantities are needed to calculate the correlation coefficient. The standard deviations of the x and y variables are found using a calculator with a statistical function. You can verify that $s_x = 281.4842$ and $s_y = 59.7592$. Finally,

$$s_{xy} = \frac{\sum x_i y_i - \dfrac{(\sum x_i)(\sum y_i)}{n}}{n - 1}$$

$$= \frac{7{,}240{,}383 - \dfrac{(20{,}980)(4043.5)}{12}}{11} = 15{,}545.19697$$

This agrees with the value given in the printout in Figure 3.8(a). Then

$$r = \frac{s_{xy}}{s_x s_y} = \frac{15,545.19697}{(281.4842)(59.7592)} = .9241$$

which also agrees with the value of the correlation coefficient given in Figure 3.8(b). (You may wish to verify the value of r using your calculator.) This value of r is fairly close to 1, which indicates that the linear relationship between these two variables is very strong. Additional information about the correlation coefficient and its role in analyzing linear relationships, along with alternative calculation formulas, can be found in Chapter 12.

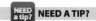
NEED A TIP?

x "explains" *y* or *y* "depends on" *x*.

x is the **explan**atory or independent variable.

y is the response or **depend**ent variable.

Sometimes the two variables, x and y, are related in a particular way. It may be that the value of y depends on the value of x; that is, the value of x in some way explains the value of y. For example, the cost of a home (y) may *depend* on its amount of floor space (x); a student's grade point average (x) may *explain* her score on an achievement test (y). In these situations, we call y the **dependent variable,** while x is called the **independent variable.**

If one of the two variables can be classified as the dependent variable y and the other as x, and if the data exhibit a straight-line pattern, it is possible to describe the relationship relating y to x using a straight line given by the equation

$$y = a + bx$$

as shown in Figure 3.9.

FIGURE 3.9

The graph of a straight line

ONLINE APPLET

How a Line Works

As you can see, a is where the line crosses or intersects the y-axis: a is called the *y-intercept.* You can also see that for every one-unit increase in x, y increases by an amount b. The quantity b determines whether the line is increasing ($b > 0$), decreasing ($b < 0$), or horizontal ($b = 0$) and is appropriately called the **slope** of the line.

When plotting the (x, y) points for two variables x and y, the points generally do not fall exactly on a straight line, but they may show a trend that could be described as a linear pattern. We can describe this trend by fitting a line as best we can through the points. This best-fitting line relating y to x, often called the **regression** or **least-squares line,** is found by minimizing the sum of the squared differences between the data points and the line itself, as shown in Figure 3.10. The formulas for computing b and a, which are derived mathematically, follow.

> ## COMPUTING FORMULAS FOR THE LEAST-SQUARES REGRESSION LINE
>
> $$b = r\left(\frac{s_y}{s_x}\right) \qquad \text{and} \qquad a = \bar{y} - b\bar{x}$$
>
> and the least-squares regression line is: $y = a + bx$

FIGURE 3.10

The best-fitting line

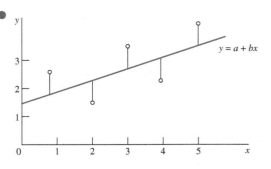

Since s_x and s_y are both positive, b and r have the same sign, so that:

 NEED A TIP?

Remember that r and b have the same sign!

- When r is positive, so is b, and the line is increasing with x.
- When r is negative, so is b, and the line is decreasing with x.
- When r is close to 0, then b is close to 0.

EXAMPLE 3.7

Find the best-fitting line relating $y = $ starting hourly wage to $x = $ number of years of work experience for the following data. Plot the line and the data points on the same graph.

x	2	3	4	5	6	7
y	$6.00	7.50	8.00	12.00	13.00	15.50

Solution Use the data entry method for your calculator to find these descriptive statistics for the bivariate data set:

$$\bar{x} = 4.5 \qquad \bar{y} = 10.333 \qquad s_x = 1.871 \qquad s_y = 3.710 \qquad r = .980$$

NEED A TIP?

Use the regression line to predict y for a given value of x.

Then

$$b = r\left(\frac{s_y}{s_x}\right) = .980\left(\frac{3.710}{1.871}\right) = 1.9432389 \approx 1.943$$

and

$$a = \bar{y} - b\bar{x} = 10.333 - 1.943(4.5) = 1.590$$

Therefore, the best-fitting line is $y = 1.590 + 1.943x$. The plot of the regression line and the actual data points are shown in Figure 3.11.

The best-fitting line can be used to estimate or predict the value of the variable y when the value of x is known. For example, if a person applying for a job has 3 years of work experience (x), what would you predict his starting hourly wage (y) to be? From the best-fitting line in Figure 3.11, the best estimate would be

$$y = a + bx = 1.590 + 1.943(3) = 7.419$$

FIGURE 3.11

Fitted line and data points for Example 3.7

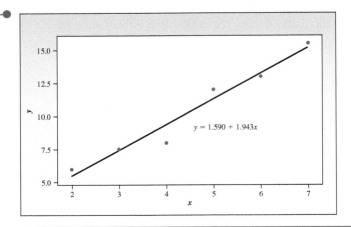

$$y = 1.590 + 1.943x$$

? NEED TO KNOW...

How to Calculate the Correlation Coefficient

1. First, create a table or use your calculator to find Σx, Σy, and Σxy.
2. Calculate the covariance, s_{xy}.
3. Use your calculator or the computing formula from Chapter 2 to calculate s_x and s_y.
4. Calculate $r = \dfrac{s_{xy}}{s_x s_y}$.

How to Calculate the Regression Line

1. First, calculate $\bar{y}$ and $\bar{x}$. Then, calculate $r = \dfrac{s_{xy}}{s_x s_y}$.

2. Find the slope, $b = r\left(\dfrac{s_y}{s_x}\right)$ and the y-intercept, $a = \bar{y} - b\bar{x}$.

3. Write the regression line by substituting the values for a and b into the equation: $y = a + bx$.

When should you describe the linear relationship between x and y using the correlation coefficient r, and when should you use the regression line $y = a + bx$? The regression approach is used when the values of x are set in advance and then the corresponding value of y is measured. The correlation approach is used when an experimental unit is selected at random and then measurements are made on both variables x and y. This technical point will be taken up in Chapter 12, which addresses regression analysis.

Most data analysts begin any data-based investigation by examining plots of the variables involved. If the relationship between two variables is of interest, data analysts can also explore bivariate plots in conjunction with numerical measures of location, dispersion, and correlation. Graphs and numerical descriptive measures are only the first of many statistical tools you will soon have at your disposal.

3.4 EXERCISES

BASIC TECHNIQUES

3.9 Suppose that the relationship between two variables x and y can be described by the regression line $y = 2.0 + 0.5x$.

a. What is the change in y for a one-unit change in x?

b. Do the values of y increase or decrease as x increases?

c. At what point does the line cross the y-axis? What is the name given to this value?

d. If $x = 2.5$, use the least squares equation to predict the value of y. What value would you predict if $x = 4.0$?

3.10 Consider this set of bivariate data: (1, 6), (3, 2), and (2, 4).

a. Calculate the covariance s_{xy}.

b. Calculate the correlation coefficient r.

c. Calculate the equation of the regression line using the computing formulas.

d. Plot the three points and the straight line on a scatterplot. Does the line pass through the middle of the three points?

 3.11 A set of bivariate data consists of these EX0311 measurements on two variables, x and y:

(3, 6) (5, 8) (2, 6) (1, 4) (4, 7) (4, 6)

a. Draw a scatterplot to describe the data.

b. Does there appear to be a relationship between x and y? If so, how do you describe it?

c. Calculate the correlation coefficient, r, using the computing formula given in this section.

d. Find the best-fitting line using the computing formulas. Graph the line on the scatterplot from part a. Does the line pass through the middle of the points?

3.12 Refer to Exercise 3.11.

a. Use the data entry method in your scientific calculator to enter the six pairs of measurements. Recall the proper memories to find the correlation coefficient, r, the y-intercept, a, and the slope, b, of the line.

b. Verify that the calculator provides the same values for r, a, and b as in Exercise 3.11.

 3.13 Consider this set of bivariate data:

EX0313

x	1	2	3	4	5	6
y	5.6	4.6	4.5	3.7	3.2	2.7

a. Draw a scatterplot to describe the data.

b. Does there appear to be a relationship between x and y? If so, how do you describe it?

c. Calculate the correlation coefficient, r. Does the value of r confirm your conclusions in part b? Explain.

 3.14 The value of a quantitative variable is EX0314 measured once a year for a 10-year period:

Year	Measurement	Year	Measurement
1	61.5	6	58.2
2	62.3	7	57.5
3	60.7	8	57.5
4	59.8	9	56.1
5	58.0	10	56.0

a. Draw a scatterplot to describe the variable as it changes over time.

b. Describe the measurements using the graph constructed in part a.

c. Use this *MINITAB* output to calculate the correlation coefficient, r:

MINITAB output for Exercise 3.14

Covariances

	x	y
x	9.16667	
y	−6.42222	4.84933

d. Find the best-fitting line using the results of part c. Verify your answer using the data entry method in your calculator.

e. Plot the best-fitting line on your scatterplot from part a. Describe the fit of the line.

APPLICATIONS

Data set EX0315 **3.15 Grocery Costs** These data relating the amount spent on groceries per week and the number of household members are from Example 3.3:

x	2	2	3	4	1	5
y	$95.75	$110.19	$118.33	$150.92	$85.86	$180.62

a. Find the best-fitting line for these data.

b. Plot the points and the best-fitting line on the same graph. Does the line summarize the information in the data points?

c. What would you estimate a household of six to spend on groceries per week? Should you use the fitted line to estimate this amount? Why or why not?

Data set EX0316 **3.16 Real Estate Prices** The data relating the square feet of living space and the selling price of 12 residential properties given in Example 3.5 are reproduced here. First, find the best-fitting line that describes these data, and then plot the line and the data points on the same graph. Comment on the goodness of the fitted line in describing the selling price of a residential property as a linear function of the square feet of living area.

Residence	x (sq. ft.)	y (in thousands)
1	1360	$278.5
2	1940	375.7
3	1750	339.5
4	1550	329.8
5	1790	295.6
6	1750	310.3
7	2230	460.5
8	1600	305.2
9	1450	288.6
10	1870	365.7
11	2210	425.3
12	1480	268.8

Data set EX0317 **3.17 Disabled Students** A social skills training program, reported in *Psychology in the Schools,* was implemented for seven students with mild handicaps in a study to determine whether the program caused improvement in pre/post measures and behavior ratings.[5] For one such test, these are the pretest and posttest scores for the seven students:

Student	Pretest	Posttest
Earl	101	113
Ned	89	89
Jasper	112	121
Charlie	105	99
Tom	90	104
Susie	91	94
Lori	89	99

a. Draw a scatterplot relating the posttest score to the pretest score.

b. Describe the relationship between pretest and posttest scores using the graph in part a. Do you see any trend?

c. Calculate the correlation coefficient and interpret its value. Does it reinforce any relationship that was apparent from the scatterplot? Explain.

Data set EX0318 **3.18 Chirping Crickets** Male crickets chirp by rubbing their front wings together, and their chirping is temperature dependent. Crickets chirp faster with increasing temperature and slower with decreasing temperatures. The table below shows the number of chirps per second for a cricket, recorded at 10 different temperatures.

Chirps per second	20	16	19	18	18	16	14	17	15	16
Temperature	88	73	91	85	82	75	69	82	69	83

a. Which of the two variables (temperature and number of chirps) is the independent variable, and which is the dependent variable?

b. Plot the data using a scatterplot. How would you describe the relationship between temperature and number of chirps?

c. Find the least-squares line relating the number of chirps to the temperature.

d. If a cricket is monitored at a temperature of 80 degrees, what would you predict his number of chirps would be?

Data set EX0319 **3.19 How to Choose a TV** As technology improves, your choice of televisions becomes more complicated. Should you choose an LCD TV, a plasma TV, and with rear or front projection? In the table below, *Consumer Reports*[6] gives the prices and screen sizes for the top 10 LCD TVs in the 46-inch and higher categories. Does the price of an LCD TV depend on the size of the screen?

Brand	Price ($)	Size	Brand	Price ($)	Size
Sony Bravia KDL-52NX800	2340	52	Sony Bravia KDL-46XBR10	2500	46
Samsung LN55C650	1600	55	Samsung UN46C8000	2200	46
Vizio VF550M	1330	55	Vizio SV472XVT	1400	47
Sony Bravia KDL-60EX700	2700	60	Samsung UN46C7000	2100	46
Sharp Aquos LED LC-52LE700UN	1620	52	LG 47LD450	900	47

a. Which of the two variables (price and size) is the independent variable, and which is the dependent variable?

b. Construct a scatterplot for the data. Does the relationship appear to be linear?

3.20 LCD TVs, continued Refer to Exercise 3.19. Suppose we assume that the relationship between x and y is linear.

a. Find the correlation coefficient, r. What does this value tell you about the strength and direction of the relationship between size and price?

b. Refer to part a. Would it be reasonable to construct a regression line used to predict the price of an LCD TV based on the size of the screen?

CHAPTER REVIEW

Key Concepts

I. **Bivariate Data**
   1. Both qualitative and quantitative variables
   2. Describing each variable separately
   3. Describing the relationship between the two variables

II. **Describing Two Qualitative Variables**
   1. Side-by-side pie charts
   2. Comparative line charts
   3. Comparative bar charts
      a. Side-by-side
      b. Stacked
   4. Relative frequencies to describe the relationship between the two variables

III. **Describing Two Quantitative Variables**
   1. Scatterplots
      a. Linear or nonlinear pattern
      b. Strength of relationship
      c. Unusual observations: clusters and outliers
   2. Covariance and correlation:

   $$\text{Covariance: } s_{xy} = \frac{\Sigma xy - \frac{(\Sigma x)(\Sigma y)}{n}}{n-1}$$

   $$\text{Correlation: } r = \frac{s_{xy}}{s_x s_y}$$

   3. The best-fitting regression line
      a. Calculating the slope and y-intercept

   $$b = r\left(\frac{s_y}{s_x}\right) \text{ and } a = \bar{y} - b\bar{x}$$

      b. Graphing the line
      c. Using the line for prediction

 TECHNOLOGY TODAY

Describing Bivariate Data in *Excel*

MS Excel provides different graphical techniques for *qualitative* and *quantitative* bivariate data, as well as commands for obtaining bivariate descriptive measures when the data are quantitative.

 EXAMPLE 3.8

(Comparative Line and Bar Charts) Suppose that the 105 students whose status was tabulated in Example 1.12 were from the University of California, Riverside, and that another 100 students from an introductory statistics class at UC Berkeley were also interviewed. Table 3.6 shows the status distribution for both sets of students.

TABLE 3.6   **Status of Students in a Statistics Class at UCR and UCB**

	Freshman	Sophomore	Junior	Senior	Grad Student
Frequency (UCR)	5	23	32	35	10
Frequency (UCB)	10	35	24	25	6

1. Enter the data into an *Excel* spreadsheet *just as it appears in the table, including the labels.* Highlight the data in the spreadsheet, click the **Insert** tab and select **Line** in the **Charts** group. In the drop-down list, you will see a variety of styles to choose from. Select the first option to produce the line chart.

2. *Editing the line chart:* Again, you can experiment with the various options in the **Chart Layout** and **Chart Styles** groups to change the look of the chart. We have chosen a design that allows a title on the vertical axis; we have added the title and have changed the "line style" of the UCR students to a "dashed" style, by double-clicking on that line. The line chart is shown in Figure 3.12(a).

FIGURE 3.12 ●

(a)   (b)

3. Once the line chart has been created, right-click on the chart area and select **Change Chart Type.** Then choose either **Stacked Column** or **Clustered Column.** The comparative bar chart (a stacked bar chart), with the same editing that you chose for the line chart, will appear as shown in Figure 3.12(b).

EXAMPLE (3.9) — **(Scatterplots, Correlation, and the Regression Line)** The data from Example 2.15 give the front and rear leg rooms (in inches) for nine different sports utility vehicles:[7]

Make and Model	Front Leg Room	Rear Leg Room
Acura MDX	41.0	28.5
Buick Enclave	41.5	30.0
Chevy TrailBlazer	40.0	25.5
Chevy Tahoe Hybrid V8 CVT	41.0	27.5
GMC Terrain 1LT 4-cyl	43.0	31.0
Honda CR-V	41.0	29.5
Hyundai Tucson	42.5	29.5
Kia Sportage	40.0	29.0
Lexus GX	42.0	30.0

1. If you did not save the *Excel* spreadsheet from Chapter 2, enter the data into the first three columns of another *Excel* spreadsheet, using the labels in the table. Highlight the front and rear leg room data (columns B and C), click the **Insert** tab and select **Scatter** in the **Charts** group, and select the first option in the drop-down list. The scatterplot appears as in Figure 3.13(a), and will need to be edited!

FIGURE 3.13

(a)　　　　　　　　　　　　　　　　　　(b)

2. *Editing the scatterplot:* With the scatterplot selected, look in the drop-down list in the **Chart Layouts** group. Find a layout that allows titles on both axes (we chose layout 1) and select it. Label the axes, remove the "legend entry" and retitle the chart as "Scatterplot of Front vs. Rear Leg Room." The scatterplot now appears in Figure 3.13(b). The plot is still not optimal, since *Excel* chooses to use zero as the lower limit of the vertical scale, causing the points to cluster at the top of the plot. To adjust this, double-click on the vertical axis. In the **Format Axis** Dialog Box, change the **Minimum** to **Fixed,** type **25** in the box, and click **Close**. (You can make a similar adjustment to the horizontal axis if needed.)

3. To plot the best-fitting line, simply right-click on one of the data points and select **Add Trendline**. In the Dialog box that opens, make sure that the radio button marked "Linear" is selected, and check the boxes marked "Display Equation on Chart" and "Display R-squared value on Chart". The final scatterplot is shown in Figure 3.14.

FIGURE 3.14

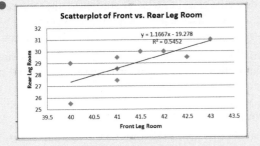

4. To find the sample correlation coefficient, *r*, you can use the command **Data ▶ Data Analysis ▶ Correlation,** selecting the two appropriate columns for the Input Range, clicking "Labels in First Row," and selecting an appropriate Output

Range. When you click **OK,** the correlation matrix will appear in the spread-sheet.

5. *(ALTERNATE PROCEDURE)* You can also place your cursor in the cell in which you want the correlation coefficient to appear. Select **Formulas ▶ More Functions ▶ Statistical ▶ CORREL** or click the "Insert Function" icon ƒₓ at the top of the spreadsheet, choosing **CORREL** fiom the **Statistical** category. Highlight or type the cell ranges for the two variables in the boxes marked "Array 1" and "Array 2" and click **OK.** For our example, the value is $r = .738$.

Describing Bivariate Data in *MINITAB*

MINITAB provides different graphical techniques for *qualitative* and *quantitative* bivariate data, as well as commands for obtaining bivariate descriptive measures when the data are quantitative.

 EXAMPLE 3.10

(Comparative Line and Bar Charts) Suppose that the 105 students whose status was tabulated in Example 1.12 were from the University of California, Riverside, and that another 100 students from an introductory statistics class at UC Berkeley were also interviewed. Table 3.7 shows the status distribution for both sets of students.

TABLE 3.7   **Status of Students in a Statistics Class at UCR and UCB**

	Freshman	Sophomore	Junior	Senior	Grad Student
Frequency (UCR)	5	23	32	35	10
Frequency (UCB)	10	35	24	25	6

1. Enter the data into a *MINITAB* worksheet as you did in Chapter 1, using your Chapter 1 project as a base if you have saved it. Column C1 will contain the 10 "Frequencies" and column C2 will contain the student "Status" corresponding to each frequency. Create a third column C3 called "College," and enter either **UCR** or **UCB** as appropriate. You can use the familiar Windows cut-and-paste commands if you like.

2. To graphically describe the UCR/UCB student data, you can use comparative pie charts—one for each school (see Chapter 1). Alternatively, you can use either stacked or side-by-side bar charts. Use **Graph ▶ Bar Chart.**

3. In the "Bar Charts" Dialog box (Figure 3.15(a)), select **Values from a Table** in the drop-down list and click either **Stack** or **Cluster** in the row marked "One Column of Values." Click **OK.** In the next Dialog box (Figure 3.15(b)), select "Frequency" for the **Graph variables** box and "Status" and "College" for the **Categorical variable for grouping** box. Click **OK.**

4. Once the bar chart is displayed (Figure 3.16), you can *right-click* on various items in the bar chart to edit. If you *right-click* on the bars and select **Update Graph Automatically,** the bar chart will automatically update when you change the data in the *MINITAB* worksheet.

FIGURE 3.15

(a) (b)

FIGURE 3.16

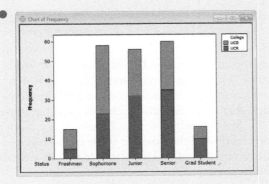

EXAMPLE 3.11

(Scatterplots, Correlation, and the Regression Line) The data from Example 2.15 give the front and rear leg rooms (in inches) for nine different sports utility vehicles:[7]

Make and Model	Front Leg Room	Rear Leg Room
Acura MDX	41.0	28.5
Buick Enclave	41.5	30.0
Chevy TrailBlazer	40.0	25.5
Chevy Tahoe Hybrid V8 CVT	41.0	27.5
GMC Terrain 1LT 4-cyl	43.0	31.0
Honda CR-V	41.0	29.5
Hyundai Tucson	42.5	29.5
Kia Sportage	40.0	29.0
Lexus GX	42.0	30.0

1. If you did not save the *MINITAB* worksheet from Chapter 2, enter the data into the first three columns of another *MINITAB* worksheet, using the labels in the table. To examine the relationship between the front and rear leg rooms, you can plot the data and numerically describe the relationship with the correlation coefficient and the best-fitting line.

2. Select **Stat ▶ Regression ▶ Fitted Line Plot,** and select "Front Leg Room" and "Rear Leg Room" for **Y** and **X,** respectively (see Figure 3.17(a)). Make sure that the radio button next to **Linear** is selected, and click **OK.** The plot of the nine data points and the best-fitting line will be generated as in Figure 3.17(b).

FIGURE 3.17

3. To calculate the correlation coefficient, use **Stat ▶ Basic Statistics ▶ Correlation,** selecting "Front Leg Room" and "Rear Leg Room" for the Variables box. To select both variables at once, hold the **Shift** key down as you highlight the variables and then click **Select.** Click **OK,** and the correlation coefficient will appear in the Session window (see Figure 3.18). Notice the relatively strong positive correlation and the positive slope of the regression line, indicating that a sports utility vehicle with a large front leg room will also tend to have a large rear leg room.

FIGURE 3.18

Supplementary Exercises

3.21 Professor Asimov Professor Isaac Asimov was one of the most prolific writers of all time. He wrote nearly 500 books during a 40-year career prior to his death in 1992. In fact, as his career progressed, he became even more productive in terms of the number of books written within a given period of time.[8] These data are the times (in months) required to write his books, in increments of 100:

Number of Books	100	200	300	400	490
Time (in months)	237	350	419	465	507

a. Plot the accumulated number of books as a function of time using a scatterplot.

b. Describe the productivity of Professor Asimov in light of the data set graphed in part a. Does the relationship between the two variables seem to be linear?

3.22 Cheese, Please! Health-conscious Americans often consult the nutritional information on food packages in an attempt to avoid foods with large amounts of fat, sodium, or cholesterol. The following information was taken from eight different brands of American cheese slices:

Brand	Fat (g)	Saturated Fat (g)	Cholesterol (mg)	Sodium (mg)	Calories
Kraft Deluxe American	7	4.5	20	340	80
Kraft Velveeta Slices	5	3.5	15	300	70
Private Selection	8	5.0	25	520	100
Ralphs Singles	4	2.5	15	340	60
Kraft 2% Milk Singles	3	2.0	10	320	50
Kraft Singles American	5	3.5	15	290	70
Borden Singles	5	3.0	15	260	60
Lake to Lake American	5	3.5	15	330	70

a. Which pairs of variables do you expect to be strongly related?

b. Draw a scatterplot for fat and saturated fat. Describe the relationship.

c. Draw a scatterplot for fat and calories. Compare the pattern to that found in part b.

d. Draw a scatterplot for fat versus sodium and another for cholesterol versus sodium. Compare the patterns. Are there any clusters or outliers?

e. For the pairs of variables that appear to be linearly related, calculate the correlation coefficients.

f. Write a paragraph to summarize the relationships you can see in these data. Use the correlations and the patterns in the four scatterplots to verify your conclusions.

3.23 Facebook Stats In Exercise 1.14, we looked at the age distribution of users of the social networking site *Facebook* (in thousands) as it changed from January 2009 to January 2010. The online article also provided insight into other demographics of *Facebook* users during this time period.[9] Two bar charts, constructed from the data, are shown below.

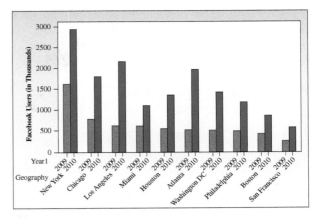

a. What variables have been measured in this study? Are the variables qualitative or quantitative?

b. Describe the populations of interest. Do these data represent populations or samples drawn from a population?

c. What type of graphical presentation has been used? What other type could have been used?

d. How would you describe the changes in the geographical and educational distributions of *Facebook* users during this 1-year period?

3.24 Cheese, again! The demand for healthy foods that are low in fats and calories has resulted in a large number of "low-fat" and "fat-free" products at the supermarket. The table shows the numbers of calories and the amounts of sodium (in milligrams) per slice for five different brands of fat-free American cheese.

Brand	Sodium (mg)	Calories
Kraft Fat Free Singles	300	30
Ralphs Fat Free Singles	300	30
Borden Fat Free	320	30
Healthy Choice Fat Free	290	30
Smart Beat American	180	25

a. Draw a scatterplot to describe the relationship between the amount of sodium and the number of calories.

b. Describe the plot in part a. Do you see any outliers? Do the rest of the points seem to form a pattern?

c. Based *only* on the relationship between sodium and calories, can you make a clear decision about which of the five brands to buy? Is it reasonable to base your choice on only these two variables? What other variables should you consider?

 3.25 Peak Current Using a chemical procedure called *differential pulse polarography,* a

EX0325

chemist measured the peak current generated (in microamperes) when a solution containing a given amount of nickel (in parts per billion) is added to a buffer. The data are shown here:

x = Ni (ppb)	y = Peak Current (μA)
19.1	.095
38.2	.174
57.3	.256
76.2	.348
95	.429
114	.500
131	.580
150	.651
170	.722

Use a graph to describe the relationship between x and y. Add any numerical descriptive measures that are appropriate. Write a paragraph summarizing your results.

3.26 Movie Money How much money do movies make on a single weekend? Does this amount in any way predict the movie's success or failure, or is the movie's total monetary success more dependent on the number of weeks that the movie remains in the movie theaters? In a recent week, the following data was collected for the top 16 movies in theaters that weekend.[10]

EX0326

TW	Title	Studio	Weekend Gross ($ millions)	Theater Count	Average ($)	Total Gross ($ millions)	Production Budget ($ millions)	Week
1	The Expendables	LGF	17.0	3270	5189	65.4	80	2
2	Vampires Suck	Fox	12.2	3233	3774	18.6	20	1
3	Eat Pray Love	Sony	12.1	3082	3930	47.2	60	2
4	Lottery Ticket	WB	10.7	1973	5399	17	17	1
5	The Other Guys	Sony	10.2	3472	2927	88.3	100	3
6	Piranha 3D	W/Dim.	10.1	2470	4092	10.1	24	1
7	The Switch	Mira.	8.4	2012	4193	8.4	—	1
8	Nanny McPhee Returns	Uni.	8.4	2784	3020	8.4	35	1
9	Inception	WB	7.8	2401	3265	262.0	160	6
10	Scott Pilgrim vs. the World	Uni.	5.2	2820	1845	20.9	60	2
11	Despicable Me	Uni.	4.7	2236	2085	231.1	69	7
12	Dinner for Schmucks	P/DW	3.5	2149	1638	65.8	69	4
13	Salt	Sony	3.4	1794	1901	109.9	110	5
14	Step Up 3-D	BV	3.2	1592	1979	36.9	30	3
15	Cats & Dogs: The Revenge of Kitty Galore	WB	1.7	1580	1077	39.7	85	4
16	Toy Story 3	BV	1.5	730	2086	403.8	200	10

a. Which pairs of variables in the table do you think will have a positive correlation? Which pairs will have a negative correlation? Explain.

b. Draw a scatterplot relating the number of weeks in release to the gross to date. How would you describe the relationship between these two variables?

c. Draw a scatterplot relating the weekend gross to the number of theaters in which the movie is being shown. How would you describe the relationship between these two variables?

d. Draw a scatterplot relating the number of theaters in which the movie is being shown to the per theater average. How would you describe the relationship between these two variables?

3.27 Movie Money, continued The data from Exercise 3.26 were entered into a *MINITAB* worksheet, and the following output was obtained.

Covariances:

	Weekend Gross	Theaters	Average	Total Gross	Week
Weekend Gross	19.6				
Theaters	2550.7	550521.6			
Average	5093.3	421558.5	1700161.7		
Total Gross	-194.1	-38787.2	-43562.8	12839.1	
Week	-7.1	-1152.8	-1852.7	278.0	6.8

a. Use the *MINITAB* output or the original data to find the correlation between the number of weeks in release and the total gross to date.

b. For the pair of variables described in part a, which of the variables would you classify as the independent variable? The dependent variable?

c. Use the *MINITAB* output or the original data to find the correlation between the weekend gross and the number of theaters in which the movie is being shown. Find the correlation between the number of theaters in which the movie is being shown and the per theater average.

d. Do the correlations found in part c confirm your answer to Exercise 3.26a? What might be the practical reasons for the direction and strength of the correlations in part c?

3.28 Heights and Gender Refer to Exercise 1.54 and data set EX0154. When the heights of these 105 students were recorded, their gender was also recorded.

a. What variables have been measured in this experiment? Are they qualitative or quantitative?

b. Look at the histogram from Exercise 1.54 along with the comparative box plots shown below. Do the box plots help to explain the two local peaks in the histogram? Explain.

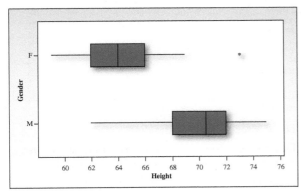

3.29 Hazardous Waste The data in Exercise 1.37 gave the number of hazardous waste sites in each of the 50 states and the District of Columbia in 2009.[11] Suspecting that there might be a relationship between the number of waste sites and the size of the state (in thousands of square miles), researchers recorded both variables and generated a scatterplot.

State	Sites	Area	State	Sites	Area	State	Sites	Area
AL	15	52	KY	14	40	ND	0	71
AK	6	663	LA	12	52	OH	41	45
AZ	9	114	ME	12	35	OK	9	70
AR	9	53	MD	19	12	OR	13	98
CA	96	164	MA	32	11	PA	97	46
CO	20	104	MI	69	97	RI	12	2
CT	15	6	MN	25	87	SC	26	32
DE	15	2	MS	6	48	SD	2	77
DC	1	0	MO	31	70	TN	15	42
FL	55	66	MT	17	147	TX	50	269
GA	16	59	NE	13	77	UT	19	85
HI	3	11	NV	1	111	VT	11	10
ID	9	84	NH	21	9	VA	31	43
IL	48	58	NJ	114	9	WA	48	71
IN	32	36	NM	14	122	WV	9	24
IA	12	56	NY	90	55	WI	39	65
KS	12	82	NC	36	54	WY	2	98

MINITAB printout for Exercise 3.29

Covariances: Sites, Area

	Sites	Area
Sites	702.776	
Area	-72.176	9346.603

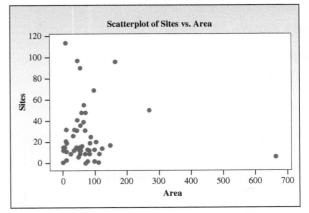

a. Is there any clear pattern in the scatterplot? Describe the relationship between number of waste sites and the size of the state.

b. Use the *MINITAB* output to calculate the correlation coefficient. Does this confirm your answer to part a?

c. Are there any outliers or clusters in the data? If so, can you explain them?

d. What other variables could you consider in trying to understand the distribution of hazardous waste sites in the United States?

3.30 Aaron Rodgers, again The number of passes completed and the total number of passing yards were recorded for Aaron Rodgers for each of the 15 regular season games that he played in the fall of 2010:[12]

Week	Completions	Total Yards	Week	Completions	Total Yards
1	19	188	9	27	289
2	19	255	11	22	301
3	34	316	12	26	344
4	12	181	13	21	298
5	27	293	14	7	46
6	18	313	16	25	404
7	21	295	17	19	229
8	15	170			

Source: www.ESPN.com

a. Draw a scatterplot to describe the relationship between number of completions and total passing yards for Aaron Rodgers.

b. Describe the plot in part a. Do you see any outliers? Do the rest of the points seem to form a pattern?

c. Calculate the correlation coefficient, r, between the number of completions and total passing yards.

d. What is the regression line for predicting total number of passing yards y based on the total number of completions x?

e. If Aaron Rodgers had 20 pass completions in his next game, what would you predict his total number of passing yards to be?

3.31 Pottery, continued In Exercise 1.59, we analyzed the percentage of aluminum oxide in 26 samples of Romano-British pottery found at four different kiln sites in the United Kingdom.[13] Since one of the sites only provided two measurements, that site is eliminated, and comparative box plots of aluminum oxide at the other three sites are shown.

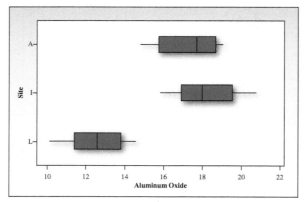

a. What two variables have been measured in this experiment? Are they qualitative or quantitative?

b. How would you compare the amount of aluminum oxide in the samples at the three sites?

3.32 Pottery, continued Here is the percentage of aluminum oxide, the percentage of iron oxide, and the percentage of magnesium oxide in five samples collected at Ashley Rails in the United Kingdom.

Sample	Al	Fe	Mg
1	17.7	1.12	0.56
2	18.3	1.14	0.67
3	16.7	0.92	0.53
4	14.8	2.74	0.67
5	19.1	1.64	0.60

a. Find the correlation coefficients describing the relationships between aluminum and iron oxide content, between iron oxide and magnesium oxide, and between aluminum oxide and magnesium oxide.

b. Write a sentence describing the relationships between these three chemicals in the pottery samples.

3.33 Gestation Times and Longevity The table below shows the gestation time in days and the average longevity in years for a variety of

mammals in captivity; the potential life span of animals is rarely attained for animals in the wild.[11]

Animal	Gestation (days)	Avg Longevity (yrs)	Animal	Gestation (days)	Avg Longevity (yrs)
Ass	365	12	Hippopotamus	238	41
Baboon	187	20	Horse	330	20
Bear (black)	219	18	Kangaroo (gray)	36	7
Bear (grizzly)	225	25	Leopard	98	12
Bear (polar)	240	20	Lion	100	15
Beaver	105	5	Monkey (rhesus)	166	15
Bison	285	15	Moose	240	12
Camel	406	12	Mouse (meadow)	21	3
Cat (domestic)	63	12	Mouse (dom.white)	19	3
Chimpanzee	230	20	Opossum (American)	13	1
Chipmunk	31	6	Pig (domestic)	112	10
Cow	284	15	Puma	90	12
Deer (whitetailed)	201	8	Rabbit (domestic)	31	5
Dog (domestic)	61	12	Rhinoceros (black)	450	15
Elephant (African)	660	35	Rhinoceros (white)	480	20
Elephant (Asian)	645	40	Sea lion (California)	350	12
Elk	250	15	Sheep (domestic)	154	12
Fox (red)	52	7	Squirrel (gray)	44	10
Giraffe	457	10	Tiger	105	16
Goat (domestic)	151	8	Wolf (maned)	63	5
Gorilla	258	20	Zebra (Grant's)	365	15
Guinea pig	68	4			

Source: The World Almanac and Book of Facts 2011

a. Draw a scatterplot for the data.

b. Describe the form, direction, and strength for the pattern in the scatterplot.

c. Are there any outliers or other unusual data points in the set? If so, to which animals do these data points correspond?

d. Remove the outliers or unusual data points from the set, and reconstruct the scatterplot. Does it appear that a straight line is appropriate for describing the data?

3.34 Housing Blues As the United States fell further and further into a recession in the years 2007–2009, the number of families who had to default on their mortgages and those who actually lost their homes grew. However, as we recover from the recession, *Moody's Economy* predicts that these numbers will fall.[14] The table below, adapted from their report, shows their approximate estimates and predictions for the years 2005–2012.

Year	Defaults (thousands)	Lost Homes (thousands)
2005	750	350
2006	800	400
2007	1350	800
2008	2600	1650
2009	3700	2000
2010	3400	2400
2011	2200	1350
2012	1000	600

a. What variables have been measured in this report? Are they qualitative or quantitative?

b. Draw side-by-side comparative bar charts to describe the number of defaults and lost homes, categorized by year.

c. Draw two line charts on the same set of axes to describe the same number over the period 2005–2012.

d. What conclusions can you draw using the two graphs in parts b and c? Which is more effective?

Data set **3.35 Armspan and Height** Leonardo
EX0335 da Vinci (1452–1519) drew a sketch of a man, indicating that a person's armspan (measuring across the back with arms outstretched to make a "T") is roughly equal to the person's height. To test this claim, we measured eight people with the following results:

Person	1	2	3	4
Armspan (inches)	68	62.25	65	69.5
Height (inches)	69	62	65	70

Person	5	6	7	8
Armspan (inches)	68	69	62	60.25
Height (inches)	67	67	63	62

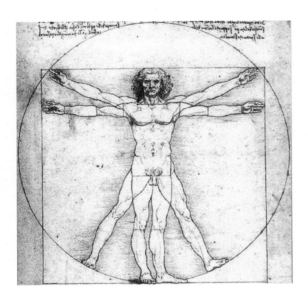

a. Draw a scatterplot for armspan and height. Use the same scale on both the horizontal and vertical axes. Describe the relationship between the two variables.

b. Calculate the correlation coefficient relating armspan and height.

c. If you were to calculate the regression line for predicting height based on a person's armspan, how would you estimate the slope of this line?

d. Find the regression line relating armspan to a person's height.

e. If a person has an armspan of 62 inches, what would you predict the person's height to be?

Data set **3.36 Midterm Scores** When a student
EX0336 performs poorly on a midterm exam, the student sometimes is convinced that their score is an anomaly and that they will do much better on the second midterm. The data below show the midterm scores (out of 100 points) for eight students in an introductory statistics class.

Student	Midterm 1	Midterm 2
1	70	88
2	58	52
3	85	84
4	82	74
5	70	80
6	40	36
7	85	48
8	85	96

a. Construct a scatterplot for the data.

b. Describe the pattern that you see in the scatterplot. Are there any clusters or outliers? If so, how would you explain them?

3.37 Midterm Scores, continued Refer to Exercise 3.36.

a. Calculate r, the correlation coefficient between the two midterm scores. How would you describe the relationship between scores on the first and second midterms?

b. Calculate the regression line for predicting a student's score on the second midterm exam based on the student's score on the first midterm.

c. Using the regression line from part b, predict a student's score on the second midterm if his score on the first midterm was 85.

Data set **3.38 Test Interviews** Of two personnel eval-
EX0338 uation techniques available, the first requires a 2-hour test-interview while the second can be completed in less than an hour. The scores for each of the eight individuals who took both tests are given in the next table.

Applicant	Test 1 (x)	Test 2 (y)
1	75	38
2	89	56
3	60	35
4	71	45
5	92	59
6	105	70
7	55	31
8	87	52

a. Construct a scatterplot for the data.

b. Describe the form, direction, and strength of the pattern in the scatterplot.

3.39 Test Interviews, continued Refer to Exercise 3.38.

a. Find the correlation coefficient, r, to describe the relationship between the two tests.

b. Would you be willing to use the second and quicker test rather than the longer test-interview to evaluate personnel? Explain.

3.40 Rain and Snow Is there a correlation EX0340 between the amount of rain and the amount of snow that falls in a particular location? The table below shows the average annual rainfall (inches) and the average annual snowfall (inches) for 10 cities in the United States.[15]

City	Rainfall (inches)	Snowfall (inches)
Billings, MT	14.77	56.9
Casper, WY	13.03	77.8
Concord, NH	37.60	64.5
Fargo, ND	21.19	40.8
Kansas City, MO	37.98	19.9
Juneau, AK	58.33	97.0
Memphis, TN	54.65	5.1
New York, NY	49.69	28.6
Portland, OR	37.07	6.5
Springfield, IL	35.56	23.2

Source: Time Almanac 2007

a. Construct a scatterplot for the data.

b. Calculate the correlation coefficient r between rainfall and snowfall. Describe the form, direction, and strength of the relationship between rainfall and snowfall.

c. Are there any outliers in the scatterplot? If so, which city does this outlier represent?

d. Remove the outlier that you found in part c from the data set and recalculate the correlation coefficient r for the remaining nine cities. Does the correlation between rainfall and snowfall change, and, if so, in what way?

3.41 Smartphones The table below shows EX0341 the prices of nine *Verizon* smartphones along with their overall score (on a scale of 0–100) in a consumer rating survey presented by *Consumer Reports*.[16]

Brand and Model	Price($)	Overall Score
Motorola Droid X	200	75
Motorola Droid	150	73
HTC Droid	200	73
LG Ally	50	72
Samsung Omnia II	50	71
HTC Imagio	100	70
Motorola Devour	80	70
Blackberry Storm2 9550	150	70
Palm Pre Plus	50	66

a. Plot the nine data points using a scatterplot. Describe the form, direction, and strength of the relationship between price and overall score.

b. Calculate r, the correlation coefficient between price and overall score.

c. Find the regression line for predicting the overall score of a smartphone based on its price.

CASE STUDY ● Are Your Dishes *Really* Clean?

 Dish-washers

Does the price of an appliance convey something about its quality? Forty-eight different dishwashers were ranked on characteristics ranging from an overall satisfaction score, washing (x_1), energy use (x_2), noise (x_3), ease of use (x_4), and cycle time (in minutes).[17] The Bosch (SHE55M1[2]UC) had the highest performance score of 82 while the GE (GLD4408R[WW]) had the lowest at 53. Ratings pictograms were converted to numerical values for $x_1, \ldots, x_4$, where 5 = Excellent, 4 = Very good, 3 = Good, 2 = Fair, and 1 = Poor. Use a statistical computer package to explore the relationships between various pairs of variables in the table.

Legend: ◉ Excellent   ◖ Very good   ○ Good   ◑ Fair   ● Poor

Brand and Model	Price $	Overall Score	Washing (x_1)	Energy Use (x_2)	Noise (x_3)	Ease of Use (x_4)	Cycle Time (min.)	x_1	x_2	x_3	x_4
Amana ADB1600AW[W]	350	61					130	4	4	2	3
Asko D3531	1600	80					145	4	5	5	4
Asko D5233XXL[HS]	1500	56	NA				180	*	4	4	4
Asko D5253XXL	1300	77					180	5	4	4	4
Bosch SHE55M1[2]UC	850	82					120	5	5	4	4
Bosch SHE6AP0[2]UC	600	75					135	5	4	3	4
Bosch SHX43P1[2]UC	800	77					115	5	5	3	4
Bosch SHX45P0[5]UC	900	79					115	5	5	4	4
Bosch SHX65P0[5]UC	1150	75					120	4	4	4	4
Bosch SHX6AP0[2]UC	700	77					110	5	5	4	4
Bosch SHX98M0[9]UC	1550	82					115	5	4	5	4
Dacor Epicure ED24[S]	1550	69					110	4	4	4	3
Electrolux Wave-Touch EWDW6505G[W]	1200	65					135	4	4	4	5
Frigidaire Gallery FGBD2431K[W]	350	70					155	5	3	3	3
Frigidaire Gallery FGBD2432K[W]	380	68					145	4	4	3	4
Frigidaire Gallery FGHD2433K[F]	500	66					135	4	4	3	4
GE GDWF100R[WW]	600	70					120	5	4	3	4
GE GLD4408R[WW]	400	53					135	3	4	2	3
GE GLD7400R[WW]	600	62					110	4	4	3	5
GE Profile PDWT500R[WW]	1300	77					110	5	4	4	5
Hotpoint HDA3600R[WW]	300	53					115	3	5	1	2
Jenn-Air JDB3200AW[W]	1100	64					125	4	5	4	3
Kenmore 1318[2]	840	79					145	5	5	4	4
Kenmore 1324[2]	410	58					125	3	4	3	3
Kenmore 1344[2]	300	60					110	4	3	2	3
Kenmore 1348[2]	500	77					120	5	4	3	3
Kenmore 1374[2]	650	80					125	5	4	3	4
Kenmore 1389[2]	500	78					135	5	4	3	4
Kenmore Elite UltraWash HE 1312[2]	780	79					140	5	5	4	4
Kenmore Elite UltraWash HE 1315[2]	1100	81					145	5	5	4	4
Kenmore Pro 1317[3]	1280	79					145	5	5	5	4
KitchenAid KUDE50CV[SS]	1200	76					125	5	5	4	4
KitchenAid KUDE60FV[WH]	1340	78					135	5	5	4	4
KitchenAid KUDE70CV[SS]	1300	81					140	5	5	5	4
KitchenAid KUDS30IV[WH]	675	77					120	5	4	3	4
KitchenAid KUDS40CV[WH]	990	79					115	5	4	4	4
LG LDF6920[WW]	700	79					125	5	4	4	4
LG Steam LDF7932[ST]	1000	81					130	5	4	4	5
Maytag MDB7609AW[W]	450	75					120	5	4	3	2
Maytag MDB8959AW[W]	750	71					120	4	4	3	5
Maytag MTB4709AW[W]	400	69					125	5	4	2	3
Miele Inspira G2142SC[WH]	1150	76					145	5	5	4	3

Legend: Excellent ◉ Very good ◓ Good ○ Fair ◒ Poor ●

Brand and Model	Price $	Overall Score	Washing (x_1)	Energy Use (x_2)	Noise (x_3)	Ease of Use (x_4)	Cycle Time (min.)	x_1	x_2	x_3	x_4
Whirlpool DU1030XTX[Q]	350	68	◉	◓	◒	○	130	5	4	2	3
Whirlpool DU1055XTV[Q]	400	76	◉	◓	○	○	125	5	4	3	3
Whirlpool DU1300XTV[Q]	420	72	◉	○	○	○	140	5	3	3	3
Whirlpool Gold GU2300XTV[Q]	550	78	◉	◓	○	◓	135	5	4	3	4
Whirlpool Gold GU2800XTV[Q]	700	77	◉	◉	○	◓	155	5	5	3	4
Whirlpool Gold GU3600XTV[Q]	800	77	◓	◉	◓	◓	145	4	5	4	4

Source: © 2007 by Consumers Union of U.S., Inc., Yonkers, NY 10703-1057, a nonprofit organization. Reprinted with permission from the September 2007 issue of *Consumer Reports*® for educational purposes only. No commercial use or reproduction permitted. www.ConsumerReports.org®.

1. Look at the variables Price, Score, and Cycle Time individually. What can you say about symmetry? About outliers?

2. Look at all the variables in pairs. Which pairs are positively correlated? Negatively correlated? Are there any pairs that exhibit little or no correlation? Are some of these results counterintuitive?

3. Does the price of an appliance, specifically a dishwasher, convey something about its quality? Which variables did you use in arriving at your answer?

Probability and Probability Distributions

4

GENERAL OBJECTIVES

Now that you have learned to describe a data set, how can you use sample data to draw conclusions about the sampled populations? The technique involves a statistical tool called *probability*. To use this tool correctly, you must first understand how it works. The first part of this chapter will present the basic concepts with simple examples.

The variables that we measured in Chapters 1 and 2 can now be redefined as random variables, whose values depend on the chance selection of the elements in the sample. Using probability as a tool, you can create probability distributions that serve as models for discrete random variables, and you can describe these random variables using a mean and standard deviation similar to those in Chapter 2.

? NEED TO KNOW...

How to Calculate the Probability of an Event
The Difference between Mutually Exclusive and Independent Events

© Tammie Arroyo/Getty Images

Probability and Decision Making in the Congo

In his exciting novel *Congo*, author Michael Crichton describes an expedition racing to find boron-coated blue diamonds in the rain forests of eastern Zaire. Can probability help the heroine Karen Ross in her search for the Lost City of Zinj? The case study at the end of this chapter involves Ross's use of probability in decision-making situations.

THE ROLE OF PROBABILITY IN STATISTICS

Probability and statistics are related in an important way. Probability is used as a *tool;* it allows you to evaluate the reliability of your conclusions about the population when you have only sample information. Consider these situations:

- When you toss a single coin, you will see either a head (H) or a tail (T). If you toss the coin repeatedly, you will generate an infinitely large number of Hs and Ts—the entire population. What does this population look like? If the coin is fair, then the population should contain 50% Hs and 50% Ts. Now toss the coin one more time. What is the chance of getting a head? Most people would say that the "probability" or chance is 1/2.

- Now suppose you are not sure whether the coin is fair; that is, you are not sure whether the makeup of the population is 50–50. You decide to perform a simple experiment. You toss the coin $n = 10$ times and observe 10 heads in a row. Can you conclude that the coin is fair? Probably not, because if the coin were fair, observing 10 heads in a row would be very *unlikely;* that is, the "probability" would be very small. It is more *likely* that the coin is biased.

As in the coin-tossing example, statisticians use probability in two ways. When the population is *known,* probability is used to describe the likelihood of observing a particular sample outcome. When the population is *unknown* and only a sample from that population is available, probability is used in making statements about the makeup of the population—that is, in making statistical inferences.

In Chapters 4–7, you will learn many different ways to calculate probabilities. You will assume that the population is *known* and calculate the probability of observing various sample outcomes. Once you begin to use probability for statistical inference in Chapter 8, the population will be *unknown* and you will use your knowledge of probability to make reliable inferences from sample information. We begin with some simple examples to help you grasp the basic concepts of probability.

EVENTS AND THE SAMPLE SPACE

Data are obtained by observing either uncontrolled events in nature or by observing events in controlled situations. We use the term **experiment** to describe either method of data collection.

Definition   An **experiment** is the process by which an observation (or measurement) is obtained.

The observation or measurement generated by an experiment may or may not produce a numerical value. Here are some examples of experiments:

- Recording a test grade
- Measuring daily rainfall
- Interviewing a householder to obtain his or her opinion on a greenbelt zoning ordinance

- Testing a printed circuit board to determine whether it is a defective product or an acceptable product
- Tossing a coin and observing the face that appears

When an experiment is performed, what we observe is an outcome called a **simple event,** often denoted by the capital E with a subscript.

Definition A **simple event** is the outcome that is observed on a single repetition of the experiment.

EXAMPLE 4.1

Experiment: Toss a die and observe the number that appears on the upper face. List the simple events in the experiment.

Solution When the die is tossed once, there are six possible outcomes. These are the simple events, listed below.

Event E_1: Observe a 1   Event E_4: Observe a 4

Event E_2: Observe a 2   Event E_5: Observe a 5

Event E_3: Observe a 3   Event E_6: Observe a 6

We can now define an **event** as a collection of simple events, often denoted by a capital letter.

Definition An **event** is a collection of simple events.

EXAMPLE
continued 4.1

We can define the events A and B for the die-tossing experiment:

A: Observe an odd number

B: Observe a number less than 4

Since event A occurs if the upper face is 1, 3, or 5, it is a collection of three simple events and we write $A = \{E_1, E_3, E_5\}$. Similarly, the event B occurs if the upper face is 1, 2, or 3 and is defined as a collection or set of these three simple events: $B = \{E_1, E_2, E_3\}$.

Sometimes when one event occurs, it means that another event cannot.

Definition Two events are **mutually exclusive** if, when one event occurs, the other cannot, and vice versa.

In the die-tossing experiment, events A and B are *not* mutually exclusive, because they have two outcomes in common—if the number on the upper face of the die is a 1 or a 3. Both events A and B will occur if either E_1 or E_3 is observed when the experiment is performed. In contrast, the six simple events $E_1, E_2, \ldots, E_6$ form a set of all mutually exclusive outcomes of the experiment. When the experiment is performed once, one and only one of these simple events can occur.

Definition The set of all simple events is called the sample space, S.

Sometimes it helps to visualize an experiment using a picture called a **Venn diagram,** shown in Figure 4.1. The white box represents the *sample space,* which contains all of the *simple events,* represented by labeled points. Since an event is a collection of one or more simple events, the appropriate points are circled and labeled with the event letter. For the die-tossing experiment, the sample space is $S = \{E_1, E_2, E_3, E_4, E_5, E_6\}$ or, more simply, $S = \{1, 2, 3, 4, 5, 6\}$. The events $A = \{1, 3, 5\}$ and $B = \{1, 2, 3\}$ are circled in the Venn diagram.

FIGURE 4.1

Venn diagram for die tossing

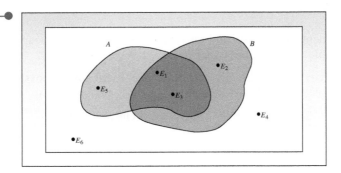

EXAMPLE 4.2

Experiment: Toss a single coin and observe the result. These are the simple events:

E_1: Observe a head (H)

E_2: Observe a tail (T)

The sample space is $S = \{E_1, E_2\}$, or, more simply, $S = \{H, T\}$.

EXAMPLE 4.3

Experiment: Record a person's blood type. The four mutually exclusive possible outcomes are these simple events:

E_1: Blood type A

E_2: Blood type B

E_3: Blood type AB

E_4: Blood type O

The sample space is $S = \{E_1, E_2, E_3, E_4\}$, or $S = \{A, B, AB, O\}$.

Some experiments can be generated in stages, and the sample space can be displayed in a **tree diagram.** Each successive level of branching on the tree corresponds to a step required to generate the final outcome.

EXAMPLE 4.4

A medical technician records a person's blood type and Rh factor. List the simple events in the experiment.

Solution   For each person, a two-stage procedure is needed to record the two variables of interest. The tree diagram is shown in Figure 4.2. The eight simple events in the tree diagram form the sample space, $S = \{A+, A-, B+, B-, AB+, AB-, O+, O-\}$.

FIGURE 4.2

Tree diagram for
Example 4.4

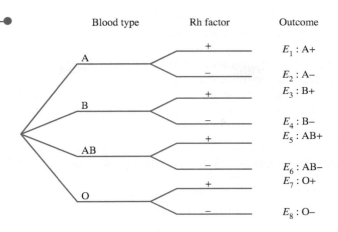

An alternative way to display the simple events is to use a **probability table,** as shown in Table 4.1. The columns and rows show the possible outcomes at the first and second stages, respectively, and the simple events are shown in the cells of the table.

TABLE 4.1

Probability Table for Example 4.4

Rh Factor	Blood Type			
	A	B	AB	O
Negative	A−	B−	AB−	O−
Positive	A+	B+	AB+	O+

CALCULATING PROBABILITIES USING SIMPLE EVENTS

4.3

The probability of an event A is a measure of our belief that the event A will occur. One practical way to interpret this measure is with the concept of *relative frequency.* Recall from Chapter 1 that if an experiment is performed n times, then the relative frequency of a particular occurrence—say, A—is

$$\text{Relative frequency} = \frac{\text{Frequency}}{n}$$

where the frequency is the number of times the event A occurred. If you let n, the number of repetitions of the experiment, become larger and larger ($n \rightarrow \infty$), you will eventually generate the entire population. In this population, the relative frequency of the event A is defined as the **probability of event A;** that is,

$$P(A) = \lim_{n \to \infty} \frac{\text{Frequency}}{n}$$

Since $P(A)$ behaves like a relative frequency, $P(A)$ must be a proportion lying between 0 and 1; $P(A) = 0$ if the event A never occurs, and $P(A) = 1$ if the event A always occurs. The closer $P(A)$ is to 1, the more likely it is that A will occur.

For example, if you tossed a balanced, six-sided die an infinite number of times, you would expect the relative frequency for any of the six values, $x = 1, 2, 3, 4, 5, 6$, to be 1/6. Needless to say, it would be very time-consuming, if not impossible, to repeat an experiment an infinite number of times. For this reason, there are alternative methods for calculating probabilities that make use of the relative frequency concept.

An important consequence of the relative frequency definition of probability involves the simple events. Since the simple events are mutually exclusive, their probabilities must satisfy two conditions.

REQUIREMENTS FOR SIMPLE-EVENT PROBABILITIES

- Each probability must lie between 0 and 1.
- The sum of the probabilities for all simple events in S equals 1.

When it is possible to write down the simple events associated with an experiment and to determine their respective probabilities, we can find the probability of an event A as follows:

Definition   The **probability of an event A** is equal to the sum of the probabilities of the simple events contained in A.

EXAMPLE   4.5

Toss two fair coins and record the outcome. Find the probability of observing exactly one head in the two tosses.

Solution   To list the simple events in the sample space, you can use a tree diagram as shown in Figure 4.3. The letters H and T mean that you observed a head or a tail, respectively, on a particular toss. To assign probabilities to each of the four simple events, you need to remember that the coins are fair. Therefore, any of the four simple events is as likely as any other. Since the sum of the four simple events must be 1, each must have probability $P(E_i) = 1/4$. The simple events in the sample space are shown in Table 4.2, along with their *equally likely probabilities*. To find $P(A) = P$(observe exactly one head), you need to find all the simple events that result in event A—namely, E_2 and E_3:

NEED A TIP?

Probabilities must lie between 0 and 1.

$$P(A) = P(E_2) + P(E_3)$$
$$= \frac{1}{4} + \frac{1}{4} = \frac{1}{2}$$

FIGURE 4.3

Tree diagram for Example 4.5

NEED A TIP?

The probabilities of all the simple events must add to 1.

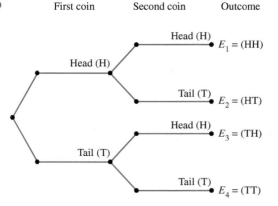

TABLE 4.2 ● **Simple Events and Their Probabilities**

Event	First Coin	Second Coin	$P(E_i)$
E_1	H	H	1/4
E_2	H	T	1/4
E_3	T	H	1/4
E_4	T	T	1/4

EXAMPLE 4.6

The proportions of blood phenotypes A, B, AB, and O in the population of all Caucasians in the United States are reported as .40, .11, .04, and .45, respectively.[1] If a single Caucasian is chosen randomly from the population, what is the probability that he or she will have either type A or type AB blood?

Solution The four simple events, A, B, AB, and O, do *not* have equally likely probabilities. Their probabilities are found using the relative frequency concept as

$$P(A) = .40 \qquad P(B) = .11 \qquad P(AB) = .04 \qquad P(O) = .45$$

The event of interest consists of two simple events, so

$$P(\text{person is either type A or type AB}) = P(A) + P(AB)$$

$$= .40 + .04 = .44$$

EXAMPLE 4.7

A candy dish contains one yellow and two red candies. You close your eyes, choose two candies one at a time from the dish, and record their colors. What is the probability that both candies are red?

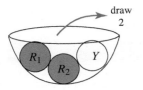
draw 2

Solution Since no probabilities are given, you must list the simple events in the sample space. The two-stage selection of the candies suggests a tree diagram, shown in Figure 4.4. There are two red candies in the dish, so you can use the letters R_1, R_2, and Y to indicate that you have selected the first red, the second red, or the yellow candy, respectively. Since you closed your eyes when you chose the candies, all six choices should be *equally likely* and are assigned probability 1/6. If A is the event that both candies are red, then

$$A = \{R_1R_2, R_2R_1\}$$

Thus,

$$P(A) = P(R_1R_2) + P(R_2R_1)$$

$$= \frac{1}{6} + \frac{1}{6} = \frac{1}{3}$$

FIGURE 4.4 ●

Tree diagram for Example 4.7

 NEED A TIP?

A tree diagram helps to find simple events.

Branch = step toward outcome

Following branches ⇒ list of simple events

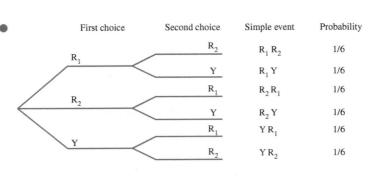

First choice	Second choice	Simple event	Probability
R_1	R_2	$R_1 R_2$	1/6
	Y	$R_1 Y$	1/6
R_2	R_1	$R_2 R_1$	1/6
	Y	$R_2 Y$	1/6
Y	R_1	$Y R_1$	1/6
	R_2	$Y R_2$	1/6

NEED TO KNOW...

How to Calculate the Probability of an Event

1. List all the simple events in the sample space.
2. Assign an appropriate probability to each simple event.
3. Determine which simple events result in the event of interest.
4. Sum the probabilities of the simple events that result in the event of interest.

In your calculation, you must always be careful that you satisfy these two conditions:

• Include all simple events in the sample space.
• Assign realistic probabilities to the simple events.

When the sample space is large, it is easy to unintentionally omit some of the simple events. If this happens, or if your assigned probabilities are wrong, your answers will not be useful in practice.

One way to determine the required number of simple events is to use the counting rules presented in the next optional section. These rules can be used to solve more complex problems, which generally involve a large number of simple events. If you need to master only the basic concepts of probability, you may choose to skip the next section.

 EXERCISES

BASIC TECHNIQUES

4.1 Tossing a Die An experiment involves tossing a single die. These are some events:

A: Observe a 2

B: Observe an even number

C: Observe a number greater than 2

D: Observe both A and B

E: Observe A or B or both

F: Observe both A and C

a. List the simple events in the sample space.

b. List the simple events in each of the events A through F.

c. What probabilities should you assign to the simple events?

d. Calculate the probabilities of the six events A through F by adding the appropriate simple-event probabilities.

4.2 A sample space S consists of five simple events with these probabilities:

$$P(E_1) = P(E_2) = .15 \qquad P(E_3) = .4$$
$$P(E_4) = 2P(E_5)$$

a. Find the probabilities for simple events E_4 and E_5.

b. Find the probabilities for these two events:

$$A = \{E_1, E_3, E_4\}$$
$$B = \{E_2, E_3\}$$

c. List the simple events that are either in event A or event B or both.

d. List the simple events that are in both event A and event B.

4.3 A sample space contains 10 simple events: E_1, E_2, . . . , E_{10}. If $P(E_1) = 3P(E_2) = .45$ and the remaining simple events are equiprobable, find the probabilities of these remaining simple events.

4.4 Free Throws A particular basketball player hits 70% of her free throws. When she tosses a pair of free throws, the four possible simple events and three of their associated probabilities are as given in the table:

Simple Event	Outcome of First Free Throw	Outcome of Second Free Throw	Probability
1	Hit	Hit	.49
2	Hit	Miss	?
3	Miss	Hit	.21
4	Miss	Miss	.09

a. Find the probability that the player will hit on the first throw and miss on the second.

b. Find the probability that the player will hit on at least one of the two free throws.

4.5 Four Coins A jar contains four coins: a nickel, a dime, a quarter, and a half-dollar. Three coins are randomly selected from the jar.

a. List the simple events in S.

b. What is the probability that the selection will contain the half-dollar?

c. What is the probability that the total amount drawn will equal 60¢ or more?

4.6 Preschool or Not? On the first day of kindergarten, the teacher randomly selects 1 of his 25 students and records the student's gender, as well as whether or not that student had gone to preschool.

a. How would you describe the experiment?

b. Construct a tree diagram for this experiment. How many simple events are there?

c. The table below shows the distribution of the 25 students according to gender and preschool experience. Use the table to assign probabilities to the simple events in part b.

	Male	Female
Preschool	8	9
No Preschool	6	2

d. What is the probability that the randomly selected student is male? What is the probability that the student is a female and did not go to preschool?

4.7 The Urn Problem A bowl contains three red and two yellow balls. Two balls are randomly selected and their colors recorded. Use a tree diagram to list the 20 simple events in the experiment, keeping in mind the order in which the balls are drawn.

4.8 The Urn Problem, continued Refer to Exercise 4.7. A ball is randomly selected from the bowl

containing three red and two yellow balls. Its color is noted, and the ball is returned to the bowl before a second ball is selected. List the additional five simple events that must be added to the sample space in Exercise 4.7.

APPLICATIONS

4.9 Need Eyeglasses? A survey classified a large number of adults according to whether they were judged to need eyeglasses to correct their reading vision and whether they used eyeglasses when reading. The proportions falling into the four categories are shown in the table. (Note that a small proportion, .02, of adults used eyeglasses when in fact they were judged not to need them.)

Judged to Need Eyeglasses	Used Eyeglasses for Reading	
	Yes	No
Yes	.44	.14
No	.02	.40

If a single adult is selected from this large group, find the probability of each event:

a. The adult is judged to need eyeglasses.

b. The adult needs eyeglasses for reading but does not use them.

c. The adult uses eyeglasses for reading whether he or she needs them or not.

4.10 Roulette The game of roulette uses a wheel containing 38 pockets. Thirty-six pockets are numbered 1, 2, . . . , 36, and the remaining two are marked 0 and 00. The wheel is spun, and a pocket is identified as the "winner." Assume that the observance of any one pocket is just as likely as any other.

a. Identify the simple events in a single spin of the roulette wheel.

b. Assign probabilities to the simple events.

c. Let A be the event that you observe either a 0 or a 00. List the simple events in the event A and find $P(A)$.

d. Suppose you placed bets on the numbers 1 through 18. What is the probability that one of your numbers is the winner?

4.11 Jury Duty Three people are randomly selected to report for jury duty. The gender of each person is noted by the county clerk.

a. Define the experiment.

b. List the simple events in S.

c. If each person is just as likely to be a man as a woman, what probability do you assign to each simple event?

d. What is the probability that only one of the three is a man?

e. What is the probability that all three are women?

4.12 Jury Duty II Refer to Exercise 4.11. Suppose that there are six prospective jurors, four men and two women, who might be impaneled to sit on the jury in a criminal case. Two jurors are randomly selected from these six to fill the two remaining jury seats.

a. List the simple events in the experiment (HINT: There are 15 simple events if you ignore the order of selection of the two jurors.)

b. What is the probability that both impaneled jurors are women?

4.13 Tea Tasters A food company plans to conduct an experiment to compare its brand of tea with that of two competitors. A single person is hired to taste and rank each of three brands of tea, which are unmarked except for identifying symbols A, B, and C.

a. Define the experiment.

b. List the simple events in S.

c. If the taster has no ability to distinguish a difference in taste among teas, what is the probability that the taster will rank tea type A as the most desirable? As the least desirable?

4.14 100-Meter Run Four equally qualified runners, John, Bill, Ed, and Dave, run a 100-meter sprint, and the order of finish is recorded.

a. How many simple events are in the sample space?

b. If the runners are equally qualified, what probability should you assign to each simple event?

c. What is the probability that Dave wins the race?

d. What is the probability that Dave wins and John places second?

e. What is the probability that Ed finishes last?

4.15 Fruit Flies In a genetics experiment, the researcher mated two *Drosophila* fruit flies and observed the traits of 300 offspring. The results are shown in the table.

Eye Color	Wing Size	
	Normal	Miniature
Normal	140	6
Vermillion	3	151

One of these offspring is randomly selected and observed for the two genetic traits.

a. What is the probability that the fly has normal eye color and normal wing size?

b. What is the probability that the fly has vermillion eyes?

c. What is the probability that the fly has either vermillion eyes or miniature wings, or both?

4.16 Creation Which of the following comes closest to your views on the origin and development of human beings? Do you believe that human beings have developed over millions of years from less advanced forms of life, but that God has guided the process? Do you think that human beings have developed over millions of years, and that God had no part in the process? Or do you believe that God created humans in their present form within the last 10,000 years or so? When asked these questions, the proportions of Americans with varying opinions are approximately as shown in the table.[2]

Opinion	Proportion
Guided by God	.36
God had no part	.13
God created in present form	.46
No opinion	.05

Source: Adapted from www.pollingreport.com

Suppose that one person is randomly selected and his or her opinion on this question is recorded.

a. What are the simple events in the experiment?

b. Are the simple events in part a equally likely? If not, what are the probabilities?

c. What is the probability that the person feels that God had some part in the creation of humans?

d. What is the probability that the person feels that God had no part in the process?

USEFUL COUNTING RULES (OPTIONAL)

4.4

Suppose that an experiment involves a large number N of simple events and you know that all the simple events are *equally likely.* Then each simple event has probability $1/N$, and the probability of an event A can be calculated as

$$P(A) = \frac{n_A}{N}$$

where n_A is the number of simple events that result in the event A. In this section, we present three simple rules that can be used to count either N, the number of simple events in the sample space, or n_A, the number of simple events in event A. Once you have obtained these counts, you can find $P(A)$ without actually listing all the simple events.

THE *mn* RULE

Consider an experiment that is performed in two stages. If the first stage can be accomplished in m ways and for each of these ways, the second stage can be accomplished in n ways, then there are mn ways to accomplish the experiment.

For example, suppose that you can order a car in one of three styles and in one of four paint colors. To find out how many options are available, you can think of first picking one of the $m = 3$ styles and then selecting one of the $n = 4$ paint colors. Using the *mn* Rule, as shown in Figure 4.5, you have $mn = (3)(4) = 12$ possible options.

FIGURE 4.5 ●
Style–color combinations

Style   Color

Tossing Dice

Two dice are tossed. How many simple events are in the sample space S?

Solution The first die can fall in one of $m = 6$ ways, and the second die can fall in one of $n = 6$ ways. Since the experiment involves two stages, forming the pairs of numbers shown on the two faces, the total number of simple events in S is

$$mn = (6)(6) = 36$$

EXAMPLE 4.9

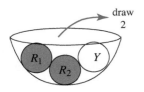

A candy dish contains one yellow and two red candies. Two candies are selected one at a time from the dish, and their colors are recorded. How many simple events are in the sample space S?

Solution   The first candy can be chosen in $m = 3$ ways. Since one candy is now gone, the second candy can be chosen in $n = 2$ ways. The total number of simple events is

$$mn = (3)(2) = 6$$

These six simple events were listed in Example 4.7.

We can extend the *mn* Rule for an experiment that is performed in more than two stages.

THE EXTENDED *mn* RULE

If an experiment is performed in k stages, with n_1 ways to accomplish the first stage, n_2 ways to accomplish the second stage, . . . , and n_k ways to accomplish the kth stage, then the number of ways to accomplish the experiment is

$$n_1 n_2 n_3 \cdots n_k$$

EXAMPLE 4.10

How many simple events are in the sample space when three coins are tossed?

Solution   Each coin can land in one of two ways. Hence, the number of simple events is

$$(2)(2)(2) = 8$$

EXAMPLE 4.11

A truck driver can take three routes from city A to city B, four from city B to city C, and three from city C to city D. If, when traveling from A to D, the driver must drive from A to B to C to D, how many possible A-to-D routes are available?

Solution   Let

n_1 = Number of routes from A to B = 3
n_2 = Number of routes from B to C = 4
n_3 = Number of routes from C to D = 3

Then the total number of ways to construct a complete route, taking one subroute from each of the three groups, (A to B), (B to C), and (C to D), is

$$n_1 n_2 n_3 = (3)(4)(3) = 36$$

A second useful counting rule follows from the *mn* Rule and involves **orderings** or **permutations.** For example, suppose you have three books, *A, B,* and *C,* but you have room for only two on your bookshelf. In how many ways can you select and arrange the two books? There are three choices for the two books—*A* and *B, A* and *C,* or *B* and *C*—but each of the pairs can be arranged in two ways on the shelf. All the permutations of the two books, chosen from three, are listed in Table 4.3. The *mn*

Rule implies that there are 6 ways, because the first book can be chosen in $m = 3$ ways and the second in $n = 2$ ways, so the result is $mn = 6$.

TABLE 4.3

Permutations of Two Books Chosen from Three

Combinations of Two	Reordering of Combinations
AB	BA
AC	CA
BC	CB

In how many ways can you arrange all three books on your bookshelf? These are the six permutations:

$$ABC \quad ACB \quad BAC$$
$$BCA \quad CAB \quad CBA$$

Since the first book can be chosen in $n_1 = 3$ ways, the second in $n_2 = 2$ ways, and the third in $n_3 = 1$ way, the total number of orderings is $n_1 n_2 n_3 = (3)(2)(1) = 6$.

Rather than applying the *mn* Rule each time, you can find the number of orderings using a general formula involving *factorial notation*.

A COUNTING RULE FOR PERMUTATIONS

The number of ways we can arrange n distinct objects, taking them r at a time, is

$$P_r^n = \frac{n!}{(n - r)!}$$

where $n! = n(n - 1)(n - 2) \cdots (3)(2)(1)$ and $0! \doteq 1$.

Since r objects are chosen, this is an *r-stage* experiment. The first object can be chosen in n ways, the second in $(n - 1)$ ways, the third in $(n - 2)$ ways, and the rth in $(n - r + 1)$ ways. We can simplify this awkward notation using the counting rule for permutations because

$$\frac{n!}{(n - r)!} = \frac{n(n - 1)(n - 2) \cdots (n - r + 1)(n - r) \cdots (2)(1)}{(n - r) \cdots (2)(1)}$$

$$= n(n - 1) \cdots (n - r + 1)$$

A SPECIAL CASE: ARRANGING *n* ITEMS

The number of ways to arrange an entire set of n distinct items is $P_n^n = n!$

EXAMPLE 4.12

Three lottery tickets are drawn from a total of 50. If the tickets will be distributed to each of three employees in the order in which they are drawn, the order will be important. How many simple events are associated with the experiment?

Solution The total number of simple events is

$$P_3^{50} = \frac{50!}{47!} = 50(49)(48) = 117,600$$

EXAMPLE (4.13) — A piece of equipment is composed of five parts that can be assembled in any order. A test is to be conducted to determine the time necessary for each order of assembly. If each order is to be tested once, how many tests must be conducted?

Solution   The total number of tests is

$$P_5^5 = \frac{5!}{0!} = 5(4)(3)(2)(1) = 120$$

When we counted the number of permutations of the two books chosen for your bookshelf, we used a systematic approach:

- First we counted the number of *combinations* or pairs of books to be chosen.
- Then we counted the number of ways to arrange the two chosen books on the shelf.

Sometimes the ordering or arrangement of the objects is not important, but only the objects that are chosen. In this case, you can use a counting rule for **combinations.** For example, you may not care in what order the books are placed on the shelf, but only which books you are able to shelve. When a five-person committee is chosen from a group of 12 students, the order of choice is unimportant because all five students will be equal members of the committee.

A COUNTING RULE FOR COMBINATIONS

The number of distinct combinations of n distinct objects that can be formed, taking them r at a time, is

$$C_r^n = \frac{n!}{r!(n-r)!}$$

The number of *combinations* and the number of *permutations* are related:

$$C_r^n = \frac{P_r^n}{r!}$$

You can see that C_r^n results when you divide the number of permutations by $r!$, the number of ways of rearranging each distinct group of r objects chosen from the total n.

EXAMPLE (4.14) — A printed circuit board may be purchased from five suppliers. In how many ways can three suppliers be chosen from the five?

Solution   Since it is important to know only which three have been chosen, not the order of selection, the number of ways is

$$C_3^5 = \frac{5!}{3!2!} = \frac{(5)(4)}{2} = 10$$

The next example illustrates the use of counting rules to solve a probability problem.

EXAMPLE (4.15) — Five manufacturers produce a certain electronic device, whose quality varies from manufacturer to manufacturer. If you were to select three manufacturers at random, what is the chance that the selection would contain exactly two of the best three?

Solution   The simple events in this experiment consist of all possible combinations of three manufacturers, chosen from a group of five. Of these five, three have been designated as "best" and two as "not best." You can think of a candy dish containing three red and two yellow candies, from which you will select three, as illustrated in Figure 4.6. The total number of simple events N can be counted as the number of ways to choose three of the five manufacturers, or

$$N = C_3^5 = \frac{5!}{3!2!} = 10$$

FIGURE 4.6

Illustration for Example 4.15

Choose 3

3 "best"

2 "not best"

Since the manufacturers are selected at random, any of these 10 simple events will be *equally likely,* with probability 1/10. But how many of these simple events result in the event?

$$A: \text{Exactly two of the "best" three}$$

You can count n_A, the number of events in A, in two steps because event A will occur when you select two of the "best" three and one of the two "not best." There are

$$C_2^3 = \frac{3!}{2!1!} = 3$$

ways to accomplish the first stage and

$$C_1^2 = \frac{2!}{1!1!} = 2$$

ways to accomplish the second stage. Applying the *mn* Rule, we find there are $n_A = (3)(2) = 6$ of the 10 simple events in event A and $P(A) = n_A/N = 6/10$.

Many other counting rules are available in addition to the three presented in this section. If you are interested in this topic, you should consult one of the many textbooks on combinatorial mathematics.

4.4 EXERCISES

BASIC TECHNIQUES

4.17 You have *two* groups of distinctly different items, 10 in the first group and 8 in the second. If you select one item from each group, how many different pairs can you form?

4.18 You have *three* groups of distinctly different items, four in the first group, seven in the second, and three in the third. If you select one item from each group, how many different triplets can you form?

4.19 Permutations Evaluate the following *permutations.* (HINT: Your scientific calculator may have a function that allows you to calculate permutations and combinations quite easily.)

a. P_3^5 **b.** P_9^{10} **c.** P_6^6 **d.** P_1^{20}

4.20 Combinations Evaluate these *combinations:*

a. C_3^5 **b.** C_9^{10} **c.** C_6^6 **d.** C_1^{20}

4.21 Choosing People In how many ways can you select five people from a group of eight if the order of selection is important?

4.22 Choosing People, again In how many ways can you select two people from a group of 20 if the order of selection is not important?

4.23 Dice Three dice are tossed. How many simple events are in the sample space?

4.24 Coins Four coins are tossed. How many simple events are in the sample space?

4.25 The Urn Problem, again Three balls are selected from a box containing 10 balls. The order of selection is not important. How many simple events are in the sample space?

APPLICATIONS

4.26 What to Wear? You own 4 pairs of jeans, 12 clean T-shirts, and 4 wearable pairs of sneakers. How many outfits (jeans, T-shirt, and sneakers) can you create?

4.27 Itineraries A businessman in New York is preparing an itinerary for a visit to six major cities. The distance traveled, and hence the cost of the trip, will depend on the order in which he plans his route. How many different itineraries (and trip costs) are possible?

4.28 Vacation Plans Your family vacation involves a cross-country air flight, a rental car, and a hotel stay in Boston. If you can choose from four major air carriers, five car rental agencies, and three major hotel chains, how many options are available for your vacation accommodations?

4.29 A Card Game Three students are playing a card game. They decide to choose the first person to play by each selecting a card from the 52-card deck and looking for the highest card in value and suit. They rank the suits from lowest to highest: clubs, diamonds, hearts, and spades.

a. If the card is replaced in the deck after each student chooses, how many possible configurations of the three choices are possible?

b. How many configurations are there in which each student picks a different card?

c. What is the probability that all three students pick exactly the same card?

d. What is the probability that all three students pick different cards?

4.30 Dinner at Gerard's A French restaurant in Riverside, California, offers a special summer menu in which, for a fixed dinner cost, you can choose from one of two salads, one of two entrees, and one of two desserts. How many different dinners are available?

4.31 Playing Poker Five cards are selected from a 52-card deck for a poker hand.

a. How many simple events are in the sample space?

b. A *royal flush* is a hand that contains the A, K, Q, J, and 10, all in the same suit. How many ways are there to get a royal flush?

c. What is the probability of being dealt a royal flush?

4.32 Poker II Refer to Exercise 4.31. You have a poker hand containing four of a kind.

a. How many possible poker hands can be dealt?

b. In how many ways can you receive four cards of the same face value *and* one card from the other 48 available cards?

c. What is the probability of being dealt four of a kind?

4.33 A Hospital Survey A study is to be conducted in a hospital to determine the attitudes of nurses toward various administrative procedures. If a sample of 10 nurses is to be selected from a total of 90, how many different samples can be selected? (HINT: Is order important in determining the makeup of the sample to be selected for the survey?)

4.34 Traffic Problems Two city council members are to be selected from a total of five to form a subcommittee to study the city's traffic problems.

a. How many different subcommittees are possible?

b. If all possible council members have an equal chance of being selected, what is the probability that members Smith and Jones are both selected?

4.35 The WNBA Professional basketball is now a reality for women basketball players in the United States. There are two conferences in the WNBA, each with six teams, as shown in the table below.[3]

Western Conference	Eastern Conference
Minnesota Lynx	Atlanta Dream
Phoenix Mercury	Indiana Fever
Tulsa Shock	New York Liberty
Los Angeles Sparks	Washington Mystics
Seattle Storm	Connecticut Sun
San Antonio Silver Stars	Chicago Sky

Source: www.espn.com

Two teams, one from each conference, are randomly selected to play an exhibition game.

a. How many pairs of teams can be chosen?

b. What is the probability that the two teams are Los Angeles and New York?

c. What is the probability that the Western Conference team is not from California?

4.36 100-Meter Run, again Refer to Exercise 4.14, in which a 100-meter sprint is run by John, Bill, Ed, and Dave. Assume that all of the runners are equally qualified, so that any order of finish is equally likely. Use the *mn* Rule or permutations to answer these questions:

a. How many orders of finish are possible?

b. What is the probability that Dave wins the sprint?

c. What is the probability that Dave wins and John places second?

d. What is the probability that Ed finishes last?

4.37 Gender Bias? A woman brought a complaint of gender discrimination to an eight-member human relations advisory board. The board, composed of five women and three men, voted 5–3 in favor of the plaintiff, the five women voting for the plaintiff and the three men against. Has the board been affected by gender bias? That is, if the vote in favor of the plaintiff was 5–3 and the board members were not biased by gender, what is the probability that the vote would split along gender lines (five women for, three men against)?

4.38 Cramming A student prepares for an exam by studying a list of 10 problems. She can solve 6 of them. For the exam, the instructor selects 5 questions at random from the list of 10. What is the probability that the student can solve all 5 problems on the exam?

4.39 Monkey Business A monkey is given 12 blocks: 3 shaped like squares, 3 like rectangles, 3 like triangles, and 3 like circles. If it draws three of each kind in order—say, 3 triangles, then 3 squares, and so on—would you suspect that the monkey associates identically shaped figures? Calculate the probability of this event.

EVENT RELATIONS AND PROBABILITY RULES

4.5

Sometimes the event of interest can be formed as a combination of several other events. Let *A* and *B* be two events defined on the sample space *S*. Here are three important relationships between events.

Definition The **union** of events *A* and *B*, denoted by $A \cup B$, is the event that either *A* or *B* or both occur.

Definition The **intersection** of events *A* and *B*, denoted by $A \cap B$, is the event that both *A* and *B* occur.[†]

Definition The **complement** of an event *A*, denoted by A^c, is the event that *A* *does not* occur.

Figures 4.7, 4.8, and 4.9 show Venn diagram representations of $A \cup B$, $A \cap B$, and A^c, respectively. Any simple event in the shaded area is a possible outcome resulting in the appropriate event. One way to find the probabilities of the union, the intersection, or the complement is to sum the probabilities of all the associated simple events.

[†]Some authors use the notation *AB*.

FIGURE 4.7

Venn diagram of $A \cup B$

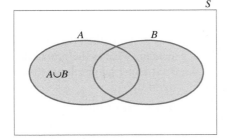

FIGURE 4.8

Venn diagram $A \cap B$

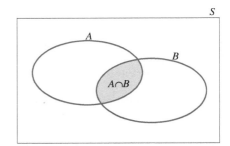

FIGURE 4.9

The complement of an event

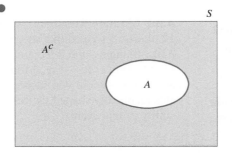

EXAMPLE 4.16

Two fair coins are tossed, and the outcome is recorded. These are the events of interest:

A: Observe at least one head

B: Observe at least one tail

Define the events A, B, $A \cap B$, $A \cup B$, and A^c as collections of simple events, and find their probabilities.

Solution   Recall from Example 4.5 that the simple events for this experiment are

E_1: HH (head on first coin, head on second)

E_2: HT

E_3: TH

E_4: TT

and that each simple event has probability 1/4. Event A, at least one head, occurs if E_1, E_2, or E_3 occurs, so that

$$A = \{E_1, E_2, E_3\} \qquad P(A) = \frac{3}{4}$$

and

$$A^c = \{E_4\} \qquad P(A^c) = \frac{1}{4}$$

Similarly,

$$B = \{E_2, E_3, E_4\} \qquad P(B) = \frac{3}{4}$$

$$A \cap B = \{E_2, E_3\} \qquad P(A \cap B) = \frac{1}{2}$$

$$A \cup B = \{E_1, E_2, E_3, E_4\} \qquad P(A \cup B) = \frac{4}{4} = 1$$

Note that $(A \cup B) = S$, the sample space, and is thus certain to occur.

The concept of unions and intersections can be extended to more than two events. For example, the union of three events A, B, and C, which is written as $A \cup B \cup C$, is the set of simple events that are in A or B or C or in any combination of those events. Similarly, the intersection of three events A, B, and C, which is written as $A \cap B \cap C$, is the collection of simple events that are common to the three events A, B, and C.

Calculating Probabilities for Unions and Complements

When we can write the event of interest in the form of a union, a complement, or an intersection, there are special probability rules that can simplify our calculations. The first rule deals with *unions* of events.

THE ADDITION RULE

Given two events, A and B, the probability of their union, $A \cup B$, is equal to

$$P(A \cup B) = P(A) + P(B) - P(A \cap B)$$

Notice in the Venn diagram in Figure 4.10 that the sum $P(A) + P(B)$ double counts the simple events that are common to both A and B. Subtracting $P(A \cap B)$ gives the correct result.

FIGURE 4.10 ●

The Addition Rule

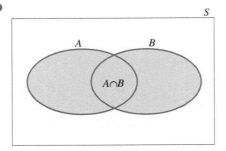

When two events A and B are **mutually exclusive** or **disjoint,** it means that when A occurs, B cannot, and vice versa. This means that the probability that they both occur, $P(A \cap B)$, must be zero. Figure 4.11 is a Venn diagram representation of two such events with no simple events in common.

FIGURE 4.11

FIGURE 4.11

Two disjoint or mutually exclusive events

 NEED A TIP?

Remember, mutually exclusive $\Leftrightarrow P(A \cap B) = 0$.

When two events A and B are **mutually exclusive,** then $P(A \cap B) = 0$ and the Addition Rule simplifies to

$$P(A \cup B) = P(A) + P(B)$$

The second rule deals with *complements* of events. You can see from the Venn diagram in Figure 4.9 that A and A^c are mutually exclusive and that $A \cup A^c = S$, the entire sample space. It follows that

$$P(A) + P(A^c) = 1 \text{ and } P(A^c) = 1 - P(A)$$

RULE FOR COMPLEMENTS

$$P(A^c) = 1 - P(A)$$

EXAMPLE 4.17 An oil-prospecting firm plans to drill two exploratory wells. Past evidence is used to assess the possible outcomes listed in Table 4.4.

TABLE 4.4 ● **Outcomes for Oil-Drilling Experiment**

Event	Description	Probability
A	Neither well produces oil or gas	.80
B	Exactly one well produces oil or gas	.18
C	Both wells produce oil or gas	.02

Find $P(A \cup B)$ and $P(B \cup C)$.

Solution By their definition, events A, B, and C are jointly mutually exclusive because the occurrence of one event precludes the occurrence of either of the other two. Therefore,

$$P(A \cup B) = P(A) + P(B) = .80 + .18 = .98$$

and

$$P(B \cup C) = P(B) + P(C) = .18 + .02 = .20$$

The event $A \cup B$ can be described as the event that *at most* one well produces oil or gas, and $B \cup C$ describes the event that *at least* one well produces gas or oil.

EXAMPLE 4.18

In a telephone survey of 1000 adults, respondents were asked their opinion about the cost of a college education. The respondents were classified according to whether they currently had a child in college and whether they thought the loan burden for most college students is too high, the right amount, or too little. The proportions responding in each category are shown in the **probability table** in Table 4.5. Suppose one respondent is chosen at random from this group.

TABLE 4.5

Probability Table

	Too High (A)	Right Amount (B)	Too Little (C)
Child in College (D)	.35	.08	.01
No Child in College (E)	.25	.20	.11

1. What is the probability that the respondent has a child in college?
2. What is the probability that the respondent does not have a child in college?
3. What is the probability that the respondent has a child in college or thinks that the loan burden is too high or both?

Solution Table 4.5 gives the probabilities for the six simple events in the cells of the table. For example, the entry in the top left corner of the table is the probability that a respondent has a child in college *and* thinks the loan burden is too high ($A \cap D$).

1. The event that a respondent has a child in college will occur regardless of his or her response to the question about loan burden. That is, event D consists of the simple events in the first row:

$$P(D) = .35 + .08 + .01 = .44$$

In general, the probabilities of *marginal* events such as D and A are found by summing the probabilities in the appropriate row or column.

2. The event that the respondent does not have a child in college is the complement of the event D denoted by D^c. The probability of D^c is found as

$$P(D^c) = 1 - P(D)$$

Using the result of part 1, we have

$$P(D^c) = 1 - .44 = .56$$

3. The event of interest is $P(A \cup D)$. Using the Addition Rule

$$P(A \cup D) = P(A) + P(D) - P(A \cap D)$$
$$= .60 + .44 - .35$$
$$= .69$$

INDEPENDENCE, CONDITIONAL PROBABILITY, AND THE MULTIPLICATION RULE

4.6

In Example 4.18, we were able to use the Addition Rule to calculate $P(A \cup D)$ because we could find $P(A \cap D)$ directly from the probability table. Sometimes, however, the intersection probability is unknown. In this situation, there is a probability rule that can be used to calculate the probability of the intersection of several events. This rule depends on the important statistical concept of **independent** or **dependent events.**

Definition   Two events, A and B, are said to be **independent** if and only if the probability of event B is not influenced or changed by the occurrence of event A, or vice versa.

Colorblindness   Suppose a researcher notes a person's gender and whether or not the person is colorblind to red and green. Does the probability that a person is colorblind change, depending on whether the person is male or not? Define two events:

   A: Person is a male

   B: Person is colorblind

In this case, since colorblindness is a male sex-linked characteristic, the probability that a man is colorblind will be greater than the probability that a person chosen from the general population will be colorblind. The probability of event B, that a person is colorblind, depends on whether or not event A, that the person is a male, has occurred. We say that A and B are *dependent events*.

Tossing Dice   On the other hand, consider tossing a single die two times, and define two events:

   A: Observe a 2 on the first toss

   B: Observe a 2 on the second toss

If the die is fair, the probability of event A is $P(A) = 1/6$. Consider the probability of event B. Regardless of whether event A has or has not occurred, the probability of observing a 2 on the second toss is still 1/6. We could write:

   $P(B$ given that A occurred$) = 1/6$

   $P(B$ given that A did not occur$) = 1/6$

Since the probability of event B is not changed by the occurrence of event A, we say that A and B are *independent events*.

The probability of an event A, given that the event B has occurred, is called the **conditional probability of A, given that B has occurred,** denoted by $P(A|B)$. The vertical bar is read "given" and the events appearing to the right of the bar are those that you know have occurred. We will use these probabilities to calculate the probability that *both A and B* occur when the experiment is performed.

THE GENERAL MULTIPLICATION RULE

The probability that *both A and B* occur when the experiment is performed is

   $P(A \cap B) = P(A)P(B|A)$

or

   $P(A \cap B) = P(B)P(A|B)$

EXAMPLE 4.19

In a color preference experiment, eight toys are placed in a container. The toys are identical except for color—two are red, and six are green. A child is asked to choose two toys *at random*. What is the probability that the child chooses the two red toys?

Solution You can visualize the experiment using a tree diagram as shown in Figure 4.12. Define the following events:

R: Red toy is chosen

G: Green toy is chosen

FIGURE 4.12

Tree diagram for Example 4.19

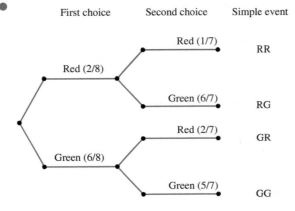

The event *A* (both toys are red) can be constructed as the intersection of two events:

$A = (\text{R on first choice}) \cap (\text{R on second choice})$

Since there are only two red toys in the container, the probability of choosing red on the first choice is 2/8. However, once this red toy has been chosen, the probability of red on the second choice is *dependent* on the outcome of the first choice (see Figure 4.12). If the first choice was red, the probability of choosing a second red toy is only 1/7 because there is only one red toy among the seven remaining. If the first choice was green, the probability of choosing red on the second choice is 2/7 because there are two red toys among the seven remaining. Using this information and the Multiplication Rule, you can find the probability of event *A*.

$P(A) = P(\text{R on first choice} \cap \text{R on second choice})$
$\qquad = P(\text{R on first choice}) \, P(\text{R on second choice}|\text{R on first})$
$\qquad = \left(\frac{2}{8}\right)\left(\frac{1}{7}\right) = \frac{2}{56} = \frac{1}{28}$

The solution in Example 4.19 was possible only because we knew $P(\text{R on second choice}|\text{R on first choice})$. If you don't know the conditional probability, $P(A|B)$, you may be able to calculate it by using the Multiplication Rule in a slightly different form. Just *rearrange the terms* in the Multiplication Rule.

CONDITIONAL PROBABILITIES

The conditional probability of event A, given that event B has occurred is

$$P(A|B) = \frac{P(A \cap B)}{P(B)} \quad \text{if} \quad P(B) \neq 0$$

The conditional probability of event B, given that event A has occurred is

$$P(B|A) = \frac{P(A \cap B)}{P(A)} \quad \text{if} \quad P(A) \neq 0$$

Notice that, in this form, you need to know $P(A \cap B)$!

Colorblindness, continued Suppose that in the general population, there are 51% men and 49% women, and that the proportions of colorblind men and women are shown in the probability table below:

	Men(B)	Women (B^C)	Total
Colorblind (A)	.04	.002	.042
Not Colorblind (A^C)	.47	.488	.958
Total	.51	.49	1.00

If a person is drawn at random from this population and is found to be a man (event B), what is the probability that the man is colorblind (event A)? If we know that the event B has occurred, we must restrict our focus to only the 51% of the population that is male. The probability of being colorblind, given that the person is male, is 4% of the 51%, or

$$P(A|B) = \frac{P(A \cap B)}{P(B)} = \frac{.04}{.51} = .078$$

What is the probability of being colorblind, given that the person is female? Now we are restricted to only the 49% of the population that is female, and

$$P(A|B^C) = \frac{P(A \cap B^C)}{P(B^C)} = \frac{.002}{.49} = .004$$

Notice that the probability of event A changed, depending on whether event B occured. This indicates that these two events are *dependent*.

When two events are **independent**—that is, if the probability of event B is the same, whether or not event A has occurred, then event A does not affect event B and

$$P(B|A) = P(B)$$

The Multiplication Rule can now be simplified.

THE MULTIPLICATION RULE FOR INDEPENDENT EVENTS

If two events A and B are independent, the probability that *both A and B* occur is

$$P(A \cap B) = P(A)P(B)$$

Similarly, if A, B, and C are mutually independent events (all pairs of events are independent), then the probability that A, B, and C all occur is

$$P(A \cap B \cap C) = P(A)P(B)P(C)$$

Coin Tosses at Football Games   A football team is involved in two over-time periods during a given game, so that there are three coin tosses. If the coin is fair, what is the probability that they lose all three tosses?

Solution   If the coin is fair, the event can be described in three steps:

  A: lose the first toss

  B: lose the second toss

  C: lose the third toss

Since the tosses are independent, and since $P(\text{win}) = P(\text{lose}) = .5$ for any of the three tosses,

$$P(A \cap B \cap C) = P(A)P(B)P(C) = (.5)(.5)(.5) = .125$$

How can you check to see if two events are independent or dependent? The easiest solution is to redefine the concept of **independence** in a more formal way.

CHECKING FOR INDEPENDENCE

Two events A and B are said to be **independent** if and only if either

  $P(A \cap B) = P(A)P(B)$

or

  $P(B|A) = P(B)$ or equivalently, $P(A|B) = P(A)$

Otherwise, the events are said to be **dependent.**

EXAMPLE 4.20

Toss two coins and observe the outcome. Define these events:

  A: Head on the first coin

  B: Tail on the second coin

Are events A and B independent?

Solution   From previous examples, you know that $S = \{HH, HT, TH, TT\}$. Use these four simple events to find

NEED A TIP?

Remember,
independence ⇔
$P(A \cap B) = P(A)P(B)$.

$$P(A) = \frac{1}{2}, \; P(B) = \frac{1}{2}, \text{ and } P(A \cap B) = \frac{1}{4}.$$

Since $P(A)P(B) = \left(\frac{1}{2}\right)\left(\frac{1}{2}\right) = \frac{1}{4}$ and $P(A \cap B) = \frac{1}{4}$, we have $P(A)P(B) = P(A \cap B)$

and the two events must be independent.

EXAMPLE 4.21

Refer to the probability table in Example 4.18, which is reproduced below.

	Too High (A)	Right Amount (B)	Too Little (C)
Child in College (D)	.35	.08	.01
No Child in College (E)	.25	.20	.11

Are events D and A independent? Explain.

Solution

1. Use the probability table to find $P(A \cap D) = .35$, $P(A) = .60$, and $P(D) = .44$. Then

$$P(A)P(D) = (.60)(.44) = .264 \text{ and } P(A \cap D) = .35$$

Since these two probabilities are not the same, events A and D are *dependent*.

2. Alternately, calculate

$$P(A|D) = \frac{P(A \cap D)}{P(D)} = \frac{.35}{.44} = .80$$

Since $P(A|D) = .80$ and $P(A) = .60$, we are again led to the conclusion that events A and D are *dependent*.

3. Equivalently,

$$P(D|A) = \frac{P(A \cap D)}{P(A)} = \frac{.35}{.60} = .58$$

while $P(D) = .44$. Again we see that A and D are dependent events.

 NEED TO KNOW...

The Difference between Mutually Exclusive and Independent Events

Many students find it hard to tell the difference between *mutually exclusive* and *independent* events.

- When two events are *mutually exclusive* or *disjoint,* they cannot both happen together when the experiment is performed. Once the event B has occurred, event A cannot occur, so that $P(A|B) = 0$, or vice versa. The occurrence of event B certainly affects the probability that event A can occur.

- Therefore, mutually exclusive events must be *dependent.*

- When two events are *mutually exclusive* or *disjoint,* $P(A \cap B) = 0$ and $P(A \cup B) = P(A) + P(B)$.

- When two events are *independent,* $P(A \cap B) = P(A)P(B)$, and $P(A \cup B) = P(A) + P(B) - P(A)P(B)$.

Using probability rules to calculate the probability of an event requires some experience and ingenuity. You need to express the event of interest as a union or intersection (or the combination of both) of two or more events whose probabilities are known or easily calculated. Often you can do this in different ways; the key is to find the right combination.

EXAMPLE (4.22) Two cards are drawn from a deck of 52 cards. Calculate the probability that the draw includes an ace and a ten.

Solution Consider the event of interest:

$$A: \text{Draw an ace and a ten}$$

Then $A = B \cup C$, where

B: Draw the ace on the first draw and the ten on the second
C: Draw the ten on the first draw and the ace on the second

Events B and C were chosen to be mutually exclusive and also to be intersections of events with known probabilities; that is,

$$B = B_1 \cap B_2 \text{ and } C = C_1 \cap C_2$$

where

B_1: Draw an ace on the first draw
B_2: Draw a ten on the second draw
C_1: Draw a ten on the first draw
C_2: Draw an ace on the second draw

Applying the Multiplication Rule, you get

$$P(B_1 \cap B_2) = P(B_1)P(B_2|B_1)$$
$$= \left(\frac{4}{52}\right)\left(\frac{4}{51}\right)$$

and

$$P(C_1 \cap C_2) = \left(\frac{4}{52}\right)\left(\frac{4}{51}\right)$$

Then, applying the Addition Rule,

$$P(A) = P(B) + P(C)$$
$$= \left(\frac{4}{52}\right)\left(\frac{4}{51}\right) + \left(\frac{4}{52}\right)\left(\frac{4}{51}\right) = \frac{8}{663}$$

Check each composition carefully to be certain that it is actually equal to the event of interest.

4.6 EXERCISES

BASIC TECHNIQUES

4.40 An experiment can result in one of five equally likely simple events, $E_1, E_2, \ldots, E_5$. Events A, B, and C are defined as follows:

A: E_1, E_3	$P(A) = .4$
B: E_1, E_2, E_4, E_5	$P(B) = .8$
C: E_3, E_4	$P(C) = .4$

Find the probabilities associated with the following events by listing the simple events in each.

a. A^c **b.** $A \cap B$ **c.** $B \cap C$
d. $A \cup B$ **e.** $B|C$ **f.** $A|B$
g. $A \cup B \cup C$ **h.** $(A \cap B)^c$

4.41 Refer to Exercise 4.40. Use the definition of a complementary event to find these probabilities:

a. $P(A^c)$ **b.** $P((A \cap B)^c)$

Do the results agree with those obtained in Exercise 4.40?

4.42 Refer to Exercise 4.40. Use the definition of conditional probability to find these probabilities:

a. $P(A|B)$ **b.** $P(B|C)$

Do the results agree with those obtained in Exercise 4.40?

4.43 Refer to Exercise 4.40. Use the Addition and Multiplication Rules to find these probabilities:

a. $P(A \cup B)$ **b.** $P(A \cap B)$ **c.** $P(B \cap C)$

Do the results agree with those obtained in Exercise 4.40?

4.44 Refer to Exercise 4.40.

a. Are events A and B independent?

b. Are events A and B mutually exclusive?

4.45 Suppose $P(A) = .1$ and $P(B) = .5$.

a. If $P(A|B) = .1$, what is $P(A \cap B)$?

b. If $P(A|B) = .1$, are A and B independent?

c. If $P(A \cap B) = 0$, are A and B independent?

d. If $P(A \cup B) = .65$, are A and B mutually exclusive?

4.46 Dice An experiment consists of tossing a single die and observing the number of dots that show on the upper face. Events A, B, and C are defined as follows:

  A: Observe a number less than 4

  B: Observe a number less than or equal to 2

  C: Observe a number greater than 3

Find the probabilities associated with the events below using either the simple event approach or the rules and definitions from this section.

a. S **b.** $A|B$ **c.** B
d. $A \cap B \cap C$ **e.** $A \cap B$ **f.** $A \cap C$
g. $B \cap C$ **h.** $A \cup C$ **i.** $B \cup C$

4.47 Refer to Exercise 4.46.

a. Are events A and B independent? Mutually exclusive?

b. Are events A and C independent? Mutually exclusive?

4.48 Two fair dice are tossed.

a. What is the probability that the sum of the number of dots shown on the upper faces is equal to 7? To 11?

b. What is the probability that you roll "doubles"— that is, both dice have the same number on the upper face?

c. What is the probability that both dice show an odd number?

4.49 Suppose that $P(A) = .4$ and $P(B) = .2$. If events A and B are independent, find these probabilities:

a. $P(A \cap B)$ **b.** $P(A \cup B)$

4.50 Suppose that $P(A) = .3$ and $P(B) = .5$. If events A and B are mutually exclusive, find these probabilities:

a. $P(A \cap B)$ **b.** $P(A \cup B)$

4.51 Suppose that $P(A) = .4$ and $P(A \cap B) = .12$.

a. Find $P(B|A)$.

b. Are events A and B mutually exclusive?

c. If $P(B) = .3$, are events A and B independent?

4.52 An experiment can result in one or both of events A and B with the probabilities shown in this probability table:

	A	A^c
B	.34	.46
B^c	.15	.05

Find the following probabilities:

a. $P(A)$ **b.** $P(B)$ **c.** $P(A \cap B)$
d. $P(A \cup B)$ **e.** $P(A|B)$ **f.** $P(B|A)$

4.53 Refer to Exercise 4.52.

a. Are events A and B mutually exclusive? Explain.

b. Are events A and B independent? Explain.

APPLICATIONS

4.54 Drug Testing Many companies are now testing prospective employees for drug use. However, opponents claim that this procedure is unfair because the tests themselves are not 100% reliable. Suppose a company uses a test that is 98% accurate—that is, it correctly identifies a person as a drug user or nonuser with probability .98—and to reduce the chance of error, each job applicant is required to take two tests. If the outcomes of the two tests on the same person are independent events, what are the probabilities of these events?

a. A nonuser fails both tests.

b. A drug user is detected (i.e., he or she fails at least one test).

c. A drug user passes both tests.

4.55 Grant Funding Suppose a group of research proposals was evaluated by a panel of experts to decide whether or not they were worthy of funding. When these same proposals were submitted to a second independent panel of experts, the decision to fund was reversed in 30% of the cases. If the probability that a proposal is judged worthy of funding by the first panel is .2, what are the probabilities of these events?

a. A worthy proposal is approved by both panels.

b. A worthy proposal is disapproved by both panels.

c. A worthy proposal is approved by one panel.

4.56 Drug Offenders A study of drug offenders who have been treated for drug abuse suggests that the likelihood of conviction within a 2-year period after treatment may depend on the offender's education. The proportions of the total number of cases that fall into four education/conviction categories are shown in the table below.

| | Status Within 2 Years After Treatment | | |
Education	Convicted	Not Convicted	Totals
10 Years or More	.10	.30	.40
9 Years or Less	.27	.33	.60
Totals	.37	.63	1.00

Suppose a single offender is selected from the treatment program. Here are the events of interest:

A: The offender has 10 or more years of education

B: The offender is convicted within 2 years after completion of treatment

Find the appropriate probabilities for these events:

a. A **b.** B **c.** $A \cap B$

d. $A \cup B$ **e.** A^c **f.** $(A \cup B)^c$

g. $(A \cap B)^c$ **h.** A given that B has occurred

i. B given that A has occurred

4.57 Use the probabilities of Exercise 4.56 to show that these equalities are true:

a. $P(A \cap B) = P(A)P(B|A)$

b. $P(A \cap B) = P(B)P(A|B)$

c. $P(A \cup B) = P(A) + P(B) - P(A \cap B)$

4.58 The Birthday Problem Two people enter a room and their birthdays (ignoring years) are recorded.

a. Identify the nature of the simple events in S.

b. What is the probability that the two people have a specific pair of birthdates?

c. Identify the simple events in event A: Both people have the same birthday.

d. Find $P(A)$. **e.** Find $P(A^c)$.

4.59 The Birthday Problem, continued If n people enter a room, find these probabilities:

A: None of the people have the same birthday

B: At least two of the people have the same birthday

Solve for

a. $n = 3$ **b.** $n = 4$

[NOTE: Surprisingly, $P(B)$ increases rapidly as n increases. For example, for $n = 20$, $P(B) = .411$; for $n = 40$, $P(B) = .891$.]

4.60 Starbucks or Peet's®? A college student frequents one of two coffee houses on campus, choosing Starbucks 70% of the time and Peet's 30% of the time. Regardless of where she goes, she buys a cafe mocha on 60% of her visits.

a. The next time she goes into a coffee house on campus, what is the probability that she goes to Starbucks and orders a cafe mocha?

b. Are the two events in part a independent? Explain.

c. If she goes into a coffee house and orders a cafe mocha, what is the probability that she is at Peet's?

d. What is the probability that she goes to Starbucks or orders a cafe mocha or both?

4.61 Inspection Lines A certain manufactured item is visually inspected by two different inspectors. When a defective item comes through the line, the probability that it gets by the first inspector is .1. Of those that get past the first inspector, the second inspector will "miss" 5 out of 10. What fraction of the defective items get by both inspectors?

4.62 Smoking and Cancer A survey of people in a given region showed that 20% were smokers. The probability of death due to lung cancer, given that a person smoked, was roughly 10 times the probability of death due to lung cancer, given that a person did not smoke. If the probability of death due to lung cancer in the region is .006, what is the probability of death due to lung cancer given that a person is a smoker?

4.63 Smoke Detectors A smoke-detector system uses two devices, A and B. If smoke is present, the probability that it will be detected by device A is .95; by device B, .98; and by both devices, .94.

a. If smoke is present, find the probability that the smoke will be detected by device A or device B or both devices.

b. Find the probability that the smoke will not be detected.

4.64 Plant Genetics In 1865, Gregor Mendel suggested a theory of inheritance based on the science of genetics. He identified heterozygous individuals for flower color that had two alleles (one r = recessive white color allele and one R = dominant red color allele). When these individuals were mated, 3/4 of the offspring were observed to have red flowers and 1/4 had white flowers. The table summarizes this mating; each parent gives one of its alleles to form the gene of the offspring.

	Parent 2	
Parent 1	r	R
r	rr	rR
R	Rr	RR

We assume that each parent is equally likely to give either of the two alleles and that, if either one or two of the alleles in a pair is dominant (R), the offspring will have red flowers.

a. What is the probability that an offspring in this mating has at least one dominant allele?

b. What is the probability that an offspring has at least one recessive allele?

c. What is the probability that an offspring has one recessive allele, given that the offspring has red flowers?

4.65 Soccer Injuries During the inaugural season of Major League Soccer in the United States, the medical teams documented 256 injuries that caused a loss of participation time to the player. The results of this investigation, reported in *The American Journal of Sports Medicine,* are shown in the table.[4]

Severity	Practice (*P*)	Game (*G*)	Total
Minor (*A*)	66	88	154
Moderate (*B*)	23	44	67
Major (*C*)	12	23	35
Total	101	155	256

If one individual is drawn at random from this group of 256 soccer players, find the following probabilities:

a. $P(A)$ **b.** $P(G)$ **c.** $P(A \cap G)$

d. $P(G|A)$ **e.** $P(G|B)$ **f.** $P(G|C)$

g. $P(C|P)$ **h.** $P(B^c)$

4.66 Choosing a Mate Men and women often disagree on how they think about selecting a mate. Suppose that a poll of 1000 individuals in their twenties gave the following responses to the question of whether it is more important for their future mate to be able to communicate their feelings (*F*) than it is for that person to make a good living (*G*).

	Feelings (*F*)	Good Living (*G*)	Totals
Men (*M*)	.35	.20	.55
Women (*W*)	.36	.09	.45
Totals	.71	.29	1.00

If an individual is selected at random from this group of 1000 individuals, calculate the following probabilities:

a. $P(F)$ **b.** $P(G)$ **c.** $P(F|M)$

d. $P(F|W)$ **e.** $P(M|F)$ **f.** $P(W|G)$

4.67 Kobe and Lamar Two stars of the *LA Lakers* are very different when it comes to making free throws. ESPN.com reports that Kobe Bryant makes 85% of his free throw shots while Lamar Odum makes 62% of his free throws.[5] Assume that the free throws are independent and that each player shoots two free throws during a team practice.

a. What is the probability that Kobe makes both of his free throws?

b. What is the probability that Lamar makes exactly one of his two free throws?

c. What is the probability that Kobe makes both of his free throws and Lamar makes neither of his?

4.68 Golfing Player *A* has entered a golf tournament but it is not certain whether player *B* will enter. Player *A* has probability 1/6 of winning the tournament if player *B* enters and probability 3/4 of winning if player *B* does not enter the tournament. If the probability that player *B* enters is 1/3, find the probability that player *A* wins the tournament.

 4.7

BAYES' RULE (OPTIONAL)

Colorblindness Let us reconsider the experiment involving colorblindness from Section 4.6. Notice that the two events

B: the person selected is a man

B^C: the person selected is a woman

taken together make up the sample space S, consisting of both men and women. Since colorblind people can be either male or female, the event A, which is that a person is colorblind, consists of both those simple events that are in A **and** B and those simple events that are in A **and** B^C. Since these two *intersections* are *mutually exclusive*, you can write the event A as

$$A = (A \cap B) \cup (A \cap B^C)$$

and

$$P(A) = P(A \cap B) + P(A \cap B^C)$$
$$= .04 + .002 = .042$$

Suppose now that the sample space can be partitioned into k subpopulations, $S_1, S_2, S_3, \ldots, S_k$, that, as in the colorblindness example, are **mutually exclusive and exhaustive;** that is, taken together they make up the entire sample space. In a similar way, you can express an event A as

$$A = (A \cap S_1) \cup (A \cap S_2) \cup (A \cap S_3) \cup \cdots \cup (A \cap S_k)$$

Then

$$P(A) = P(A \cap S_1) + P(A \cap S_2) + P(A \cap S_3) + \cdots + P(A \cap S_k)$$

This is illustrated for $k = 3$ in Figure 4.13.

FIGURE 4.13

Decomposition of event A

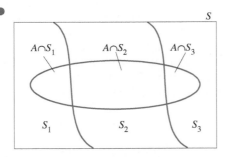

You can go one step further and use the Multiplication Rule to write $P(A \cap S_i)$ as $P(S_i)P(A|S_i)$, for $i = 1, 2, \ldots, k$. The result is known as the **Law of Total Probability.**

LAW OF TOTAL PROBABILITY

Given a set of events $S_1, S_2, S_3, \ldots, S_k$ that are mutually exclusive and exhaustive and an event A, the probability of the event A can be expressed as

$$P(A) = P(S_1)P(A|S_1) + P(S_2)P(A|S_2) + P(S_3)P(A|S_3) + \cdots + P(S_k)P(A|S_k)$$

EXAMPLE 4.23

Sneakers are no longer just for the young. In fact, most adults own multiple pairs of sneakers. Table 4.6 gives the fraction of U.S. adults 20 years of age and older who own five or more pairs of wearable sneakers, along with the fraction of the U.S. adult population 20 years or older in each of five age groups.[6] Use the Law of Total Probability to determine the unconditional probability of an adult 20 years and older owning five or more pairs of wearable sneakers.

TABLE 4.6 ● **Probability Table**

	Groups and Ages				
	G_1 20–24	G_2 25–34	G_3 35–49	G_4 50–64	G_5 ≥ 65
Fraction with ≥ 5 Pairs	.26	.20	.13	.18	.14
Fraction of U.S. Adults 20 and Older	.09	.18	.30	.25	.18

Solution   Let A be the event that a person chosen at random from the U.S. adult population 20 years of age and older owns five or more pairs of wearable sneakers. Let $G_1, G_2, \ldots, G_5$ represent the event that the person selected belongs to each of the five age groups, respectively. Since the five groups are *exhaustive,* you can write the event A as

$$A = (A \cap G_1) \cup (A \cap G_2) \cup (A \cap G_3) \cup (A \cap G_4) \cup (A \cap G_5)$$

Using the Law of Total Probability, you can find the probability of A as

$$P(A) = P(A \cap G_1) + P(A \cap G_2) + P(A \cap G_3) + P(A \cap G_4) + P(A \cap G_5)$$
$$= P(G_1)P(A|G_1) + P(G_2)P(A|G_2) + P(G_3)P(A|G_3)$$
$$+ P(G_4)P(A|G_4) + P(G_5)P(A|G_5)$$

From the probabilities in Table 4.6,

$$P(A) = (.09)(.26) + (.18)(.20) + (.30)(.13) + (.25)(.18) + (.18)(.14)$$
$$= .0234 + .0360 + .0390 + .0450 + .0252 = .1686$$

The *unconditional probability* that a person selected at random from the population of U.S. adults 20 years of age and older owns at least five pairs of wearable sneakers is about .17. Notice that the Law of Total Probability is a weighted average of the probabilities within each group, with weights .09, .18, .30, .25, and .18, which reflect the relative sizes of the groups.

Often you need to find the conditional probability of an event B, given that an event A has occurred. One such situation occurs in screening tests, which used to be associated primarily with medical diagnostic tests but are now finding applications in a variety of fields. Automatic test equipment is routinely used to inspect parts in high-volume production processes. Steroid testing of athletes, home pregnancy tests, and AIDS testing are some other applications. Screening tests are evaluated on the probability of a false negative or a false positive, and both of these are *conditional probabilities.*

A **false positive** is the event that the test is positive for a given condition, given that the person does not have the condition. A **false negative** is the event that the test is negative for a given condition, given that the person has the condition. You can evaluate these conditional probabilities using a formula derived by the probabilist Thomas Bayes.

The experiment involves selecting a sample from one of k subpopulations that are mutually exclusive and exhaustive. Each of these subpopulations, denoted by $S_1, S_2, \ldots, S_k$, has a selection probability $P(S_1), P(S_2), P(S_3), \ldots, P(S_k)$, called *prior probabilities.* An event A is observed in the selection. What is the probability that the sample came from subpopulation S_i, given that A has occurred?

You know from Section 4.6 that $P(S_i|A) = [P(A \cap S_i)]/P(A)$, which can be rewritten as $P(S_i|A) = [P(S_i)P(A|S_i)]/P(A)$. Using the Law of Total Probability to rewrite $P(A)$, you have

$$P(S_i|A) = \frac{P(S_i)P(A|S_i)}{P(S_1)P(A|S_1) + P(S_2)P(A|S_2) + P(S_3)P(A|S_3) + \cdots + P(S_k)P(A|S_k)}$$

These new probabilities are often referred to as *posterior probabilities*—that is, probabilities of the subpopulations (also called *states of nature*) that have been updated after observing the sample information contained in the event A. Bayes suggested that if the prior probabilities are unknown, they can be taken to be $1/k$, which implies that each of the events S_1 through S_k is equally likely.

BAYES' RULE

Let $S_1, S_2, \ldots, S_k$ represent k mutually exclusive and exhaustive subpopulations with prior probabilities $P(S_1), P(S_2), \ldots, P(S_k)$. If an event A occurs, the posterior probability of S_i given A is the conditional probability

$$P(S_i|A) = \frac{P(S_i)P(A|S_i)}{\displaystyle\sum_{j=1}^{k} P(S_j)P(A|S_j)}$$

for $i = 1, 2, \ldots, k$.

EXAMPLE (4.24)

Refer to Example 4.23. Find the probability that the person selected was 65 years of age or older, given that the person owned at least five pairs of wearable sneakers.

Solution   You need to find the conditional probability given by

$$P(G_5|A) = \frac{P(A \cap G_5)}{P(A)}$$

You have already calculated $P(A) = .1686$ using the Law of Total Probability. Therefore,

$P(G_5|A) =$

$$\frac{P(G_5)P(A|G_5)}{P(G_1)P(A|G_1) + P(G_2)P(A|G_2) + P(G_3)P(A|G_3) + P(G_4P(A|G_4) + P(G_5)P(A|G_5)}$$

$$= \frac{(.18)(.14)}{(.09)(.26) + (.18)(.20) + (.30)(.13) + (.25)(.18) + (.18)(.14)}$$

$$= \frac{.0252}{.1686} = .1495$$

In this case, the posterior probability of .15 is somewhat larger than the prior probability of .13 (from Table 4.6). This group *a priori* was the second smallest, and only a small proportion of this segment had five or more pairs of wearable sneakers.

What is the posterior probability for those aged 35 to 49? For this group of adults, we have

$$P(G_3|A) = \frac{(.30)(.13)}{(.09)(.26) + (.18)(.20) + (.30)(.13) + (.25)(.18) + (.18)(.14)}$$

$$= \frac{.0390}{.1686} = .2313$$

This posterior probability of .23 is substantially less than the prior probability of .30. In effect, this group was *a priori* the largest segment of the population sampled, but at the same time, the proportion of individuals in this group who had at least five pairs of wearable sneakers was the smallest of any of the groups. These two facts taken together cause a downward adjustment of almost one-third in the *a priori* probability of .30.

 4.7 **EXERCISES**

BASIC TECHNIQUES

4.69 Bayes' Rule A sample is selected from one of two populations, S_1 and S_2, with probabilities $P(S_1) = .7$ and $P(S_2) = .3$. If the sample has been selected from S_1, the probability of observing an event A is $P(A|S_1) = .2$. Similarly, if the sample has been selected from S_2, the probability of observing A is $P(A|S_2) = .3$.

a. If a sample is randomly selected from one of the two populations, what is the probability that event A occurs?

b. If the sample is randomly selected and event A is observed, what is the probability that the sample was selected from population S_1? From population S_2?

4.70 Bayes' Rule II If an experiment is conducted, one and only one of three mutually exclusive events S_1, S_2, and S_3 can occur, with these probabilities:

$$P(S_1) = .2 \qquad P(S_2) = .5 \qquad P(S_3) = .3$$

The probabilities of a fourth event A occurring, given that event S_1, S_2, or S_3 occurs, are

$$P(A|S_1) = .2 \quad P(A|S_2) = .1 \quad P(A|S_3) = .3$$

If event A is observed, find $P(S_1|A)$, $P(S_2|A)$, and $P(S_3|A)$.

4.71 Law of Total Probability A population can be divided into two subgroups that occur with probabilities 60% and 40%, respectively. An event A occurs 30% of the time in the first subgroup and 50% of the time in the second subgroup. What is the unconditional probability of the event A, regardless of which subgroup it comes from?

APPLICATIONS

4.72 Violent Crime City crime records show that 20% of all crimes are violent and 80% are nonviolent, involving theft, forgery, and so on. Ninety percent of violent crimes are reported versus 70% of nonviolent crimes.

a. What is the overall reporting rate for crimes in the city?

b. If a crime in progress is reported to the police, what is the probability that the crime is violent? What is the probability that it is nonviolent?

c. Refer to part b. If a crime in progress is reported to the police, why is it more likely that it is a nonviolent crime? Wouldn't violent crimes be more likely to be reported? Can you explain these results?

4.73 Worker Error A worker-operated machine produces a defective item with probability .01 if the worker follows the machine's operating instructions exactly, and with probability .03 if he does not. If the worker follows the instructions 90% of the time, what proportion of all items produced by the machine will be defective?

4.74 Airport Security Suppose that, in a particular city, airport A handles 50% of all airline traffic, and airports B and C handle 30% and 20%, respectively.

The detection rates for weapons at the three airports are .9, .8, and .85, respectively. If a passenger at one of the airports is found to be carrying a weapon through the boarding gate, what is the probability that the passenger is using airport A? Airport C?

4.75 Football Strategies A particular football team is known to run 30% of its plays to the left and 70% to the right. A linebacker on an opposing team notes that the right guard shifts his stance most of the time (80%) when plays go to the right and that he uses a balanced stance the remainder of the time. When plays go to the left, the guard takes a balanced stance 90% of the time and the shift stance the remaining 10%. On a particular play, the linebacker notes that the guard takes a balanced stance.

a. What is the probability that the play will go to the left?
b. What is the probability that the play will go to the right?
c. If you were the linebacker, which direction would you prepare to defend if you saw the balanced stance?

4.76 No Pass, No Play Under the "no pass, no play" rule for athletes, an athlete who fails a course is disqualified from participating in sports activities during the next grading period. Suppose the probability that an athlete who has not previously been disqualified will be disqualified is .15 and the probability that an athlete who has been disqualified will be disqualified again in the next time period is .5. If 30% of the athletes have been disqualified before, what is the unconditional probability that an athlete will be disqualified during the next grading period?

4.77 Medical Diagnostics Medical case histories indicate that different illnesses may produce identical symptoms. Suppose a particular set of symptoms, which we will denote as event H, occurs only when any one of three illnesses—A, B, or C—occurs. (For the sake of simplicity, we will assume that illnesses A, B, and C are mutually exclusive.) Studies show these probabilities of getting the three illnesses:

$$P(A) = .01$$
$$P(B) = .005$$
$$P(C) = .02$$

The probabilities of developing the symptoms H, given a specific illness, are

$$P(H|A) = .90$$
$$P(H|B) = .95$$
$$P(H|C) = .75$$

Assuming that an ill person shows the symptoms H, what is the probability that the person has illness A?

4.78 Cheating on Your Taxes? Suppose 5% of all people filing the long income tax form seek deductions that they know are illegal, and an additional 2% incorrectly list deductions because they are unfamiliar with income tax regulations. Of the 5% who are guilty of cheating, 80% will deny knowledge of the error if confronted by an investigator. If the filer of the long form is confronted with an unwarranted deduction and he or she denies the knowledge of the error, what is the probability that he or she is guilty?

4.79 Screening Tests Suppose that a certain disease is present in 10% of the population, and that there is a screening test designed to detect this disease if present. The test does not always work perfectly. Sometimes the test is negative when the disease is present, and sometimes it is positive when the disease is absent. The table below shows the proportion of times that the test produces various results.

	Test Is Positive (P)	Test Is Negative (N)
Disease Present (D)	.08	.22
Disease Absent (D^c)	.05	.85

a. Find the following probabilities from the table: $P(D)$, $P(D^c)$, $P(N|D^c)$, $P(N|D)$.
b. Use Bayes' Rule and the results of part a to find $P(D|N)$.
c. Use the definition of conditional probability to find $P(D|N)$. (Your answer should be the same as the answer to part b.)
d. Find the probability of a false positive, that the test is positive, given that the person is disease-free.
e. Find the probability of a false negative, that the test is negative, given that the person has the disease.
f. Are either of the probabilities in parts d or e large enough that you would be concerned about the reliability of this screening method? Explain.

DISCRETE RANDOM VARIABLES AND THEIR PROBABILITY DISTRIBUTIONS

4.8

In Chapter 1, *variables* were defined as characteristics that change or vary over time and/or for different individuals or objects under consideration. *Quantitative variables* generate numerical data, whereas *qualitative variables* generate categorical data. However, even qualitative variables can generate numerical data if the categories are numerically coded to form a scale. For example, if you toss a single coin, the qualitative outcome could be recorded as "0" if a head and "1" if a tail.

Random Variables

A numerically valued variable x will vary or change depending on the particular outcome of the experiment being measured. For example, suppose you toss a die and measure x, the number observed on the upper face. The variable x can take on any of six values—1, 2, 3, 4, 5, 6—depending on the *random* outcome of the experiment. For this reason, we refer to the variable x as a **random variable.**

Definition A variable x is a **random variable** if the value that it assumes, corresponding to the outcome of an experiment, is a chance or random event.

You can think of many examples of random variables:

- x = Number of defects on a *randomly selected* piece of furniture
- x = SAT score for a *randomly selected* college applicant
- x = Number of telephone calls received by a crisis intervention hotline during a *randomly selected* time period

As in Chapter 1, quantitative random variables are classified as either *discrete* or *continuous,* according to the values that x can assume. It is important to distinguish between discrete and continuous random variables because different techniques are used to describe their distributions. We focus on discrete random variables in the remainder of this chapter; continuous random variables are the subject of Chapter 6.

Probability Distributions

In Chapters 1 and 2, you learned how to construct the *relative frequency distribution* for a set of numerical measurements on a variable x. The distribution gave this information about x:

- What values of x occurred
- How often each value of x occurred

You also learned how to use the mean and standard deviation to measure the center and variability of this data set.

In this chapter, we defined *probability* as the limiting value of the relative frequency as the experiment is repeated over and over again. Now we define the **probability distribution** for a random variable x as the *relative frequency distribution* constructed for the entire population of measurements.

Definition The **probability distribution** for a discrete random variable is a formula, table, or graph that gives the possible values of x, and the probability $p(x)$ associated with each value of x.

The values of x represent mutually exclusive numerical events. Summing $p(x)$ over all values of x is equivalent to adding the probabilities of all simple events and therefore equals 1.

REQUIREMENTS FOR A DISCRETE PROBABILITY DISTRIBUTION

- $0 \leq p(x) \leq 1$
- $\Sigma\, p(x) = 1$

EXAMPLE 4.25

Toss two fair coins and let x equal the number of heads observed. Find the probability distribution for x.

Solution   The simple events for this experiment with their respective probabilities are listed in Table 4.7. Since $E_1 = $ HH results in two heads, this simple event results in the value $x = 2$. Similarly, the value $x = 1$ is assigned to E_2, and so on.

TABLE 4.7   ● **Simple Events and Probabilities in Tossing Two Coins**

Simple Event	Coin 1	Coin 2	$P(E_i)$	x
E_1	H	H	1/4	2
E_2	H	T	1/4	1
E_3	T	H	1/4	1
E_4	T	T	1/4	0

For each value of x, you can calculate $p(x)$ by adding the probabilities of the simple events in that event. For example, when $x = 0$, simple event E_4 occurs, so that

$$p(0) = P(E_4) = \frac{1}{4}$$

and when $x = 1$,

$$p(1) = P(E_2) + P(E_3) = \frac{1}{2}$$

The values of x and their respective probabilities, $p(x)$, are listed in Table 4.8. Notice that the probabilities add to 1.

TABLE 4.8   ● **Probability Distribution for x (x = Number of Heads)**

x	Simple Events in x	$p(x)$
0	E_4	1/4
1	E_2, E_3	1/2
2	E_1	1/4
		$\Sigma\, p(x) = 1$

The probability distribution in Table 4.8 can be graphed using the methods of Section 1.5 to form the **probability histogram** in Figure 4.14.[†] The three values of the random variable x are located on the horizontal axis, and the probabilities $p(x)$ are

---

[†]The probability distribution in Table 4.8 can also be presented using a formula, which is given in Section 5.2.

located on the vertical axis (replacing the relative frequencies used in Chapter 1). Since the width of each bar is 1, the area under the bar is the probability of observing the particular value of x and the total area equals 1.

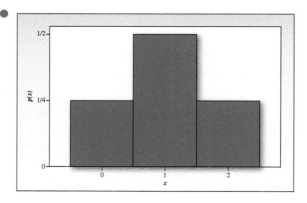

The Mean and Standard Deviation for a Discrete Random Variable

The probability distribution for a discrete random variable looks very similar to the relative frequency distribution discussed in Chapter 1. The difference is that the relative frequency distribution describes a *sample* of n measurements, whereas the probability distribution is constructed as a model for the *entire population* of measurements. Just as the mean $\bar{x}$ and the standard deviation s measured the center and spread of the sample data, you can calculate similar measures to describe the center and spread of the population.

The population mean, which measures the average value of x in the population, is also called the **expected value** of the random variable x and is written as $E(x)$. It is the value that you would *expect* to observe on *average* if the experiment is repeated over and over again. The formula for calculating the population mean is easier to understand by example. Toss those two fair coins again, and let x be the number of heads observed. We constructed this probability distribution for x:

x	0	1	2
$p(x)$	1/4	1/2	1/4

Suppose the experiment is repeated a large number of times—say, $n = 4,000,000$ times. Intuitively, you would expect to observe approximately 1 million zeros, 2 million ones, and 1 million twos. Then the average value of x would equal

$$\frac{\text{Sum of measurements}}{n} = \frac{1,000,000(0) + 2,000,000(1) + 1,000,000(2)}{4,000,000}$$

$$= \left(\frac{1}{4}\right)(0) + \left(\frac{1}{2}\right)(1) + \left(\frac{1}{4}\right)(2)$$

Note that the first term in this sum is $(0)p(0)$, the second is equal to $(1)p(1)$, and the third is $(2)p(2)$. The average value of x, then, is

$$\Sigma xp(x) = 0 + \frac{1}{2} + \frac{2}{4} = 1$$

This result provides some intuitive justification for the definition of the expected value of a discrete random variable x.

Definition   Let x be a discrete random variable with probability distribution $p(x)$. The mean or **expected value of x** is given as

$$\mu = E(x) = \Sigma xp(x)$$

where the elements are summed over all values of the random variable x.

We could use a similar argument to justify the formulas for the **population variance** σ^2 and the **population standard deviation** σ. These numerical measures describe the spread or variability of the random variable using the "average" or "expected value" of $(x - \mu)^2$, the squared deviations of the x-values from their mean μ.

Definition   Let x be a discrete random variable with probability distribution $p(x)$ and mean μ. The **variance of x** is

$$\sigma^2 = E[(x - \mu)^2] = \Sigma(x - \mu)^2 p(x)$$

where the summation is over all values of the random variable x.[†]

Definition   The **standard deviation σ of a random variable x** is equal to the positive square root of its variance.

EXAMPLE   4.26

An electronics store sells a particular model of a laptop computer. There are only four laptops in stock, and the manager wonders what today's demand for this particular model will be. She learns from the marketing department that the probability distribution for x, the daily demand for the laptop, is as shown in the table. Find the mean, variance, and standard deviation of x. Is it likely that five or more customers will want to buy the laptop today?

x	0	1	2	3	4	5
$p(x)$	.10	.40	.20	.15	.10	.05

Solution   Table 4.9 shows the values of x and $p(x)$, along with the individual terms used in the formulas for μ and σ^2. The sum of the values in the third column is

$$\mu = \Sigma xp(x) = (0)(.10) + (1)(.40) + \cdots + (5)(.05) = 1.90$$

while the sum of the values in the fifth column is

$$\sigma^2 = \Sigma(x - \mu)^2 p(x)$$
$$= (0 - 1.9)^2(.10) + (1 - 1.9)^2(.40) + \cdots + (5 - 1.9)^2(.05) = 1.79$$

and

$$\sigma = \sqrt{\sigma^2} = \sqrt{1.79} = 1.34$$

[†]It can be shown (proof omitted) that $\sigma^2 = \Sigma(x - \mu)^2 p(x) = \Sigma x^2 p(x) - \mu^2$. This result is analogous to the computing formula for the sum of squares of deviations given in Chapter 2.

TABLE 4.9 ● **Calculations for Example 4.26**

x	$p(x)$	$xp(x)$	$(x - \mu)^2$	$(x - \mu)^2 p(x)$
0	.10	.00	3.61	.361
1	.40	.40	.81	.324
2	.20	.40	.01	.002
3	.15	.45	1.21	.1815
4	.10	.40	4.41	.441
5	.05	.25	9.61	.4805
Totals	1.00	$\mu = 1.90$		$\sigma^2 = 1.79$

The graph of the probability distribution is shown in Figure 4.15. Since the distribution is approximately mound-shaped, approximately 95% of all measurements should lie within *two* standard deviations of the mean—that is,

$$\mu \pm 2\sigma \Rightarrow 1.90 \pm 2(1.34) \qquad \text{or } -.78 \text{ to } 4.58$$

Since $x = 5$ lies outside this interval, you can say it is unlikely that five or more customers will want to buy the laptop today. In fact, $P(x \geq 5)$ is exactly .05, or 1 time in 20.

FIGURE 4.15 ●

Probability distribution for Example 4.26

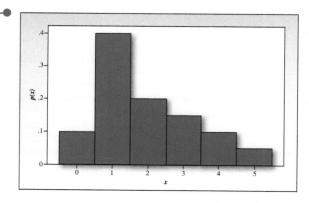

EXAMPLE **4.27**

In a lottery conducted to benefit a local charity, 8000 tickets are to be sold at $10 each. The prize is a $24,000 compact car. If you purchase two tickets, what is your expected gain?

Solution Your gain x may take one of two values. You will either lose $20 (i.e., your "gain" will be $-$20) or win $23,980, with probabilities 7998/8000 and 2/8000, respectively. The probability distribution for the gain x is shown in the table:

x	$p(x)$
$-$20	7998/8000
$23,980	2/8000

The expected gain will be

$$\mu = \Sigma \, xp(x)$$

$$= (-\$20)\left(\frac{7998}{8000}\right) + (\$23,980)\left(\frac{2}{8000}\right) = -\$14$$

Recall that the expected value of x is the average of the theoretical population that would result if the lottery were repeated an infinitely large number of times. If this were done, your average or expected gain per lottery ticket would be a loss of $14.

EXAMPLE (4.28) Determine the yearly premium for a $10,000 insurance policy covering an event that, over a long period of time, has occurred at the rate of 2 times in 100. Let x equal the yearly financial gain to the insurance company resulting from the sale of the policy, and let C equal the unknown yearly premium. Calculate the value of C such that the expected gain $E(x)$ will equal zero. Then C is the premium required to break even. To this, the company would add administrative costs and profit.

Solution The first step in the solution is to determine the values that the gain x may take and then to determine $p(x)$. If the event does not occur during the year, the insurance company will gain the premium of $x = C$ dollars. If the event does occur, the gain will be negative; that is, the company will lose $10,000 less the premium of C dollars already collected. Then $x = -(10,000 - C)$ dollars. The probabilities associated with these two values of x are 98/100 and 2/100, respectively. The probability distribution for the gain is shown in the table:

$x = $ Gain	$p(x)$
C	98/100
$-(10,000 - C)$	2/100

Since the company wants the insurance premium C such that, in the long run (for many similar policies), the mean gain will equal zero, you can set the expected value of x equal to zero and solve for C. Then

$$\mu = E(x) = \Sigma x p(x)$$

$$= C\left(\frac{98}{100}\right) + [-10,000 + C]\left(\frac{2}{100}\right) = 0$$

or

$$\frac{98}{100}C + \frac{2}{100}C - 200 = 0$$

Solving this equation for C, you obtain $C = 200. Therefore, if the insurance company charged a yearly premium of $200, the average gain calculated for a large number of similar policies would equal zero. The actual premium would equal $200 plus administrative costs and profit.

The method for calculating the expected value of x for a continuous random variable is similar to what you have done, but in practice it involves the use of calculus. Nevertheless, the basic results concerning expectations are the same for continuous and discrete random variables. For example, regardless of whether x is continuous or discrete, $\mu = E(x)$ and $\sigma^2 = E[(x - \mu)^2]$.

4.8 EXERCISES

BASIC TECHNIQUES

4.80 Discrete or Continuous? Identify the following as discrete or continuous random variables:

a. Total number of points scored in a football game

b. Shelf life of a particular drug

c. Height of the ocean's tide at a given location

d. Length of a 2-year-old black bass

e. Number of aircraft near-collisions in a year

4.81 Discrete or Continuous? II Identify the following as discrete or continuous random variables:

a. Increase in length of life attained by a cancer patient as a result of surgery

b. Tensile breaking strength (in pounds per square inch) of 1-inch-diameter steel cable

c. Number of deer killed per year in a state wildlife preserve

d. Number of overdue accounts in a department store at a particular time

e. Your blood pressure

4.82 Probability Distribution I A random variable x has this probability distribution:

x	0	1	2	3	4	5
$p(x)$	.1	.3	.4	.1	?	.05

a. Find $p(4)$.

b. Construct a probability histogram to describe $p(x)$.

c. Find μ, σ^2, and σ.

d. Locate the interval $\mu \pm 2\sigma$ on the x-axis of the histogram. What is the probability that x will fall into this interval?

e. If you were to select a very large number of values of x from the population, would most fall into the interval $\mu \pm 2\sigma$? Explain.

4.83 Probability Distribution II A random variable x can assume five values: 0, 1, 2, 3, 4. A portion of the probability distribution is shown here:

x	0	1	2	3	4
$p(x)$	.1	.3	.3	?	.1

a. Find $p(3)$.

b. Construct a probability histogram for $p(x)$.

c. Calculate the population mean, variance, and standard deviation.

d. What is the probability that x is greater than 2?

e. What is the probability that x is 3 or less?

4.84 Dice Let x equal the number observed on the throw of a single balanced die.

a. Find and graph the probability distribution for x.

b. What is the average or expected value of x?

c. What is the standard deviation of x?

d. Locate the interval $\mu \pm 2\sigma$ on the x-axis of the graph in part a. What proportion of all the measurements would fall into this range?

4.85 Grocery Visits Let x represent the number of times a customer visits a grocery store in a 1-week period. Assume this is the probability distribution of x:

x	0	1	2	3
$p(x)$	.1	.4	.4	.1

Find the expected value of x, the average number of times a customer visits the store.

4.86 If you toss a pair of dice, the sum T of the numbers appearing on the upper faces of the dice can assume the value of an integer in the interval $2 \le T \le 12$.

a. Find the probability distribution for T. Display this probability distribution in a table.

b. Construct a probability histogram for $P(T)$. How would you describe the shape of this distribution?

APPLICATIONS

4.87 RU Texting? The proportion of adults (18 years or more) who admit to texting while driving is 47%.[7] Suppose you randomly select three adult drivers and ask if they text while driving.

a. Find the probability distribution for x, the number of drivers in the sample who admit to texting while driving.

b. Construct a probability histogram for $p(x)$.

c. What is the probability that exactly one of the three drivers texts while driving?

d. What are the population mean and standard deviation for the random variable x?

4.88 Which Key Fits? A key ring contains four office keys that are identical in appearance, but only one will open your office door. Suppose you randomly select one key and try it. If it does not fit, you randomly select one of the three remaining keys. If it does not fit, you randomly select one of the last two. Each different sequence that could occur in selecting the keys represents one of a set of equiprobable simple events.

a. List the simple events in S and assign probabilities to the simple events.

b. Let x equal the number of keys that you try before you find the one that opens the door $(x = 1, 2, 3, 4)$. Then assign the appropriate value of x to each simple event.

c. Calculate the values of $p(x)$ and display them in a table.

d. Construct a probability histogram for $p(x)$.

4.89 Gender Bias? A company has five applicants for two positions: two women and three men. Suppose that the five applicants are equally qualified and that no preference is given for choosing either gender. Let x equal the number of women chosen to fill the two positions.

a. Find $p(x)$.

b. Construct a probability histogram for x.

4.90 Defective Equipment A piece of electronic equipment contains six computer chips, two of which are defective. Three chips are selected at random, removed from the piece of equipment, and inspected. Let x equal the number of defectives observed, where $x = 0$, 1, or 2. Find the probability distribution for x. Express the results graphically as a probability histogram.

4.91 Drilling Oil Wells Past experience has shown that, on the average, only 1 in 10 wells drilled hits oil. Let x be the number of drillings until the first success (oil is struck). Assume that the drillings represent independent events.

a. Find $p(1)$, $p(2)$, and $p(3)$.

b. Give a formula for $p(x)$.

c. Graph $p(x)$.

4.92 Tennis, Anyone? Two tennis professionals, A and B, are scheduled to play a match; the winner is the first player to win three sets in a total that cannot exceed five sets. The event that A wins any one set is independent of the event that A wins any other, and the probability that A wins any one set is equal to .6. Let x equal the total number of sets in the match; that is, $x = 3$, 4, or 5. Find $p(x)$.

4.93 Tennis, again In Exercise 4.92 you found the probability distribution for x, the number of sets required to play a best-of-five-sets match, given that the probability that A wins any one set—call this $P(A)$—is .6.

a. Find the expected number of sets required to complete the match for $P(A) = .6$.

b. Find the expected number of sets required to complete the match when the players are of equal ability—that is, $P(A) = .5$.

c. Find the expected number of sets required to complete the match when the players differ greatly in ability—that is, say, $P(A) = .9$.

d. What is the relationship between $P(A)$ and $E(x)$, the expected number of sets required to complete the match?

4.94 The PGA One professional golfer plays best on short-distance holes. Experience has shown that the numbers x of shots required for 3-, 4-, and 5-par holes have the probability distributions shown in the table:

Par-3 Holes		Par-4 Holes		Par-5 Holes	
x	$p(x)$	x	$p(x)$	x	$p(x)$
2	.12	3	.14	4	.04
3	.80	4	.80	5	.80
4	.06	5	.04	6	.12
5	.02	6	.02	7	.04

What is the golfer's expected score on these holes?

a. A par-3 hole

b. A par-4 hole

c. A par-5 hole

4.95 Insuring Your Diamonds You can insure a $50,000 diamond for its total value by paying a premium of D dollars. If the probability of loss in a given year is estimated to be .01, what premium should the insurance company charge if it wants the expected gain to equal $1000?

4.96 FDA Testing The maximum patent life for a new drug is 17 years. Subtracting the length of time required by the FDA for testing and approval of the drug provides the actual patent life of the drug—that is, the length of time that a company has to recover research and development costs and make a profit. Suppose the distribution of the lengths of patent life for new drugs is as shown here:

Years, x	3	4	5	6	7	8
$p(x)$	.03	.05	.07	.10	.14	.20

Years, x	9	10	11	12	13
$p(x)$	.18	.12	.07	.03	.01

a. Find the expected number of years of patent life for a new drug.

b. Find the standard deviation of x.

c. Find the probability that x falls into the interval $\mu \pm 2\sigma$.

4.97 Coffee Breaks Most coffee drinkers take a little time each day for their favorite beverage, and many take more than one coffee break every day. The table below, adapted from a Snapshot in *USA Today*, shows the probability distribution for x, the number of daily coffee breaks taken per day by coffee drinkers.[8]

x	0	1	2	3	4	5
$p(x)$	.28	.37	.17	.12	.05	.01

a. What is the probability that a randomly selected coffee drinker would take no coffee breaks during the day?

b. What is the probability that a randomly selected coffee drinker would take more than two coffee breaks during the day?

c. Calculate the mean and standard deviation for the random variable x.

d. Find the probability that x falls into the interval $\mu \pm 2\sigma$.

4.98 Shipping Charges From experience, a shipping company knows that the cost of delivering a small package within 24 hours is $14.80. The company charges $15.50 for shipment but guarantees to refund the charge if delivery is not made within

24 hours. If the company fails to deliver only 2% of its packages within the 24-hour period, what is the expected gain per package?

4.99 Actuaries A CEO is considering buying an insurance policy to cover possible losses incurred by marketing a new product. If the product is a complete failure, a loss of $800,000 would be incurred; if it is only moderately successful, a loss of $250,000 would be incurred. Insurance actuaries have determined that the probabilities that the product will be a failure or only moderately successful are .01 and .05, respectively. Assuming that the CEO is willing to ignore all other possible losses, what premium should the insurance company charge for a policy in order to break even?

CHAPTER REVIEW

Key Concepts and Formulas

I. Experiments and the Sample Space

1. Experiments, events, mutually exclusive events, simple events

2. The sample space

3. Venn diagrams, tree diagrams, probability tables

II. Probabilities

1. Relative frequency definition of probability

2. Properties of probabilities

   a. Each probability lies between 0 and 1.

   b. Sum of all simple-event probabilities equals 1.

3. $P(A)$, the sum of the probabilities for all simple events in A

III. Counting Rules

1. mn Rule; extended mn Rule

2. Permutations: $P_r^n = \dfrac{n!}{(n-r)!}$

3. Combinations: $C_r^n = \dfrac{n!}{r!(n-r)!}$

IV. Event Relations

1. Unions and intersections

2. Events

   a. Disjoint or mutually exclusive:
      $P(A \cap B) = 0$

   b. Complementary: $P(A) = 1 - P(A^c)$

3. Conditional probability: $P(A|B) = \dfrac{P(A \cap B)}{P(B)}$

4. Independent and dependent events

5. Addition Rule:
   $P(A \cup B) = P(A) + P(B) - P(A \cap B)$

6. Multiplication Rule: $P(A \cap B) = P(A)P(B|A)$

7. Law of Total Probability

8. Bayes' Rule

V. Discrete Random Variables and Probability Distributions

1. Random variables, discrete and continuous

2. Properties of probability distributions

   a. $0 \le p(x) \le 1$

   b. $\Sigma p(x) = 1$

3. Mean or expected value of a discrete random variable: $\mu = \Sigma x p(x)$

4. Variance and standard deviation of a discrete random variable: $\sigma^2 = \Sigma (x - \mu)^2 p(x)$ and $\sigma = \sqrt{\sigma^2}$

TECHNOLOGY TODAY

Discrete Probability Distributions in *MS Excel*

Although *MS Excel* cannot help you solve the types of general probability problems presented in this chapter, it is useful for calculating the mean, variance, and standard deviation of the random variable x. In Chapters 5 and 6, we will use *Excel* to calculate exact probabilities for three special cases: the binomial, the Poisson, and the normal random variables.

EXAMPLE Suppose you have this general discrete probability distribution:

x	0	1	3	5
$p(x)$	.25	.35	.25	.15

1. Enter the values of x and $p(x)$ into columns A and B of a new *Excel* spreadsheet. Then create two columns—column C (named "$x*p(x)$") and column D (named "$xsq*p(x)$"). You can now use the **Function** f_x command to fill in columns C and D. In *Excel*, an "equals" sign indicates that you are going to type an equation (or insert a function). Hence, in cell C2, we type:

   =A2*B2

   Then, to copy this formula to the remaining three cells in column C, simply click on cell C2, grab the square in the lower right corner of the cell with your mouse, and drag to cell C5 to copy.

2. To fill in column D, type the following equation into cell D2: = A2*A2*B2 and then copy this formula to the remaining three cells in column D as explained above.

3. Finally, use the first three cells in column F to type the names "Mean," "Variance," and "Std Dev." Again, use the equation (or insert function) commands. In cell G1 (Mean), type: =SUM(C2:C5); in cell G2 (Variance), type: =SUM(D2:D5)-(G1*G1); and in cell G3 (Standard Deviation), type: =SQRT(G2). The resulting spreadsheet is shown in Figure 4.16.

FIGURE 4.16

	A	B	C	D	E	F	G
1	x	p(x)	x*p(x)	xsq*p(x)		Mean	1.85
2	0	0.25	0	0		Variance	2.9275
3	1	0.35	0.35	0.35		Std Dev	1.710994
4	3	0.25	0.75	2.25			
5	5	0.15	0.75	3.75			

Discrete Probability Distributions in *MINITAB*

Although *MINITAB* cannot help you solve the types of general probability problems presented in this chapter, it is useful for graphing the probability distribution $p(x)$ for a general discrete random variable x when the probabilities are known, and for calculating the mean, variance, and standard deviation of the random variable x. In Chapters 5 and 6, we will use *MINITAB* to calculate exact probabilities for three special cases: the binomial, the Poisson, and the normal random variables.

EXAMPLE (4.30)

Suppose you have this general probability distribution:

x	0	1	3	5
$p(x)$	.25	.35	.25	.15

1. Enter the values of x and $p(x)$ into columns C1 and C2 of a new *MINITAB* worksheet. In the gray boxes just below C3, C4, and C5, respectively, type the names "Mean," "Variance," and "Std Dev." You can now use the **Calc ▶ Calculator** command to calculate μ, σ^2, and σ and to store the results in columns C3–C5 of the worksheet.

2. Use the same approach for all three parameters. In the Calculator dialog box, select "Mean" as the column in which to store μ. In the Expression box, use the Functions list, the calculator keys, and the variables list on the left to highlight, select, and create the expression for the mean (see Figure 4.17(a)):

SUM('x'*'p(x)')

MINITAB will multiply each row element in C1 times the corresponding row element in C2, sum the resulting products, and store the result in C3! You can check the result by hand if you like.

3. The formulas for the variance and standard deviation are selected in a similar way:

Variance: SUM(('x' − 'Mean')**2*'p(x)')
Std Dev: SQRT('Variance')

4. To see the tabular form of the probability distribution and the three parameters, use **Data ▶ Display Data** and select all five columns. Click **OK** and the results will be displayed in the Session window, as shown in Figure 4.17(b).

FIGURE 4.17

(a)

(b)

The probability histogram can be plotted using the *MINITAB* command **Graph ▶ Scatterplot ▶ Simple ▶ OK.** In the Scatterplot dialog box, select 'p(x)' for **Y variables** and 'x' for **X variables.** To display the discrete probability bars, click on **Data View,** uncheck the box marked "Symbols," and check the box marked "Project Lines." Click **OK** twice to see the plot. You will see a single straight line projected at each of the four values of x. If you want the plot to look more like the discrete probability histograms in Section 4.8, position your cursor on one of the lines, right-click the mouse and choose "Edit Project Lines." Under the "Attributes" tab, select **Custom** and change the line size to **75**. Click **OK**. If the bar width is not satisfactory, you can readjust the line size. Finally, right-click on the X-axis, choose "Edit X Scale" and select **−.5** and **5.5** for the minimum and maximum **Scale Ranges**. Click **OK**. The probability histogram is shown in Figure 4.18.

Locate the mean on the graph. Is it at the center of the distribution? If you mark off two standard deviations on either side of the mean, do most of the possible values of x fall into this interval?

FIGURE 4.18

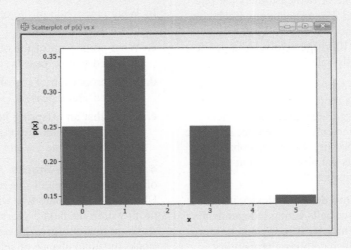

Supplementary Exercises

Starred (*) exercises are optional.

4.100 Playing the Slots A slot machine has three slots; each will show a cherry, a lemon, a star, or a bar when spun. The player wins if all three slots show the same three items. If each of the four items is equally likely to appear on a given spin, what is your probability of winning?

4.101 Whistle Blowers Although there is legal protection for "whistle blowers"—employees who report illegal or unethical activities in the workplace—it has been reported that approximately 23% of those who reported fraud suffered reprisals such as demotion or poor performance ratings. Suppose the probability that an employee will fail to report a case of fraud is .69. Find the probability that an employee who observes a case of fraud will report it and will subsequently suffer some form of reprisal.

4.102 Aspirin Two cold tablets are unintentionally placed in a box containing two aspirin tablets. The

four tablets are identical in appearance. One tablet is selected at random from the box and is swallowed by the first patient. A tablet is then selected at random from the three remaining tablets and is swallowed by the second patient. Define the following events as specific collections of simple events:

a. The sample space S

b. The event A that the first patient obtained a cold tablet

c. The event B that exactly one of the two patients obtained a cold tablet

d. The event C that neither patient obtained a cold tablet

4.103 Refer to Exercise 4.102. By summing the probabilities of simple events, find $P(A)$, $P(B)$, $P(A \cap B)$, $P(A \cup B)$, $P(C)$, $P(A \cap C)$, and $P(A \cup C)$.

4.104 DVRs A retailer sells two styles of high-priced digital video recorders (DVR) that experience indicates are in equal demand. (Fifty percent of all potential customers prefer style 1, and 50% favor style 2.) If the retailer stocks four of each, what is the probability that the first four customers seeking a DVR all purchase the same style?

4.105 Interstate Commerce A shipping container contains seven complex electronic systems. Unknown to the purchaser, three are defective. Two of the seven are selected for thorough testing and are then classified as defective or nondefective. What is the probability that no defectives are found?

4.106 Heavy Equipment A heavy-equipment sales-man can contact either one or two customers per day with probabilities 1/3 and 2/3, respectively. Each con-tact will result in either no sale or a $50,000 sale with probabilities 9/10 and 1/10, respectively. What is the expected value of his daily sales?

4.107 Fire Insurance A county containing a large number of rural homes is thought to have 60% of those homes insured against fire. Four rural homeowners are chosen at random from the entire population, and x are found to be insured against fire. Find the probability distribution for x. What is the probability that at least three of the four will be insured?

4.108 Fire Alarms A fire-detection device uses three temperature-sensitive cells acting independently of one another in such a manner that any one or more can activate the alarm. Each cell has a probability $p = .8$ of activating the alarm when the temperature reaches 135°F or higher. Let x equal the number of cells activating the alarm when the temperature reaches 135°F.

a. Find the probability distribution of x.

b. Find the probability that the alarm will function when the temperature reaches 135°F.

c. Find the expected value and the variance for the random variable x.

4.109 Roulette Exercise 4.10 described the game of roulette. Suppose you bet $5 on a single number—say, the number 18. The payoff on this type of bet is usu-ally 35 to 1. What is your expected gain?

4.110 Plant Genetics Refer to the experiment conducted by Gregor Mendel in Exercise 4.64. Suppose you are interested in following two indepen-dent traits in snap peas—seed texture (S = smooth, s = wrinkled) and seed color (Y = yellow, y = green)—in a second-generation cross of heterozy-gous parents. Remember that the capital letter repre-sents the dominant trait. Complete the table with the gene pairs for both traits. All possible pairings are equally likely.

	Seed Color			
Seed Texture	yy	yY	Yy	YY
ss	(ss yy)	(ss yY)		
sS				
Ss				
SS				

a. What proportion of the offspring from this cross will have smooth yellow peas?

b. What proportion of the offspring will have smooth green peas?

c. What proportion of the offspring will have wrinkled yellow peas?

d. What proportion of the offspring will have wrinkled green peas?

e. Given that an offspring has smooth yellow peas, what is the probability that this offspring carries one s allele? One s allele *and* one y allele?

4.111 Profitable Stocks An investor has the option of investing in three of five recommended stocks. Unknown to her, only two will show a substantial profit within the next 5 years. If she selects the three stocks at random (giving every combination of three stocks an equal chance of selection), what is the proba-bility that she selects the two profitable stocks? What is the probability that she selects only one of the two profitable stocks?

4.112 Racial Bias? Four union men, two from a minority group, are assigned to four distinctly different one-man jobs, which can be ranked in order of desir-ability.

a. Define the experiment.

b. List the simple events in S.

c. If the assignment to the jobs is unbiased—that is, if any one ordering of assignments is as probable as any other—what is the probability that the two men from the minority group are assigned to the least desirable jobs?

4.113 A Reticent Salesman A salesperson figures that the probability of her making a sale during the first contact with a client is .4 but improves to .55 on the second contact if the client did not buy during the first contact. Suppose this salesperson makes one and only one callback to any client. If she contacts a client, calculate the probabilities for these events:

a. The client will buy.

b. The client will not buy.

4.114 Bus or Subway A man takes either a bus or the subway to work with probabilities .3 and .7, respectively. When he takes the bus, he is late 30% of the days. When he takes the subway, he is late 20% of the days. If the man is late for work on a particular day, what is the probability that he took the bus?

4.115 Guided Missiles The failure rate for a guided missile control system is 1 in 1000. Suppose that a duplicate, but completely independent, control system is installed in each missile so that, if the first fails, the second can take over. The reliability of a missile is the probability that it does not fail. What is the reliability of the modified missile?

4.116 Rental Trucks A rental truck agency services its vehicles on a regular basis, routinely checking for mechanical problems. Suppose that the agency has six moving vans, two of which need to have new brakes. During a routine check, the vans are tested one at a time.

a. What is the probability that the last van with brake problems is the fourth van tested?

b. What is the probability that no more than four vans need to be tested before both brake problems are detected?

c. Given that one van with bad brakes is detected in the first two tests, what is the probability that the remaining van is found on the third or fourth test?

4.117 Pennsylvania Lottery Probability played a role in the rigging of the April 24, 1980, Pennsylvania state lottery. To determine each digit of the three-digit winning number, each of the numbers 0, 1, 2, . . . , 9 is written on a Ping-Pong ball, the 10 balls are blown into a compartment, and the number selected for the digit is the one on the ball that floats to the top of the machine. To alter the odds, the conspirators injected a liquid into all balls used in the game except those numbered 4 and 6, making it almost certain that the lighter balls would be selected and determine the digits in the winning number. They then proceeded to buy lottery tickets bearing the potential winning numbers. How many potential winning numbers were there (666 was the eventual winner)?

*4.118 Lottery, continued Refer to Exercise 4.117. Hours after the rigging of the Pennsylvania state lottery was announced on September 19, 1980, Connecticut state lottery officials were stunned to learn that *their* winning number for the day was 666.

a. All evidence indicates that the Connecticut selection of 666 was pure chance. What is the probability that a 666 would be drawn in Connecticut, given that a 666 had been selected in the April 24, 1980, Pennsylvania lottery?

b. What is the probability of drawing a 666 in the April 24, 1980, Pennsylvania lottery (remember, this drawing was rigged) *and* a 666 on the September 19, 1980, Connecticut lottery?

*4.119 ACL/MCL Tears *The American Journal of Sports Medicine* published a study of 810 women collegiate rugby players with two common knee injuries: medial cruciate ligament (MCL) sprains and anterior cruciate ligament (ACL) tears.[9] For backfield players, it was found that 39% had MCL sprains and 61% had ACL tears. For forwards, it was found that 33% had MCL sprains and 67% had ACL tears. Since a rugby team consists of eight forwards and seven backs, you can assume that 47% of the players with knee injuries are backs and 53% are forwards.

a. Find the unconditional probability that a rugby player selected at random from this group of players has experienced an MCL sprain.

b. Given that you have selected a player who has an MCL sprain, what is the probability that the player is a forward?

c. Given that you have selected a player who has an ACL tear, what is the probability that the player is a back?

4.120 MRIs An article in *The American Journal of Sports Medicine* compared the results of magnetic resonance imaging (MRI) evaluation with arthroscopic surgical evaluation of cartilage tears at two sites in the knees of 35 patients. The $2 \times 35 = 70$ examinations produced the classifications shown in the table.[10] Actual tears were confirmed by arthroscopic surgical examination.

	Tears	No Tears	Total
MRI Positive	27	0	27
MRI Negative	4	39	43
Total	31	39	70

a. What is the probability that a site selected at random has a tear and has been identified as a tear by MRI?

b. What is the probability that a site selected at random has no tear and has been identified as having a tear?

c. What is the probability that a site selected at random has a tear and has not been identified by MRI?

d. What is the probability of a positive MRI, given that there is a tear?

e. What is the probability of a false negative—that is, a negative MRI, given that there is a tear?

4.121 The Match Game Two men each toss a coin. They obtain a "match" if either both coins are heads or both are tails. Suppose the tossing is repeated three times.

a. What is the probability of three matches?

b. What is the probability that all six tosses (three for each man) result in tails?

c. Coin tossing provides a model for many practical experiments. Suppose that the coin tosses represent the answers given by two students for three specific true–false questions on an examination. If the two students gave three matches for answers, would the low probability found in part a suggest collusion?

4.122 Contract Negotiations Experience has shown that, 50% of the time, a particular union–management contract negotiation led to a contract settlement within a 2-week period, 60% of the time the union strike fund was adequate to support a strike, and 30% of the time both conditions were satisfied. What is the probability of a contract settlement given that the union strike fund is adequate to support a strike? Is settlement of a contract within a 2-week period dependent on whether the union strike fund is adequate to support a strike?

4.123 Work Tenure Suppose the probability of remaining with a particular company 10 years or longer is 1/6. A man and a woman start work at the company on the same day.

a. What is the probability that the man will work there less than 10 years?

b. What is the probability that both the man and the woman will work there less than 10 years? (Assume they are unrelated and their lengths of service are independent of each other.)

c. What is the probability that one or the other or both will work 10 years or longer?

4.124 Accident Insurance Accident records collected by an automobile insurance company give the following information: The probability that an insured driver has an automobile accident is .15; if an accident has occurred, the damage to the vehicle amounts to 20% of its market value with probability .80, 60% of

its market value with probability .12, and a total loss with probability .08. What premium should the company charge on a $22,000 car so that the expected gain by the company is zero?

4.125 Waiting Times Suppose that at a particular supermarket the probability of waiting 5 minutes or longer for checkout at the cashier's counter is .2. On a given day, a man and his wife decide to shop individually at the market, each checking out at different cashier counters. They both reach cashier counters at the same time.

a. What is the probability that the man will wait less than 5 minutes for checkout?

b. What is probability that both the man and his wife will be checked out in less than 5 minutes? (Assume that the checkout times for the two are independent events.)

c. What is the probability that one or the other or both will wait 5 minutes or longer?

4.126 Quality Control A quality-control plan calls for accepting a large lot of crankshaft bearings if a sample of seven is drawn and none are defective. What is the probability of accepting the lot if none in the lot are defective? If 1/10 are defective? If 1/2 are defective?

4.127 Mass Transit Only 40% of all people in a community favor the development of a mass transit system. If four citizens are selected at random from the community, what is the probability that all four favor the mass transit system? That none favors the mass transit system?

4.128 Blood Pressure Meds A research physician compared the effectiveness of two blood pressure drugs A and B by administering the two drugs to each of four pairs of identical twins. Drug A was given to one member of a pair; drug B to the other. If, in fact, there is no difference in the effects of the drugs, what is the probability that the drop in the blood pressure reading for drug A exceeds the corresponding drop in the reading for drug B for all four pairs of twins? Suppose drug B created a greater drop in blood pressure than drug A for each of the four pairs of twins. Do you think this provides sufficient evidence to indicate that drug B is more effective in lowering blood pressure than drug A?

4.129 Blood Tests To reduce the cost of detecting a disease, blood tests are conducted on a pooled sample of blood collected from a group of n people. If no indication of the disease is present in the pooled blood sample, none have the disease. If analysis of the pooled blood sample indicates that the disease is present, the blood of each individual must be tested. The individual tests are conducted in sequence. If, among a group of five people, one person has the disease, what is the probability that six blood tests (including the pooled test) are required to detect the single diseased person? If two people have the disease, what is the probability that six tests are required to locate both diseased people?

4.130 Tossing a Coin How many times should a coin be tossed to obtain a probability equal to or greater than .9 of observing at least one head?

4.131 Flextime A survey to determine the availability of flextime schedules in the California workplace provided the following information for 220 firms located in two California cities.

| | Flextime Schedule | | |
City	Available	Not Available	Total
A	39	75	114
B	25	81	106
Totals	64	156	220

A company is selected at random from this pool of 220 companies.

a. What is the probability that the company is located in city A?

b. What is the probability that the company is located in city B and offers flextime work schedules?

c. What is the probability that the company does not have flextime schedules?

d. What is the probability that the company is located in city B, given that the company has flextime schedules available?

4.132 A Color Recognition Experiment An experiment is run as follows—the colors red, yellow, and blue are each flashed on a screen for a short period of time. A subject views the colors and is asked to choose the one he feels was flashed for the longest time. The experiment is repeated three times with the same subject.

a. If all the colors were flashed for the same length of time, find the probability distribution for x, the number of times that the subject chose the color red. Assume that his three choices are independent.

b. Construct the probability histogram for the random variable x.

4.133 Pepsi™ or Coke™? A taste-testing experiment is conducted at a local supermarket, where passing shoppers are asked to taste two soft-drink samples—one Pepsi and one Coke—and state their preference. Suppose that four shoppers are chosen at random and asked to participate in the experiment, and that there is actually no difference in the taste of the two brands.

a. What is the probability that all four shoppers choose Pepsi?

b. What is the probability that exactly one of the four shoppers chooses Pepsi?

4.134 Viruses A certain virus afflicted the families in three adjacent houses in a row of 12 houses. If houses were randomly chosen from a row of 12 houses, what is the probability that the three houses would be adjacent? Is there reason to believe that this virus is contagious?

4.135 Orchestra Politics The board of directors of a major symphony orchestra has voted to create a committee for the purpose of handling employee complaints. The committee will consist of the president and vice president of the symphony board and two orchestra representatives. The two orchestra representatives will be randomly selected from a list of six volunteers, consisting of four men and two women.

a. Find the probability distribution for x, the number of women chosen to be orchestra representatives.

b. What is the probability that both orchestra representatives will be women?

c. Find the mean and variance for the random variable x.

4.136 Independence and Mutually Exclusive Suppose that $P(A) = .3$ and $P(B) = .4$.

a. If $P(A \cap B) = .12$ are A and B independent? Justify your answer.

b. If $P(A \cup B) = .7$ what is $P(A \cap B)$? Justify your answer.

c. If A and B are independent, what is $P(A|B)$?

d. If A and B are mutually exclusive, what is $P(A|B)$?

4.137 Bringing Home the Bacon The following information reflects the results of a survey reported by Mya Frazier in an *Ad Age Insights* white paper.[11] Working spouses were asked "Who is the household breadwinner?" Suppose that one person is selected at random from these 200 individuals.

	You	Spouse or Significant Other	About Equal	Totals
Men	64	16	20	100
Women	32	45	23	100
Totals	96	61	43	200

a. What is the probability that this person will identify his/herself as the household breadwinner?

b. What is the probability that the person selected will be a man who indicates that he and his spouse/significant other are equal breadwinners?

c. If the person selected indicates that the spouse or significant other is the breadwinner, what is the probability that the person is a man?

CASE STUDY

Probability and Decision Making in the Congo

In his exciting novel *Congo,* Michael Crichton describes a search by Earth Resources Technology Service (ERTS), a geological survey company, for deposits of boron-coated blue diamonds, diamonds that ERTS believes to be the key to a new generation of optical computers.[12] In the novel, ERTS is racing against an international consortium to find the Lost City of Zinj, a city that thrived on diamond mining and existed several thousand years ago (according to African fable), deep in the rain forests of eastern Zaire.

After the mysterious destruction of its first expedition, ERTS launches a second expedition under the leadership of Karen Ross, a 24-year-old computer genius who is accompanied by Professor Peter Elliot, an anthropologist; Amy, a talking gorilla; and the famed mercenary and expedition leader, "Captain" Charles Munro. Ross's efforts to find the city are blocked by the consortium's offensive actions, by the deadly rain forest, and by hordes of "talking" killer gorillas whose perceived mission is to defend the diamond mines. Ross overcomes these obstacles by using space-age computers to evaluate the probabilities of success for all possible circumstances and all possible actions that the expedition might take. At each stage of the expedition, she is able to quickly evaluate the chances of success.

At one stage in the expedition, Ross is informed by her Houston headquarters that their computers estimate that she is 18 hours and 20 minutes behind the competing Euro-Japanese team, instead of 40 hours ahead. She changes plans and decides to have the 12 members of her team—Ross, Elliot, Munro, Amy, and eight native porters—parachute into a volcanic region near the estimated location of Zinj. As Crichton relates, "Ross had double-checked outcome probabilities from the Houston computer, and the results were unequivocal. The probability of a successful jump was .7980, meaning that there was approximately one chance in five that someone would be badly hurt. However, given a successful jump, the probability of expedition success was .9943, making it virtually certain that they would beat the consortium to the site."

Keeping in mind that this is an excerpt from a novel, let us examine the probability, .7980, of a successful jump. If you were one of the 12-member team, what is the probability that you would successfully complete your jump? In other words, if the probability of a successful jump by all 12 team members is .7980, what is the probability that a single member could successfully complete the jump?

Several Useful Discrete Distributions

© Kim Steele/Photodisc/Getty Images

NEED TO KNOW...

How to Use Table 1 to Calculate Binomial Probabilities
How to Use Table 2 to Calculate Poisson Probabilities

A Mystery: Cancers Near a Reactor

Is the Pilgrim I nuclear reactor responsible for an increase in cancer cases in the surrounding area? A political controversy was set off when the Massachusetts Department of Public Health found an unusually large number of cases in a 4-mile-wide coastal strip just north of the nuclear reactor in Plymouth, Massachusetts. The case study at the end of this chapter examines how this question can be answered using one of the discrete probability distributions presented here.

INTRODUCTION

Examples of *discrete random variables* can be found in a variety of everyday situations and across most academic disciplines. However, there are three discrete probability distributions that serve as *models* for a large number of these applications. In this chapter we study the binomial, the Poisson, and the hypergeometric probability distributions and discuss their usefulness in different physical situations.

THE BINOMIAL PROBABILITY DISTRIBUTION

A coin-tossing experiment is a simple example of an important discrete random variable called the **binomial random variable.** Many practical experiments result in data similar to the head or tail outcomes of the coin toss. For example, consider the political polls used to predict voter preferences in elections. Each sampled voter can be compared to a coin because the voter may be in favor of our candidate— a "head"—or not—a "tail." In most cases, the proportion of voters who favor our candidate does not equal 1/2; that is, the coin is not fair. In fact, the proportion of voters who favor our candidate is exactly what the poll is designed to measure!

Here are some other situations that are similar to the coin-tossing experiment:

- A sociologist is interested in the proportion of elementary school teachers who are men.
- A soft drink marketer is interested in the proportion of cola drinkers who prefer her brand.
- A geneticist is interested in the proportion of the population who possess a gene linked to Alzheimer's disease.

Each sampled person is analogous to tossing a coin, but the probability of a "head" is not necessarily equal to 1/2. Although these situations have different practical objectives, they all exhibit the common characteristics of the **binomial experiment.**

Definition A **binomial experiment** is one that has these five characteristics:

1. The experiment consists of n identical trials.
2. Each trial results in one of two outcomes. For lack of a better name, the one outcome is called a success, S, and the other a failure, F.
3. The probability of success on a single trial is equal to p and remains the same from trial to trial. The probability of failure is equal to $(1 - p) = q$.
4. The trials are independent.
5. We are interested in x, the number of successes observed during the n trials, for $x = 0, 1, 2, \ldots, n$.

EXAMPLE 5.1

Suppose there are approximately 1,000,000 adults in a county and an unknown proportion p favors term limits for politicians. A sample of 1000 adults will be chosen in such a way that every one of the 1,000,000 adults has an equal chance of being selected, and each adult is asked whether he or she favors term limits. (The ultimate objective of this survey is to estimate the unknown proportion p, a problem that we will discuss in Chapter 8.) Is this a binomial experiment?

Solution Does the experiment have the five binomial characteristics?

1. A "trial" is the choice of a single adult from the 1,000,000 adults in the county. This sample consists of $n = 1000$ identical trials.

2. Since each adult will either favor or not favor term limits, there are two outcomes that represent the "successes" and "failures" in the binomial experiment.[†]

3. The probability of success, p, is the probability that an adult favors term limits. Does this probability remain the same for each adult in the sample? For all practical purposes, the answer is *yes*. For example, if 500,000 adults in the population favor term limits, then the probability of a "success" when the first adult is chosen is 500,000/1,000,000 = 1/2. When the second adult is chosen, the probability p changes slightly, depending on the first choice. That is, there will be either 499,999 or 500,000 successes left among the 999,999 adults. In either case, p is still approximately equal to 1/2.

4. The independence of the trials is guaranteed because of the large group of adults from which the sample is chosen. The probability of an adult favoring term limits does not change depending on the responses of previously chosen people.

5. The random variable x is the number of adults in the sample who favor term limits.

Because the survey satisfies the five characteristics reasonably well, for all practical purposes it can be viewed as a binomial experiment.

EXAMPLE 5.2

A patient fills a prescription for a 10-day regimen of 2 pills daily. Unknown to the pharmacist and the patient, the 20 tablets consist of 18 pills of the prescribed medication and 2 pills that are the generic equivalent of the prescribed medication. The patient selects two pills at random for the first day's dosage. If we check the selection and record the number of pills that are generic, is this a binomial experiment?

Solution Again, check the sampling procedure for the characteristics of a binomial experiment.

1. A "trial" is the selection of a pill from the 20 in the prescription. This experiment consists of $n = 2$ trials.

2. Each trial results in one of two outcomes. Either the pill is generic (call this a "success") or not (a "failure").

3. Since the pills in a prescription bottle can be considered randomly "mixed," the unconditional probability of drawing a generic pill on a given trial would be 2/20.

4. The condition of independence between trials is *not* satisfied, because the probability of drawing a generic pill on the second trial is dependent on the first trial. For example, if the first pill drawn is generic, then there is only 1 generic pill in the remaining 19. Therefore,

$$P(\text{generic on trial 2}|\text{generic on trial 1}) = 1/19$$

[†]Although it is traditional to call the two possible outcomes of a trial "success" and "failure," they could have been called "head" and "tail," "red" and "white," or any other pair of words. Consequently, the outcome called a "success" does not need to be viewed as a success in the ordinary use of the word.

If the first selection *does not* result in a generic pill, then there are still 2 generic pills in the remaining 19, and the probability of a "success" (a generic pill) changes to

$$P(\text{generic on trial 2}|\text{no generic on trial 1}) = 2/19$$

Therefore, the trials are dependent and the sampling does not represent a binomial experiment.

Think about the difference between these two examples. When the sample (the n identical trials) came from a large population, the probability of success p stayed about the same from trial to trial. When the population size N was small, the probability of success p changed quite dramatically from trial to trial, and the experiment *was not* binomial.

RULE OF THUMB

If the sample size is large relative to the population size—in particular, if $n/N \geq .05$—then the resulting experiment is not binomial.

In Chapter 4, we tossed two fair coins and constructed the probability distribution for x, the number of heads—a binomial experiment with $n = 2$ and $p = .5$. The general binomial probability distribution is constructed in the same way, but the procedure gets complicated as n gets large. Fortunately, the probabilities $p(x)$ follow a general pattern. This allows us to use a single formula to find $p(x)$ for any given value of x.

THE BINOMIAL PROBABILITY DISTRIBUTION

A binomial experiment consists of n identical trials with probability of success p on each trial. The probability of k successes in n trials is

$$P(x = k) = C_k^n p^k q^{n-k} = \frac{n!}{k!(n-k)!} p^k q^{n-k}$$

for values of $k = 0, 1, 2, \ldots, n$. The symbol C_k^n equals

$$\frac{n!}{k!(n-k)!}$$

where $n! = n(n-1)(n-2) \cdots (2)(1)$ and $0! \equiv 1$.

The general formulas for μ, σ^2, and σ given in Chapter 4 can be used to derive the following simpler formulas for the binomial mean and standard deviation.

MEAN AND STANDARD DEVIATION FOR THE BINOMIAL RANDOM VARIABLE

The random variable x, the number of successes in n trials, has a probability distribution with this center and spread:

Mean: $\mu = np$
Variance: $\sigma^2 = npq$
Standard deviation: $\sigma = \sqrt{npq}$

EXAMPLE **5.3**

Find $P(x = 2)$ for a binomial random variable with $n = 10$ and $p = .1$.

Solution $P(x = 2)$ is the probability of observing 2 successes and 8 failures in a sequence of 10 trials. You might observe the 2 successes first, followed by 8 consecutive failures:

S, S, F, F, F, F, F, F, F, F

Since p is the probability of success and q is the probability of failure, this particular sequence has probability

$$ppqqqqqqqq = p^2 q^8$$

However, many *other* sequences also result in $x = 2$ successes. The binomial formula uses C_2^{10} to count the number of sequences and gives the exact probability when you use the binomial formula with $k = 2$:

$$P(x = 2) = C_2^{10}(.1)^2(.9)^{10-2}$$

$$= \frac{10!}{2!(10 - 2)!}(.1)^2(.9)^8 = \frac{10(9)}{2(1)}(.01)(.430467) = .1937$$

> **NEED A TIP?**
>
> $n! =$
> $n(n - 1)(n - 2) \dots (2)(1)$
> For example,
> $5! = 5(4)(3)(2)(1) = 120$
> and $0! \equiv 1$.

You could repeat the procedure in Example 5.3 for each value of x—0, 1, 2, . . . , 10—and find all the values of $p(x)$ necessary to construct a probability histogram for x. This would be a long and tedious job, but the resulting graph would look like Figure 5.1(a). You can check the height of the bar for $x = 2$ and find $p(2) = P(x = 2) = .1937$. The graph is skewed right; that is, most of the time you will observe small values of x. The mean or "balancing point" is around $x = 1$; in fact, you can use the formula to find the exact mean:

$$\mu = np = 10(.1) = 1$$

Figures 5.1(b) and 5.1(c) show two other binomial distributions with $n = 10$ but with different values of p. Look at the shapes of these distributions. When $p = .5$, the distribution is exactly symmetric about the mean, $\mu = np = 10(.5) = 5$. When $p = .9$, the distribution is the "mirror image" of the distribution for $p = .1$ and is skewed to the left.

FIGURE 5.1

Binomial probability distributions

EXAMPLE 5.4

Over a long period of time, it has been observed that a professional basketball player can make a free throw on a given trial with probability equal to .8. Suppose he shoots four free throws.

1. What is the probability that he will make exactly two free throws?
2. What is the probability that he will make at least one free throw?

Solution A "trial" is a single free throw, and you can define a "success" as a basket and a "failure" as a miss, so that $n = 4$ and $p = .8$. If you assume that the player's chance of making the free throw does not change from shot to shot, then the number x of times that he makes the free throw is a *binomial random variable*.

1. $P(x = 2) = C_2^4(.8)^2(.2)^2$

$$= \frac{4!}{2!2!}(.64)(.04) = \frac{4(3)(2)(1)}{2(1)(2)(1)}(.64)(.04) = .1536$$

The probability is .1536 that he will make exactly two free throws.

2. $P(\text{at least one}) = P(x \geq 1) = p(1) + p(2) + p(3) + p(4)$
$$= 1 - p(0)$$
$$= 1 - C_0^4(.8)^0(.2)^4$$
$$= 1 - .0016 = .9984.$$

Although you could calculate $P(x = 1)$, $P(x = 2)$, $P(x = 3)$, and $P(x = 4)$ to find this probability, using the complement of the event makes your job easier; that is,

$$P(x \geq 1) = 1 - P(x < 1) = 1 - P(x = 0).$$

Can you think of any reason your assumption of independent trials might be wrong? If the player learns from his previous attempt (i.e., he adjusts his shooting according to his last attempt), then his probability p of making the free throw may change, possibly increase, from shot to shot. The trials would *not* be independent and the experiment would *not* be binomial.

NEED A TIP?

Use Table 1 in Appendix I rather than the binomial formula whenever possible. *This is an easier way!*

Calculating binomial probabilities can become tedious even for relatively small values of n. As n gets larger, it becomes almost impossible without the help of a calculator or computer. Fortunately, both of these tools are available to us. Computer-generated tables of **cumulative binomial probabilities** are given in Table 1 of Appendix I for values of n ranging from 2 to 25 and for selected values of p. These probabilities can also be generated using *MINITAB, MS Excel,* or the Java applets on the CourseMate Web site.

Cumulative binomial probabilities differ from the *individual* binomial probabilities that you calculated with the binomial formula. Once you find the column of probabilities for the correct values of n and p in Table 1, the row marked k gives the sum of all the binomial probabilities from $x = 0$ to $x = k$. Table 5.1 shows part of Table 1 for $n = 5$ and $p = .6$. If you look in the row marked $k = 3$, you will find

$$P(x \leq 3) = p(0) + p(1) + p(2) + p(3) = .663$$

TABLE 5.1 ● **Portion of Table 1 in Appendix I for _n_ = 5**

							p							
k	.01	.05	.10	.20	.30	.40	.50	.60	.70	.80	.90	.95	.99	k
0	—	—	—	—	—	—	—	.010	—	—	—	—	—	0
1	—	—	—	—	—	—	—	.087	—	—	—	—	—	1
2	—	—	—	—	—	—	—	.317	—	—	—	—	—	2
3	—	—	—	—	—	—	—	.663	—	—	—	—	—	3
4	—	—	—	—	—	—	—	.922	—	—	—	—	—	4
5	—	—	—	—	—	—	—	1.000	—	—	—	—	—	5

If the probability you need to calculate is not in this form, you will need to think of a way to rewrite your probability to make use of the tables!

EXAMPLE 5.5 Use the cumulative binomial table for $n = 5$ and $p = .6$ to find the probabilities of these events:

1. Exactly three successes
2. Three or more successes

Solution

1. When $k = 3$ in Table 5.1, the tabled value is

$$P(x \leq 3) = p(0) + p(1) + p(2) + p(3)$$

Since you want only $P(x = 3) = p(3)$, you must subtract out the unwanted probability:

$$P(x \leq 2) = p(0) + p(1) + p(2)$$

which is found in Table 5.1 with $k = 2$. Then

$$P(x = 3) = P(x \leq 3) - P(x \leq 2)$$
$$= .663 - .317 = .346$$

2. To find $P(\text{three or more successes}) = P(x \geq 3)$ using Table 5.1, you must use the complement of the event of interest. Write

$$P(x \geq 3) = 1 - P(x < 3) = 1 - P(x \leq 2)$$

You can find $P(x \leq 2)$ in Table 5.1 with $k = 2$. Then

$$P(x \geq 3) = 1 - P(x \leq 2)$$
$$= 1 - .317 = .683$$

EXAMPLE 5.6 ── Refer to Example 5.5 and the binomial random variable x with $n = 5$ and $p = .6$. Use the cumulative binomial table to find the remaining binomial probabilities, $p(0)$, $p(1)$, $p(2)$, $p(4)$, and $p(5)$. Construct the probability histogram for the random variable x and describe its shape and location.

Solution

1. You can find $P(x = 0)$ directly from Table 5.1 with $k = 0$. That is, $p(0) = .010$.
2. The other probabilities can be found by subtracting successive entries in Table 5.1. Then

$$P(x = 1) = P(x \leq 1) - P(x = 0) = .087 - .010 = .077$$
$$P(x = 2) = P(x \leq 2) - P(x \leq 1) = .317 - .087 = .230$$
$$P(x = 4) = P(x \leq 4) - P(x \leq 3) = .922 - .663 = .259$$
$$P(x = 5) = P(x \leq 5) - P(x \leq 4) = 1.000 - .922 = .078$$

The probability histogram is shown in Figure 5.2. The distribution is relatively mound-shaped, with a center around 3.

FIGURE 5.2

Binomial probability distribution for Example 5.6.

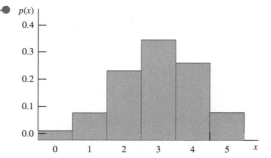

NEED TO KNOW...

How to Use Table 1 to Calculate Binomial Probabilities

1. Find the necessary values of n and p. Isolate the appropriate column in Table 1.
2. Table 1 gives $P(x \leq k)$ in the row marked k. Rewrite the probability you need so that it is in this form.
   - List the values of x in your event.
   - From the list, write the event as either the difference of two probabilities:
   $$P(x \leq a) - P(x \leq b) \quad \text{for } a > b$$
   or the complement of the event:
   $$1 - P(x \leq a)$$
   or just the event itself:
   $$P(x \leq a) \text{ or } P(x < a) = P(x \leq a - 1)$$

EXAMPLE 5.7

Calculating Binomial
Probabilities

A regimen consisting of a daily dose of vitamin C was tested to determine its effectiveness in preventing the common cold. Ten people who were following the prescribed regimen were observed for a period of 1 year. Eight survived the winter without a cold. Suppose the probability of surviving the winter without a cold is .5 when the vitamin C regimen is not followed. What is the probability of observing eight or more survivors, given that the regimen is ineffective in increasing resistance to colds?

Solution If you assume that the vitamin C regimen is ineffective, then the probability p of surviving the winter without a cold is .5. The probability distribution for x, the number of survivors, is

$$p(x) = C_x^{10}(.5)^x(.5)^{10-x}$$

You have learned several ways to find $P(8 \text{ or more survivors}) = P(x \geq 8)$. You will get the same results with any of these methods; choose the most convenient method for your particular problem.

1. *The binomial formula:*

$$
\begin{aligned}
P(8 \text{ or more}) &= p(8) + p(9) + p(10) \\
&= C_8^{10}(.5)^{10} + C_9^{10}(.5)^{10} + C_{10}^{10}(.5)^{10} \\
&= .055
\end{aligned}
$$

2. *The cumulative binomial tables:* Find the column corresponding to $p = .5$ in the table for $n = 10$:

$$
\begin{aligned}
P(8 \text{ or more}) &= P(x \geq 8) = 1 - P(x \leq 7) \\
&= 1 - .945 = .055
\end{aligned}
$$

3. *Output from MINITAB or MS Excel:* The outputs shown in Figures 5.3(a) and 5.3(b) give the **cumulative distribution function,** which are the same probabilities you found in the cumulative binomial tables. The **probability density function** gives the individual binomial probabilities, which you found using the binomial formula.

FIGURE 5.3(a)
MINITAB output for
Example 5.7

Cumulative Distribution Function

Binomial with n = 10 and p = 0.5

x	P(X <= x)
0	0.00098
1	0.01074
2	0.05469
3	0.17187
4	0.37695
5	0.62305
6	0.82813
7	0.94531
8	0.98926
9	0.99902
10	1.00000

Probability Density Function

Binomial with n = 10 and p = 0.5

x	P(X = x)
0	0.000977
1	0.009766
2	0.043945
3	0.117188
4	0.205078
5	0.246094
6	0.205078
7	0.117188
8	0.043945
9	0.009766
10	0.000977

FIGURE 5.3(b)

Excel output for
Example 5.7

x	P(X<= x)	P(X = x)
0	0.000977	0.000977
1	0.010742	0.009766
2	0.054688	0.043945
3	0.171875	0.117188
4	0.376953	0.205078
5	0.623047	0.246094
6	0.828125	0.205078
7	0.945313	0.117188
8	0.989258	0.043945
9	0.999023	0.009766
10	1	0.000977

Using the cumulative distribution function, calculate

$$P(x \geq 8) = 1 - P(x \leq 7)$$
$$= 1 - .94531 = .05469$$

Or, using the probability density function, calculate

$$P(x \geq 8) = p(8) + p(9) + p(10)$$
$$= .043945 + .009766 + .000977 = .05469$$

EXAMPLE 5.8

Would you rather take a multiple-choice or a full recall test? If you have absolutely no knowledge of the material, you will score zero on a full recall test. However, if you are given five choices for each question, you have at least one chance in five of guessing correctly! If a multiple-choice exam contains 100 questions, each with five possible answers, what is the expected score for a student who is guessing on each question? Within what limits will the "no-knowledge" scores fall?

Solution If x is the number of correct answers on the 100-question exam, the probability of a correct answer, p, is one in five, so that $p = .2$. Since the student is randomly selecting answers, the $n = 100$ answers are independent, and the expected score for this binomial random variable is

$$\mu = np = 100(.2) = 20 \quad \text{correct answers}$$

To evaluate the spread or variability of the scores, you can calculate

$$\sigma = \sqrt{npq} = \sqrt{100(.2)(.8)} = 4$$

Then, using your knowledge of variation from Tchebysheff's Theorem and the Empirical Rule, you can make these statements:

- A large proportion of the scores will lie within two standard deviations of the mean, or from $20 - 8 = 12$ to $20 + 8 = 28$.
- Almost all the scores will lie within three standard deviations of the mean, or from $20 - 12 = 8$ to $20 + 12 = 32$.

The "guessing" option gives the student a better score than the zero score on the full recall test, but the student still will not pass the exam. What other options does the student have?

5.2 EXERCISES

BASIC TECHNIQUES

5.1 Consider a binomial random variable with $n = 8$ and $p = .7$. Let x be the number of successes in the sample.

a. Find the probability that x is 3 or less.
b. Find the probability that x is 3 or more.
c. Find $P(x < 3)$.
d. Find $P(x = 3)$.
e. Find $P(3 \leq x \leq 5)$.

5.2 Consider a binomial random variable with $n = 9$ and $p = .3$. Let x be the number of successes in the sample.

a. Find the probability that x is exactly 2.
b. Find the probability that x is less than 2.
c. Find $P(x > 2)$.
d. Find $P(2 \leq x \leq 4)$.

5.3 Evaluate these binomial probabilities:

a. $C_2^8(.3)^2(.7)^6$ **b.** $C_0^4(.05)^0(.95)^4$
c. $C_3^{10}(.5)^3(.5)^7$ **d.** $C_1^7(.2)^1(.8)^6$

5.4 Evaluate these binomial probabilities:

a. $C_0^8(.2)^0(.8)^8$ **b.** $C_1^8(.2)^1(.8)^7$ **c.** $C_2^8(.2)^2(.8)^6$
d. $P(x \leq 1)$ when $n = 8$, $p = .2$
e. $P(\text{two or fewer successes})$

5.5 Let x be a binomial random variable with $n = 7$, $p = .3$. Find these values:

a. $P(x = 4)$ **b.** $P(x \leq 1)$ **c.** $P(x > 1)$
d. $\mu = np$ **e.** $\sigma = \sqrt{npq}$

5.6 Use the formula for the binomial probability distribution to calculate the values of $p(x)$ and construct the probability histogram for x when $n = 6$ and $p = .2$. [HINT: Calculate $P(x = k)$ for seven different values of k.]

5.7 Refer to Exercise 5.6. Construct the probability histogram for a binomial random variable x with $n = 6$ and $p = .8$. Use the results of Exercise 5.6; do not recalculate all the probabilities.

5.8 If x has a binomial distribution with $p = .5$, will the shape of the probability distribution be symmetric, skewed to the left, or skewed to the right?

5.9 Let x be a binomial random variable with $n = 10$ and $p = .4$. Find these values:

a. $P(x = 4)$ **b.** $P(x \geq 4)$ **c.** $P(x > 4)$
d. $P(x \leq 4)$ **e.** $\mu = np$ **f.** $\sigma = \sqrt{npq}$

5.10 Use Table 1 in Appendix I to find the sum of the binomial probabilities from $x = 0$ to $x = k$ for these cases:

a. $n = 10$, $p = .1$, $k = 3$
b. $n = 15$, $p = .6$, $k = 7$
c. $n = 25$, $p = .5$, $k = 14$

5.11 Use Table 1 in Appendix I to evaluate the following probabilities for $n = 6$ and $p = .8$:

a. $P(x \geq 4)$ **b.** $P(x = 2)$
c. $P(x < 2)$ **d.** $P(x > 1)$

Verify these answers using the values of $p(x)$ calculated in Exercise 5.7.

5.12 Find $P(x \leq k)$ for each of the following cases:

a. $n = 20$, $p = .05$, $k = 2$
b. $n = 15$, $p = .7$, $k = 8$
c. $n = 10$, $p = .9$, $k = 9$

5.13 Use Table 1 in Appendix I to find the following:

a. $P(x < 12)$ for $n = 20$, $p = .5$
b. $P(x \leq 6)$ for $n = 15$, $p = .4$
c. $P(x > 4)$ for $n = 10$, $p = .4$
d. $P(x \geq 6)$ for $n = 15$, $p = .6$
e. $P(3 < x < 7)$ for $n = 10$, $p = .5$

5.14 Find the mean and standard deviation for a binomial distribution with these values:

a. $n = 1000$, $p = .3$ **b.** $n = 400$, $p = .01$
c. $n = 500$, $p = .5$ **d.** $n = 1600$, $p = .8$

5.15 Find the mean and standard deviation for a binomial distribution with $n = 100$ and these values of p:

a. $p = .01$ **b.** $p = .9$ **c.** $p = .3$
d. $p = .7$ **e.** $p = .5$

5.16 In Exercise 5.15, the mean and standard deviation for a binomial random variable were calculated for a fixed sample size, $n = 100$, and for different values of p. Graph the values of the standard deviation for the five values of p given in Exercise 5.15. For what value of p does the standard deviation seem to be a maximum?

5.17 Let x be a binomial random variable with $n = 20$ and $p = .1$.

a. Calculate $P(x \leq 4)$ using the binomial formula.

b. Calculate $P(x \leq 4)$ using Table 1 in Appendix I.

c. Use the *Excel* output below to calculate $P(x \leq 4)$. Compare the results of parts a, b, and c.

d. Calculate the mean and standard deviation of the random variable x.

e. Use the results of part d to calculate the intervals $\mu \pm \sigma$, $\mu \pm 2\sigma$, and $\mu \pm 3\sigma$. Find the probability that an observation will fall into each of these intervals.

f. Are the results of part e consistent with Tchebysheff's Theorem? With the Empirical Rule? Why or why not?

Excel output for Exercise 5.17: Binomial with $n = 20$ and $p = .1$

x	p(x)	x	p(x)
0	0.1216	11	7E-07
1	0.2702	12	5E-08
2	0.2852	13	4E-09
3	0.1901	14	2E-10
4	0.0898	15	9E-12
5	0.0319	16	3E-13
6	0.0089	17	8E-15
7	0.0020	18	2E-16
8	0.0004	19	2E-18
9	0.0001	20	1E-20
10	0.0000		

APPLICATIONS

5.18 The Urn Problem A jar contains five balls: three red and two white. Two balls are randomly selected without replacement from the jar, and the number x of red balls is recorded. Explain why x is or is not a binomial random variable. (HINT: Compare the characteristics of this experiment with the characteristics of a binomial experiment given in this section.) If the experiment is binomial, give the values of n and p.

5.19 The Urn Problem, continued Refer to Exercise 5.18. Assume that the sampling was conducted with replacement. That is, assume that the first ball was selected from the jar, observed, and then replaced, and that the balls were then mixed before the second ball was selected. Explain why x, the number of red balls observed, is or is not a binomial random variable.

If the experiment is binomial, give the values of n and p.

5.20 Chicago Weather A meteorologist in Chicago recorded the number of days of rain during a 30-day period. If the random variable x is defined as the number of days of rain, does x have a binomial distribution? If not, why not? If so, are both values of n and p known?

5.21 Telemarketers A market research firm hires operators to conduct telephone surveys. The computer randomly dials a telephone number, and the operator asks the respondent whether or not he has time to answer some questions. Let x be the number of telephone calls made until the first respondent is willing to answer the operator's questions. Is this a binomial experiment? Explain.

5.22 SAT Scores In 2010, the average overall SAT score (Critical Reading, Math, and Writing) for college-bound students in the United States was 1509 out of 2400. Suppose that 45% of all high school graduates took this test, and that 100 high school graduates are randomly selected from throughout the United States.[1] Which of the following random variables has an approximate binomial distribution? If possible, give the values for n and p.

a. The number of students who took the SAT.

b. The scores of the 100 students on the SAT.

c. The number of students who scored above average on the SAT.

d. The amount of time it took the students to complete the SAT.

5.23 Security Systems A home security system is designed to have a 99% reliability rate. Suppose that nine homes equipped with this system experience an attempted burglary. Find the probabilities of these events:

a. At least one of the alarms is triggered.

b. More than seven of the alarms are triggered.

c. Eight or fewer alarms are triggered.

5.24 Blood Types In a certain population, 85% of the people have Rh-positive blood. Suppose that two people from this population get married. What is the probability that they are both Rh-negative, thus making it inevitable that their children will be Rh-negative?

5.25 Car Colors Car color preferences change over the years and according to the particular model that the customer selects. In a recent year, suppose that 10% of all luxury cars sold were black. If 25 cars of that year and type are randomly selected, find the following probabilities:

a. At least five cars are black.

b. At most six cars are black.

c. More than four cars are black.

d. Exactly four cars are black.

e. Between three and five cars (inclusive) are black.

f. More than 20 cars are not black.

5.26 O Canada! The National Hockey League (NHL) has 80% of its players born outside the United States, and of those born outside the United States, 50% are born in Canada.[2] Suppose that $n = 12$ NHL players were selected at random. Let x be the number of players in the sample who were born outside of the United States so that $p = .8$. Find the following probabilities:

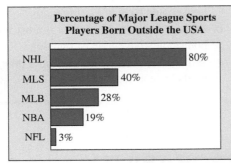

Just over 50% of the foreign players in the NHL are from Canada

Percentage of Major League Sports Players Born Outside the USA

NHL	80%
MLS	40%
MLB	28%
NBA	19%
NFL	3%

a. At least five or more of the sampled players were born outside the United States.

b. Exactly seven of the players were born outside the United States.

c. Fewer than six were born outside the United States.

5.27 Medical Bills Records show that 30% of all patients admitted to a medical clinic fail to pay their bills and that eventually the bills are forgiven. Suppose $n = 4$ new patients represent a random selection from the large set of prospective patients served by the clinic. Find these probabilities:

a. All the patients' bills will eventually have to be forgiven.

b. One will have to be forgiven.

c. None will have to be forgiven.

5.28 Medical Bills II Refer to Exercise 5.27 where 30% of all admitted patients fail to pay their bills and the debts are eventually forgiven. Suppose that the clinic treats 2000 different patients over a period of 1 year, and let x be the number of forgiven debts.

a. What is the mean (expected) number of debts that have to be forgiven?

b. Find the variance and standard deviation of x.

c. What can you say about the probability that x will exceed 700? (HINT: Use the values of μ and σ, along with Tchebysheff's Theorem.)

5.29 Whitefly Infestation Suppose that 10% of the fields in a given agricultural area are infested with the sweet potato whitefly. One hundred fields in this area are randomly selected and checked for whitefly.

a. What is the average number of fields sampled that are infested with whitefly?

b. Within what limits would you expect to find the number of infested fields, with probability approximately 95%?

c. What might you conclude if you found that $x = 25$ fields were infested? Is it possible that one of the characteristics of a binomial experiment is not satisfied in this experiment? Explain.

5.30 Color Preferences in Mice In a psychology experiment, the researcher designs a maze in which a mouse must choose one of two paths, colored either red or blue, at each of 10 intersections. At the end of the maze, the mouse is given a food reward. The researcher counts the number of times the mouse chooses the red path. If you were the researcher, how would you use this count to decide whether the mouse has any preference for color?

5.31 Back Pain Six in 10 adults say lower back pain substantially limits their athletic activities.[3] A random sample of $n = 8$ adults were asked if lower back pain was a limiting factor in their athletic activities. The printout below shows the cumulative and individual probabilities for a binomial random variable with $n = 8$ and $p = .6$.

MINITAB Output for Exercise 5.31

Cumulative Distribution Function		Probability Density Function	
Binomial with n = 8 and p = 0.6		Binomial with n = 8 and p = 0.6	
x	P(X <= x)	x	P(X = x)
0	0.00066	0	0.000655
1	0.00852	1	0.007864
2	0.04981	2	0.041288
3	0.17367	3	0.123863
4	0.40591	4	0.232243
5	0.68461	5	0.278692
6	0.89362	6	0.209019
7	0.98320	7	0.089580
8	1.00000	8	0.016796

a. Use the binomial formula to find the probability that all eight indicate that lower back pain was a limiting factor in their athletic activities.

b. Confirm the results of part a using the printout.

c. What is the probability that at most seven individuals give lower back pain as a limiting factor in their athletic activities?

5.32 Fast Food and Gas Stations Forty percent of all Americans who travel by car look for gas stations and food outlets that are close to or visible from the highway. Suppose a random sample of $n = 25$ Americans who travel by car are asked how they determine where to stop for food and gas. Let x be the number in the sample who respond that they look for gas stations and food outlets that are close to or visible from the highway.

a. What are the mean and variance of x?

b. Calculate the interval $\mu \pm 2\sigma$. What values of the binomial random variable x fall into this interval?

c. Find $P(6 \leq x \leq 14)$. How does this compare with the fraction in the interval $\mu \pm 2\sigma$ for any distribution? For mound-shaped distributions?

5.33 Taste Test for PTC The taste test for PTC (phenylthiocarbamide) is a favorite exercise for every human genetics class. It has been established that a single gene determines the characteristic, and that 70% of Americans are "tasters," while 30% are "non-tasters." Suppose that 20 Americans are randomly chosen and are tested for PTC.

a. What is the probability that 17 or more are "tasters"?

b. What is the probability that 15 or fewer are "tasters"?

5.34 Man's Best Friend According to the Humane Society of the United States, there are approximately 77.5 million owned dogs in the United States, and approximately 40% of all U.S. households own at least one dog.[4] Suppose that the 40% figure is correct and that 15 households are randomly selected for a pet ownership survey.

a. What is the probability that exactly eight of the households have at least one dog?

b. What is the probability that at most four of the households have at least one dog?

c. What is the probability that more than 10 households have at least one dog?

THE POISSON PROBABILITY DISTRIBUTION

5.3

Another discrete random variable that has numerous practical applications is the **Poisson random variable.** Its probability distribution provides a good model for data that represent the number of occurrences of a specified event in a given unit of time or space. Here are some examples of experiments for which the random variable x can be modeled by the Poisson random variable:

- The number of calls received by a technical support specialist during a given period of time
- The number of bacteria per small volume of fluid
- The number of customer arrivals at a checkout counter during a given minute
- The number of machine breakdowns during a given day
- The number of traffic accidents on a section of freeway during a given time period

In each example, **x represents the number of events that occur in a period of time or space during which an average of μ such events can be expected to occur.** The only assumptions needed when one uses the Poisson distribution to model experiments such as these are that the counts or events occur **randomly and independently** of one another. The formula for the Poisson probability distribution, as well as its mean and variance, are given next.

THE POISSON PROBABILITY DISTRIBUTION

Let μ be the average number of times that an event occurs in a certain period of time or space. The probability of k occurrences of this event is

$$P(x = k) = \frac{\mu^k e^{-\mu}}{k!}$$

for values of $k = 0, 1, 2, 3, \ldots$. The mean and standard deviation of the Poisson random variable x are

Mean: μ Standard deviation: $\sigma = \sqrt{\mu}$

The symbol $e = 2.71828\ldots$ is evaluated using your scientific calculator, which should have a function such as e^x. For each value of k, you can obtain the individual probabilities for the Poisson random variable, just as you did for the binomial random variable.

NEED A TIP?

Use either the Poisson formula or Table 2 to calculate Poisson probabilities.

EXAMPLE 5.9

The average number of traffic accidents on a certain section of highway is two per week. Assume that the number of accidents follows a Poisson distribution with $\mu = 2$.

1. Find the probability of no accidents on this section of highway during a 1-week period.
2. Find the probability of at most three accidents on this section of highway during a 2-week period.

Solution

1. The average number of accidents per week is $\mu = 2$. Therefore, the probability of no accidents on this section of highway during a given week is

$$P(x = 0) = p(0) = \frac{2^0 e^{-2}}{0!} = e^{-2} = .135335$$

2. During a 2-week period, the average number of accidents on this section of highway is $2(2) = 4$. The probability of at most three accidents during a 2-week period is

$$P(x \leq 3) = p(0) + p(1) + p(2) + p(3)$$

where

$$p(0) = \frac{4^0 e^{-4}}{0!} = .018316 \qquad p(2) = \frac{4^2 e^{-4}}{2!} = .146525$$

$$p(1) = \frac{4^1 e^{-4}}{1!} = .073263 \qquad p(3) = \frac{4^3 e^{-4}}{3!} = .195367$$

Therefore,

$$P(x \leq 3) = .018316 + .073263 + .146525 + .195367 = .433471$$

Once the values for $p(x)$ have been calculated, you can use them to construct a probability histogram for the random variable x. Graphs of the Poisson probability distribution for $\mu = .5, 2,$ and 4 are shown in Figure 5.4.

FIGURE 5.4 ●

Poisson probability distributions for $\mu = .5, 2,$ and 4

Alternatively, you can use **cumulative Poisson tables** (Table 2 in Appendix I) or the cumulative or individual probabilities generated by *MINITAB* or *MS Excel*. All of these options are usually more convenient than hand calculation. The procedures are similar to those used for the binomial random variable.

NEED TO KNOW...

How to Use Table 2 to Calculate Poisson Probabilities

1. Find the necessary value of μ. Isolate the appropriate column in Table 2.

2. Table 2 gives $P(x \le k)$ in the row marked k. Rewrite the probability you need so that it is in this form.

   • List the values of x in your event.

   • From the list, write the event as either the difference of two probabilities:

   $$P(x \le a) - P(x \le b) \text{ for } a > b$$

   or the complement of the event:

   $$1 - P(x \le a)$$

   or just the event itself:

   $$P(x \le a) \text{ or } P(x < a - 1)$$

EXAMPLE 5.10

Refer to Example 5.9, where we calculated probabilities for a Poisson distribution with $\mu = 2$ and $\mu = 4$. Use the cumulative Poisson table to find the probabilities of these events:

1. No accidents during a 1-week period.
2. At most three accidents during a 2-week period.

Solution
A portion of Table 2 in Appendix I is shown in Figure 5.5.

FIGURE 5.5

Portion of Table 2 in Appendix I

			μ		
k	2.0	2.5	3.0	3.5	4.0
0	.135	.082	.055	.033	.018
1	.406	.287	.199	.136	.092
2	.677	.544	.423	.321	.238
3	.857	.758	.647	.537	.433
4	.947	.891	.815	.725	.629
5	.983	.958	.916	.858	.785
6	.995	.986	.966	.935	.889
7	.999	.996	.988	.973	.949
8	1.000	.999	.996	.990	.979
9		1.000	.999	.997	.992
10			1.000	.999	.997
11				1.000	.999
12					1.000

1. From Example 5.9, the average number of accidents in a 1-week period is $\mu = 2.0$. Therefore, the probability of no accidents in a 1-week period can be read directly from Table 2 in the column marked "2.0" as $P(x = 0) = p(0) = .135$.

2. The average number of accidents in a 2-week period is $2(2) = 4$. Therefore, the probability of at most three accidents in a 2-week period is found in Table 2, indexing $\mu = 4.0$ and $k = 3$ as $P(x \leq 3) = .433$.

Both of these probabilities match the calculations done in Example 5.9, correct to three decimal places.

In Section 5.2, we used the cumulative binomial tables to simplify the calculation of binomial probabilities. Unfortunately, in practical situations, n is often large and no tables are available.

THE POISSON APPROXIMATION TO THE BINOMIAL DISTRIBUTION

 NEED A TIP?

You can estimate binomial probabilities with the Poisson when n is large and p is small.

The Poisson probability distribution provides a simple, easy-to-compute, and accurate approximation to binomial probabilities when n is large and $\mu = np$ is small, preferably with $np < 7$. An approximation suitable for larger values of $\mu = np$ will be given in Chapter 6.

EXAMPLE 5.11

Suppose a life insurance company insures the lives of 5000 men aged 42. If actuarial studies show the probability that any 42-year-old man will die in a given year to be .001, find the exact probability that the company will have to pay $x = 4$ claims during a given year.

Solution The exact probability is given by the binomial distribution as

$$P(x = 4) = p(4) = \frac{5000!}{4!4996!}(.001)^4(.999)^{4996}$$

for which binomial tables are not available. To compute $P(x = 4)$ without the aid of a scientific calculator or a computer would be very time-consuming, but the Poisson distribution can be used to provide a good approximation to $P(x = 4)$. Computing $\mu = np = (5000)(.001) = 5$ and substituting into the formula for the Poisson probability distribution, we have

$$p(4) \approx \frac{\mu^4 e^{-\mu}}{4!} = \frac{5^4 e^{-5}}{4!} = \frac{(625)(.006738)}{24} = .175$$

The value of $p(4)$ could also be obtained using Table 2 in Appendix I with $\mu = 5$ as

$$p(4) = P(x \le 4) - P(x \le 3) = .440 - .265 = .175$$

EXAMPLE 5.12

A manufacturer of power lawn mowers buys 1-horsepower, two-cycle engines in lots of 1000 from a supplier. She then equips each of the mowers produced by her plant with one of the engines. History shows that the probability of any one engine from that supplier proving unsatisfactory is .001. In a shipment of 1000 engines, what is the probability that none is defective? Three are? Four are?

Solution This is a binomial experiment with $n = 1000$ and $p = .001$. The expected number of defectives in a shipment of $n = 1000$ engines is $\mu = np = (1000)(.001) = 1$. Since this is a binomial experiment with $np < 7$, the probability of x defective engines in the shipment may be approximated by

$$P(x = k) = p(k) = \frac{\mu^k e^{-\mu}}{k!} = \frac{1^k e^{-1}}{k!} = \frac{e^{-1}}{k!}$$

Therefore,

$$p(0) \approx \frac{e^{-1}}{0!} = \frac{.368}{1} = .368$$

$$p(3) \approx \frac{e^{-1}}{3!} = \frac{.368}{6} = .061$$

$$p(4) \approx \frac{e^{-1}}{4!} = \frac{.368}{24} = .015$$

The individual Poisson probabilities for $\mu = 1$ along with the individual binomial probabilities for $n = 1000$ and $p = .001$ and $x = 0, 1, \ldots, 10$ were generated by *MS Excel* and are shown in Figure 5.6. The individual probabilities, even though they are computed with totally different formulas, are almost the same. The exact binomial probabilities are in the left section of Figure 5.6, and the Poisson approximations are on the right.

FIGURE 5.6 ●

Excel output of binomial and Poisson probabilities

x	Binomial p(x)	x	Poisson p(x)
0	0.3677	0	0.3679
1	0.3681	1	0.3679
2	0.1840	2	0.1839
3	0.0613	3	0.0613
4	0.0153	4	0.0153
5	0.0030	5	0.0031
6	0.0005	6	0.0005
7	0.0001	7	0.0001
8	0.0000	8	0.0000
9	0.0000	9	0.0000
10	0.0000	10	0.0000

5.3 EXERCISES

BASIC TECHNIQUES

5.35 Consider a Poisson random variable with $\mu = 2.5$. Use the Poisson formula to calculate the following probabilities:

a. $P(x = 0)$ b. $P(x = 1)$

c. $P(x = 2)$ d. $P(x \leq 2)$

5.36 Consider a Poisson random variable with $\mu = 3$. Use the Poisson formula to calculate the following probabilities:

a. $P(x = 0)$ b. $P(x = 1)$

c. $P(x > 1)$

5.37 Consider a Poisson random variable with $\mu = 3$. Use Table 2 to find the following probabilities:

a. $P(x \leq 3)$ b. $P(x > 3)$

c. $P(x = 3)$ d. $P(3 \leq x \leq 5)$

5.38 Consider a Poisson random variable with $\mu = 0.8$. Use Table 2 to find the following probabilities:

a. $P(x = 0)$ b. $P(x \leq 2)$

c. $P(x > 2)$ d. $P(2 \leq x \leq 4)$

5.39 Let x be a Poisson random variable with mean $\mu = 2$. Calculate these probabilities:

a. $P(x = 0)$ b. $P(x = 1)$

c. $P(x > 1)$ d. $P(x = 5)$

5.40 Let x be a Poisson random variable with mean $\mu = 2.5$. Use Table 2 in Appendix I to calculate these probabilities:

a. $P(x \geq 5)$ b. $P(x < 6)$

c. $P(x = 2)$ d. $P(1 \leq x \leq 4)$

5.41 Poisson vs. Binomial Let x be a binomial random variable with $n = 20$ and $p = .1$.

a. Calculate $P(x \leq 2)$ using Table 1 in Appendix I to obtain the exact binomial probability.

b. Use the Poisson approximation to calculate $P(x \leq 2)$.

c. Compare the results of parts a and b. Is the approximation accurate?

5.42 Poisson vs. Binomial II To illustrate how well the Poisson probability distribution approximates the binomial probability distribution, calculate the Poisson approximate values for $p(0)$ and $p(1)$ for a binomial probability distribution with $n = 25$ and $p = .05$. Compare the answers with the exact values obtained from Table 1 in Appendix I.

APPLICATIONS

5.43 Airport Safety The increased number of small commuter planes in major airports has heightened concern over air safety. An eastern airport has recorded a monthly average of five near misses on landings and takeoffs in the past 5 years.

a. Find the probability that during a given month there are no near misses on landings and takeoffs at the airport.

b. Find the probability that during a given month there are five near misses.

c. Find the probability that there are at least five near misses during a particular month.

5.44 Intensive Care The number x of people entering the intensive care unit at a particular hospital on any one day has a Poisson probability distribution with mean equal to five persons per day.

a. What is the probability that the number of people entering the intensive care unit on a particular day is two? Less than or equal to two?

b. Is it likely that x will exceed 10? Explain.

5.45 Accident Prone According to a study conducted by the Department of Pediatrics at the University of California, San Francisco, children who are injured two or more times tend to sustain these injuries during a relatively limited time, usually 1 year or less. If the average number of injuries per year for school-age children is two, what are the probabilities of these events?

a. A school-age child will sustain two injuries during the year.

b. A school-age child will sustain two or more injuries during the year.

c. A school-age child will sustain at most one injury during the year.

5.46 Accident Prone, continued Refer to Exercise 5.45.

a. Calculate the mean and standard deviation for x, the number of injuries per year sustained by a school-age child.

b. Within what limits would you expect the number of injuries per year to fall?

5.47 Bacteria in Water Samples If a drop of water is placed on a slide and examined under a microscope, the number x of a specific type of bacteria present has been found to have a Poisson probability distribution. Suppose the maximum permissible count per water specimen for this type of bacteria is five. If the mean count for your water supply is two and you test a single specimen, is it likely that the count will exceed the maximum permissible count? Explain.

5.48 E. coli Outbreaks An outbreak of *E. coli* infections in August of 2010 occurred in three Washington state day care centers. There were eight confirmed and six suspected cases of *E. coli*, with over 70 children awaiting test results.[5] Outbreaks of *E. coli* infections for 2009 are reported to be less than 1 per 100,000, down from 2.5 per 100,000 reported earlier.[6] Using the rate of 1 per 100,000, find the following probabilities.

a. What is the probability that at most two outbreaks per 100,000 are reported across the United States this year?

b. What is the probability that more than three outbreaks per 100,000 are reported across the United States this year?

THE HYPERGEOMETRIC PROBABILITY DISTRIBUTION

5.4

Suppose you are selecting a sample of elements from a population and you record whether or not each element possesses a certain characteristic. You are recording the typical "success" or "failure" data found in the binomial experiment. The sample survey of Example 5.1 and the sampling for defectives of Example 5.2 are practical illustrations of these sampling situations.

If the number of elements in the population is large relative to the number in the sample (as in Example 5.1), the probability of selecting a success on a single trial is equal to the proportion p of successes in the population. Because the population is large in relation to the sample size, this probability will remain constant (for all practical purposes) from trial to trial, and the number x of successes in the sample will follow a binomial probability distribution. However, if the number of elements in the population is small in relation to the sample size ($n/N \geq .05$), the probability of a success for a given trial is dependent on the outcomes of preceding trials. Then the number x of successes follows what is known as a **hypergeometric probability distribution.**

It is easy to visualize the **hypergeometric random variable x** by thinking of a bowl containing M red balls and $N - M$ white balls, for a *total of N* balls in the bowl. You select n balls from the bowl and record x, the number of red balls that you see. If you now define a "success" to be a red ball, you have an example of the hypergeometric random variable x.

The formula for calculating the probability of exactly k successes in n trials is given next.

THE HYPERGEOMETRIC PROBABILITY DISTRIBUTION

A population contains M successes and $N - M$ failures. The probability of exactly k successes in a random sample of size n is

$$P(x = k) = \frac{C_k^M C_{n-k}^{N-M}}{C_n^N}$$

for values of k that depend on N, M, and n with

$$C_n^N = \frac{N!}{n!(N-n)!}$$

The mean and variance of a hypergeometric random variable are very similar to those of a binomial random variable with a correction for the finite population size:

$$\mu = n\left(\frac{M}{N}\right)$$

$$\sigma^2 = n\left(\frac{M}{N}\right)\left(\frac{N-M}{N}\right)\left(\frac{N-n}{N-1}\right)$$

EXAMPLE 5.13

A case of wine has 12 bottles, 3 of which contain spoiled wine. A sample of 4 bottles is randomly selected from the case.

1. Find the probability distribution for x, the number of bottles of spoiled wine in the sample.
2. What are the mean and variance of x?

Solution For this example, $N = 12$, $n = 4$, $M = 3$, and $(N - M) = 9$. Then

$$p(x) = \frac{C_x^3 C_{4-x}^9}{C_4^{12}}$$

1. The possible values for x are 0, 1, 2, and 3, with probabilities

$$p(0) = \frac{C_0^3 C_4^9}{C_4^{12}} = \frac{1(126)}{495} = .25$$

$$p(1) = \frac{C_1^3 C_3^9}{C_4^{12}} = \frac{3(84)}{495} = .51$$

$$p(2) = \frac{C_2^3 C_2^9}{C_4^{12}} = \frac{3(36)}{495} = .22$$

$$p(3) = \frac{C_3^3 C_1^9}{C_4^{12}} = \frac{1(9)}{495} = .02$$

2. The mean is given by

$$\mu = 4\left(\frac{3}{12}\right) = 1$$

NEED A TIP?

Draw 4

3S
9G

and the variance is

$$\sigma^2 = 4\left(\frac{3}{12}\right)\left(\frac{9}{12}\right)\left(\frac{12-4}{11}\right) = .5455$$

EXAMPLE 5.14

A particular industrial product is shipped in lots of 20. Testing to determine whether an item is defective is costly; hence, the manufacturer samples production rather than using a 100% inspection plan. A sampling plan constructed to minimize the number of defectives shipped to customers calls for sampling five items from each lot and rejecting the lot if more than one defective is observed. (If the lot is rejected, each item in the lot is then tested.) If a lot contains four defectives, what is the probability that it will be accepted?

Solution Let x be the number of defectives in the sample. Then $N = 20$, $M = 4$, $(N - M) = 16$, and $n = 5$. The lot will be rejected if $x = 2$, 3, or 4. Then

$$P(\text{accept the lot}) = P(x \le 1) = p(0) + p(1) = \frac{C_0^4 C_5^{16}}{C_5^{20}} + \frac{C_1^4 C_4^{16}}{C_5^{20}}$$

$$= \frac{\left(\frac{4!}{0!4!}\right)\left(\frac{16!}{5!11!}\right)}{\frac{20!}{5!15!}} + \frac{\left(\frac{4!}{1!3!}\right)\left(\frac{16!}{4!12!}\right)}{\frac{20!}{5!15!}}$$

$$= \frac{91}{323} + \frac{455}{969} = .2817 + .4696 = .7513$$

5.4 EXERCISES

BASIC TECHNIQUES

5.49 Evaluate these probabilities:

a. $\dfrac{C_1^2 C_1^1}{C_2^3}$ **b.** $\dfrac{C_0^4 C_2^2}{C_2^6}$ **c.** $\dfrac{C_2^2 C_1^2}{C_3^4}$

5.50 Let x be the number of successes observed in a sample of $n = 4$ items selected from a population of $N = 8$. Suppose that of the $N = 8$ items, 5 are considered "successes."

a. Find the probability of observing all successes.

b. Find the probability of observing one success.

c. Find the probability of observing at most two successes.

5.51 Evaluate these probabilities:

a. $\dfrac{C_1^3 C_1^2}{C_2^5}$ **b.** $\dfrac{C_2^4 C_1^3}{C_3^7}$ **c.** $\dfrac{C_4^5 C_0^3}{C_4^8}$

5.52 Let x be the number of successes observed in a sample of $n = 5$ items selected from $N = 10$. Suppose that, of the $N = 10$ items, 6 are considered "successes."

a. Find the probability of observing no successes.

b. Find the probability of observing at least two successes.

c. Find the probability of observing exactly two successes.

5.53 Let x be a hypergeometric random variable with $N = 15$, $n = 3$, and $M = 4$.

a. Calculate $p(0)$, $p(1)$, $p(2)$, and $p(3)$.

b. Construct the probability histogram for x.

c. Use the formulas given in Section 5.4 to calculate $\mu = E(x)$ and σ^2.

d. What proportion of the population of measurements fall into the interval $(\mu \pm 2\sigma)$? Into the interval

($\mu \pm 3\sigma$)? Do these results agree with those given by Tchebysheff's Theorem?

5.54 Candy Choices A candy dish contains five blue and three red candies. A child reaches up and selects three candies without looking.

a. What is the probability that there are two blue and one red candies in the selection?

b. What is the probability that the candies are all red?

c. What is the probability that the candies are all blue?

APPLICATIONS

5.55 Defective Computer Chips A piece of electronic equipment contains six computer chips, two of which are defective. Three computer chips are randomly chosen for inspection, and the number of defective chips is recorded. Find the probability distribution for x, the number of defective computer chips. Compare your results with the answers obtained in Exercise 4.90.

5.56 Gender Bias? A company has five applicants for two positions: two women and three men. Suppose that the five applicants are equally qualified and that no preference is given for choosing either gender. Let x equal the number of women chosen to fill the two positions.

a. Write the formula for $p(x)$, the probability distribution of x.

b. What are the mean and variance of this distribution?

c. Construct a probability histogram for x.

5.57 Teaching Credentials In southern California, a growing number of persons pursuing a teaching credential are choosing paid internships over traditional student teaching programs. A group of eight candidates for three teaching positions consisted of five paid interns and three traditional student teachers. Let us assume that all eight candidates are equally qualified for the positions. Let x represent the number of paid interns who are hired for these three positions.

a. Does x have a binomial distribution or a hypergeometric distribution? Support your answer.

b. Find the probability that three paid interns are hired for these positions.

c. What is the probability that none of the three hired was a paid intern?

d. Find $P(x \le 1)$.

5.58 Seed Treatments Seeds are often treated with a fungicide for protection in poor-draining, wet environments. In a small-scale trial prior to a large-scale experiment to determine what dilution of the fungicide to apply, five treated seeds and five untreated seeds were planted in clay soil and the number of plants emerging from the treated and untreated seeds were recorded. Suppose the dilution was not effective and only four plants emerged. Let x represent the number of plants that emerged from treated seeds.

a. Find the probability that $x = 4$.

b. Find $P(x \le 3)$.

c. Find $P(2 \le x \le 3)$.

CHAPTER REVIEW

Key Concepts and Formulas

I. The Binomial Random Variable

1. **Five characteristics:** n identical independent trials, each resulting in either *success* (S) or *failure* (F); probability of success is p and remains constant from trial to trial; and x is the number of successes in n trials

2. **Calculating binomial probabilities**

   a. Formula: $P(x = k) = C_k^n p^k q^{n-k}$

   b. Cumulative binomial tables

   c. Individual and cumulative probabilities using *MINITAB* and *MS Excel*

3. Mean of the binomial random variable:
   $\mu = np$

4. Variance and standard deviation: $\sigma^2 = npq$
   and $\sigma = \sqrt{npq}$

II. The Poisson Random Variable

1. The number of events that occur in a period of time or space, during which an average of μ such events are expected to occur

2. **Calculating Poisson probabilities**

   a. Formula: $P(x = k) = \dfrac{\mu^k e^{-\mu}}{k!}$

b. Cumulative Poisson tables

c. Individual and cumulative probabilities using *MINITAB* and *MS Excel*

3. Mean of the Poisson random variable:
$E(x) = \mu$

4. Variance and standard deviation: $\sigma^2 = \mu$ and $\sigma = \sqrt{\mu}$

5. Binomial probabilities can be approximated with Poisson probabilities when $np < 7$, using $\mu = np$.

III. The Hypergeometric Random Variable

1. The number of successes in a sample of size n from a finite population containing M successes and $N - M$ failures

2. Formula for the probability of k successes in n trials:
$$P(x = k) = \frac{C_k^M C_{n-k}^{N-M}}{C_n^N}$$

3. Mean of the hypergeometric random variable:
$$\mu = n\left(\frac{M}{N}\right)$$

4. Variance and standard deviation:
$$\sigma^2 = n\left(\frac{M}{N}\right)\left(\frac{N-M}{N}\right)\left(\frac{N-n}{N-1}\right) \text{ and } \sigma = \sqrt{\sigma^2}$$

 TECHNOLOGY TODAY

Binomial and Poisson Probabilities in *Microsoft Excel*

For a random variable that has either a binomial or a Poisson probability distribution, *MS Excel* has been programmed to calculate either exact probabilities—$P(x = k)$—for a given value of k or cumulative probabilities—$P(x \le k)$—for a given value of k. You must specify which distribution you are using and the necessary parameters: n and p for the binomial distribution and μ for the Poisson distribution.

Binomial Probabilities

1. Consider a binomial distribution with $n = 10$ and $p = .25$. The *value* of p does not appear in Table 1 of Appendix I, but you can use *Excel* to generate the entire probability distribution as well as the cumulative probabilities by entering the numbers 0–10 in column A.

One way to quickly enter a set of consecutive integers in a column is to do the following:

- Name columns A, B, and C as "x," "P(x = k)", and P(x<=k), respectively.
- Enter the first two values of x—0 and 1—to create a pattern in column A.
- Use your mouse to highlight the first two integers. Then grab the square handle in the lower right corner of the highlighted area. Drag the handle down to continue the pattern.
- As you drag, you will see an integer appear in a small rectangle. Release the mouse when you have the desired number of integers—in this case, 10.

2. Once the necessary values of x have been entered, place your cursor in the cell corresponding to $p(0)$, cell B2 in the spreadsheet. Select the **Insert Function** icon f_x. In the drop-down list, select the **Statistical** category, select the

BINOM.DIST function and click **OK**. (NOTE: This function is called BINOMDIST in *Excel 2007* and earlier versions.) The Dialog box shown in Figure 5.7 will appear.

FIGURE 5.7

3. You must type in or select numbers or cell locations for each of the four boxes. When you place your cursor in the box, you will see a description of the necessary input for that box. Enter the address of the cell corresponding to $x = 0$ (cell A2) in the first box, the value of n in the second box, the value of p in the third box, and the word FALSE in the fourth box to calculate $P(x = k)$.

4. The resulting probability is marked as "Formula result = .056313515" at the bottom of the box, and when you click **OK**, the probability $P(x = 0)$ will appear in cell B2. To obtain the other probabilities, simply place your cursor in cell B2, grab the square handle in the lower right corner of the cell and drag the handle down to copy the formula into the other nine cells. *MS Excel* will automatically adjust the cell location in the formula as you copy.

5. If you want to generate the cumulative probabilities, $P(x \leq k)$, place your cursor in the cell corresponding to $P(x \leq 0)$, cell C2 in the spreadsheet. Then select **Insert Function ▶ Statistical ▶ BINOM.DIST,** and click **OK**. Continue as in steps 3 and 4, but type TRUE in the fourth line of the Dialog box to calculate $P(x \leq k)$. The resulting output is shown in Figure 5.8.

FIGURE 5.8

	A	B	C
1	x	P(x = k)	P(x <= k)
2	0	0.0563	0.0563
3	1	0.1877	0.2440
4	2	0.2816	0.5256
5	3	0.2503	0.7759
6	4	0.1460	0.9219
7	5	0.0584	0.9803
8	6	0.0162	0.9965
9	7	0.0031	0.9996
10	8	0.0004	1.0000
11	9	0.0000	1.0000
12	10	0.0000	1.0000

6. What value k is such that only 5% of the values of x exceed this value (and 95% are less than or equal to k)? Place your cursor in an empty cell, select **Insert Function ▶ Statistical ▶ BINOM.INV**, and click **OK**. (NOTE: This function is new to *Excel 2010*.) The resulting Dialog box will calculate what is sometimes called the **inverse cumulative probability.** Type **10** in the first box, **.25** in the second box, and **.95** in the third box. When you click **OK,** the number **5** will

appear in the empty cell. This is the smallest value of k for which $P(x \leq k)$ is greater than or equal to .95. Refer to Figure 5.8 and notice that $P(x \leq 5) = .9803$ so that $P(x > 5) = 1 - .9803 = .0197$. Hence, if you observed a value of $x = 5$, this would be an unusual observation.

Poisson Probabilities

1. The procedures for calculating individual or cumulative probabilities and probability distributions for the Poisson random variable are similar to those used for the binomial distribution.

2. To find Poisson probabilities $P(x = k)$ or $P(x \leq k)$ select **Insert Function ▶ Statistical ▶ POISSON.DIST,** and click **OK.** (NOTE: This function is called **POISSON** in *Excel 2007* and earlier versions.) Enter the values for k, μ, and **FALSE/TRUE** before clicking **OK.**

3. There is no **inverse cumulative probability** as there was for the binomial distribution.

Binomial and Poisson Probabilities in *MINITAB*

For a random variable that has either a binomial or a Poisson probability distribution, *MINITAB* has been programmed to calculate either exact probabilities—$P(X = x)$—for a given value of x or the cumulative probabilities—$P(X \leq x)$—for a given value of x. (NOTE: *MINITAB* uses the notation "X" for the random variable and "x" for a particular value of the random variable.) You must specify which distribution you are using and the necessary parameters: n and p for the binomial distribution and μ for the Poisson distribution.

Binomial Probabilities

1. Consider a binomial distribution with $n = 10$ and $p = .25$. The *value* of p does not appear in Table 1 of Appendix I, but you can use *MINITAB* to generate the entire probability distribution as well as the cumulative probabilities by entering the numbers 0 to 10 in column A.

2. One way to quickly enter a set of consecutive integers in a column is to do the following:

   • Name columns C1, C2, and C3 as "x," "P(X = x)," and P(X<=x), respectively.

   • Enter the first two values of x—0 and 1—to create a pattern in column C1.

   • Use your mouse to highlight the first two integers. Then grab the square handle in the lower right corner of the highlighted area. Drag the handle down to continue the pattern.

   • As you drag, you will see an integer appear in a small rectangle. Release the mouse when you have the desired number of integers—in this case, $\boxed{10}$.

3. Once the necessary values of x have been entered, use **Calc ▶ Probability Distributions ▶ Binomial** to generate the Dialog box shown in Figure 5.9.

FIGURE 5.9

4. Type the number of trials and the value of p (Event probability) in the appropriate boxes, select C1 ('x') for the input column, and select C2 ('P(X = x)') for the Optional storage. Make sure that the radio button marked "Probability" is selected. The probability distribution for x will appear in column C2 when you click **OK.** (NOTE: If you do not select a column for the Optional storage, the results will be displayed in the Session window.)

5. If you want to generate the cumulative probabilities, $P(x \leq k)$, again use **Calc ▶ Probability Distributions ▶ Binomial** to generate the Dialog box. This time, select the radio button marked "Cumulative probability" and select C3 (P(X<=x)) for the Optional storage in the Dialog box (Figure 5.9). The cumulative probability distribution will appear in column C3 when you click **OK.** You can display both distributions in the Session window using **Data ▶ Display Data,** selecting columns C1–C3 and clicking **OK.** The results are shown in Figure 5.10(a).

(a)

FIGURE 5.10

(b)

6. What value x is such that only 5% of the values of the random variable X exceed this value (and 95% are less than or equal to x)? Again, use **Calc ▶ Probability Distributions ▶ Binomial** to generate the Dialog box. This time, select the radio button marked "Inverse cumulative probability" and enter the probability **.95** into the "Input constant" box (Figure 5.9). When you click **OK,** the values of x on either side of the ".95 mark" will appear in the Session window as shown in Figure 5.10(b). Hence, if you observed a value of $x = 5$, this would be an unusual observation, because $P(x > 5) = 1 - .980272 = .019728$.

Poisson Probabilities

1. The procedures for calculating individual or cumulative probabilities and probability distributions for the Poisson random variable are similar to those used for the binomial distribution.

2. To find Poisson probabilities $P(X = x)$ or $P(X \leq x)$, use **Calc ▶ Probability Distributions ▶ Poisson** to generate the Dialog box. Enter the value for the mean μ, choose the appropriate radio button, and input either a column or a constant to indicate the value(s) of X for which you want to calculate a probability before clicking **OK.**

3. The **inverse cumulative probability** calculates the values of x such that $P(X \leq x) = C$, where C is a constant probability, between 0 and 1. Follow the steps described for the binomial random variable in step 6 above.

Supplementary Exercises

5.59 List the five identifying characteristics of the binomial experiment.

5.60 Under what conditions can the Poisson random variable be used to approximate the probabilities associated with the binomial random variable? What application does the Poisson distribution have other than to estimate certain binomial probabilities?

5.61 Under what conditions would you use the hypergeometric probability distribution to evaluate the probability of x successes in n trials?

5.62 Tossing a Coin A balanced coin is tossed three times. Let x equal the number of heads observed.

a. Use the formula for the binomial probability distribution to calculate the probabilities associated with $x = 0, 1, 2,$ and 3.

b. Construct the probability distribution.

c. Find the mean and standard deviation of $x,$ using these formulas:

$$\mu = np$$
$$\sigma = \sqrt{npq}$$

d. Using the probability distribution in part b, find the fraction of the population measurements lying within one standard deviation of the mean. Repeat for two standard deviations. How do your results agree with Tchebysheff's Theorem and the Empirical Rule?

5.63 Coins, continued Refer to Exercise 5.62. Suppose the coin is definitely unbalanced and the probability of a head is equal to $p = .1.$ Follow the instructions in parts a, b, c, and d. Note that the probability

distribution loses its symmetry and becomes skewed when p is not equal to 1/2.

5.64 Cancer Survivor Rates The 10-year survival rate for bladder cancer is approximately 50%. If 20 people who have bladder cancer are properly treated for the disease, what is the probability that:

a. At least 1 will survive for 10 years?

b. At least 10 will survive for 10 years?

c. At least 15 will survive for 10 years?

5.65 Garbage Collection A city commissioner claims that 80% of all people in the city favor private garbage collection in contrast to collection by city employees. To check the 80% claim, you randomly sample 25 people and find that $x,$ the number of people who support the commissioner's claim, is 22.

a. What is the probability of observing at least 22 who support the commissioner's claim if, in fact, $p = .8?$

b. What is the probability that x is exactly equal to 22?

c. Based on the results of part a, what would you conclude about the claim that 80% of all people in the city favor private collection? Explain.

5.66 Integers If a person is given the choice of an integer from 0 to 9, is it more likely that he or she will choose an integer near the middle of the sequence than one at either end?

a. If the integers are equally likely to be chosen, find the probability distribution for $x,$ the number chosen.

b. What is the probability that a person will choose a 4, 5, or 6?

c. What is the probability that a person will not choose a 4, 5, or 6?

5.67 Integers II Refer to Exercise 5.66. Twenty people are asked to select a number from 0 to 9. Eight of them choose a 4, 5, or 6.

a. If the choice of any one number is as likely as any other, what is the probability of observing eight or more choices of the numbers 4, 5, or 6?

b. What conclusions would you draw from the results of part a?

5.68 Checking In Fewer Americans are really getting away while on vacation. In fact, among small business owners, more than half (51%) say they check in with the office at least once a day while on vacation; only 27% say they cut the cord completely.[7] If 20 small business owners are randomly selected, and we assume that exactly half check in with the office at least once a day, then $n = 20$ and $p = .5$. Find the following probabilities.

a. Exactly 16 say that they check in with the office at least once a day while on vacation.

b. Between 15 and 18 (inclusive) say they check in with the office at least once a day while on vacation.

c. Five or fewer say that they check in with the office at least once a day while on vacation. Would this be an unlikely occurrence?

5.69 Psychosomatic Problems A psychiatrist believes that 80% of all people who visit doctors have problems of a psychosomatic nature. She decides to select 25 patients at random to test her theory.

a. Assuming that the psychiatrist's theory is true, what is the expected value of x, the number of the 25 patients who have psychosomatic problems?

b. What is the variance of x, assuming that the theory is true?

c. Find $P(x \leq 14)$. (Use tables and assume that the theory is true.)

d. Based on the probability in part c, if only 14 of the 25 sampled had psychosomatic problems, what conclusions would you make about the psychiatrist's theory? Explain.

5.70 Student Fees A student government states that 80% of all students favor an increase in student fees to subsidize a new recreational area. A random sample of $n = 25$ students produced 15 in favor of increased fees. What is the probability that 15 or fewer in the sample would favor the issue if student government is correct? Do the data support the student government's assertion, or does it appear that the percentage favoring an increase in fees is less than 80%?

5.71 Gray Hair on Campus College campuses are graying! According to a recent article, one in four college students is aged 30 or older. Assume that the 25% figure is accurate, that your college is representative of colleges at large, and that you sample $n = 200$ students, recording x, the number of students age 30 or older.

a. What are the mean and standard deviation of x?

b. If there are 35 students in your sample who are age 30 or older, would you be willing to assume that the 25% figure is representative of your campus? Explain.

5.72 Probability of Rain To check the accuracy of a particular weather forecaster, records were checked only for those days when the forecaster predicted rain "with 30% probability." A check of 25 of those days indicated that it rained on 10 of the 25.

a. If the forecaster is accurate, what is the appropriate value of p, the probability of rain on one of the 25 days?

b. What are the mean and standard deviation of x, the number of days on which it rained, assuming that the forecaster is accurate?

c. Calculate the z-score for the observed value, $x = 10$. [HINT: Recall from Section 2.6 that $z\text{-score} = (x - \mu)/\sigma$.]

d. Do these data disagree with the forecast of a "30% probability of rain"? Explain.

5.73 What's for Breakfast? A packaging experiment is conducted by placing two different package designs for a breakfast food side by side on a supermarket shelf. On a given day, 25 customers purchased a package of the breakfast food from the supermarket. Let x equal the number of buyers who choose the second package design.

a. If there is no preference for either of the two designs, what is the value of p, the probability that a buyer chooses the second package design?

b. If there is no preference, use the results of part a to calculate the mean and standard deviation of x.

c. If 5 of the 25 customers choose the first package design and 20 choose the second design, what do you conclude about the customers' preference for the second package design?

5.74 Plant Density One model for plant competition assumes that there is a zone of resource depletion around each plant seedling. Depending on the size of the zones and the density of the plants, the zones of resource depletion may overlap with those of other seedlings in the vicinity. When the seeds are randomly dispersed over a wide area, the number of neighbors that a seedling may have usually follows a Poisson distribution with a mean equal to the density of seedlings per unit area. Suppose that the density of seedlings is four per square meter (m^2).

a. What is the probability that a given seedling has no neighbors within 1 m^2?

b. What is the probability that a seedling has at most three neighbors per m^2?

c. What is the probability that a seedling has five or more neighbors per m^2?

d. Use the fact that the mean and variance of a Poisson random variable are equal to find the proportion of neighbors that would fall into the interval $\mu \pm 2\sigma$. Comment on this result.

5.75 Plant Genetics A peony plant with red petals was crossed with another plant having streaky petals. The probability that an offspring from this cross has red flowers is .75. Let x be the number of plants with red petals resulting from ten seeds from this cross that were collected and germinated.

a. Does the random variable x have a binomial distribution? If not, why not? If so, what are the values of n and p?

b. Find $P(x \geq 9)$.

c. Find $P(x \leq 1)$.

d. Would it be unusual to observe one plant with red petals and the remaining nine plants with streaky petals? If these experimental results actually occurred, what conclusions could you draw?

5.76 Dominant Traits The alleles for black (B) and white (b) feather color in chickens show incomplete dominance; individuals with the gene pair Bb have "blue" feathers. When one individual that is homozygous dominant (BB) for this trait is mated with an individual that is homozygous recessive (bb) for this trait, 1/2 will carry the gene pair Bb. Let x be the

number of chicks with "blue" feathers in a sample of $n = 20$ chicks resulting from this type of cross.

a. Does the random variable x have a binomial distribution? If not, why not? If so, what are the values of n and p?

b. What is the mean number of chicks with "blue" feathers in the sample?

c. What is the probability of observing fewer than five chicks with "blue" feathers?

d. What is the probability that the number of chicks with "blue" feathers is greater than or equal to 10 but less than or equal to 12?

5.77 Football Coin Tosses During the 1992 football season, the Los Angeles Rams (now the St. Louis Rams) had a bizarre streak of coin-toss losses. In fact, they lost the call 11 weeks in a row.[8]

a. The Rams' computer system manager said that the odds against losing 11 straight tosses are 2047 to 1. Is he correct?

b. After these results were published, the Rams lost the call for the next two games, for a total of 13 straight losses. What is the probability of this happening if, in fact, the coin was fair?

5.78 Diabetes in Children Insulin-dependent diabetes (IDD) among children occurs most frequently in persons of northern European descent. The incidence ranges from a low of 1–2 cases per 100,000 per year to a high of more than 40 per 100,000 in parts of Finland.[9] Let us assume that an area in Europe has an incidence of 5 cases per 100,000 per year.

a. Can the distribution of the number of cases of IDD in this area be approximated by a Poisson distribution? If so, what is the mean?

b. What is the probability that the number of cases of IDD in this area is less than or equal to 3 per 100,000?

c. What is the probability that the number of cases is greater than or equal to 3 but less than or equal to 7 per 100,000?

d. Would you expect to observe 10 or more cases of IDD per 100,000 in this area in a given year? Why or why not?

5.79 Problems with Your New Smartphone? A new study by Square Trade indicates that smartphones are 50% more likely to malfunction than simple phones over a 3-year period.[10] Of smartphone failures, 30% are related to internal components not working, and overall, there is a 31% chance of having your smartphone fail over 3 years. Suppose that smartphones are shipped in cartons of $N = 50$ phones. Before shipment $n = 10$ phones are selected from each carton and the carton is shipped if none of the selected phones are defective. If one or more are found to be defective, the whole carton is tested.

a. What is the probability distribution of x, the number of defective phones related to internal components not working in the sample of $n = 10$ phones?

b. What is the probability that the carton will be shipped if two of the $N = 50$ smartphones in the carton have defective internal components?

c. What is the probability that the carton will be shipped if it contains four defectives? Six defectives?

5.80 Dark Chocolate Despite reports that dark chocolate is beneficial to the heart, 47% of adults still prefer milk chocolate to dark chocolate.[11] Suppose a random sample of $n = 5$ adults is selected and asked whether they prefer milk chocolate to dark chocolate.

a. What is the probability that all five adults say that they prefer milk chocolate to dark chocolate?

b. What is the probability that exactly three of the five adults say they prefer milk chocolate to dark chocolate?

c. What is the probability that at least one adult prefers milk chocolate to dark chocolate?

5.81 Tay–Sachs Disease Tay–Sachs disease is a genetic disorder that is usually fatal in young children. If both parents are carriers of the disease, the probability that their offspring will develop the disease is approximately .25. Suppose a husband and wife are both carriers of the disease and the wife is pregnant on three different occasions. If the occurrence of Tay–Sachs in any one offspring is independent of the occurrence in any other, what are the probabilities of these events?

a. All three children will develop Tay–Sachs disease.

b. Only one child will develop Tay–Sachs disease.

c. The third child will develop Tay–Sachs disease, given that the first two did not.

5.82 Conservative Spenders A Snapshot in *USA Today* shows that 60% of consumers say they have become more conservative spenders.[12] When asked "What would you do first if you won $1 million

tomorrow?" the answers had to do with somewhat conservative measures like "hire a financial advisor," or "pay off my credit card," or "pay off my mortgage." Suppose a random sample of $n = 15$ consumers is selected and the number x of those who say they have become conservative spenders recorded.

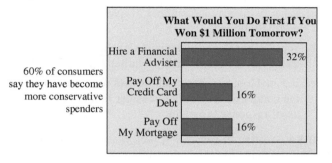

a. What is the probability that more than six consumers say they have become conservative spenders?

b. What is the probability that fewer than five of those sampled have become conservative spenders?

c. What is the probability that exactly nine of those sampled are now conservative spenders.

5.83 The Triangle Test A procedure often used to control the quality of name-brand food products utilizes a panel of five "tasters." Each member of the panel tastes three samples, two of which are from batches of the product known to have the desired taste and the other from the latest batch. Each taster selects the sample that is different from the other two. Assume that the latest batch does have the desired taste, and that there is no communication between the tasters.

a. If the latest batch tastes the same as the other two batches, what is the probability that the taster picks it as the one that is different?

b. What is the probability that exactly one of the tasters picks the latest batch as different?

c. What is the probability that at least one of the tasters picks the latest batch as different?

5.84 Do You Return Your Questionnaires? A public opinion research firm claims that approximately 70% of those sent questionnaires respond by returning the questionnaire. Twenty such questionnaires are sent out, and assume that the president's claim is correct.

a. What is the probability that exactly 10 of the questionnaires are filled out and returned?

b. What is the probability that at least 12 of the questionnaires are filled out and returned?

c. What is the probability that at most 10 of the questionnaires are filled out and returned?

5.85 Questionnaires, continued Refer to Exercise 5.84. If $n = 20$ questionnaires are sent out,

a. What is the average number of questionnaires that will be returned?

b. What is the standard deviation of the number of questionnaires that will be returned?

c. If $x = 10$ of the 20 questionnaires are returned to the company, would you consider this to be an unusual response? Explain.

5.86 Poultry Problems A preliminary investigation reported that approximately 30% of locally grown poultry were infected with an intestinal parasite that, though not harmful to those consuming the poultry, decreased the usual weight growth rates in the birds. A diet supplement believed to be effective against this parasite was added to the bird's food. Twenty-five birds were examined after having the supplement for at least two weeks, and three birds were still found to be infested with the parasite.

a. If the diet supplement is ineffective, what is the probability of observing three or fewer birds infected with the intestinal parasite?

b. If in fact the diet supplement was effective and reduced the infection rate to 10%, what is the probability observing three or fewer infected birds?

5.87 Machine Breakdowns In a food processing and packaging plant, there are, on the average, two packaging machine breakdowns per week. Assume the weekly machine breakdowns follow a Poisson distribution.

a. What is the probability that there are no machine breakdowns in a given week?

b. Calculate the probability that there are no more than two machine breakdowns in a given week.

5.88 Safe Drivers? Evidence shows that the probability that a driver will be involved in a serious automobile accident during a given year is .01. A particular corporation employs 100 full-time traveling sales reps. Based on this evidence, use the Poisson approximation to the binomial distribution to find the probability that exactly two of the sales reps will be involved in a serious automobile accident during the coming year.

5.89 Stressed Out A subject is taught to do a task in two different ways. Studies have shown that when subjected to mental strain and asked to perform the task, the subject most often reverts to the method first learned, regardless of whether it was easier or more difficult. If the probability that a subject returns to the first method learned is .8 and six subjects are tested, what is the probability that at least five of the subjects revert to their first learned method when asked to perform their task under stress?

5.90 Enrolling in College A West Coast university has found that about 90% of its accepted applicants for enrollment in the freshman class will actually enroll. In 2012, 1360 applicants were accepted to the university. Within what limits would you expect to find the size of the freshman class at this university in the fall of 2012?

5.91 Earthquakes! Suppose that one out of every 10 homeowners in the state of California has invested in earthquake insurance. If 15 homeowners are randomly chosen to be interviewed,

a. What is the probability that at least one had earthquake insurance?

b. What is the probability that four or more have earthquake insurance?

c. Within what limits would you expect the number of homeowners insured against earthquakes to fall?

5.92 Bad Wiring Improperly wired control panels were mistakenly installed on two of eight large automated machine tools. It is uncertain which of the machine tools have the defective panels, and a sample of four tools is randomly chosen for inspection. What is the probability that the sample will include no defective panels? Both defective panels?

5.93 Eating on the Run How do you survive when there's no time to eat—fast food, no food, a protein bar, candy? A Snapshot in *USA Today* indicates that 36% of women aged 25–55 say that, when they are too busy to eat, they get fast food from a drive-thru.[13] A random sample of 100 women aged 25–55 is selected.

a. What is the average number of women who say they eat fast food when they're too busy to eat?

b. What is the standard deviation for the number of women who say they eat fast food when they're too busy to eat?

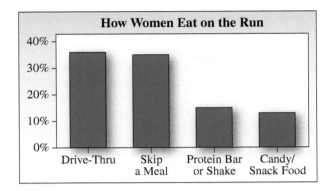

How Women Eat on the Run

c. If 49 of the women in the sample said they eat fast food when they're too busy to eat, would this be an unusual occurrence? Explain.

5.94 Drivable Streets According to a *USA Today* Snapshot, drivers say fixing or repaving streets is the best way to make their communities drivable—better than building new roads or adding lanes.[14] Suppose that $n = 15$ drivers are randomly selected and x is the number who say that improved road conditions would make their communities more drivable. Let $p = .4$ when finding probabilities associated with any following outcomes:

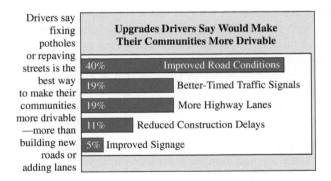

Drivers say fixing potholes or repaving streets is the best way to make their communities more drivable —more than building new roads or adding lanes

Upgrades Drivers Say Would Make Their Communities More Drivable

- 40% Improved Road Conditions
- 19% Better-Timed Traffic Signals
- 19% More Highway Lanes
- 11% Reduced Construction Delays
- 5% Improved Signage

a. What is the probability distribution for x?

b. What is $P(x \leq 4)$?

c. Find the probability that x exceeds 5.

d. What is the largest value of c for which $P(x \leq c) \leq .5$?

5.95 Credit Problems? The recession has caused many people to use their credit cards far less. In fact, in the United States, 60% of consumers say they are committed to living with fewer credit cards.[15] A sample of $n = 400$ consumers with credit cards are randomly selected.

a. What is the average number of consumers in the sample who said they are committed to living with fewer credit cards?

b. What is the standard deviation of the number in the sample who said they are committed to living with fewer credit cards?

c. Within what range would you expect to find the number in the sample who said they are committed to living with fewer credit cards?

d. If only 200 of the sample of consumers said they were committed to living with fewer credit cards, would you consider this unusual? Explain. What conclusion might you draw from this sample information?

5.96 Successful Surgeries A new surgical procedure is said to be successful 80% of the time. Suppose the operation is performed five times and the results are assumed to be independent of one another. What are the probabilities of these events?

a. All five operations are successful.

b. Exactly four are successful.

c. Less than two are successful.

5.97 Surgery, continued Refer to Exercise 5.96. If less than two operations were successful, how would you feel about the performance of the surgical team?

5.98 Engine Failure Suppose the four engines of a commercial aircraft are arranged to operate independently and that the probability of in-flight failure of a single engine is .01. What is the probability of the following events on a given flight?

a. No failures are observed.

b. No more than one failure is observed.

5.99 McDonald's or Burger King? Suppose that 50% of all young adults prefer McDonald's to Burger King when asked to state a preference. A group of 10 young adults were randomly selected and their preferences recorded.

a. What is the probability that more than 6 preferred McDonald's?

b. What is the probability that between 4 and 6 (inclusive) preferred McDonald's?

c. What is the probability that between 4 and 6 (inclusive) preferred Burger King?

5.100 Vacation Destinations High gas prices may keep some American vacationers closer to home. However, when given a choice of getaway spots, 66% of U.S. leisure travelers indicated that they would like to visit national parks.[16] A random sample of $n = 100$ leisure travelers is selected.

a. What is the average of x, the number of travelers in the sample who indicate they would like to visit national parks? What is the standard deviation of x?

b. Would it be unlikely to find only 50 or fewer of those sampled who indicated they would like to visit national parks?

CASE STUDY

A Mystery: Cancers Near a Reactor

How safe is it to live near a nuclear reactor? Men who lived in a coastal strip that extends 20 miles north from a nuclear reactor in Plymouth, Massachusetts, developed some forms of cancer at a rate 50% higher than the statewide rate, according to a study endorsed by the Massachusetts Department of Public Health and reported in the May 21, 1987, edition of the *New York Times*.[17]

The cause of the cancers is a mystery, but it was suggested that the cancer was linked to the Pilgrim I reactor, which had been shut down for 13 months because of management problems. Boston Edison, the owner of the reactor, acknowledged radiation releases in the mid-1970s that were just above permissible levels. If the reactor was in fact responsible for the excessive cancer rate, then the currently acknowledged level of radiation required to cause cancer would have to change. However, confounding the mystery was the fact that women in this same area were seemingly unaffected.

In his report, Dr. Sidney Cobb, an epidemiologist, noted the connection between the radiation releases at the Pilgrim I reactor and 52 cases of hematopoietic cancers. The report indicated that this unexpectedly large number might be attributable to airborne radioactive effluents from Pilgrim I, concentrated along the coast by wind patterns and not dissipated, as assumed by government regulators. How unusual was this number of cancer cases? That is, statistically speaking, is 52 a highly improbable number of cases? If the answer is yes, then either some external factor (possibly radiation) caused this unusually large number, or we have observed a very rare event!

The Poisson probability distribution provides a good approximation to the distributions of variables such as the number of deaths in a region due to a rare disease, the number of accidents in a manufacturing plant per month, or the number of airline crashes per month. Therefore, it is reasonable to assume that the Poisson distribution provides an appropriate model for the number of cancer cases in this instance.

1. If the 52 reported cases represented a rate 50% higher than the statewide rate, what is a reasonable estimate of μ, the average number of such cancer cases statewide?

2. Based on your estimate of μ, what is the estimated standard deviation of the number of cancer cases statewide?

3. What is the z-score for the $x = 52$ observed cases of cancer? How do you interpret this z-score in light of the concern about an elevated rate of hematopoietic cancers in this area?

The Normal Probability Distribution

Laurence Gough/Shutterstock.com

GENERAL OBJECTIVES

In Chapters 4 and 5, you learned about discrete random variables and their probability distributions. In this chapter, you will learn about continuous random variables and their probability distributions and about one very important continuous random variable—the normal. You will learn how to calculate normal probabilities and, under certain conditions, how to use the normal probability distribution to approximate the binomial probability distribution. Then, in Chapter 7 and in the chapters that follow, you will see how the normal probability distribution plays a central role in statistical inference.

CHAPTER INDEX

 NEED TO KNOW...

How to Use Table 3 to Calculate Probabilities under the Standard Normal Curve
How to Calculate Binomial Probabilities Using the Normal Approximation

"Are You Going to Curve the Grades?"

"Curving the grades" doesn't necessarily mean that you will receive a higher grade on a test, although many students would like to think so! Curving actually refers to a method of assigning the letter grades A, B, C, D, or F using fixed proportions of the grades corresponding to each of the letter grades. One such curving technique assumes that the distribution of the grades is approximately normal and uses these proportions.

Letter Grade	A	B	C	D	F
Proportion of grades	10%	20%	40%	20%	10%

In the case study at the end of this chapter, we will examine this and other assigned proportions for curving grades.

PROBABILITY DISTRIBUTIONS FOR CONTINUOUS RANDOM VARIABLES

6.1

When a random variable x is discrete, you can assign a positive probability to each value that x can take and get the probability distribution for x. The sum of all the probabilities associated with the different values of x is 1. However, not all experiments result in random variables that are discrete.

Continuous random variables, such as heights and weights, length of life of a particular product, or experimental laboratory error, can assume the infinitely many values corresponding to points on a line interval. If you try to assign a positive probability to each of these uncountable values, the probabilities will no longer sum to 1, as with discrete random variables. Therefore, you must use a different approach to generate the probability distribution for a continuous random variable.

Suppose you have a set of measurements on a continuous random variable, and you create a relative frequency histogram to describe their distribution. For a small number of measurements, you could use a small number of classes; then as more and more measurements are collected, you can use more classes and reduce the class width. The outline of the histogram will change slightly, for the most part becoming less and less irregular, as shown in Figure 6.1. As the number of measurements becomes very large and the class widths become very narrow, the relative frequency histogram appears more and more like the smooth curve shown in Figure 6.1(d). This smooth curve describes the **probability distribution of the continuous random variable.**

FIGURE 6.1

Relative frequency histograms for increasingly large sample sizes

(a)

(b)

(c)

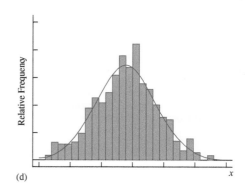

(d)

How can you create a model for this probability distribution? A continuous random variable can take on any of an infinite number of values on the real line, much like the infinite number of grains of sand on a beach. The probability distribution is created by distributing one unit of probability along the line, much as you might distribute a handful of sand. The probability—grains of sand or measurements—will pile up in certain places, and the result is the probability distribution shown in Figure 6.2. The depth or **density** of the probability, which varies with x, may be described by a mathematical formula $f(x)$, called the **probability distribution** or **probability density function** for the random variable x.

FIGURE 6.2

The probability distribution $f(x)$; $P(a < x < b)$ is equal to the shaded area under the curve

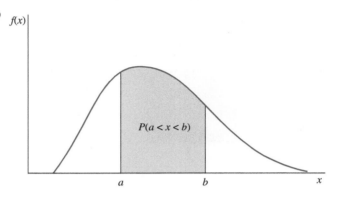

Remember that, for discrete random variables, (1) the sum of all the probabilities $p(x)$ equals 1 and (2) the probability that x falls into a certain interval is the sum of all the probabilities in that interval. Continuous random variables have some parallel characteristics listed next.

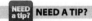

- The area under a continuous probability distribution is equal to 1.
- The probability that x will fall into a particular interval—say, from a to b—is equal to the area under the curve between the two points a and b. This is the shaded area in Figure 6.2.

There is also one important difference between discrete and continuous random variables. Consider the probability that x equals some particular value—say, a. Since there is no area above a single point—say, $x = a$—in the probability distribution for a continuous random variable, our definition implies that the probability is 0.

- $P(x = a) = 0$ for continuous random variables.
- This implies that $P(x \geq a) = P(x > a)$ and $P(x \leq a) = P(x < a)$.
- This is *not* true in general for discrete random variables.

How do you choose the model—that is, the probability distribution $f(x)$—appropriate for a given experiment? Many types of continuous curves are available for modeling. Some are mound-shaped, like the one in Figure 6.1(d), but others are not. In general, try to pick a model that meets these criteria:

- It fits the accumulated body of data.
- It allows you to make the best possible inferences using the data.

EXAMPLE 6.1

The *uniform random variable** is used to model the behavior of a continuous random variable whose values are uniformly or evenly distributed over a given interval. For example, the error x introduced by rounding an observation to the nearest inch would probably have a uniform distribution over the interval from $-.5$ to $.5$. The probability density function $f(x)$ would be "flat" as shown in Figure 6.3. The height of the rectangle is set at 1, so that the total area under the probability distribution is 1.

FIGURE 6.3

A uniform probability distribution

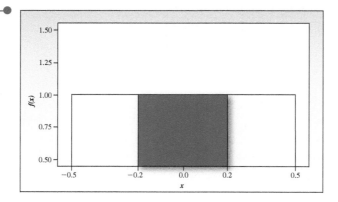

What is the probability that the rounding error is less than .2 in magnitude?

Solution   This probability corresponds to the area under the distribution between $x = -.2$ and $x = .2$. Since the height of the rectangle is 1,

$$P(-.2 < x < .2) = [.2 - (-.2)] \times 1 = .4$$

EXAMPLE 6.2

The *exponential random variable** is used to model continuous random variables such as waiting times or lifetimes associated with electronic components. For example, the waiting time at a supermarket checkout counter has an exponential distribution with an average waiting time of 5 minutes. The probability density function $f(x) = .2e^{-.2x}$ is shown in Figure 6.4. To find areas under this curve, you can use the fact that $P(x > a) = e^{-.2a}$ for $a > 0$. What is the probability that you have to wait 10 minutes or more at the checkout counter?

FIGURE 6.4

An exponential probability distribution

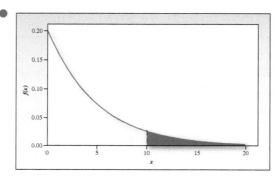

*The probability density function (*pdf*) for the *general uniform random variable x* is $f(x) = 1/(b - a)$ for $a \le x \le b$ and with mean $\mu = (a + b)/2$ and $\sigma^2 = (b - a)^2/12$. For the *exponential random variable x*, the *pdf* is $f(x) = \lambda e^{-\lambda x}$ for $x \ge 0$, $\lambda > 0$, with $\mu = 1/\lambda$ and $\sigma^2 = 1/\lambda^2$.

Solution   The probability to be calculated is the area shaded in Figure 6.4. Use the general formula for $P(x > a)$ to find

$$P(x > 10) = e^{-.2(10)} = .135$$

Your model may not always fit the experimental situation perfectly, but you should try to choose a model that *best fits* the population relative frequency histogram. The better the model approximates reality, the better your inferences will be. Fortunately, many continuous random variables have mound-shaped frequency distributions, such as the data in Figure 6.1(d). The **normal probability distribution** provides a good model for describing this type of data.

THE NORMAL PROBABILITY DISTRIBUTION

6.2

Continuous probability distributions can assume a variety of shapes. However, a large number of random variables observed in nature possess a frequency distribution that is approximately mound-shaped or, as the statistician would say, is approximately a normal probability distribution. The formula or probability density function (pdf) that generates this distribution is shown next.

NORMAL PROBABILITY DISTRIBUTION

$$f(x) = \frac{1}{\sigma\sqrt{2\pi}}\, e^{-(x-\mu)^2/(2\sigma^2)} \qquad -\infty < x < \infty$$

The symbols e and π are mathematical constants given approximately by 2.7183 and 3.1416, respectively; μ and σ ($\sigma > 0$) are parameters that represent the population mean and standard deviation, respectively.

The graph of a normal probability distribution with mean μ and standard deviation σ is shown in Figure 6.5. The mean μ locates the *center* of the distribution, and the distribution is *symmetric* about its mean μ. Since the total area under the normal probability distribution is equal to 1, this symmetry implies that the area to the right of μ is .5 and the area to the left of μ is also .5.

The *shape* of the distribution is determined by σ, the population standard deviation. As you can see in Figure 6.6, large values of σ reduce the height of the curve and increase the spread; small values of σ increase the height of the curve

FIGURE 6.5

Normal probability
distribution

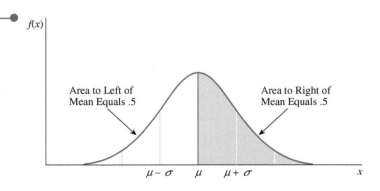

FIGURE 6.6

Normal probability
distributions with differing
values of μ and σ

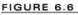 **ONLINE APPLET**

Visualizing Normal
Curves

and reduce the spread. Figure 6.6 shows three normal probability distributions with different means and standard deviations. Notice the differences in shape and location.

You rarely find a variable with values that are infinitely small ($-\infty$) or infinitely large ($+\infty$). Even so, many *positive* random variables (such as heights, weights, and times) have distributions that are well approximated by a normal distribution. According to the Empirical Rule, almost all values of a normal random variable lie in the interval $\mu \pm 3\sigma$. As long as the values within three standard deviations of the mean are *positive*, the normal distribution provides a good model to describe the data.

TABULATED AREAS OF THE NORMAL PROBABILITY DISTRIBUTION

6.3

To find the probability that a normal random variable x lies in the interval from a to b, we need to find the area under the normal curve between the points a and b (see Figure 6.2). However (see Figure 6.6), there are an infinitely large number of normal distributions—one for each different mean and standard deviation. A separate table of areas for each of these curves is obviously impractical. Instead, we use a standardization procedure that allows us to use the same table for all normal distributions.

The Standard Normal Random Variable

A normal random variable x is **standardized** by expressing its value as the number of standard deviations (σ) it lies to the left or right of its mean μ. This is really just a change in the units of measure that we use, as if we were measuring in inches rather than in centimeters! The standardized normal random variable, z, is defined as

$$z = \frac{x - \mu}{\sigma}$$

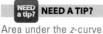 **NEED A TIP?**

Area under the z-curve
equals 1.

or equivalently,

$$x = \mu + z\sigma$$

From the formula for z, we can draw these conclusions:

- When x is less than the mean μ, the value of z is negative.
- When x is greater than the mean μ, the value of z is positive.
- When $x = \mu$, the value of $z = 0$.

The probability distribution for z, shown in Figure 6.7, is called the **standardized normal distribution** because its mean is 0 and its standard deviation is 1. Values of z on the left side of the curve are negative, while values on the right side are positive. The area under the standard normal curve to the left of a specified value of z—say, z_0—is the probability $P(z \le z_0)$. This **cumulative area** is recorded in Table 3 of Appendix I and is shown as the shaded area in Figure 6.7. An abbreviated version of Table 3 is given in Table 6.1. Notice that the table contains both positive and negative values of z. The left-hand column of the table gives the value of z correct to the tenths place; the second decimal place for z, corresponding to hundredths, is given across the top row.

FIGURE 6.7

Standardized normal distribution

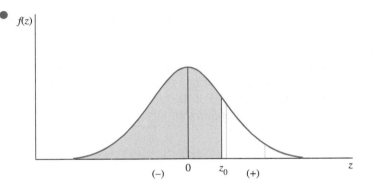

TABLE 6.1

Abbreviated Version of Table 3 in Appendix I
Table 3. Areas Under the Normal Curve

z	.00	.01	.02	.03	. . .	.09
−3.4	.0003	.0003	.0003	.0003		
−3.3	.0005	.0005	.0005	.0004		
−3.2	.0007	.0007	.0006	.0006		
−3.1	.0010	.0009	.0009	.0009		
−3.0	.0013	.0013	.0013	.0012	. . .	.0010
−2.9	.0019	.	.	.		
−2.8	.0026	.	.	.		
−2.7	.0035	.	.	.		
−2.6	.0047					
−2.5	.0062					
.	.					
.	.					
−2.0	.0228					
.	.					
.	.					
.	.					

Table 3. Areas Under the Normal Curve (*continued*)

z	.00	.01	.02	.03	.04	. . .	.09
0.0	.5000	.5040	.5080	.5120	.5160		
0.1	.5398	.5438	.5478	.5517	.5557		
0.2	.5793	.5832	.5871	.5910	.5948		
0.3	.6179	.6217	.6255	.6293	.6331		
0.4	.6554	.6591	.6628	.6664	.6700	. . .	.6879
0.5	.6915	.	.	.			
0.6	.7257	.	.	.			
0.7	.7580	.	.	.			
0.8	.7881						
0.9	.8159						
.	.						
.	.						
.	.						
2.0	.9772						

EXAMPLE 6.3

NEED A TIP?

$P(z \le 1.63) = P(z < 1.63)$

Find $P(z \le 1.63)$. This probability corresponds to the area to the left of a point $z = 1.63$ standard deviations to the right of the mean (see Figure 6.8).

Solution The area is shaded in Figure 6.8. Since Table 3 in Appendix I gives areas under the normal curve to the left of a specified value of z, you simply need to find the tabled value for $z = 1.63$. Proceed down the left-hand column of the table to $z = 1.6$ and across the top of the table to the column marked .03. The intersection of this row and column combination gives the area .9484, which is $P(z \le 1.63)$.

FIGURE 6.8

Area under the standard normal curve for Example 6.3

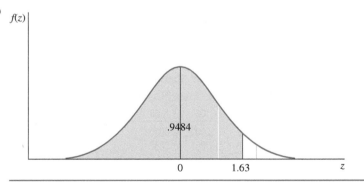

Areas to the left of $z = 0$ are found using negative values of z.

EXAMPLE 6.4

Find $P(z \ge -.5)$. This probability corresponds to the area to the *right* of a point $z = -.5$ standard deviation to the left of the mean (see Figure 6.9).

FIGURE 6.9

Area under the standard normal curve for Example 6.4

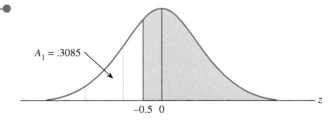

Solution   The area given in Table 3 in Appendix I is the area to the left of a specified value of z. Indexing $z = -.5$ in Table 3, we can find the area A_1 to the *left* of $-.5$ to be .3085.

Since the area under the curve is 1, we find

$$P(z \geq -.5) = 1 - A_1 = 1 - .3085 = .6915.$$

EXAMPLE 6.5   Find $P(-.5 \leq z \leq 1.0)$. This probability is the area between $z = -.5$ and $z = 1.0$, as shown in Figure 6.10.

FIGURE 6.10

Area under the standard normal curve for Example 6.5

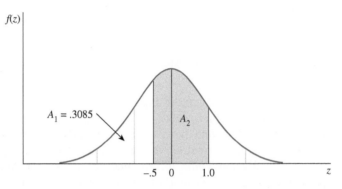

Solution   The area required is the shaded area A_2 in Figure 6.10. From Table 3 in Appendix I, you can find the area to the left of $z = -.5$ ($A_1 = .3085$) and the area to the left of $z = 1.0$ ($A_1 + A_2 = .8413$). To find the area marked A_2, we subtract the two entries:

$$A_2 = (A_1 + A_2) - A_1 = .8413 - .3085 = .5328$$

That is, $P(-.5 \leq z \leq 1.0) = .5328$.

? NEED TO KNOW...

How to Use Table 3 to Calculate Probabilities under the Standard Normal Curve

- To calculate the area to the left of a z-value, find the area directly from Table 3.

- To calculate the area to the right of a z-value, find the area in Table 3, and subtract from 1.

- To calculate the area between two values of z, find the two areas in Table 3, and subtract one area from the other.

EXAMPLE 6.6   Find the probability that a normally distributed random variable will fall within these ranges:

1. One standard deviation of its mean
2. Two standard deviations of its mean

Solution

1. Since the standard normal random variable z measures the distance from the mean in units of standard deviations, you need to find

$$P(-1 \leq z \leq 1) = .8413 - .1587 = .6826$$

   Remember that you calculate the area between two z-values by subtracting the tabled entries for the two values.

2. As in part 1, $P(-2 \leq z \leq 2) = .9772 - .0228 = .9544$.

These probabilities agree with the approximate values of 68% and 95% in the Empirical Rule from Chapter 2.

EXAMPLE **6.7**

Find the value of z—say z_0—such that .95 of the area is within $\pm z_0$ standard deviations of the mean.

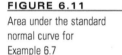

We know the area. Work from the inside of the table out.

Solution The shaded area in Figure 6.11 is the area within $\pm z_0$ standard deviations of the mean, which needs to be equal to .95. The "tail areas" under the curve are not shaded, and have a combined area of $1 - .95 = .05$. Because of the symmetry of the normal curve, these two tail areas have the same area, so that $A_1 = .05/2 = .025$ in Figure 6.11. Thus, the entire *cumulative area* to the left of $-z_0$ equals $A_1 = .025$. This area is found in the interior of Table 3 in Appendix I in the row corresponding to $z = -1.9$ and the .06 column. Hence, $-z_0 = -1.96$ or $z_0 = 1.96$. Note that this result is very close to the approximate value, $z = 2$, used in the Empirical Rule.

FIGURE 6.11

Area under the standard normal curve for Example 6.7

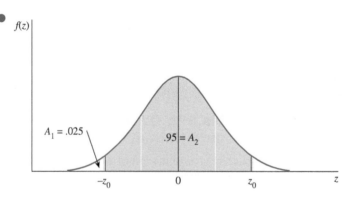

Calculating Probabilities for a General Normal Random Variable

Most of the time, the probabilities you are interested in will involve x, a normal random variable with mean μ and standard deviation σ. You must then *standardize* the interval of interest, writing it as the equivalent interval in terms of z, the standard normal random variable. Once this is done, the probability of interest is the area that you find using the *standard normal probability distribution*.

EXAMPLE 6.8

Let x be a normally distributed random variable with a mean of 10 and a standard deviation of 2. Find the probability that x lies between 11 and 13.6.

Solution The interval from $x = 11$ to $x = 13.6$ must be standardized using the formula for z. When $x = 11$,

NEED A TIP?

Always draw a picture—it helps!

$$z = \frac{x - \mu}{\sigma} = \frac{11 - 10}{2} = .5$$

and when $x = 13.6$,

$$z = \frac{x - \mu}{\sigma} = \frac{13.6 - 10}{2} = 1.8$$

The desired probability is therefore $P(.5 \leq z \leq 1.8)$, the area lying between $z = .5$ and $z = 1.8$, as shown in Figure 6.12. From Table 3 in Appendix I, you find that the area to the left of $z = .5$ is .6915, and the area to the left of $z = 1.8$ is .9641. The desired probability is the difference between these two probabilities, or

$$P(.5 \leq z \leq 1.8) = .9641 - .6915 = .2726$$

FIGURE 6.12

Area under the standard normal curve for Example 6.8

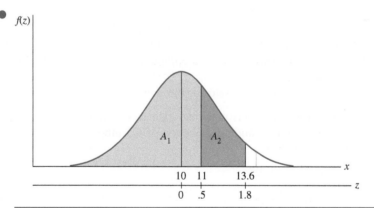

ONLINE APPLET

Normal Probability Distributions & Normal Probabilities and z-Scores?

EXAMPLE 6.9

Studies show that gasoline use for compact cars sold in the United States is normally distributed, with a mean of 35.5 miles per gallon (mpg) and a standard deviation of 4.5 mpg. What percentage of compacts get 40 mpg or more?

Solution The proportion of compacts that get 40 mpg or more is given by the shaded area in Figure 6.13. To solve this problem, you must first find the z-value corresponding to $x = 40$ by calculating.

$$z = \frac{x - \mu}{\sigma} = \frac{40 - 35.5}{4.5} = 1.0$$

From Table 3 in Appendix I, the area A_1 to the left of $z = 1.0$ is .8413. Then the proportion of compacts that get 40 mpg or more is equal to

$$P(x \geq 40) = 1 - P(z < 1) = 1 - .8413 = .1587$$

The percentage exceeding 40 mpg is

$$100(.1587) = 15.87\%$$

FIGURE 6.13

Area under the standard
normal curve for
Example 6.9

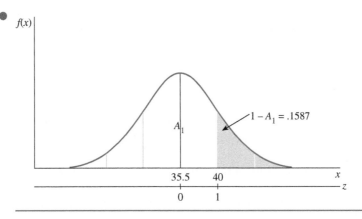

EXAMPLE 6.10

Refer to Example 6.9. In times of scarce energy resources, a competitive advantage is given to an automobile manufacturer who can produce a car that has substantially better fuel economy than the competitors' cars. If a manufacturer wishes to develop a compact car that outperforms 95% of the current compacts in fuel economy, what must the gasoline use rate for the new car be?

Solution   The gasoline use rate x has a normal distribution with a mean of 35.5 mpg and a standard deviation of 4.5 mpg. You need to find a particular value—say, x_0—such that

$$P(x \leq x_0) = .95$$

This is the 95th percentile of the distribution of gasoline use rate x. Since the only information you have about normal probabilities is in terms of the standard normal random variable z, start by standardizing the value of x_0:

$$z_0 = \frac{x_0 - 35.5}{4.5}$$

Since the value of z_0 corresponds to x_0, it must *also* have area .95 to its left, as shown in Figure 6.14. If you look in the interior of Table 3 in Appendix I, you will find that the area .9500 is exactly halfway between the areas for $z = 1.64$ and $z = 1.65$. Thus, we take z_0 to be halfway between 1.64 and 1.65, or

$$z_0 = \frac{x_0 - 35.5}{4.5} = 1.645$$

Solving for x_0, you obtain

$$x_0 = \mu + z_0\sigma = 35.5 + (1.645)(4.5) = 42.9$$

FIGURE 6.14 ●

Area under the standard
normal curve for
Example 6.10

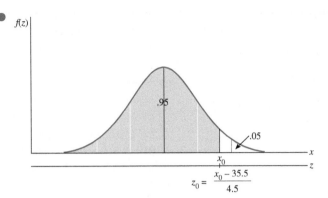

$z_0 = \dfrac{x_0 - 35.5}{4.5}$

The manufacturer's new compact car must therefore get 42.9 mpg to outperform 95% of the compact cars currently available on the U.S. market.

6.3 EXERCISES

BASIC TECHNIQUES

6.1 Consider a standard normal random variable with $\mu = 0$ and standard deviation $\sigma = 1$. Use Table 3 to find the following probabilities:

a. $P(z < 2)$ **b.** $P(z > 1.16)$
c. $P(-2.33 < z < 2.33)$ **d.** $P(z < 1.88)$

6.2 Find these probabilities associated with the standard normal random variable z:

a. $P(z > 5)$ **b.** $P(-3 < z < 3)$
c. $P(z < 2.81)$ **d.** $P(z > 2.81)$

6.3 Calculate the area under the standard normal curve to the left of these values:

a. $z = 1.6$ **b.** $z = 1.83$
c. $z = .90$ **d.** $z = 4.18$

6.4 Calculate the area under the standard normal curve between these values:

a. $z = -1.4$ and $z = 1.4$ **b.** $z = -3.0$ and $z = 3.0$

6.5 Find the following probabilities for the standard normal random variable z:

a. $P(-1.43 < z < .68)$ **b.** $P(.58 < z < 1.74)$
c. $P(-1.55 < z < -.44)$ **d.** $P(z > 1.34)$
e. $P(z < -4.32)$

6.6 Find these probabilities for the standard normal random variable z:

a. $P(z < 2.33)$ **b.** $P(z < 1.645)$
c. $P(z > 1.96)$ **d.** $P(-2.58 < z < 2.58)$

6.7 a. Find a z_0 such that $P(z > z_0) = .025$.
b. Find a z_0 such that $P(z < z_0) = .9251$.

6.8 Find a z_0 such that $P(-z_0 < z < z_0) = .8262$.

6.9 a. Find a z_0 that has area .9505 to its left.
b. Find a z_0 that has area .05 to its left.

6.10 a. Find a z_0 such that $P(-z_0 < z < z_0) = .90$.
b. Find a z_0 such that $P(-z_0 < z < z_0) = .99$.

6.11 Find the following *percentiles* for the standard normal random variable z:

a. 90th percentile   **b.** 95th percentile
c. 98th percentile   **d.** 99th percentile

6.12 A normal random variable x has mean $\mu = 10$ and standard deviation $\sigma = 2$. Find the probability associated with each of the following intervals.

a. $x > 13.5$ **b.** $x < 8.2$ **c.** $9.4 < x < 10.6$

6.13 A normal random variable x has mean $\mu = 1.2$ and standard deviation $\sigma = .15$. Find the probability associated with each of the following intervals.

a. $1.00 < x < 1.10$ **b.** $x > 1.38$
c. $1.35 < x < 1.50$

6.14 A normal random variable x has an unknown mean μ and standard deviation $\sigma = 2$. If the probability that x exceeds 7.5 is .8023, find μ.

6.15 A normal random variable x has mean 35 and standard deviation 10. Find a value of x that has area

.01 to its right. This is the *99th percentile* of this normal distribution.

6.16 A normal random variable x has mean 50 and standard deviation 15. Would it be unusual to observe the value $x = 0$? Explain your answer.

6.17 A normal random variable x has an unknown mean and standard deviation. The probability that x exceeds 4 is .9772, and the probability that x exceeds 5 is .9332. Find μ and σ.

APPLICATIONS

6.18 Hamburger Meat The meat department at a local supermarket specifically prepares its "1-pound" packages of ground beef so that there will be a variety of weights, some slightly more and some slightly less than 1 pound. Suppose that the weights of these "1-pound" packages are normally distributed with a mean of 1.00 pound and a standard deviation of .15 pound.

a. What proportion of the packages will weigh more than 1 pound?

b. What proportion of the packages will weigh between .95 and 1.05 pounds?

c. What is the probability that a randomly selected package of ground beef will weigh less than .80 pound?

d. Would it be unusual to find a package of ground beef that weighs 1.45 pounds? How would you explain such a large package?

6.19 Human Heights Human heights are one of many biological random variables that can be modeled by the normal distribution. Assume that the heights of American men have a mean of 69.5 inches and a standard deviation of 3.5 inches.

a. What proportion of all men will be taller than 6'0"? (HINT: Convert the measurements to inches.)

b. What is the probability that a randomly selected man will be between 5'8" and 6'1" tall?

c. President Barack Obama is 6'1". Is this an unusual height?

d. Of the 43 presidents elected from 1789 through 2008, 18 were 6'0" or taller.[1] Would you consider this to be unusual, given the proportion found in part a?

6.20 Christmas Trees The diameters of Douglas firs grown at a Christmas tree farm are normally distributed with a mean of 4 inches and a standard deviation of 1.5 inches.

a. What proportion of the trees will have diameters between 3 and 5 inches?

b. What proportion of the trees will have diameters less than 3 inches?

c. Your Christmas tree stand will expand to a diameter of 6 inches. What proportion of the trees will not fit in your Christmas tree stand?

6.21 Cerebral Blood Flow Cerebral blood flow (CBF) in the brains of healthy people is normally distributed with a mean of 74 and a standard deviation of 16.

a. What proportion of healthy people will have CBF readings between 60 and 80?

b. What proportion of healthy people will have CBF readings above 100?

c. If a person has a CBF reading below 40, he is classified as at risk for a stroke. What proportion of healthy people will mistakenly be diagnosed as "at risk"?

6.22 Braking Distances For a car traveling 30 miles per hour (mph), the distance required to brake to a stop is normally distributed with a mean of 50 feet and a standard deviation of 8 feet. Suppose you are traveling 30 mph in a residential area and a car moves abruptly into your path at a distance of 60 feet.

a. If you apply your brakes, what is the probability that you will brake to a stop within 40 feet or less? Within 50 feet or less?

b. If the only way to avoid a collision is to brake to a stop, what is the probability that you will avoid the collision?

6.23 Elevator Capacities Suppose that you must establish regulations concerning the maximum number of people who can occupy an elevator. A study indicates that if eight people occupy the elevator, the probability distribution of the total weight of the eight people is approximately normally distributed with a mean equal to 1200 pounds and a standard deviation of 99 pounds. What is the probability that the total weight of eight people exceeds 1300 pounds? 1500 pounds?

6.24 A Phosphate Mine The discharge of suspended solids from a phosphate mine is normally distributed, with a mean daily discharge of 27 milligrams per liter (mg/l) and a standard deviation of 14 mg/l. On what proportion of days will the daily discharge exceed 50 mg/l?

6.25 Sunflowers An experimenter publishing in the *Annals of Botany* investigated whether the stem diameters of the dicot sunflower would change depending on whether the plant was left to sway freely in the wind or was artificially supported.[2] Suppose that the unsupported stem diameters at the base of a particular species of sunflower plant have a normal distribution with an average diameter of 35 millimeters (mm) and a standard deviation of 3 mm.

a. What is the probability that a sunflower plant will have a basal diameter of more than 40 mm?

b. If two sunflower plants are randomly selected, what is the probability that both plants will have a basal diameter of more than 40 mm?

c. Within what limits would you expect the basal diameters to lie, with probability .95?

d. What diameter represents the 90th percentile of the distribution of diameters?

6.26 Breathing Rates The number of times x an adult human breathes per minute when at rest has a probability distribution that is approximately normal, with the mean equal to 16 and the standard deviation equal to 4. If a person is selected at random and the number x of breaths per minute while at rest is recorded, what is the probability that x will exceed 22?

6.27 Economic Forecasts One method of arriving at economic forecasts is to use a consensus approach. A forecast is obtained from each of a large number of analysts, and the average of these individual forecasts is the consensus forecast. Suppose the individual 2013 January prime interest rate forecasts of economic analysts are approximately normally distributed with the mean equal to 4.75% and a standard deviation equal to 0.2%. If a single analyst is randomly selected from among this group, what is the probability that the analyst's forecast of the prime rate will take on these values?

a. Exceed 4.25%

b. Be less than 4.375%

6.28 Tax Audits How does the IRS decide on the percentage of income tax returns to audit for each state? Suppose they do it by randomly selecting 50 values from a normal distribution with a mean equal to 1.55% and a standard deviation equal to .45%. (Computer programs are available for this type of sampling.)

a. What is the probability that a particular state will have more than 2.5% of its income tax returns audited?

b. What is the probability that a state will have less than 1% of its income tax returns audited?

6.29 Bacteria in Drinking Water Suppose the numbers of a particular type of bacteria in samples of 1 milliliter (ml) of drinking water tend to be approximately normally distributed, with a mean of 85 and a standard deviation of 9. What is the probability that a given 1-ml sample will contain more than 100 bacteria?

6.30 Loading Grain A grain loader can be set to discharge grain in amounts that are normally distributed, with mean μ bushels and standard deviation equal to 25.7 bushels. If a company wishes to use the loader to fill containers that hold 2000 bushels of grain and wants to overfill only one container in 100, at what value of μ should the company set the loader?

6.31 How Many Words? A publisher has discovered that the number of words contained in new manuscripts is normally distributed, with a mean equal to 20,000 words in excess of that specified in the author's contract and a standard deviation of 10,000 words. If the publisher wants to be almost certain (say, with a probability of .95) that the manuscript will have less than 100,000 words, what number of words should the publisher specify in the contract?

6.32 Tennis Anyone? A stringer of tennis rackets has found that the actual string tension achieved for any individual racket will vary as much as 6 pounds per square inch from the desired tension set on the stringing machine. If the stringer wishes to string at a tension lower than that specified by a customer only 5% of the time, how much above or below the customer's specified tension should the stringer set the stringing machine? (NOTE: Assume that the distribution of string tensions produced by the stringing machine is normally distributed, with a mean equal to the tension set on the machine and a standard deviation equal to 2 pounds per square inch.)

6.33 Mall Rats An article in *American Demographics* claims that more than twice as many shoppers are out shopping on the weekends than during the week.[3] Not only that, such shoppers also spend more money on their purchases on Saturdays and Sundays! Suppose that the amount of money spent at shopping centers between 4 P.M. and 6 P.M. on Sundays has a normal distribution with mean $85 and with a standard deviation of $20. A shopper is randomly selected on a Sunday between 4 P.M. and 6 P.M. and asked about his spending patterns.

a. What is the probability that he has spent more than $95 at the mall?

b. What is the probability that he has spent between $95 and $115 at the mall?

c. If two shoppers are randomly selected, what is the probability that both shoppers have spent more than $115 at the mall?

6.34 Pulse Rates What's a *normal* pulse rate? That depends on a variety of factors. Pulse rates between 60 and 100 beats per minute are considered normal for children over 10 and adults.[4] Suppose that these pulse

rates are approximately normally distributed with a mean of 78 and a standard deviation of 12.

a. What proportion of adults will have pulse rates between 60 and 100?

b. What is the 95th percentile for the pulse rates of adults?

c. Would a pulse rate of 110 be considered unusual? Explain.

THE NORMAL APPROXIMATION TO THE BINOMIAL PROBABILITY DISTRIBUTION (OPTIONAL)

In Chapter 5, you learned three ways to calculate probabilities for the binomial random variable x:

- Using the binomial formula, $P(x = k) = C_k^n p^k q^{n-k}$
- Using the cumulative binomial tables
- Using *MS Excel* and *MINITAB*

The binomial formula produces lengthy calculations, and the tables are available for only certain values of n and p. There is another option available when $np < 7$; the Poisson probabilities can be used to approximate $P(x = k)$. When this approximation *does not work* and n is large, the normal probability distribution provides another approximation for binomial probabilities.

> ## THE NORMAL APPROXIMATION TO THE BINOMIAL PROBABILITY DISTRIBUTION
>
> Let x be a binomial random variable with n trials and probability p of success. The probability distribution of x is approximated using a normal curve with
>
> $$\mu = np \quad \text{and} \quad \sigma = \sqrt{npq}$$
>
> This approximation is adequate as long as n is large and p is not too close to 0 or 1.

Since the normal distribution is continuous, the area under the curve at any single point is equal to 0. Keep in mind that this result applies only to continuous random variables. Because the binomial random variable x is a discrete random variable, the probability that x takes some specific value—say, $x = 11$—will not necessarily equal 0.

Figures 6.15 and 6.16 show the binomial probability histograms for $n = 25$ with $p = .5$ and $p = .1$, respectively. The distribution in Figure 6.15 is exactly symmetric. If you superimpose a normal curve with the same mean, $\mu = np$, and the same standard deviation, $\sigma = \sqrt{npq}$, over the top of the bars, it "fits" quite well; that is, the

FIGURE 6.15 ●

The binomial probability distribution for $n = 25$ and $p = .5$ and the approximating normal distribution with $\mu = 12.5$ and $\sigma = 2.5$

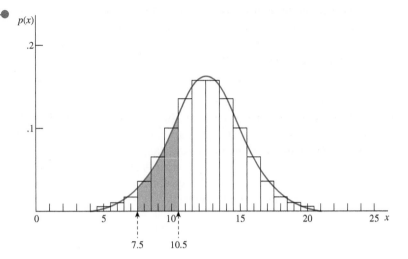

FIGURE 6.16 ●

The binomial probability distribution and the approximating normal distribution for $n = 25$ and $p = .1$

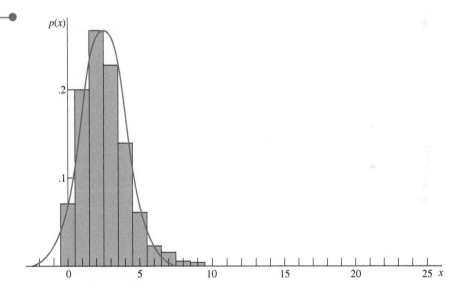

areas under the curve are almost the same as the areas under the bars. However, when the probability of success, p, gets small and the distribution is skewed, as in Figure 6.16, the symmetric normal curve no longer fits very well. If you try to use the normal curve areas to approximate the area under the bars, your approximation will not be very good.

EXAMPLE 6.11

Use the normal curve to approximate the probability that $x = 8$, 9, or 10 for a binomial random variable with $n = 25$ and $p = .5$. Compare this approximation to the exact binomial probability.

Solution You can find the exact binomial probability for this example because there are cumulative binomial tables for $n = 25$. From Table 1 in Appendix I,

$$P(x = 8, 9, \text{ or } 10) = P(x \le 10) - P(x \le 7) = .212 - .022 = .190$$

To use the normal approximation, first find the appropriate mean and standard deviation for the normal curve:

$$\mu = np = 25(.5) = 12.5$$

$$\sigma = \sqrt{npq} = \sqrt{25(.5)(.5)} = 2.5$$

NEED A TIP?

Only use the continuity correction if *x* has a **binomial** distribution!

The probability that you need corresponds to the area of the three rectangles lying over $x = 8$, 9, and 10. The equivalent area under the normal curve lies between $x = 7.5$ (the lower edge of the rectangle for $x = 8$) and $x = 10.5$ (the upper edge of the rectangle for $x = 10$). This area is shaded in Figure 6.15.

To find the normal probability, follow the procedures of Section 6.3. First you standardize each interval endpoint:

$$z = \frac{x - \mu}{\sigma} = \frac{7.5 - 12.5}{2.5} = -2.0$$

$$z = \frac{x - \mu}{\sigma} = \frac{10.5 - 12.5}{2.5} = -.8$$

Then the approximate probability (shaded in Figure 6.17) is found from Table 3 in Appendix I:

$$P(-2.0 < z < -.8) = .2119 - .0228 = .1891$$

You can compare the approximation, .1891, to the actual probability, .190. They are quite close!

FIGURE 6.17

Area under the normal curve for Example 6.11

ONLINE APPLET

Normal Approximation to Binomial Probabilities

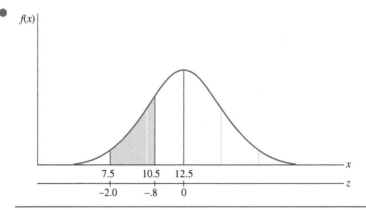

You must be careful not to exclude half of the two extreme probability rectangles when you use the normal approximation to the binomial probability distribution. This adjustment, called the **continuity correction,** helps account for the fact that you are approximating a *discrete random variable* with a *continuous* one. If you forget the correction, your approximation will not be very good! Use this correction only for *binomial probabilities;* do not try to use it when the random variable is already continuous, such as a height or weight.

How can you tell when it is appropriate to use the normal approximation to binomial probabilities? The normal approximation works well when the binomial histogram is roughly symmetric. This happens when the binomial distribution is not "bunched up" near 0 or n—that is, when it can spread out at least two standard deviations from its mean without exceeding its limits, 0 and n. Using this criterion, you can derive this simple rule of thumb:

RULE OF THUMB

The normal approximation to the binomial probabilities will be adequate if both

$$np > 5 \quad \text{and} \quad nq > 5$$

 NEED TO KNOW...

How to Calculate Binomial Probabilities Using the Normal Approximation

- Find the necessary values of n and p. Calculate $\mu = np$ and $\sigma = \sqrt{npq}$.
- Write the probability you need in terms of x and locate the appropriate area on the curve.
- Correct the value of x by $\pm.5$ to include the entire block of probability for that value. This is the *continuity correction*.
- Convert the necessary x-values to z-values using

$$z = \frac{x \pm .5 - np}{\sqrt{npq}}$$

- Use Table 3 in Appendix I to calculate the approximate probability.

 EXAMPLE 6.12

The reliability of an electrical fuse is the probability that a fuse, chosen at random from production, will function under its designed conditions. A random sample of 1000 fuses was tested and $x = 27$ defectives were observed. Calculate the approximate probability of observing 27 or more defectives, assuming that the fuse reliability is .98.

Solution   The probability of observing a defective when a single fuse is tested is $p = .02$, given that the fuse reliability is .98. Then

$$\mu = np = 1000(.02) = 20$$
$$\sigma = \sqrt{npq} = \sqrt{1000(.02)(.98)} = 4.43$$

The probability of 27 or more defective fuses, given $n = 1000$, is

NEED A TIP?

If np and nq are both greater than 5, you can use the normal approximation.

$$P(x \geq 27) = p(27) + p(28) + p(29) + \cdots + p(999) + p(1000)$$

It is appropriate to use the normal approximation to the binomial probability because

$$np = 1000(.02) = 20 \quad \text{and} \quad nq = 1000(.98) = 980$$

are both greater than 5. The normal area used to approximate $P(x \geq 27)$ is the area under the normal curve to the right of 26.5, so that the entire rectangle for $x = 27$ is included. Then, the z-value corresponding to $x = 26.5$ is

$$z = \frac{x - \mu}{\sigma} = \frac{26.5 - 20}{4.43} = \frac{6.5}{4.43} = 1.47$$

and the area to the left of $z = 1.47$ is equal to .9292, as shown in Figure 6.18. Since the total area under the curve is 1, you have

$$P(x \geq 27) \approx P(z \geq 1.47) = 1 - .9292 = .0708$$

FIGURE 6.18 ●

Normal approximation to the binomial for Example 6.12

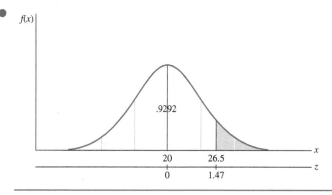

EXAMPLE 6.13

A producer of soft drinks was fairly certain that her brand had a 10% share of the soft drink market. In a market survey involving 2500 consumers of soft drinks, $x = 211$ expressed a preference for her brand. If the 10% figure is correct, find the probability of observing 211 or fewer consumers who prefer her brand of soft drink.

Solution If the producer is correct, then the probability that a consumer prefers her brand of soft drink is $p = .10$. Calculate

$$\mu = np = 2500(.10) = 250$$
$$\sigma = \sqrt{npq} = \sqrt{2500(.10)(.90)} = 15$$

The probability of observing 211 or fewer who prefer her brand is

$$P(x \leq 211) = p(0) + p(1) + \cdots + p(210) + p(211)$$

The normal approximation to this probability is the area to the left of 211.5 under a normal curve with a mean of 250 and a standard deviation of 15. First calculate

$$z = \frac{x - \mu}{\sigma} = \frac{211.5 - 250}{15} = -2.57$$

Then

$$P(x \leq 211) \approx P(z < -2.57) = .0051$$

The probability of observing a sample value of 211 or less when $p = .10$ is so small that you can conclude that one of two things has occurred: Either you have observed an unusual sample even though really $p = .10$, *or* the sample reflects that the actual value of p is less than .10 and perhaps closer to the observed sample proportion, $211/2500 = .08$.

6.4 EXERCISES

BASIC TECHNIQUES

6.35 Consider a binomial random variable x with $n = 25$ and $p = .6$.

a. Can the normal approximation be used to approximate probabilities in this case? Why or why not?

b. What are the mean and standard deviation of x?

c. Using the correction for continuity, approximate $P(x > 9)$.

6.36 Consider a binomial random variable x with $n = 45$ and $p = .05$.

a. Are np and nq both larger than 5?

b. Based on your answer to part a, can we use the normal approximation to approximate the binomial probabilities associated with x? If not, is there another possible approximation we could use?

6.37 Let x be a binomial random variable with $n = 25$ and $p = .3$.

a. Is the normal approximation appropriate for this binomial random variable?

b. Find the mean and standard deviation for x.

c. Use the normal approximation to find $P(6 \leq x \leq 9)$.

d. Use Table 1 in Appendix I to find the exact probability $P(6 \leq x \leq 9)$. Compare the results of parts c and d. How close was your approximation?

6.38 Let x be a binomial random variable with $n = 15$ and $p = .5$.

a. Is the normal approximation appropriate?

b. Find $P(x \geq 6)$ using the normal approximation.

c. Find $P(x > 6)$ using the normal approximation.

d. Find the exact probabilities for parts b and c, and compare these with your approximations.

6.39 Let x be a binomial random variable with $n = 100$ and $p = .2$. Find approximations to these probabilities:

a. $P(x > 22)$ **b.** $P(x \geq 22)$

c. $P(20 < x < 25)$ **d.** $P(x \leq 25)$

6.40 Let x be a binomial random variable with $n = 25$, $p = .2$.

a. Use Table 1 in Appendix I to calculate $P(4 \leq x \leq 6)$.

b. Find μ and σ for the binomial probability distribution, and use the normal distribution to approximate the probability $P(4 \leq x \leq 6)$. Note that this value

is a good approximation to the exact value of $P(4 \leq x \leq 6)$ even though $np = 5$.

6.41 Suppose the random variable x has a binomial distribution corresponding to $n = 20$ and $p = .30$. Use Table 1 of Appendix I to calculate these probabilities:

a. $P(x = 5)$ **b.** $P(x \geq 7)$

6.42 Refer to Exercise 6.41. Use the normal approximation to calculate $P(x = 5)$ and $P(x \geq 7)$. Compare with the exact values obtained from Table 1 in Appendix I.

6.43 Consider a binomial experiment with $n = 20$ and $p = .4$. Calculate $P(x \geq 10)$ using each of these methods:

a. Table 1 in Appendix I

b. The normal approximation to the binomial probability distribution

6.44 Find the normal approximation to $P(355 \leq x \leq 360)$ for a binomial probability distribution with $n = 400$ and $p = .9$.

APPLICATIONS

6.45 How Can I Help? Are you helping to save the environment? A *USA Today* Snapshot found that about 78% of Americans believe that recycling trash makes the biggest difference in protecting the environment.[5]

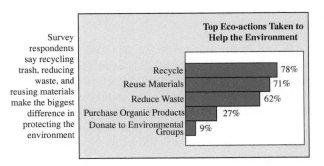

Suppose a random sample of $n = 50$ adults are polled and asked if they believed that recycling made the biggest difference in protecting our environment. Let us assume that the 78% figure is, in fact, correct. What are the probabilities for the following events?

a. Fewer than 30 individuals believe that recycling makes the biggest difference?

b. More than 40 individuals believe that recycling makes the biggest difference?

c. More than 10 individuals believe that recycling *does not* make the biggest difference?

6.46 Genetic Defects Data collected over a long period of time show that a particular genetic defect occurs in 1 of every 1000 children. The records of a medical clinic show $x = 60$ children with the defect in a total of 50,000 examined. If the 50,000 children were a random sample from the population of children represented by past records, what is the probability of observing a value of x equal to 60 or more? Would you say that the observation of $x = 60$ children with genetic defects represents a rare event?

6.47 No Shows Airlines and hotels often grant reservations in excess of capacity to minimize losses due to no-shows. Suppose the records of a hotel show that, on the average, 10% of their prospective guests will not claim their reservation. If the hotel accepts 215 reservations and there are only 200 rooms in the hotel, what is the probability that all guests who arrive to claim a room will receive one?

6.48 Lung Cancer Compilation of large masses of data on lung cancer shows that approximately 1 of every 40 adults acquires the disease. Workers in a certain occupation are known to work in an air-polluted environment that may cause an increased rate of lung cancer. A random sample of $n = 400$ workers shows 19 with identifiable cases of lung cancer. Do the data provide sufficient evidence to indicate a higher rate of lung cancer for these workers than for the national average?

6.49 Tall or Short? Do Americans tend to vote for the taller of the two major candidates in a presidential election? In 49 of our presidential elections for which the heights of all the major-party candidates are known, 26 of the winners were taller than their opponents.[1] Assume that Americans are not biased by a candidate's height and that the winner is just as likely to be taller or shorter than his opponent.

a. Is the observed number of taller winners in the U.S. presidential election unusual? Find the approximate probability of finding 26 or more of the 49 pairs in which the taller candidate wins.

b. Based on your answer to part a, can you conclude that Americans might consider a candidate's height when casting their ballot?

6.50 The Rh Factor In a certain population, 15% of the people have Rh-negative blood. A blood bank serving this population receives 92 blood donors on a particular day.

a. What is the probability that 10 or fewer are Rh-negative?

b. What is the probability that 15 to 20 (inclusive) of the donors are Rh-negative?

c. What is the probability that more than 80 of the donors are Rh-positive?

6.51 Pepsi's Market Share Two of the biggest soft drink rivals, Pepsi and Coke, are very concerned about their market shares. The pie chart that follows claims that PepsiCo's share of the beverage market is 25%.[6] Assume that this proportion will be close to the probability that a person selected at random indicates a preference for a Pepsi product when choosing a soft drink.

U.S. Refreshment Beverage Market Share

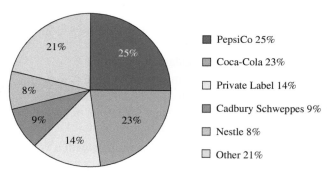

A group of $n = 500$ consumers is selected and the number preferring a Pepsi product is recorded. Use the normal curve to approximate the following binomial probabilities.

a. Exactly 150 consumers prefer a Pepsi product.

b. Between 120 and 150 consumers (inclusive) prefer a Pepsi product.

c. Fewer than 150 consumers prefer a Pepsi product.

d. Would it be unusual to find that 232 of the 500 consumers preferred a Pepsi product? If this were to occur, what conclusions would you draw?

6.52 Ready, Set, Relax! In a study conducted for the Center for a New American Dream, *Time* magazine reports that 60% of Americans felt pressure to work too much, and 80% wished for more family time.[7] Assume that these percentages are correct for all

Americans, and that a random sample of 25 Americans is selected.

a. Use Table 1 in Appendix I to find the probability that more than 20 felt pressure to work too much.

b. Use the normal approximation to the binomial distribution to aproximate the probability in part a. Compare your answer with the exact value from part a.

c. Use Table 1 in Appendix I to find the probability that between 15 and 20 (inclusive) wished for more family time.

d. Use the normal approximation to the binomial distribution to approximate the probability in part c. Compare your answer with the exact value from part c.

6.53 We said, "Relax!" The article in *Time* magazine[7] (Exercise 6.52) also reported that 80% of men and 62% of women put in more than 40 hours a week on the job. Assume that these percentages are correct for all Americans, and that a random sample of 50 working women is selected.

a. What is the average number of women who put in more than 40 hours a week on the job?

b. What is the standard deviation for the number of women who put in more than 40 hours a week on the job?

c. Suppose that in our sample of 50 working women, there are 25 who work more than 40 hours a week. Would you consider this to be an unusual occurrence? Explain.

CHAPTER REVIEW

Key Concepts and Formulas

I. Continuous Probability Distributions

   1. Continuous random variables

   2. Probability distributions or probability density functions

     a. Curves are smooth.

     b. Area under the curve equals 1.

     c. The area under the curve between a and b represents the probability that x falls between a and b.

     d. $P(x = a) = 0$ for continuous random variables.

II. The Normal Probability Distribution

   1. Symmetric about its mean μ

   2. Shape determined by its standard deviation σ

III. The Standard Normal Distribution

   1. The standard normal random variable z has mean 0 and standard deviation 1.

   2. Any normal random variable x can be transformed to a standard normal random variable using

$$z = \frac{x - \mu}{\sigma}$$

   3. Convert necessary values of x to z.

   4. Use Table 3 in Appendix I to compute standard normal probabilities.

   5. Several important z-values have right-tail areas as follows:

Right-Tail Area	.005	.01	.025	.05	.10
z-Value	2.58	2.33	1.96	1.645	1.28

TECHNOLOGY TODAY

Normal Probabilities in *Microsoft Excel*

When the random variable of interest has a normal probability distribution, you can generate the following probabilities using the following functions:

- **NORM.DIST** and **NORM.S.DIST:** Generate cumulative probabilities— $P(x \le x_0)$ for a general normal random variable or $P(z \le z_0)$ for a standard normal random variable. (NOTE: These functions are called **NORMDIST** and **NORMSDIST** in *Excel 2007* and earlier versions.)

- **NORM.INV** and **NORM.S.INV:** Generate inverse cumulative probabilities— the value x_0 such that the area to its left under the general normal probability distribution is equal to a, or the value z_0 such that the area to its left under the standard normal probability distribution is equal to a. (NOTE: These functions are called **NORMINV** and **NORMSINV** in *Excel 2007* and earlier versions.)

You must specify which normal distribution you are using and, if it is a general normal random variable, the values for the mean μ and the standard deviation σ. As in Chapter 5, you must also specify the values for x_0, z_0, or a, depending on the function you are using.

EXAMPLE 6.14

For a standard normal random variable z, find $P(1.2 < z < 1.96)$. Find the value z_0 with area .025 to its right.

1. Name columns A and B of an *Excel* spreadsheet as "z0", and "P(z<=z0)", respectively. Then enter the two values for z_0 (1.2 and 1.96) in cells A2 and A3. To generate cumulative probabilities for these two values, first place your cursor in cell B2. Select **Insert Function** f_x ▶ **Statistical** ▶ **NORM.S.DIST** and click **OK.** The Dialog box shown in Figure 6.19 will appear.

FIGURE 6.19

2. Enter the location of first value of z_0 (cell A2) into the first box and the word TRUE into the second box. The resulting probability is marked as "Formula result = .8849" at the bottom of the box, and when you click **OK,** the probability $P(z \le 1.2)$ will appear in cell B2. To obtain the other probability, simply place your cursor in cell B2, grab the square handle in the lower right corner of the cell and drag the handle down to copy the formula into the other cell and obtain $P(z \le 1.96) = .9750$. *MS Excel* has automatically adjusted the cell location in the formula as you copied.

3. To find $P(1.2 < z < 1.96)$, remember that the cumulative probability is the area to the left of the given value of z. Hence,

$$P(1.2 < z < 1.96) = P(z < 1.96) - P(z < 1.2) = .9750 - .8849 = .0901.$$

You can check this calculation using Table 3 in Appendix I if you wish!

4. To calculate inverse cumulative probabilities, place your cursor in an empty cell, select **Insert Function ▶ Statistical ▶ NORM.S.INV** and click **OK.** We need a value z_0 with area .025 to its right, or area .975 to its left. Enter **.975** in the box marked "Probability" and notice the "Formula Result = 1.959963985," which will appear in the empty cell when you click **OK.** This value, when rounded to two decimal places, is the familiar $z_0 = 1.96$ used in Example 6.7.

EXAMPLE 6.15

Suppose that the average birth weights of babies born at hospitals owned by a major health maintenance organization (HMO) are approximately normal with mean 6.75 pounds and standard deviation 0.54 pound. What proportion of babies born at these hospitals weigh between 6 and 7 pounds? Find the 95th percentile of these birth weights.

1. Name columns A and B of an *Excel* spreadsheet as "x0", and "P(x<=x0)", respectively. Then enter the two values for x_0 (6 and 7) in cells A2 and A3. Proceed as in Example 6.14, this time selecting **Insert Function ▶ Statistical ▶ NORM.DIST** and clicking **OK.** The Dialog box shown in Figure 6.20 will appear.

FIGURE 6.20

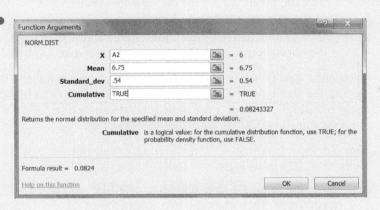

2. Enter the location of first value of x_0 (cell A2) into the first box, the appropriate mean and standard deviation in the second and third boxes, and the word TRUE into the fourth box. The resulting probability is marked as "Formula result = .0824" at the bottom of the box, and when you click **OK,** the probability $P(x \leq 6)$ will appear in cell B2. To obtain the other probability, simply place your cursor in cell B2, grab the square handle in the lower right corner of the cell and drag the handle down to copy the formula into the other cell and obtain $P(x \leq 7) = .6783$.

3. Finally, use the values calculated by *Excel* to calculate

$$P(6 < x < 7) = P(x < 7) - P(x < 6) = .6783 - .0824 = .5959.$$

4. To calculate the 95th percentile, place your cursor in an empty cell, select **Insert Function ▶ Statistical ▶ NORM.INV** and click **OK.** We need a value x_0 with area .95 to its left. Enter **.95** in the box marked "Probability," the appropriate mean and standard deviation (see Figure 6.21), and notice the "Formula Result = 7.638220959," which will appear in the empty cell when you click **OK.**

FIGURE 6.21

That is, 95% of all babies born at these hospitals weigh 7.638 pounds or less. Would you consider a baby who weighs 9 pounds to be unusually large?

Normal Probabilities in *MINITAB*

When the random variable of interest has a normal probability distribution, you can generate the following probabilities:

- Cumulative probabilities—$P(X \leq x)$ for a given value of x. (NOTE: *MINITAB* uses the notation "X" for the random variable and "x" for a particular value of the random variable.)

- Inverse cumulative probabilities—the value x such that the area to its left under the normal probability distribution is equal to a.

You must specify which normal distribution you are using and the values for the mean μ and the standard deviation σ. As in Chapter 5, you have the option of specifying only one single value of x (or a) or several values of x (or a), which should be stored in a column (say, C1) of the *MINITAB* worksheet.

EXAMPLE 6.16

For a standard normal random variable z, find $P(1.2 < z < 1.96)$. Find the value z_0 with area .025 to its right.

1. Name columns C1 and C2 of a *MINITAB* worksheet as "x", and "P(X<=x)", respectively. Then enter the two values for x (1.2 and 1.96) in the first two cells of column C1. To generate cumulative probabilities for these two values, select **Calc ▶ Probability Distributions ▶ Normal** and the Dialog box shown in Figure 6.22 will appear.

FIGURE 6.22

```
Normal Distribution                                    x

C1  x          ○ Probability density
C2  P(X<=x)    ● Cumulative probability
               ○ Inverse cumulative probability

               Mean: [0.0    ]
               Standard deviation: [1.0    ]

               ● Input column:    [x     ]
                 Optional storage: [      ]

               ○ Input constant:  [      ]
   Select        Optional storage: [      ]

   Help              OK       Cancel
```

2. By default, *MINITAB* chooses $\mu = 0$ and $\sigma = 1$ as the mean and standard deviation of the standard normal z distribution, so you need only to enter the Input column (C1) and make sure that the radio button marked "Cumulative probability" is selected. If you do not specify a column for "Optional storage," *MINITAB* will display the results in the Session window, shown in Figure 6.23(a).

FIGURE 6.23

(a)

(b)

3. To find $P(1.2 < z < 1.96)$, remember that the cumulative probability is the area to the left of the given value of z. Hence,

$$P(1.2 < z < 1.96) = P(z < 1.96) - P(z < 1.2) = .975002 - .884930 = .090072.$$

You can check this calculation using Table 3 in Appendix I if you wish!

4. To calculate inverse cumulative probabilities, select **Calc ▶ Probability Distributions ▶ Normal**, and click the radio button marked "Inverse cumulative probability," shown in Figure 6.22. We need a value z_0 with area .025 to its right, or area .975 to its left. Enter **.975** in the box marked "Input constant" and click **OK.** The value of z_0 will appear in the Session window, shown in Figure 6.23(b). This value, when rounded to two decimal places, is the familiar $z_0 = 1.96$ used in Example 6.7.

EXAMPLE 6.17

Suppose that the average birth weights of babies born at hospitals owned by a major health maintenance organization (HMO) are approximately normal with mean 6.75 pounds and standard deviation 0.54 pound. What proportion of babies born at these hospitals weigh between 6 and 7 pounds? Find the 95th percentile of these birth weights.

1. Enter the two values for x (6 and 7) in the first two cells of column C1. Proceed as in Example 6.16, again selecting **Calc ▶ Probability Distributions ▶ Normal.** This time, enter the values for the mean ($\mu = 6.75$) and standard deviation ($\sigma = .54$) in the appropriate boxes, and select column C1 ("x") for the Input column. Make sure that the radio button marked "Cumulative probability" is selected and click **OK.** In the Session window, you will see that $P(x \leq 7) = .678305$ and $P(x \leq 6) = .082433$.

2. Finally, use the values calculated by *MINITAB* to calculate

$$P(6 < x < 7) = P(x < 7) - P(x < 6) = .678305 - .082433 = .595872.$$

3. To calculate the 95th percentile, selecting **Calc ▶ Probability Distributions ▶ Normal,** enter the values for the mean ($\mu = 6.75$) and standard deviation ($\sigma = .54$) in the appropriate boxes, and make sure that the radio button marked "Inverse

cumulative probability" is selected. We need a value x_0 with area .95 to its left. Enter **.95** in the box marked "Input constant" and click **OK.** In the Session window, you will see the 95th percentile, as shown in Figure 6.24.

FIGURE 6.24

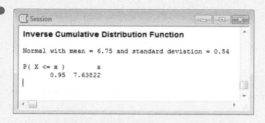

That is, 95% of all babies born at these hospitals weigh 7.63822 pounds or less. Would you consider a baby who weighs 9 pounds to be unusually large?

Supplementary Exercises

6.54 Calculate the area under the standard normal curve to the left of these values:
a. $z = -.90$ **b.** $z = 2.34$
c. $z = 5.4$

6.55 Calculate the area under the standard normal curve between these values:
a. $z = -2.0$ and $z = 2.0$ **b.** $z = -2.3$ and $z = -1.5$

6.56 Find the following probabilities for the standard normal random variable z:
a. $P(-1.96 \le z \le 1.96)$ **b.** $P(z > 1.96)$
c. $P(z < -1.96)$

6.57 a. Find a z_0 such that $P(z > z_0) = .9750$.
b. Find a z_0 such that $P(z > z_0) = .3594$.

6.58 a. Find a z_0 such that $P(-z_0 \le z \le z_0) = .95$.
b. Find a z_0 such that $P(-z_0 \le z \le z_0) = .98$.
c. Find a z_0 such that $P(-z_0 \le z \le z_0) = .90$.
d. Find a z_0 such that $P(-z_0 \le z \le z_0) = .99$.

6.59 A normal random variable x has mean $\mu = 5$ and standard deviation $\sigma = 2$. Find the probabilities associated with the following intervals:
a. $1.2 < x < 10$ **b.** $x > 7.5$ **c.** $x \le 0$

6.60 Let x be a binomial random variable with $n = 36$ and $p = .54$. Use the normal approximation to find:
a. $P(x \le 25)$ **b.** $P(15 \le x \le 20)$ **c.** $P(x > 30)$

6.61 Using Table 3 in Appendix I, calculate the area under the standard normal curve to the left of the following:
a. $z = 1.2$ **b.** $z = -.9$
c. $z = 1.46$ **d.** $z = -.42$

6.62 Find the following probabilities for the standard normal random variable:
a. $P(.3 < z < 1.56)$ **b.** $P(-.2 < z < .2)$

6.63 a. Find the probability that z is greater than $-.75$.
b. Find the probability that z is less than 1.35.

6.64 Find z_0 such that $P(z > z_0) = .5$.

6.65 Find the probability that z lies between $z = -1.48$ and $z = 1.48$.

6.66 Find z_0 such that $P(-z_0 < z < z_0) = .5$. What percentiles do $-z_0$ and z_0 represent?

6.67 Drill Bits It is estimated that the mean life span of oil-drilling bits is 75 hours. Suppose an oil exploration company purchases drill bits that have a life span that is approximately normally distributed with a mean equal to 75 hours and a standard deviation equal to 12 hours.
a. What proportion of the company's drill bits will fail before 60 hours of use?
b. What proportion will last at least 60 hours?
c. What proportion will have to be replaced after more than 90 hours of use?

6.68 Faculty Ages The influx of new ideas into a college or university, introduced primarily by new young faculty, is becoming a matter of concern because of the increasing ages of faculty members. That is, the distribution of faculty ages is shifting upward; there is a shortage of vacant positions and an oversupply of PhDs. If the retirement age at most universities is 65, would you expect the distribution of faculty ages to be normal? Explain.

6.69 Bearing Diameters A machine operation produces bearings whose diameters are normally distributed, with mean and standard deviation equal to .498 and .002, respectively. If specifications require that the bearing diameter equal .500 inch $\pm$ .004 inch, what fraction of the production will be unacceptable?

6.70 Used Cars A used-car dealership has found that the length of time before a major repair is required on the cars it sells is normally distributed with a mean equal to 10 months and a standard deviation of 3 months. If the dealer wants only 5% of the cars to fail before the end of the guarantee period, for how many months should the cars be guaranteed?

6.71 Restaurant Sales The daily sales total (excepting Saturday) at a small restaurant has a probability distribution that is approximately normal, with a mean μ equal to $1230 per day and a standard deviation σ equal to $120.

a. What is the probability that the sales will exceed $1400 for a given day?

b. The restaurant must have at least $1000 in sales per day to break even. What is the probability that on a given day the restaurant will not break even?

6.72 Washers The life span of a type of automatic washer is approximately normally distributed with mean and standard deviation equal to 10.5 and 3.0 years, respectively. If this type of washer is guaranteed for a period of 5 years, what fraction will need to be repaired and/or replaced?

6.73 Garage Door Openers Most users of automatic garage door openers activate their openers at distances that are normally distributed with a mean of 30 feet and a standard deviation of 11 feet. To minimize interference with other remote-controlled devices, the manufacturer is required to limit the operating distance to 50 feet. What percentage of the time will users attempt to operate the opener outside its operating limit?

6.74 How Long Is the Test? The average length of time required to complete a college achievement test was found to equal 70 minutes with a standard deviation of 12 minutes. When should the test be terminated if you wish to allow sufficient time for 90% of the students to complete the test? (Assume that the time required to complete the test is normally distributed.)

6.75 Servicing Automobiles The length of time required to run a 5000-mile check and to service an automobile has a mean equal to 1.4 hours and a standard deviation of .7 hour. Suppose that the service department plans to service 50 automobiles per 8-hour day and that, in order to do so, it must spend no more than an average of 1.6 hours per automobile. What proportion of all days will the service department have to work overtime?

6.76 TV Viewers An advertising agency has stated that 20% of all television viewers watch a given program. In a random sample of 1000 viewers, $x = 184$ viewers were watching the program. Do these data present sufficient evidence to contradict the advertiser's claim?

6.77 Forecasting Earnings A researcher notes that senior corporation executives are not very accurate forecasters of their own annual earnings. He states that his studies of a large number of company executive forecasts "showed that the average estimate missed the mark by 15%."

a. Suppose the distribution of these forecast errors has a mean of 15% and a standard deviation of 10%. Is it likely that the distribution of forecast errors is approximately normal?

b. Suppose the probability is .5 that a corporate executive's forecast error exceeds 15%. If you were to sample the forecasts of 100 corporate executives, what is the probability that more than 60 would be in error by more than 15%?

6.78 Filling Soda Cups A soft drink machine can be regulated to discharge an average of μ ounces per cup. If the ounces of fill are normally distributed, with standard deviation equal to .3 ounce, give the setting for μ so that 8-ounce cups will overflow only 1% of the time.

6.79 Light Bulbs A manufacturing plant uses light bulbs whose life spans are normally distributed, with mean and standard deviation equal to 500 and 50 hours, respectively. In order to minimize the number of bulbs that burn out during operating hours, all the bulbs are replaced after a given period of operation. How often should the bulbs be replaced if we wish no more than

1% of the bulbs to burn out between replacement periods?

6.80 The Freshman Class The admissions office of a small college is asked to accept deposits from a number of qualified prospective freshmen so that, with probability about .95, the size of the freshman class will be less than or equal to 120. Suppose the applicants constitute a random sample from a population of applicants, 80% of whom would actually enter the freshman class if accepted.

a. How many deposits should the admissions counselor accept?

b. If applicants in the number determined in part a are accepted, what is the probability that the freshman class size will be less than 105?

6.81 No Shows An airline finds that 5% of the persons making reservations on a certain flight will not show up for the flight. If the airline sells 160 tickets for a flight that has only 155 seats, what is the probability that a seat will be available for every person holding a reservation and planning to fly?

6.82 Long Distance It is known that 30% of all calls coming into a telephone exchange are long-distance calls. If 200 calls come into the exchange, what is the probability that at least 50 will be long-distance calls?

6.83 Plant Genetics In Exercise 5.75, a cross between two peony plants—one with red petals and one with streaky petals—produced offspring plants with red petals 75% of the time. Suppose that 100 seeds from this cross were collected and germinated, and x, the number of plants with red petals, was recorded.

a. What is the exact probability distribution for x?

b. Is it appropriate to approximate the distribution in part a using the normal distribution? Explain.

c. Use an appropriate method to find the approximate probability that between 70 and 80 (inclusive) offspring plants have red flowers.

d. What is the probability that 53 or fewer offspring plants had red flowers? Is this an unusual occurrence?

e. If you actually observed 53 of 100 offspring plants with red flowers, and if you were certain that the genetic ratio 3:1 was correct, what other explanation could you give for this unusual occurrence?

6.84 Suppliers A or B? A purchaser of electric relays buys from two suppliers, A and B. Supplier A supplies two of every three relays used by the company. If 75 relays are selected at random from those in use by the company, find the probability that at most

48 of these relays come from supplier A. Assume that the company uses a large number of relays.

6.85 Snacking and TV Psychologists believe that excessive eating may be associated with emotional states (being upset or bored) and environmental cues (watching television, reading, and so on). To test this theory, suppose you randomly selected 60 persons and matched them by weight and gender in pairs. For a period of 2 weeks, one of each pair is required to spend evenings reading novels of interest to him or her, while the other spends each evening watching television. The calorie count for all snack and drink intake for the evenings is recorded for each person, and you record $x = 19$, the number of pairs for which the television watchers' calorie intake exceeded the intake of the readers. If there is no difference in the effects of television and reading on calorie intake, the probability p that the calorie intake of one member of a pair exceeds that of the other member is .5. Do these data provide sufficient evidence to indicate a difference between the effects of television watching and reading on calorie intake? (HINT: Calculate the z-score for the observed value, $x = 19$.)

6.86 Gestation Times *The Biology Data Book* reports that the gestation time for human babies averages 278 days with a standard deviation of 12 days.[8] Suppose that these gestation times are normally distributed.

a. Find the upper and lower quartiles for the gestation times.

b. Would it be unusual to deliver a baby after only 6 months of gestation? Explain.

6.87 Tax Audits In Exercise 6.28, we suggested that the IRS assigns auditing rates per state by randomly selecting 50 auditing percentages from a normal distribution with a mean equal to 1.55% and a standard deviation of .45%.

a. What is the probability that a particular state would have more than 2% of its tax returns audited?

b. What is the expected value of x, the number of states that will have more than 2% of their income tax returns audited?

c. Is it likely that as many as 15 of the 50 states will have more than 2% of their income tax returns audited?

6.88 Your Favorite Sport Among the 10 most popular sports, men include competition-type sports—pool and billiards, basketball, and softball—whereas women include aerobics, running, hiking, and calisthenics. However, the top recreational activity for men was still the relaxing sport of fishing, with 41% of those surveyed indicating that they had fished during the year. Suppose 180 randomly selected men are asked whether they had fished in the past year.

a. What is the probability that fewer than 50 had fished?

b. What is the probability that between 50 and 75 (inclusive) had fished?

c. If the 180 men selected for the interview were selected by the marketing department of a sporting-goods company based on information obtained from their mailing lists, what would you conclude about the reliability of their survey results?

6.89 Introvert or Extrovert? A psychological introvert–extrovert test produced scores that had a normal distribution with a mean and standard deviation of 75 and 12, respectively. If we wish to designate the *highest* 15% as extroverts, what would be the proper score to choose as the cutoff point?

6.90 Normal Distribution? The chest measurements for 5738 Scottish militiamen in the early 19th century are given below.[9] Chest sizes are measured in inches, and each observation reports the number of soldiers with that chest size.

Count	Chest	Count	Chest
3	33	934	41
18	34	658	42
81	35	370	43
185	36	92	44
420	37	50	45
749	38	21	46
1073	39	4	47
1079	40	1	48

Notice the approximate normality of the histogram of the 5738 chest measurements.

Histogram of Chest Measurements

a. The mean of this distribution is $\bar{x} = 39.83$ and the standard deviation is $s = 2.05$. What is the 95th percentile of this distribution based on a normal curve with $\mu = 39.83$ and $\sigma = 2.05$?

b. Find the empirical estimate of the 95th percentile and compare with your answer in part a. (HINT: The 95th percentile will be in position $.95(n + 1) = .95 \times 5739 = 5452.05$ from the left tail of the distribution or in position $5738 - 5452.05 = 285.95$ from the right tail of the distribution.)

c. Find the 90th percentile of this distribution based on a normal curve with $\mu = 39.83$ and $\sigma = 2.05$. What is the value of the empirical 90th percentile? How does it compare with the value assuming normality?

6.91 Normal Distribution? continued Assume that the chest measurements in Exercise 6.90 are normally distributed with a mean of $\mu = 39.83$ and standard deviation of $\sigma = 2.05$.

a. What proportion of the observations would lie between 36.5 and 43.5 inches?

b. Between what two measurements would 95% of the observations lie?

c. What are the actual proportions for parts a and b using the data directly? Comment on the accuracy of the proportions found using assumed normality of the chest measurements.

6.92 Normal Temperatures In Exercise 1.67, Allen Shoemaker derived a distribution of human body temperatures, which has a distinct mound-shape.[10] Suppose we assume that the temperatures of healthy humans is approximately normal with a mean of 98.6 degrees and a standard deviation of 0.8 degrees.

a. If a healthy person is selected at random, what is the probability that the person has a temperature above 99.0 degrees?

b. What is the 95th percentile for the body temperatures of healthy humans?

6.93 Cellphone Etiquette A Snapshot in *USA Today* indicates that 51% of Americans say the average person is not very considerate of others when talking on a cellphone.[11] Suppose that 100 Americans are randomly selected. Find the approximate probability that 60 or more Americans would indicate that the average person is not very considerate of others when talking on a cellphone.

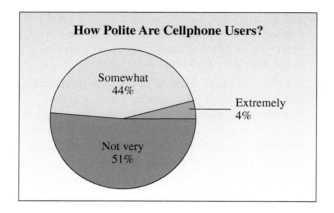

How Polite Are Cellphone Users?

Somewhat 44%

Extremely 4%

Not very 51%

6.94 Stamps Philatelists (stamp collectors) often buy stamps at or near retail prices, but, when they sell, the price is considerably lower. For example, depending on the mix of a collection, condition, demand, economic conditions, etc., a collection may sell at $x\%$ of the retail price, where x is normally distributed with a mean equal to 45% and a standard deviation of 4.5%. If a philatelist has a collection to sell that has a retail value of $30,000, what is the probability that the philatelist receives these amounts for the collection?
a. More than $15,000   **b.** Less than $15,000
c. Less than $12,000

6.95 Test Scores The scores on a national achievement test were approximately normally distributed, with a mean of 540 and a standard deviation of 110.

a. If you achieved a score of 680, how far, in standard deviations, did your score depart from the mean?

b. What percentage of those who took the examination scored higher than you?

6.96 Faculty Salaries Although faculty salaries at colleges and universities in the United States continue to rise, they do not always keep pace with the cost of living nor with salaries in the private sector. In 2010, the National Center for Educational Statistics indicated that the average salary for Assistant Professors at public 4-year colleges was $63,441.[12] Suppose that these salaries are normally distributed with a standard deviation of $4000.

a. What proportion of assistant professors at public 4-year colleges will have salaries less than $55,000?

b. What proportion of these professors will have salaries between $55,000 and $65,000?

6.97 Transplanting Cells Briggs and King developed the technique of nuclear transplantation, in which the nucleus of a cell from one of the later stages of the development of an embryo is transplanted into a zygote (a single-cell fertilized egg) to see whether the nucleus can support normal development. If the probability that a single transplant from the early gastrula stage will be successful is .65, what is the probability that more than 70 transplants out of 100 will be successful?

CASE STUDY **"Are You Going to Curve the Grades?"**

Very often, at the end of an exam that seemed particularly difficult, students will ask the professor, "Are you going to curve the grades?" Unfortunately, "curving the grades" doesn't necessarily mean that you will receive a higher grade on a test, although you might like to think so! Curving grades is actually a technique whereby a fixed proportion of the highest grades receive As (even if the highest grade is a failing grade on a percentage basis), and a fixed proportion of the lowest grades receive Fs (even if the lowest score is a passing grade on a percentage basis). The B, C, and D grades are also assigned according to fixed proportion. One such allocation uses the following proportions.

Letter Grade	A	B	C	D	F
Proportion of grades	10%	20%	40%	20%	10%

1. If the average C grade is centered on the average grade for all students, and if we assume that the grades are normally distributed, how many standard deviations on each side of the mean will designate the C grades?

2. How many standard deviations on either side of the mean will be the cutoff points for the B and D grades?

A histogram of the grades for an introductory Statistics class together with summary statistics follows.

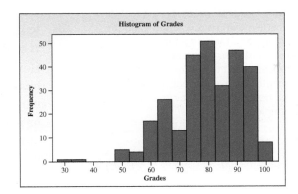

Descriptive Statistics: Grades

Variable	N	N*	Mean	StDev	Minimum	Q1	Median	Q3	Maximum
Grades	290	0	79.972	12.271	31.000	73.000	82.000	88.000	100.000

For ease of calculation, round the number of standard deviations for C grades to ±.5 standard deviations and for B and D grades to ±1.5 standard deviations.

3. Find the cutoff points for A, B, C, D, and F grades corresponding to these rounded values.

4. If you had a score of 92 on the exam and you had the choice of curving the grades or using the absolute standard of 90–100 for an A, 80–89 for a B, 70–79 for a C, and so on, what would be your choice? Explain your reasoning. Is the skewness of the distribution of grades a problem?

© PictureNet/CORBIS

Sampling Distributions

GENERAL OBJECTIVES

In the past several chapters, we studied *populations* and the *parameters* that describe them. These populations were either discrete or continuous, and we used *probability* as a tool for determining how likely certain sample outcomes might be. In this chapter, our focus changes as we begin to study *samples* and the *statistics* that describe them. These sample statistics are used to make inferences about the corresponding population parameters. This chapter involves sampling and sampling distributions, which describe the behavior of sample statistics in repeated sampling.

CHAPTER INDEX

? NEED TO KNOW...

When the Sample Size is Large Enough to Use the Central Limit Theorem
How to Calculate Probabilities for the Sample Mean $\bar{x}$
How to Calculate Probabilities for the Sample Proportion $\hat{p}$

Sampling the Roulette at Monte Carlo

How would you like to try your hand at gambling without the risk of losing? You could do it by simulating the gambling process, making imaginary bets, and observing the results. This technique, called a Monte Carlo procedure, is the topic of the case study at the end of this chapter.

7.1 INTRODUCTION

In the previous three chapters, you have learned a lot about probability distributions, such as the binomial and normal distributions. The shape of the normal distribution is determined by its mean μ and its standard deviation σ, whereas the shape of the binomial distribution is determined by p. These numerical descriptive measures—called **parameters**—are needed to calculate the probability of observing sample results.

In practical situations, you may be able to decide which *type* of probability distribution to use as a model, but the values of the *parameters* that specify its *exact form* are unknown. Here are two examples:

- A pollster is sure that the responses to his "agree/disagree" questions will follow a binomial distribution, but p, the proportion of those who "agree" in the population, is unknown.
- An agronomist believes that the yield per acre of a variety of wheat is approximately normally distributed, but the mean μ and standard deviation σ of the yields are unknown.

In these cases, you must rely on the *sample* to learn about these parameters. The proportion of those who "agree" in the pollster's sample provides information about the actual value of p. The mean and standard deviation of the agronomist's sample approximate the actual values of μ and σ. If you want the sample to provide *reliable information* about the population, however, you must select your sample in a certain way!

7.2 SAMPLING PLANS AND EXPERIMENTAL DESIGNS

The way a sample is selected is called the **sampling plan** or **experimental design** and determines the quantity of information in the sample. Knowing the sampling plan used in a particular situation will often allow you to measure the reliability or goodness of your inference.

Simple random sampling is a commonly used sampling plan in which every sample of size n has the same chance of being selected. For example, suppose you want to select a sample of size $n = 2$ from a population containing $N = 4$ objects. If the four objects are identified by the symbols x_1, x_2, x_3, and x_4, there are six distinct pairs that could be selected, as listed in Table 7.1. If the sample of $n = 2$ observations is selected so that each of these six samples has the same chance of selection, given by 1/6, then the resulting sample is called a **simple random sample,** or just a **random sample.**

TABLE 7.1 ● **Ways of Selecting a Sample of Size 2 from 4 Objects**

Sample	Observations in Sample
1	x_1, x_2
2	x_1, x_3
3	x_1, x_4
4	x_2, x_3
5	x_2, x_4
6	x_3, x_4

Definition   If a sample of *n* elements is selected from a population of *N* elements using a sampling plan in which each of the possible samples has the same chance of selection, then the sampling is said to be **random** and the resulting sample is a **simple random sample.**

Perfect random sampling is difficult to achieve in practice. If the size of the population *N* is small, you might write each of *N* numbers on a poker chip, mix the chips, and select a sample of *n* chips. The numbers that you select correspond to the *n* measurements that appear in the sample. Since this method is not always very practical, a simpler and more reliable method uses **random numbers**—digits generated so that the values 0 to 9 occur randomly and with equal frequency. These numbers can be generated by computer or may even be available on your scientific calculator. Alternatively, Table 10 in Appendix I is a table of random numbers that you can use to select a *random sample.*

EXAMPLE   7.1

A computer database at a downtown law firm contains files for $N = 1000$ clients. The firm wants to select $n = 5$ files for review. Select a simple random sample of five files from this database.

Solution   You must first label each file with a number from 1 to 1000. Perhaps the files are stored alphabetically, and the computer has already assigned a number to each. Then generate a sequence of 10 three-digit random numbers. If you are using Table 10 of Appendix I, select a random starting point and use a portion of the table similar to the one shown in Table 7.2. The random starting point ensures that you will not use the same sequence over and over again.

The first three digits of Table 7.2 indicate the number of the first file to be reviewed. The random number 001 corresponds to file #1, and the last file, #1000, corresponds to the random number 000. Using Table 7.2, you would choose the five files numbered 155, 450, 32, 882, and 350 for review. Alternately, you might choose to read across the lines, and choose files 155, 350, 989, 450, and 369 for review.

TABLE 7.2 ● **Portion of a Table of Random Numbers**

15574	35026	98924
45045	36933	28630
03225	78812	50856
88292	26053	21121

The situation described in Example 7.1 is called an **observational study** because the data already existed before you decided to *observe* or describe their characteristics. Most sample surveys, in which information is gathered with a questionnaire, fall into this category. Computer databases make it possible to assign identification numbers to each element even when the population is large and to select a simple random sample. However, when conducting a sample survey, you must be careful to watch for these frequently occurring problems:

- **Nonresponse:** You have carefully selected your random sample and sent out your questionnaires, but only 50% of those surveyed return their questionnaires. Are the responses you received still representative of the entire population, or are they **biased** because only those people who were particularly opinionated about the subject chose to respond?

- **Undercoverage:** You have selected your random sample using land-line telephone records as a database. Does the database you used systematically exclude certain segments of the population—perhaps those who use only cell phones or have unlisted numbers?
- **Wording bias:** Your questionnaire may have questions that are too complicated or tend to confuse the reader. Possibly the questions are sensitive in nature—for example, "Have you ever used drugs?" or "Have you ever cheated on your income tax?"—and the respondents will not answer truthfully.

Methods have been devised to solve some of these problems, but only if you know that they exist. If your survey is *biased* by any of these problems, then your conclusions will not be very reliable, even though you did select a random sample!

Some research involves **experimentation,** in which an experimental condition or *treatment* is imposed on the *experimental units.* Selecting a simple random sample is more difficult in this situation.

EXAMPLE 7.2

A research chemist is testing a new method for measuring the amount of titanium (Ti) in ore samples. She chooses 10 ore samples of the same weight for her experiment. Five of the samples will be measured using a standard method, and the other five using the new method. Use random numbers to assign the 10 ore samples to the new and standard groups. Do these data represent a simple random sample from the population?

Solution There are really two populations in this experiment. They consist of titanium measurements, using either the new or standard method, for *all possible* ore samples of this weight. These populations do not exist in fact; they are **hypothetical populations,** envisioned in the mind of the researcher. Thus, it is impossible to select a simple random sample using the methods of Example 7.1. Instead, the researcher selects what she believes are 10 *representative* ore samples and hopes that these samples will *behave as if* they had been randomly selected from the two populations.

The researcher can, however, randomly select the five samples to be measured with each method. Number the samples from 1 to 10. The five samples selected for the new method may correspond to 5 one-digit random numbers. Use this sequence of random digits generated on a scientific calculator:

$$948247817184610$$

Since you cannot select the same ore sample twice, you must skip any digit that has already been chosen. Ore samples 9, 4, 8, 2, and 7 will be measured using the new method. The other samples—1, 3, 5, 6, and 10—will be measured using the standard method.

In addition to *simple random sampling,* there are other sampling plans that involve randomization and therefore provide a probabilistic basis for inference making. Three such plans are based on *stratified, cluster,* and *systematic sampling.*

When the population consists of two or more subpopulations, called **strata,** a sampling plan that ensures that each subpopulation is represented in the sample is called a **stratified random sample.**

Definition **Stratified random sampling** involves selecting a simple random sample from each of a given number of subpopulations, or **strata.**

Citizens' opinions about the construction of a performing arts center could be collected using a stratified random sample with city voting wards as strata. National polls usually involve some form of stratified random sampling with states as strata.

Another form of random sampling is used when the available sampling units are groups of elements, called **clusters.** For example, a household is a *cluster* of individuals living together. A city block or a neighborhood might be a convenient sampling unit and might be considered a *cluster* for a given sampling plan.

Definition A **cluster sample** is a simple random sample of clusters from the available clusters in the population.

When a particular cluster is included in the sample, a census of every element in the cluster is taken.

Sometimes the population to be sampled is ordered, such as an alphabetized list of people with driver's licenses, a list of utility users arranged by service addresses, or a list of customers by account numbers. In these and other situations, one element is chosen at random from the first k elements, and then every kth element thereafter is included in the sample.

Definition A **1-in-k systematic random sample** involves the random selection of one of the first k elements in an ordered population, and then the systematic selection of every kth element thereafter.

NEED A TIP?

All sampling plans used for making inferences must involve *randomization*!

Not all sampling plans, however, involve random selection. You have probably heard of the nonrandom telephone polls in which those people who wish to express support for a question call one "900 number" and those opposed call a second "900 number." Each person must pay for his or her call. It is obvious that those people who call do not represent the population at large. This type of sampling plan is one form of a **convenience sample**—a sample that can be easily and simply obtained without random selection. Advertising for subjects who will be paid a fee for participating in an experiment produces a convenience sample. **Judgment sampling** allows the sampler to decide who will or will not be included in the sample. **Quota sampling,** in which the makeup of the sample must reflect the makeup of the population on some preselected characteristic, often has a nonrandom component in the selection process. **Remember that nonrandom samples can be described but cannot be used for making inferences!**

7.2 EXERCISES

BASIC TECHNIQUES

7.1 A population consists of $N = 500$ experimental units. Use a random number table to select a random sample of $n = 20$ experimental units. (HINT: Since you need to use three-digit numbers, you can assign 2 three-digit numbers to each of the sampling units in the manner shown in the table.) What is the probability that each experimental unit is selected for inclusion in the sample?

Experimental Units	Random Numbers
1	001, 501
2	002, 502
3	003, 503
4	004, 504
.	.
.	.
.	.
499	499, 999
500	500, 000

7.2 A political analyst wishes to select a sample of $n = 20$ people from a population of 2000. Use the random number table to identify the people to be included in the sample.

7.3 A population contains 50,000 voters. Use the random number table to identify the voters to be included in a random sample of $n = 15$.

7.4 A small city contains 20,000 voters. Use the random number table to identify the voters to be included in a random sample of $n = 15$.

7.5 Every 10th Person A random sample of public opinion in a small town was obtained by selecting every 10th person who passed by the busiest corner in the downtown area. Will this sample have the characteristics of a random sample selected from the town's citizens? Explain.

7.6 Parks and Recreation A questionnaire was mailed to 1000 registered municipal voters selected at random. Only 500 questionnaires were returned, and of the 500 returned, 360 respondents were strongly opposed to a surcharge proposed to support the city Parks and Recreation Department. Are you willing to accept the 72% figure as a valid estimate of the percentage in the city who are opposed to the surcharge? Why or why not?

7.7 DMV Lists In many states, lists of possible jurors are assembled from voter registration lists and Department of Motor Vehicles records of licensed drivers and car owners. In what ways might this list not cover certain sectors of the population adequately?

7.8 Sex and Violence One question on a survey questionnaire is phrased as follows: "Don't you agree that there is too much sex and violence during prime TV viewing hours?" Comment on possible problems with the responses to this question. Suggest a better way to pose the question.

APPLICATIONS

7.9 Omega-3 Fats Contrary to current thought about omega-3 fatty acids, new research shows that the beneficial fats may not help reduce second heart attacks in heart attack survivors. The study included 4837 men and women being treated for heart disease. The experimental group received an additional 400 mg of the fats daily.[1] Suppose that this experiment was repeated with 50 individuals in the control group and 50 individuals in the experimental group. Determine a randomization scheme to assign the 100 individuals to the two groups.

7.10 Racial Bias? Does the race of an interviewer matter? This question was investigated by Chris Gilberg and colleagues and reported in an issue of *Chance* magazine.[2] The interviewer asked, "Do you feel that affirmative action should be used as an occupation selection criteria?" with possible answers of yes or no.

a. What problems might you expect with responses to this question when asked by interviewers of different ethnic origins?

b. When people were interviewed by an African-American, the response was about 70% in favor of affirmative action, approximately 35% when interviewed by an Asian, and approximately 25% when interviewed by a Caucasian. Do these results support your answer in part a?

7.11 Native American Youth The *American Journal of Human Biology* reported on a study of a dietary assessment tool for use in the population of urban Native American youth.[3] The subjects were Native American youth attending an after-school program in Minneapolis, MN. All 61 children between the ages of 9 and 13 who satisfied the requirements of the study objectives were included in the experiment.

a. Describe the sampling plan used to select study participants.

b. What chance mechanism was used to select this sample of 61 Native American 9- to 13-year-old individuals?

c. Can valid inferences be made using the results of this study? Why or why not?

d. If you had to devise an alternative sampling plan, what would you change?

7.12 Tai Chi and Fibromyalgia A small new study shows that tai chi, an ancient Chinese practice of exercise and meditation, may relieve symptoms of chronic painful fibromyalgia. The study assigned 66 fibromyalgia patients to take either a 12-week tai chi class, or attend a wellness education class.[4]

a. Provide a randomization scheme to assign 66 subjects to the two groups.

b. Will your randomization scheme result in equal-sized groups? Explain.

7.13 Going to the Moon Two different Gallup Polls were conducted for *CNN/USA Today,* both of which involved people's feelings about the U.S.

space program.[5] Here is a question from each poll, along with the responses of the sampled Americans:

Space Exploration

CNN/USA Today/Gallup Poll. Nationwide:

"Would you favor or oppose a new U.S. space program that would send astronauts to the moon?" Form A (N = 510, MoE ± 5)

	Favor %	Oppose %	No Opinion %
12/03	53	45	2

"Would you favor or oppose the U.S. government spending billions of dollars to send astronauts to the moon?" Form B (N = 494, MoE ± 5)

	Favor %	Oppose %	No Opinion %
12/03	31	67	2

---

a. Read the two poll questions. Which of the two wordings is more unbiased? Explain.

b. Look at the responses for the two different polls. How would you explain the large differences in the percentages either favoring or opposing the new program?

7.14 Ask America A 2003 nationwide policy survey titled "Ask America" was sent by the National Republican Congressional Committee to voters in the Forty-fourth Congressional District, asking for opinions on a variety of political issues.[6] Here are some questions from the survey:

- In recent years has the federal government grown more or less intrusive in your personal and business affairs?
- Is President Bush right in trying to rein in the size and scope of the federal government against the wishes of the big government Democrats?
- Do you believe the death penalty is a deterrrent to crime?
- Do you agree that the obstructionist Democrats should not be allowed to gain control of the U.S. Congress in the upcoming elections?

Comment on the effect of wording bias on the responses gathered using this survey.

7.3 STATISTICS AND SAMPLING DISTRIBUTIONS

When you select a random sample from a population, the numerical descriptive measures you calculate from the sample are called **statistics.** These statistics vary or change for each different random sample you select; that is, they are *random variables*. The probability distributions for statistics are called **sampling distributions** because, in repeated sampling, they provide this information:

- What values of the statistic can occur.
- How often each value occurs.

Definition The **sampling distribution of a statistic** is the probability distribution for the possible values of the statistic that results when random samples of size *n* are repeatedly drawn from the population.

There are three ways to find the sampling distribution of a statistic:

1. Derive the distribution *mathematically* using the laws of probability.
2. Use a *simulation* to approximate the distribution. That is, draw a large number of samples of size *n*, calculating the value of the statistic for each sample, and tabulate the results in a relative frequency histogram. When the number of

samples is large, the histogram will be very close to the theoretical sampling distribution.

3. Use *statistical theorems* to derive exact or approximate sampling distributions.

The next example demonstrates how to derive the sampling distributions of two statistics for a very small population.

EXAMPLE (7.3) A population consists of $N = 5$ numbers: 3, 6, 9, 12, 15. If a random sample of size $n = 3$ is selected without replacement, find the sampling distributions for the sample mean $\bar{x}$ and the sample median m.

Solution You are sampling from the population shown in Figure 7.1. It contains five distinct numbers and each is equally likely, with probability $p(x) = 1/5$. You can easily find the population mean and median as

$$\mu = \frac{3 + 6 + 9 + 12 + 15}{5} = 9 \quad \text{and} \quad M = 9$$

FIGURE 7.1

Probability histogram for the $N = 5$ population values in Example 7.3

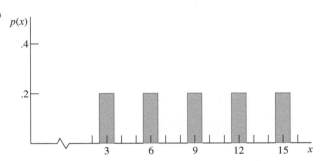

NEED A TIP?

Sampling distributions can be either discrete or continuous.

There are 10 possible random samples of size $n = 3$ and each is equally likely, with probability 1/10. These samples, along with the calculated values of $\bar{x}$ and m for each, are listed in Table 7.3. You will notice that some values of $\bar{x}$ are more likely than others because they occur in more than one sample. For example,

$$P(\bar{x} = 8) = \frac{2}{10} = .2 \quad \text{and} \quad P(m = 6) = \frac{3}{10} = .3$$

The values in Table 7.3 are tabulated, and the sampling distributions for $\bar{x}$ and m are shown in Table 7.4 and Figure 7.2.

Since the population of $N = 5$ values is symmetric about the value $x = 9$, both the *population mean* and the *median* equal 9. It would seem reasonable, therefore, to consider using either $\bar{x}$ or m as a possible estimator of $M = \mu = 9$. Which estimator would you choose? From Table 7.3, you see that, in using m as an estimator, you would be in error by $9 - 6 = 3$ with probability .3 or by $9 - 12 = -3$ with probability .3. That is, the error in estimation using m would be 3 with probability .6. In using $\bar{x}$, however, an error of 3 would occur with probability only .2. On these grounds alone, you may wish to use $\bar{x}$ as an estimator in preference to m.

TABLE 7.3

Values of $\bar{x}$ and m for Simple Random Sampling when $n = 3$ and $N = 5$

Sample	Sample Values	$\bar{x}$	m
1	3, 6, 9	6	6
2	3, 6, 12	7	6
3	3, 6, 15	8	6
4	3, 9, 12	8	9
5	3, 9, 15	9	9
6	3, 12, 15	10	12
7	6, 9, 12	9	9
8	6, 9, 15	10	9
9	6, 12, 15	11	12
10	9, 12, 15	12	12

TABLE 7.4

Sampling Distributions for (a) the Sample Mean and (b) the Sample Median

(a) $\bar{x}$	$p(\bar{x})$	(b) m	$p(m)$
6	.1	6	.3
7	.1	9	.4
8	.2	12	.3
9	.2		
10	.2		
11	.1		
12	.1		

FIGURE 7.2

Probability histograms for the sampling distributions of the sample mean, $\bar{x}$, and the sample median, m, in Example 7.3

 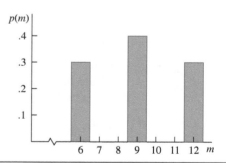

NEED A TIP?

Almost every statistic has a mean and a standard deviation (or *standard error*) describing its center and spread.

It was not too difficult to derive these sampling distributions in Example 7.3 because the number of elements in the population was very small. When this is not the case, you may need to use one of these methods:

- Use a simulation to approximate the sampling distribution empirically.
- Rely on statistical theorems and theoretical results.

One important statistical theorem that describes the sampling distribution of statistics that are sums or averages is presented in the next section.

THE CENTRAL LIMIT THEOREM

7.4

The **Central Limit Theorem** states that, under rather general conditions, sums and means of random samples of measurements drawn from a population tend to have an approximately normal distribution. Suppose you toss a balanced die $n = 1$ time. The random variable x is the number observed on the upper face. This familiar random variable can take six values, each with probability 1/6, and its probability distribution is shown in Figure 7.3. The shape of the distribution is *flat* or *uniform* and symmetric about the mean $\mu = 3.5$, with a standard deviation $\sigma = 1.71$. (See Section 4.8 and Exercise 4.84.)

FIGURE 7.3 ●

Probability distribution for *x*, the number appearing on a single toss of a die

Now, take a sample of size $n = 2$ from this population; that is, toss two dice and record the sum of the numbers on the two upper faces, $\Sigma x_i = x_1 + x_2$. Table 7.5 shows the 36 possible outcomes, each with probability 1/36. The sums are tabulated, and each of the possible sums is divided by $n = 2$ to obtain an average. The result is the **sampling distribution** of $\bar{x} = \Sigma x_i/n$, shown in Figure 7.4. You should notice the dramatic difference in the shape of the sampling distribution. It is now roughly mound-shaped but still symmetric about the mean $\mu = 3.5$.

TABLE 7.5 ● **Sums of the Upper Faces of Two Dice**

	First Die					
Second Die	1	2	3	4	5	6
1	2	3	4	5	6	7
2	3	4	5	6	7	8
3	4	5	6	7	8	9
4	5	6	7	8	9	10
5	6	7	8	9	10	11
6	7	8	9	10	11	12

FIGURE 7.4

Sampling distribution of $\bar{x}$ for $n = 2$ dice

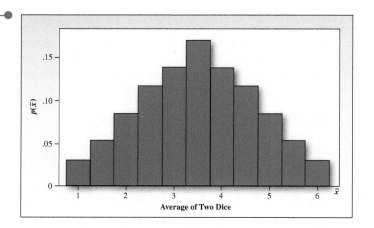

Using a similar procedure, we generated the sampling distributions of $\bar{x}$ when $n = 3$ and $n = 4$. For $n = 3$, the sampling distribution in Figure 7.5 clearly shows the mound shape of the normal probability distribution, still centered at $\mu = 3.5$. Notice also that the spread of the distribution is slowly *decreasing* as the sample size *n increases*. Figure 7.6 dramatically shows that the distribution of $\bar{x}$ is approximately normally distributed based on a sample as small as $n = 4$. This phenomenon is the result of an important statistical theorem called the **Central Limit Theorem (CLT).**

FIGURE 7.5

Sampling distribution of $\bar{x}$ for $n = 3$ dice

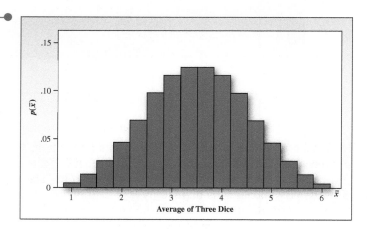

FIGURE 7.6

Sampling distribution of $\bar{x}$ for $n = 4$ dice

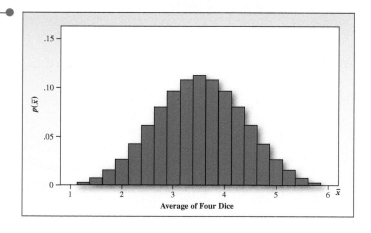

Central Limit Theorem

If random samples of n observations are drawn from a nonnormal population with finite mean μ and standard deviation σ, then, when n is large, the sampling distribution of the sample mean $\bar{x}$ is approximately normally distributed, with mean μ and standard deviation

$$\frac{\sigma}{\sqrt{n}}$$

The approximation becomes more accurate as n becomes large.

Regardless of its shape, the sampling distribution of $\bar{x}$ always has a mean identical to the mean of the sampled population and a standard deviation equal to the population standard deviation σ divided by $\sqrt{n}$. Consequently, *the spread of the distribution of sample means is considerably less than the spread of the sampled population.*

The Central Limit Theorem can be restated to apply to the **sum of the sample measurements Σx_i,** which, as n becomes large, also has an approximately normal distribution with mean $n\mu$ and standard deviation $\sigma\sqrt{n}$.

NEED A TIP?

The sampling distribution of $\bar{x}$ always has a mean μ and standard deviation $\sigma/\sqrt{n}$. The CLT helps describe its **shape.**

The important contribution of the Central Limit Theorem is in statistical inference. Many estimators that are used to make inferences about population parameters are sums or averages of the sample measurements. When the sample size is sufficiently large, you can expect these estimators to have sampling distributions that are approximately normal. You can then use the normal distribution to describe the behavior of these estimators in repeated sampling and evaluate the probability of observing certain sample results. As in Chapter 6, these probabilities are calculated using the standard normal random variable

$$z = \frac{\text{Estimator} - \text{Mean}}{\text{Standard deviation}}$$

As you reread the Central Limit Theorem, you may notice that the approximation is valid as long as the sample size n is "large"—but how large is "large"? Unfortunately, there is no clear answer to this question. The appropriate value of n depends on the shape of the population from which you sample as well as on how you want to use the approximation. However, these guidelines will help:

❓ NEED TO KNOW...

When the Sample Size is Large Enough to Use the Central Limit Theorem

- If the sampled population is **normal,** then the sampling distribution of $\bar{x}$ will also be normal, no matter what sample size you choose. This result can be proven theoretically, but it should not be too difficult for you to accept without proof.
- When the sampled population is approximately **symmetric,** the sampling distribution of $\bar{x}$ becomes approximately normal for relatively small values of n. Remember how rapidly the "flat" distribution in the dice example became mound-shaped ($n = 3$).
- When the sampled population is **skewed,** the sample size n must be larger, with n at least 30 before the sampling distribution of $\bar{x}$ becomes approximately normal.

These guidelines suggest that, for many populations, the sampling distribution of $\bar{x}$ will be approximately normal for moderate sample sizes; an exception to this rule occurs in sampling a binomial population when either p or $q = (1 - p)$ is very small. As specific applications of the Central Limit Theorem arise, we will give you the appropriate sample size n.

THE SAMPLING DISTRIBUTION OF THE SAMPLE MEAN

7.5

If the population mean μ is unknown, you might choose several *statistics* as an estimator; the sample mean $\bar{x}$ and the sample median m are two that readily come to mind. Which should you use? Consider these criteria in choosing the estimator for μ:

- Is it easy or hard to calculate?
- Does it produce estimates that are generally too high or too low?
- Is it more or less variable than other possible estimators?

The sampling distributions for $\bar{x}$ and m with $n = 3$ for the small population in Example 7.3 showed that, in terms of these criteria, the sample mean performed better than the sample median as an estimator of μ. In many situations, the sample mean $\bar{x}$ has desirable properties as an estimator that are not shared by other competing estimators; therefore, it is more widely used.

THE SAMPLING DISTRIBUTION OF THE SAMPLE MEAN, $\bar{x}$

- If a random sample of n measurements is selected from a population with mean μ and standard deviation σ, the sampling distribution of the sample mean $\bar{x}$ will have mean μ and standard deviation[*]

$$\frac{\sigma}{\sqrt{n}}$$

- If the population has a *normal* distribution, the sampling distribution of $\bar{x}$ will be *exactly* normally distributed, *regardless of the sample size, n.*
- If the population distribution is *nonnormal,* the sampling distribution of $\bar{x}$ will be *approximately* normally distributed for large samples (by the Central Limit Theorem). Conservatively, we require $n \geq 30$.

---

[*]When repeated samples of size n are randomly selected from a *finite* population with N elements whose mean is μ and whose variance is σ^2, the standard deviation of $\bar{x}$ is

$$\frac{\sigma}{\sqrt{n}} \sqrt{\frac{N - n}{N - 1}}$$

where σ^2 is the population variance. When N is large relative to the sample size n, $\sqrt{(N - n)(N - 1)}$ is approximately equal to 1, and the standard deviation of $\bar{x}$ is

$$\frac{\sigma}{\sqrt{n}}$$

Standard Error

Definition The standard deviation of a statistic used as an estimator of a population parameter is also called the **standard error of the estimator** (abbreviated **SE**) because it refers to the precision of the estimator. Therefore, the standard deviation of $\bar{x}$—given by $\sigma/\sqrt{n}$—is referred to as the **standard error of the mean** (abbreviated as SE($\bar{x}$), SEM, or sometimes just SE).

NEED TO KNOW...

How to Calculate Probabilities for the Sample Mean $\bar{x}$

If you know that the sampling distribution of $\bar{x}$ is *normal* or *approximately normal*, you can describe the behavior of the sample mean $\bar{x}$ by calculating the probability of observing certain values of $\bar{x}$ in repeated sampling.

1. Find μ and calculate SE $(\bar{x}) = \sigma/\sqrt{n}$.
2. Write down the event of interest in terms of $\bar{x}$, and locate the appropriate area on the normal curve.
3. Convert the necessary values of $\bar{x}$ to z-values using

$$z = \frac{\bar{x} - \mu}{\sigma/\sqrt{n}}$$

4. Use Table 3 in Appendix I to calculate the probability.

EXAMPLE 7.4

The duration of Alzheimer's disease from the onset of symptoms until death ranges from 3 to 20 years; the average is 8 years with a standard deviation of 4 years. The administrator of a large medical center randomly selects the medical records of 30 deceased Alzheimer's patients from the medical center's database, and records the average duration. Find the approximate probabilities for these events:

1. The average duration is less than 7 years.
2. The average duration exceeds 7 years.
3. The average duration lies within 1 year of the population mean $\mu = 8$.

NEED A TIP?

If x is normal, $\bar{x}$ is normal for any n.

If x is not normal, $\bar{x}$ is approximately normal for large n.

Solution Since the administrator has selected a random sample from the database at this medical center, he can draw conclusions about only past, present, or future patients with Alzheimer's disease at this medical center. If, on the other hand, this medical center can be considered representative of other medical centers in the country, it may be possible to draw more far-reaching conclusions.

What can you say about the shape of the sampled population? It is not symmetric, because the mean $\mu = 8$ does not lie halfway between the maximum and minimum values. Since the mean is closer to the minimum value, the distribution is skewed to the right, with a few patients living a long time after the onset of the disease. Regardless of the shape of the population distribution, however, the sampling distribution of $\bar{x}$ has a mean $\mu = 8$ and standard deviation $\sigma/\sqrt{n} = 4/\sqrt{30} = .73$. In addition, because the sample size is $n = 30$, the Central Limit Theorem ensures the *approximate normality* of the sampling distribution of $\bar{x}$.

1. The probability that $\bar{x}$ is less than 7 is given by the shaded area in Figure 7.7. To find this area, you need to calculate the value of z corresponding to $\bar{x} = 7$:

$$z = \frac{\bar{x} - \mu}{\sigma/\sqrt{n}} = \frac{7 - 8}{.73} = -1.37$$

From Table 3 in Appendix I, you can find the cumulative area corresponding to $z = -1.37$ and

$$P(\bar{x} < 7) = P(z < -1.37) = .0853$$

FIGURE 7.7

The probability that $\bar{x}$
is less than 7 for
Example 7.4

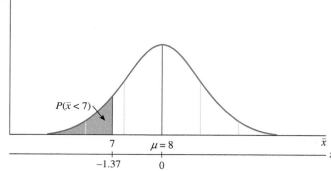

(NOTE: You must use $\sigma/\sqrt{n}$ (not σ) in the formula for z because you are finding an area under the sampling distribution for $\bar{x}$, not under the probability distribution for x.)

2. The event that $\bar{x}$ exceeds 7 is the complement of the event that $\bar{x}$ is less than 7. Thus, the probability that $\bar{x}$ exceeds 7 is

$$P(\bar{x} > 7) = 1 - P(\bar{x} \leq 7)$$
$$= 1 - .0853 = .9147$$

 NEED A TIP?

Remember that for
continuous random
variables, there is no
probability
assigned to a single
point. Therefore,
$P(\bar{x} \leq 7) = P(\bar{x} < 7)$.

3. The probability that $\bar{x}$ lies within 1 year of $\mu = 8$ is the shaded area in Figure 7.8. The z-value corresponding to $\bar{x} = 7$ is $z = -1.37$, from part 1, and the z-value for $\bar{x} = 9$ is

$$z = \frac{\bar{x} - \mu}{\sigma/\sqrt{n}} = \frac{9 - 8}{.73} = 1.37$$

The probability of interest is

$$P(7 < \bar{x} < 9) = P(-1.37 < z < 1.37)$$
$$= .9147 - .0853 = .8294$$

FIGURE 7.8

The probability that $\bar{x}$ lies within 1 year of $\mu = 8$ for Example 7.4

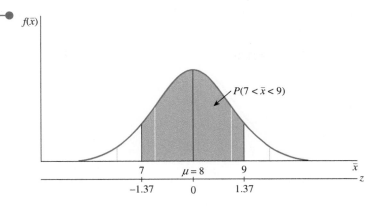

EXAMPLE 7.5

To avoid difficulties with the Federal Trade Commission or state and local consumer protection agencies, a beverage bottler must make reasonably certain that 12-ounce bottles actually contain 12 ounces of beverage. To determine whether a bottling machine is working satisfactorily, one bottler randomly samples 10 bottles per hour and measures the amount of beverage in each bottle. The mean $\bar{x}$ of the 10 fill measurements is used to decide whether to readjust the amount of beverage delivered per bottle by the filling machine.

If records show that the amount of fill per bottle is normally distributed, with a standard deviation of .2 ounce, and if the bottling machine is set to produce a mean fill per bottle of 12.1 ounces, what is the approximate probability that the sample mean $\bar{x}$ of the 10 test bottles is less than 12 ounces?

Solution The mean of the sampling distribution of the sample mean $\bar{x}$ is identical to the mean of the population of bottle fills—namely, $\mu = 12.1$ ounces—and the standard error of $\bar{x}$ is

$$\text{SE} = \frac{\sigma}{\sqrt{n}} = \frac{.2}{\sqrt{10}} = .063$$

(NOTE: σ is the standard deviation of the population of bottle fills, and n is the number of bottles in the sample.) Since the amount of fill is normally distributed, $\bar{x}$ is also normally distributed, as shown in Figure 7.9.

To find the probability that $\bar{x}$ is less than 12 ounces, express the value $\bar{x} = 12$ in units of standard deviations:

$$z = \frac{\bar{x} - \mu}{\sigma/\sqrt{n}} = \frac{12 - 12.1}{.063} = -1.59$$

Then

$$P(\bar{x} < 12) = P(z < -1.59) = .0559 \approx .056$$

Thus, if the machine is set to deliver an average fill of 12.1 ounces, the mean fill $\bar{x}$ of a sample of 10 bottles will be less than 12 ounces with a probability equal to .056.

FIGURE 7.9

Probability distribution of $\bar{x}$, the mean of the $n = 10$ bottle fills, for Example 7.5

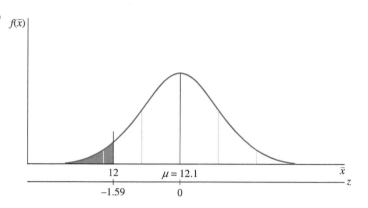

When this danger signal occurs ($\bar{x}$ is less than 12), the bottler takes a larger sample to recheck the setting of the filling machine.

7.5   EXERCISES

BASIC TECHNIQUES

7.15 Random samples of size n were selected from populations with the means and variances given here. Find the mean and standard deviation of the sampling distribution of the sample mean in each case:

a. $n = 36$, $\mu = 10$, $\sigma^2 = 9$

b. $n = 100$, $\mu = 5$, $\sigma^2 = 4$

c. $n = 8$, $\mu = 120$, $\sigma^2 = 1$

7.16 Refer to Exercise 7.15.

a. If the sampled populations are normal, what is the sampling distribution of $\bar{x}$ for parts a, b, and c?

b. According to the Central Limit Theorem, if the sampled populations are *not* normal, what can be said about the sampling distribution of $\bar{x}$ for parts a, b, and c?

Data set **7.17** A population consists of $N = 5$ numbers: EX0717 1, 3, 5, 6, and 7. It can be shown that the mean and standard deviation for this population are $\mu = 4.4$ and $\sigma = 2.15$, respectively.

a. Construct a probability histogram for this population.

b. Use the random number table, Table 10 in Appendix I, to select a random sample of size $n = 10$ with replacement from the population. Calculate the sample mean, $\bar{x}$. Repeat this procedure, calculating the sample mean $\bar{x}$ for your second sample.

(HINT: Assign the random digits 0 and 1 to the measurement $x = 1$; assign digits 2 and 3 to the measurement $x = 3$, and so on.)

c. To simulate the sampling distribution of $\bar{x}$, we have selected 50 more samples of size $n = 10$ with replacement, and have calculated the corresponding sample means. Construct a relative frequency histogram for these 50 values of $\bar{x}$. What is the shape of this distribution?

4.8	4.2	4.2	4.5	4.3	4.3	5.0	4.0	3.3	4.7
3.0	5.9	5.7	4.2	4.4	4.8	5.0	5.1	4.8	4.2
4.6	4.1	3.4	4.9	4.1	4.0	3.7	4.3	4.3	4.5
5.0	4.6	4.1	5.1	3.4	5.9	5.0	4.3	4.5	3.9
4.4	4.2	4.2	5.2	5.4	4.8	3.6	5.0	4.5	4.9

7.18 Refer to Exercise 7.17.

a. Use the data entry method in your calculator to find the mean and standard deviation of the 50 values of $\bar{x}$ given in Exercise 7.17, part c.

b. Compare the values calculated in part a to the theoretical mean μ and the theoretical standard deviation $\sigma/\sqrt{n}$ for the sampling distribution of $\bar{x}$. How close do the values calculated from the 50 measurements come to the theoretical values?

7.19 A random sample of n observations is selected from a population with standard deviation $\sigma = 1$. Calculate the standard error of the mean (SE) for these values of n:

a. $n = 1$ **b.** $n = 2$ **c.** $n = 4$

d. $n = 9$ **e.** $n = 16$ **f.** $n = 25$
g. $n = 100$

7.20 Refer to Exercise 7.19. Plot the standard error of the mean (SE) versus the sample size n and connect the points with a smooth curve. What is the effect of increasing the sample size on the standard error?

7.21 A random sample of size $n = 49$ is selected from a population with mean $\mu = 53$ and standard deviation $\sigma = 21$.

a. What will be the approximate shape of the sampling distribution of $\bar{x}$?

b. What will be the mean and standard deviation of the sampling distribution of $\bar{x}$?

7.22 Refer to Exercise 7.21. Find the probability that the sample mean is greater than 55.

7.23 A random sample of size $n = 40$ is selected from a population with mean $\mu = 100$ and standard deviation $\sigma = 20$.

a. What will be the approximate shape of the sampling distribution of $\bar{x}$?

b. What will be the mean and standard deviation of the sampling distribution of $\bar{x}$?

7.24 Refer to Exercise 7.23. Find the probability that the sample mean is between 105 and 110.

7.25 Suppose a random sample of $n = 25$ observations is selected from a population that is normally distributed with mean equal to 106 and standard deviation equal to 12.

a. Give the mean and the standard deviation of the sampling distribution of the sample mean $\bar{x}$.

b. Find the probability that $\bar{x}$ exceeds 110.

c. Find the probability that the sample mean deviates from the population mean $\mu = 106$ by no more than 4.

APPLICATIONS

7.26 Faculty Salaries Suppose that college faculty with the rank of professor at public 2-year institutions earn an average of $71,802 per year[7] with a standard deviation of $4000. In an attempt to verify this salary level, a random sample of 60 professors was selected from a personnel database for all 2-year institutions in the United States.

a. Describe the sampling distribution of the sample mean $\bar{x}$.

b. Within what limits would you expect the sample average to lie, with probability .95?

c. Calculate the probability that the sample mean $\bar{x}$ is greater than $73,000?

d. If your random sample actually produced a sample mean of $73,000, would you consider this unusual? What conclusion might you draw?

7.27 Measurement Error When research chemists perform experiments, they may obtain slightly different results on different replications, even when the experiment is performed identically each time. These differences are due to a phenomenon called "measurement error."

a. List some variables in a chemical experiment that might cause some small changes in the final response measurement.

b. If you want to make sure that your measurement error is small, you can replicate the experiment and take the sample average of all the measurements. To decrease the amount of variability in your average measurement, should you use a large or a small number of replications? Explain.

7.28 Tomatoes Explain why the weight of a package of one dozen tomatoes should be approximately normally distributed if the dozen tomatoes represent a random sample.

7.29 Bacteria in Water Use the Central Limit Theorem to explain why a Poisson random variable—say, the number of a particular type of bacteria in a cubic foot of water—has a distribution that can be approximated by a normal distribution when the mean μ is large. (HINT: One cubic foot of water contains 1728 cubic inches of water.)

7.30 Paper Strength A paper manufacturer requires a minimum strength of 20 pounds per square inch. To check on the quality of the paper, a random sample of 10 pieces of paper is selected each hour from the previous hour's production and a strength measurement is recorded for each. Assume that the strength measurements are normally distributed with a standard deviation $\sigma = 2$ pounds per square inch.

a. What is the approximate sampling distribution of the sample mean of $n = 10$ test pieces of paper?

b. If the mean of the population of strength measurements is 21 pounds per square inch, what is the approximate probability that, for a random sample of $n = 10$ test pieces of paper, $\bar{x} < 20$?

c. What value would you select for the mean paper strength μ in order that $P(\bar{x} < 20)$ be equal to .001?

7.31 Potassium Levels The normal daily human potassium requirement is in the range of 2000 to 6000 milligrams (mg), with larger amounts required during hot summer weather. The amount of potassium in food varies, but bananas are often associated with high potassium, with approximately 422 mg in a medium-sized banana[8]. Suppose the distribution of potassium in a banana is normally distributed, with mean equal to 422 mg and standard deviation equal to 13 mg per banana. You eat $n = 3$ bananas per day, and T is the total number of milligrams of potassium you receive from them.

a. Find the mean and standard deviation of T.

b. Find the probability that your total daily intake of potassium from the three bananas will exceed 1300 mg. (HINT: Note that T is the sum of three random variables, x_1, x_2, and x_3, where x_1 is the amount of potassium in banana number 1, etc.)

7.32 Deli Sales The total daily sales, x, in the deli section of a local market is the sum of the sales generated by a fixed number of customers who make purchases on a given day.

a. What kind of probability distribution do you expect the total daily sales to have? Explain.

b. For this particular market, the average sale per customer in the deli section is $8.50 with $\sigma = 2.50. If 30 customers make deli purchases on a given day, give the mean and standard deviation of the probability distribution of the total daily sales, x.

7.33 Normal Temperatures In Exercise 1.67, Allen Shoemaker derived a distribution of human body temperatures with a distinct mound shape.[9] Suppose we assume that the temperatures of healthy humans are approximately normal with a mean of 98.6° and a standard deviation of 0.8°.

a. If 130 healthy people are selected at random, what is the probability that the average temperature for these people is 98.25° or lower?

b. Would you consider an average temperature of 98.25° to be an unlikely occurrence, given that the true average temperature of healthy people is 98.6°? Explain.

7.34 Sports and Achilles Tendon Injuries Sports that involve a significant amount of running, jumping, or hopping put participants at risk for Achilles tendinopathy (AT), an inflammation and thickening of the Achilles tendon. A study in *The American Journal of Sports Medicine* looked at the diameter (in mm) of the affected and nonaffected tendons for patients who participated in these types of sports activities.[10] Suppose that the Achilles tendon diameters in the general population have a mean of 5.97 millimeters (mm) with a standard deviation of 1.95 mm.

a. What is the probability that a randomly selected sample of 31 patients would produce an average diameter of 6.5 mm or less for the nonaffected tendon?

b. When the diameters of the affected tendon were measured for a sample of 31 patients, the average diameter was 9.80. If the average tendon diameter in the population of patients with AT is no different than the average diameter of the nonaffected tendons (5.97 mm), what is the probability of observing an average diameter of 9.80 or higher?

c. What conclusions might you draw from the results of part b?

THE SAMPLING DISTRIBUTION OF THE SAMPLE PROPORTION

7.6

There are many practical examples of the binomial random variable x. One common application involves consumer preference or opinion polls, in which we use a random sample of n people to estimate the proportion p of people in the population who have a specified characteristic. If x of the sampled people have this characteristic, then the sample proportion

$$\hat{p} = \frac{x}{n}$$

can be used to estimate the population proportion p.[†]

The binomial random variable x has a probability distribution $p(x)$, described in Chapter 5, with mean np and standard deviation $\sqrt{npq}$. Since $\hat{p}$ is simply the value of x, expressed as a proportion $\left(\hat{p} = \dfrac{x}{n}\right)$, the sampling distribution of $\hat{p}$ is identical to the probability distribution of x, except that it has a new scale along the horizontal axis (Figure 7.10).

FIGURE 7.10

Sampling distribution of the binomial random variable x and the sample proportion $\hat{p}$

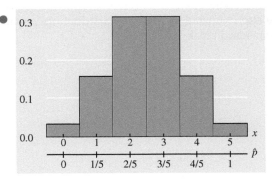

Because of this change of scale, the mean and standard deviation of $\hat{p}$ are also rescaled, so that the mean of the sampling distribution of $\hat{p}$ is p, and its standard error is

$$\text{SE}(\hat{p}) = \sqrt{\frac{pq}{n}} \qquad \text{where } q = 1 - p$$

Finally, just as we can approximate the probability distribution of x with a normal distribution when the sample size n is large, we can do the same with the sampling distribution of $\hat{p}$.

PROPERTIES OF THE SAMPLING DISTRIBUTION OF THE SAMPLE PROPORTION, $\hat{p}$

- If a random sample of n observations is selected from a binomial population with parameter p, then the sampling distribution of the sample proportion

$$\hat{p} = \frac{x}{n}$$

will have a mean

$$p$$

and a standard deviation

$$\text{SE}(\hat{p}) = \sqrt{\frac{pq}{n}} \qquad \text{where } q = 1 - p$$

- When the sample size n is large, the sampling distribution of $\hat{p}$ can be approximated by a normal distribution. The approximation will be adequate if $np > 5$ and $nq > 5$.

[†]A "hat" placed over the symbol of a population parameter denotes a statistic used to estimate the population parameter. For example, the symbol $\hat{p}$ denotes the sample proportion.

EXAMPLE (7.6)

In a survey, 500 mothers and fathers were asked about the importance of sports for boys and girls. Of the parents interviewed, 60% agreed that the genders are equal and should have equal opportunities to participate in sports. Describe the sampling distribution of the sample proportion $\hat{p}$ of parents who agree that the genders are equal and should have equal opportunities.

Solution You can assume that the 500 parents represent a random sample of the parents of all boys and girls in the United States and that the true proportion in the population is equal to some unknown value that you can call p. The sampling distribution of $\hat{p}$ can be approximated by a normal distribution,[††] with mean equal to p (see Figure 7.11) and standard error

$$SE(\hat{p}) = \sqrt{\frac{pq}{n}}$$

FIGURE 7.11

The sampling distribution for $\hat{p}$ based on a sample of $n = 500$ parents for Example 7.6

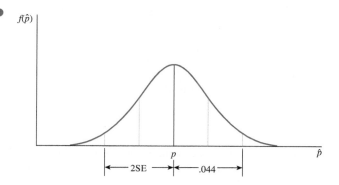

You can see from Figure 7.11 that the sampling distribution of $\hat{p}$ is centered over its mean p. Even though you do not know the exact value of p (the sample proportion $\hat{p} = .60$ may be larger or smaller than p), an approximate value for the standard deviation of the sampling distribution can be found using the sample proportion $\hat{p} = .60$ to approximate the unknown value of p. Thus,

$$SE = \sqrt{\frac{pq}{n}} \approx \sqrt{\frac{\hat{p}\hat{q}}{n}}$$

$$= \sqrt{\frac{(.60)(.40)}{500}} = .022$$

Therefore, approximately 95% of the time, $\hat{p}$ will fall within $2SE \approx .044$ of the (unknown) value of p.

[††]Checking the conditions that allow the normal approximation to the distribution of $\hat{p}$, you can see that $n = 500$ is adequate for values of p near .60 because $n\hat{p} = 300$ and $n\hat{q} = 200$ are both greater than 5.

NEED TO KNOW...

How to Calculate Probabilities for the Sample Proportion $\hat{p}$

1. Find the necessary values of n and p.
2. Check whether the normal approximation to the binomial distribution is appropriate ($np > 5$ and $nq > 5$).
3. Write down the event of interest in terms of $\hat{p}$, and locate the appropriate area on the normal curve.
4. Convert the necessary values of $\hat{p}$ to z-values using

$$z = \frac{\hat{p} - p}{\sqrt{\dfrac{pq}{n}}}$$

5. Use Table 3 in Appendix I to calculate the probability.

EXAMPLE **7.7**

Refer to Example 7.6. Suppose the proportion p of parents in the population is actually equal to .55. What is the probability of observing a sample proportion as large as or larger than the observed value $\hat{p} = .60$?

Solution Figure 7.12 shows the sampling distribution of $\hat{p}$ when $p = .55$, with the observed value $\hat{p} = .60$ located on the horizontal axis. The probability of observing a sample proportion $\hat{p}$ equal to or larger than .60 is approximated by the shaded area in the upper tail of this normal distribution with

$$p = .55$$

and

$$SE = \sqrt{\frac{pq}{n}} = \sqrt{\frac{(.55)(.45)}{500}} = .0222$$

FIGURE 7.12

The sampling distribution of $\hat{p}$ for $n = 500$ and $p = .55$ for Example 7.7

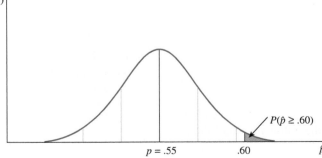

To find this shaded area, first calculate the z-value corresponding to $\hat{p} = .60$:

$$z = \frac{\hat{p} - p}{\sqrt{pq/n}} = \frac{.60 - .55}{.0222} = 2.25$$

Using Table 3 in Appendix I, you find

$$P(\hat{p} > .60) \approx P(z > 2.25) = 1 - .9878 = .0122$$

That is, if you were to select a random sample of $n = 500$ observations from a population with proportion p equal to .55, the probability that the sample proportion $\hat{p}$ would be as large as or larger than .60 is only .0122.

When the normal distribution was used in Chapter 6 to approximate the binomial probabilities associated with x, a correction of $\pm.5$ was applied to improve the approximation. The equivalent correction here is $\pm(.5/n)$. For example, for $\hat{p} = .60$ the value of z with the correction is

$$z_1 = \frac{(.60 - .001) - .55}{\sqrt{\dfrac{(.55)(.45)}{500}}} = 2.20$$

with $P(\hat{p} > .60) \approx .0139$. To two-decimal-place accuracy, this value agrees with the earlier result. When n is large, the effect of using the correction is generally negligible. You should solve problems in this and the remaining chapters *without* the correction factor unless you are specifically instructed to use it.

7.6   EXERCISES

BASIC TECHNIQUES

7.35 Random samples of size n were selected from binomial populations with population parameters p given here. Find the mean and the standard deviation of the sampling distribution of the sample proportion $\hat{p}$ in each case:

a. $n = 100, p = .3$

b. $n = 400, p = .1$

c. $n = 250, p = .6$

7.36 Is it appropriate to use the normal distribution to approximate the sampling distribution of $\hat{p}$ in the following circumstances?

a. $n = 50, p = .05$

b. $n = 75, p = .1$

c. $n = 250, p = .99$

7.37 Random samples of size $n = 75$ were selected from a binomial population with $p = .4$. Use the normal distribution to approximate the following probabilities:

a. $P(\hat{p} \le .43)$

b. $P(.35 \le \hat{p} \le .43)$

7.38 Random samples of size $n = 500$ were selected from a binomial population with $p = .1$.

a. Is it appropriate to use the normal distribution to approximate the sampling distribution of $\hat{p}$? Check to make sure the necessary conditions are met.

Using the results of part a, find these probabilities:

b. $\hat{p} > .12$

c. $\hat{p} < .10$

d. $\hat{p}$ lies within .02 of p

7.39 Calculate $SE(\hat{p})$ for $n = 100$ and these values of p:

a. $p = .01$ **b.** $p = .10$ **c.** $p = .30$

d. $p = .50$ **e.** $p = .70$ **f.** $p = .90$

g. $p = .99$

h. Plot $SE(\hat{p})$ versus p on graph paper and sketch a smooth curve through the points. For what value of p is the standard deviation of the sampling distribution of $\hat{p}$ a maximum? What happens to the standard error when p is near 0 or near 1.0?

7.40 A random sample of size $n = 50$ is selected from a binomial distribution with population proportion $p = .7$.

a. What will be the approximate shape of the sampling distribution of $\hat{p}$?

b. What will be the mean and standard deviation (or standard error) of the sampling distribution of $\hat{p}$?

c. Find the probability that the sample proportion $\hat{p}$ is less than .8.

7.41 A random sample of size $n = 80$ is selected from a binomial distribution with population proportion $p = .25$.

a. What will be the approximate shape of the sampling distribution of $\hat{p}$?

b. What will be the mean and standard deviation (or standard error) of the sampling distribution of $\hat{p}$?

c. Find the probability that the sample proportion $\hat{p}$ is between .18 and .44.

7.42 **a.** Is the normal approximation to the sampling distribution of $\hat{p}$ appropriate when $n = 400$ and $p = .8$?

b. Use the results of part a to find the probability that $\hat{p}$ is greater than .83.

c. Use the results of part a to find the probability that $\hat{p}$ lies between .76 and .84.

APPLICATIONS

7.43 **Eco-Friendly** Recycling trash, reducing waste, and reusing materials are eco-actions that will help the environment. According to a *USA Today* snapshot (Exercise 6.45), 78% of respondents list recycling as the leading way to help our environment.[11] Suppose that a random sample of $n = 100$ adults is selected and that the 78% figure is correct.

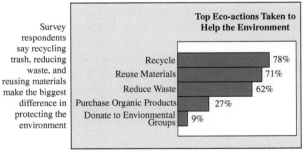

Survey respondents say recycling trash, reducing waste, and reusing materials make the biggest difference in protecting the environment

Top Eco-actions Taken to Help the Environment

Recycle 78%
Reuse Materials 71%
Reduce Waste 62%
Purchase Organic Products 27%
Donate to Enviornmental Groups 9%

a. Does the distribution of $\hat{p}$, the sample proportion of adults who list recycling as the leading way to help the environment have an approximate normal distribution? If so, what is its mean and standard deviation?

b. What is the probability that the sample proportion $\hat{p}$ is less than 75%?

c. What is the probability that $\hat{p}$ lies in the interval .7 to .75?

d. What might you conclude about p if the sample proportion were less than .65?

7.44 **Road Trip!** Parents with children list a GPS system (28%) and a DVD player (28%) as "must have" accessories for a road trip.[12] Suppose a sample of $n = 1000$ parents are randomly selected and asked what devices they would like to have for a family road trip. Let $\hat{p}$ be the proportion of parents in the sample who choose either a GPS system or a DVD player.

Must have accessories on family road trips

To surive road trips with the family, parents consider a GPS navigation device and a DVD player essential.

GPS 28% | DVD player 28%

Extra power supplies 7%
Don't know 7%
Roof rack 8%
Extra cup holders 14%
Folding/removable seating 8%

a. If $p = .28 + .28 = .56$, what is the exact distribution of $\hat{p}$? How can you approximate the distribution of $\hat{p}$?

b. What is the probability that $\hat{p}$ exceeds .6?

c. What is the probability that $\hat{p}$ lies between .5 and .6?

d. Would a sample percentage of $\hat{p} = .7$ contradict the reported value of .56?

7.45 **M&M'S** An advertiser claims that the average percentage of brown M&M'S candies in a package of milk chocolate M&M'S is 13%. Suppose you randomly select a package of milk chocolate M&M'S that contains 55 candies and determine the proportion of brown candies in the package.

a. What is the approximate distribution of the sample proportion of brown candies in a package that contains 55 candies?

b. What is the probability that the sample percentage of brown candies is less than 20%?

c. What is the probability that the sample percentage exceeds 35%?

d. Within what range would you expect the sample proportion to lie about 95% of the time?

7.46 **Fido's in the Car** It seems that driving with a pet in the car is the third worst driving distraction, behind talking on the phone and texting.

Source: USA Today, 19 August 2010, p. 8A

According to an American Automobile Association study, 80% of drivers admit to driving with a pet in the car, and of those, 20% allow their dogs to sit on their laps.[13] Suppose that you randomly select a sample of 100 drivers who have admitted to driving with a pet in their car.

a. What is the probability that 25% or more of the drivers allow their dogs to sit on their laps?

b. What is the probability that 10% or fewer of the drivers allow their dogs to sit on their laps?

c. Would it be unusual to find that 35% of the drivers allow their dogs to sit on their laps?

7.47 Oh, Nuts! Are you a chocolate "purist," or do you like other ingredients in your chocolate? *American Demographics* reports that almost 75% of consumers like traditional ingredients such as nuts or caramel in their chocolate. They are less enthusiastic about the taste of mint or coffee that provide more distinctive flavors.[14] A random sample of 200 consumers is selected and the number who like nuts or caramel in their chocolate is recorded.

a. What is the approximate sampling distribution for the sample proportion $\hat{p}$? What are the mean and standard deviation for this distribution?

b. What is the probability that the sample percentage is greater than 80%?

c. Within what limits would you expect the sample proportion to lie about 95% of the time?

A SAMPLING APPLICATION: STATISTICAL PROCESS CONTROL (OPTIONAL)

Statistical process control (SPC) methodology was developed to monitor, control, and improve products and services. For example, steel bearings must conform to size and hardness specifications, industrial chemicals must have a low prespecified level of impurities, and accounting firms must minimize and ultimately eliminate incorrect bookkeeping entries. It is often said that statistical process control consists of 10% statistics, 90% engineering and common sense. We can statistically monitor a process mean and tell when the mean falls outside preassigned limits, but we cannot tell *why* it is out of control. Answering this last question requires knowledge of the process and problem-solving ability—the other 90%!

Product quality is usually monitored using statistical control charts. Measurements on a process variable change or vary over time. The cause of this change is said to be *assignable* if it can be found and corrected. Other variation—small haphazard changes due to alteration in the production environment—that is not controllable is regarded as *random variation.* If the variation in a process variable is solely random, the process is said to be *in control.*

The first objective in statistical process control is to eliminate assignable causes of variation in the process variable and then get the process in control. The next step is to reduce variation and get the measurements on the process variable within *specification limits,* the limits within which the measurements on usable items or services must fall.

Once a process is in control and is producing a satisfactory product, the process variables are monitored with **control charts.** Samples of n items are drawn from the process at specified intervals of time, and a sample statistic is computed. These statistics are plotted on the control chart, so that the process can be checked for shifts in the process variable that might indicate control problems.

A Control Chart for the Process Mean: The $\bar{x}$ Chart

Assume that n items are randomly selected from a production process at equal intervals and that measurements are recorded on the process variable. If the process is in control, the sample means should vary about the population mean μ in a random manner. Moreover, according to the Central Limit Theorem, the sampling distribution of $\bar{x}$ should be approximately normal, so that almost all of the values of $\bar{x}$ fall into the interval $(\mu \pm 3 \text{ SE}) = \mu \pm 3(\sigma/\sqrt{n})$. Although the exact values of μ and σ are unknown, you can obtain accurate estimates by using the sample measurements.

Every control chart has a *centerline* and *control limits.* The centerline for the $\bar{x}$ **chart** is the estimate of μ, the grand average of all the sample statistics calculated from the measurements on the process variable. The upper and lower *control limits* are placed three standard deviations above and below the centerline. If you monitor the process mean based on k samples of size n taken at regular intervals, the centerline is $\bar{\bar{x}}$, the average of the sample means, and the control limits are at $\bar{\bar{x}} \pm 3(\sigma/\sqrt{n})$, with σ estimated by s, the standard deviation of the nk measurements.

EXAMPLE 7.8

A statistical process control monitoring system samples the inside diameters of $n = 4$ bearings each hour. Table 7.6 provides the data for $k = 25$ hourly samples. Construct an $\bar{x}$ chart for monitoring the process mean.

Solution The sample mean was calculated for each of the $k = 25$ samples. For example, the mean for sample 1 is

$$\bar{x} = \frac{.992 + 1.007 + 1.016 + .991}{4} = 1.0015$$

The sample means are shown in the last column of Table 7.6. The centerline is located at the average of the sample means, or

$$\bar{\bar{x}} = \frac{24.9675}{25} = .9987$$

The calculated value of s, the sample standard deviation of all $nk = 4(25) = 100$ observations, is $s = .011458$, and the estimated standard error of the mean of $n = 4$ observations is

$$\frac{s}{\sqrt{n}} = \frac{.011458}{\sqrt{4}} = .005729$$

TABLE 7.6 ● **25 Hourly Samples of Bearing Diameters, $n = 4$ Bearings per Sample**

Sample	Sample Measurements				Sample Mean, $\bar{x}$
1	.992	1.007	1.016	.991	1.00150
2	1.015	.984	.976	1.000	.99375
3	.988	.993	1.011	.981	.99325
4	.996	1.020	1.004	.999	1.00475
5	1.015	1.006	1.002	1.001	1.00600
6	1.000	.982	1.005	.989	.99400
7	.989	1.009	1.019	.994	1.00275
8	.994	1.010	1.009	.990	1.00075
9	1.018	1.016	.990	1.011	1.00875
10	.997	1.005	.989	1.001	.99800
11	1.020	.986	1.002	.989	.99925
12	1.007	.986	.981	.995	.99225
13	1.016	1.002	1.010	.999	1.00675
14	.982	.995	1.011	.987	.99375
15	1.001	1.000	.983	1.002	.99650
16	.992	1.008	1.001	.996	.99925
17	1.020	.988	1.015	.986	1.00225
18	.993	.987	1.006	1.001	.99675
19	.978	1.006	1.002	.982	.99200
20	.984	1.009	.983	.986	.99050
21	.990	1.012	1.010	1.007	1.00475
22	1.015	.983	1.003	.989	.99750
23	.983	.990	.997	1.002	.99300
24	1.011	1.012	.991	1.008	1.00550
25	.987	.987	1.007	.995	.99400

The upper and lower control limits are found as

$$\text{UCL} = \bar{\bar{x}} + 3\frac{s}{\sqrt{n}} = .9987 + 3(.005729) = 1.015887$$

and

$$\text{LCL} = \bar{\bar{x}} - 3\frac{s}{\sqrt{n}} = .9987 - 3(.005729) = .981513$$

Figure 7.13 shows the $\bar{x}$ chart constructed from the data. If you assume that the samples used to construct the $\bar{x}$ chart were collected when the process was in con-

FIGURE 7.13 ●

$\bar{x}$ chart for
Example 7.8

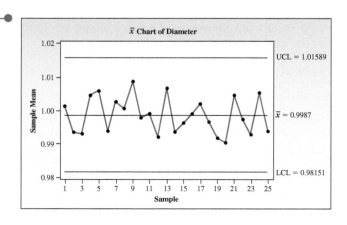

trol, the chart can now be used to detect changes in the process mean. Sample means are plotted periodically, and if a sample mean falls outside the control limits, the process should be checked to locate the cause of the unusually large or small mean.

A Control Chart for the Proportion Defective: The *p* Chart

Sometimes the observation made on an item is simply whether or not it meets specifications; thus, it is judged to be defective or nondefective. If the fraction defective produced by the process is p, then x, the number of defectives in a sample of n items, has a binomial distribution.

To monitor a process for defective items, samples of size n are selected at periodic intervals and the sample proportion $\hat{p}$ is calculated. When the process is in control, $\hat{p}$ should fall into the interval $p \pm 3SE$, where p is the proportion of defectives in the population (or the process fraction defective) with standard error

$$SE = \sqrt{\frac{pq}{n}} = \sqrt{\frac{p(1-p)}{n}}$$

The process fraction defective is unknown but can be estimated by the average of the k sample proportions:

$$\bar{p} = \frac{\Sigma \hat{p}_i}{k}$$

and the standard error is estimated by

$$SE = \sqrt{\frac{\bar{p}(1-\bar{p})}{n}}$$

The centerline for the **p chart** is located at $\bar{p}$, and the upper and lower control limits are

$$UCL = \bar{p} + 3\sqrt{\frac{\bar{p}(1-\bar{p})}{n}}$$

and

$$LCL = \bar{p} - 3\sqrt{\frac{\bar{p}(1-\bar{p})}{n}}$$

EXAMPLE 7.9

A manufacturer of ballpoint pens randomly samples 400 pens per day and tests each to see whether the ink flow is acceptable. The proportions of pens judged defective each day over a 40-day period are listed in Table 7.7. Construct a control chart for the proportion $\hat{p}$ defective in samples of $n = 400$ pens selected from the process.

Solution The estimate of the process proportion defective is the average of the $k = 40$ sample proportions in Table 7.7. Therefore, the centerline of the control chart is located at

$$\bar{p} = \frac{\Sigma \hat{p}_i}{k} = \frac{.0200 + .0125 + \cdots + .0225}{40} = \frac{.7600}{40} = .019$$

TABLE 7.7 ● **Proportions of Defectives in Samples of $n = 400$ Pens**

Day	Proportion	Day	Proportion	Day	Proportion	Day	Proportion
1	.0200	11	.0100	21	.0300	31	.0225
2	.0125	12	.0175	22	.0200	32	.0175
3	.0225	13	.0250	23	.0125	33	.0225
4	.0100	14	.0175	24	.0175	34	.0100
5	.0150	15	.0275	25	.0225	35	.0125
6	.0200	16	.0200	26	.0150	36	.0300
7	.0275	17	.0225	27	.0200	37	.0200
8	.0175	18	.0100	28	.0250	38	.0150
9	.0200	19	.0175	29	.0150	39	.0150
10	.0250	20	.0200	30	.0175	40	.0225

An estimate of SE, the standard error of the sample proportions, is

$$\sqrt{\frac{\bar{p}(1 - \bar{p})}{n}} = \sqrt{\frac{(.019)(.981)}{400}} = .00683$$

and $3SE = (3)(.00683) = .0205$. Therefore, the upper and lower control limits for the p chart are located at

$$UCL = \bar{p} + 3SE = .0190 + .0205 = .0395$$

and

$$LCL = \bar{p} - 3SE = .0190 - .0205 = -.0015$$

Or, since p cannot be negative, $LCL = 0$.

The p control chart is shown in Figure 7.14. Note that all 40 sample proportions fall within the control limits. If a sample proportion collected at some time in the future falls outside the control limits, the manufacturer should be concerned about an increase in the defective rate. He should take steps to look for the possible causes of this increase.

FIGURE 7.14 ●

p chart for
Example 7.9

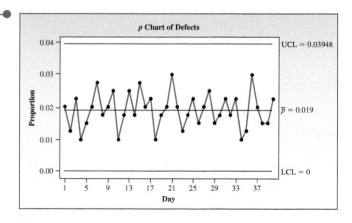

Other commonly used control charts are the *R chart,* which is used to monitor variation in the process variable by using the sample range, and the *c chart,* which is used to monitor the number of defects per item.

7.7 EXERCISES

BASIC TECHNIQUES

7.48 The sample means were calculated for 30 samples of size $n = 10$ for a process that was judged to be in control. The means of the 30 $\bar{x}$-values and the standard deviation of the combined 300 measurements were $\bar{\bar{x}} = 20.74$ and $s = .87$, respectively.

a. Use the data to determine the upper and lower control limits for an $\bar{x}$ chart.

b. What is the purpose of an $\bar{x}$ chart?

c. Construct an $\bar{x}$ chart for the process and explain how it can be used.

7.49 The sample means were calculated for 40 samples of size $n = 5$ for a process that was judged to be in control. The means of the 40 values and the standard deviation of the combined 200 measurements were $\bar{\bar{x}} = 155.9$ and $s = 4.3$, respectively.

a. Use the data to determine the upper and lower control limits for an $\bar{x}$ chart.

b. Construct an $\bar{x}$ chart for the process and explain how it can be used.

7.50 Explain the difference between an $\bar{x}$ chart and a p chart.

7.51 Samples of $n = 100$ items were selected hourly over a 100-hour period, and the sample proportion of defectives was calculated each hour. The mean of the 100 sample proportions was .035.

a. Use the data to find the upper and lower control limits for a p chart.

b. Construct a p chart for the process and explain how it can be used.

7.52 Samples of $n = 200$ items were selected hourly over a 100-hour period, and the sample proportion of defectives was calculated each hour. The mean of the 100 sample proportions was .041.

a. Use the data to find the upper and lower control limits for a p chart.

b. Construct a p chart for the process and explain how it can be used.

APPLICATIONS

7.53 Black Jack A gambling casino records and plots the mean daily gain or loss from five blackjack tables on an $\bar{x}$ chart. The overall mean of the sample means and the standard deviation of the combined data over 40 weeks were $\bar{\bar{x}} = \$10,752$ and $s = \$1605$, respectively.

a. Construct an $\bar{x}$ chart for the mean daily gain per blackjack table.

b. How can this $\bar{x}$ chart be of value to the manager of the casino?

7.54 Brass Rivets A producer of brass rivets randomly samples 400 rivets each hour and calculates the proportion of defectives in the sample. The mean sample proportion calculated from 200 samples was equal to .021. Construct a control chart for the proportion of defectives in samples of 400 rivets. Explain how the control chart can be of value to a manager.

 7.55 Lumber Specs The manager of a building-supplies company randomly samples EX0755 incoming lumber to see whether it meets quality specifications. From each shipment, 100 pieces of 2×4 lumber are inspected and judged according to whether they are first (acceptable) or second (defective) grade. The proportions of second-grade 2×4s recorded for 30 shipments were as follows:

.14	.21	.19	.18	.23	.20	.25	.19	.22	.17
.21	.15	.23	.12	.19	.22	.15	.26	.22	.21
.14	.20	.18	.22	.21	.13	.20	.23	.19	.26

Construct a control chart for the proportion of second-grade 2×4s in samples of 100 pieces of lumber. Explain how the control chart can be of use to the manager of the building-supplies company.

7.56 Coal-Burning Power Plant A coal-burning power plant tests and measures three specimens of coal each day to monitor the percentage of ash in the coal. The overall mean of 30 daily sample means and the combined standard deviation of all the data were $\bar{\bar{x}} = 7.24$ and $s = .07$, respectively. Construct an $\bar{x}$ chart for the process and explain how it can be of value to the manager of the power plant.

7.57 Nuclear Power Plant The data in the EX0757 table are measures of the radiation in air particulates at a nuclear power plant. Four measurements were recorded at weekly intervals over a 26-week period. Use the data to construct an $\bar{x}$ chart and plot the 26 values of $\bar{x}$. Explain how the chart can be used.

Week	Radiation			
1	.031	.032	.030	.031
2	.025	.026	.025	.025
3	.029	.029	.031	.030
4	.035	.037	.034	.035
5	.022	.024	.022	.023
6	.030	.029	.030	.030
7	.019	.019	.018	.019
8	.027	.028	.028	.028
9	.034	.032	.033	.033
10	.017	.016	.018	.018
11	.022	.020	.020	.021
12	.016	.018	.017	.017
13	.015	.017	.018	.017
14	.029	.028	.029	.029
15	.031	.029	.030	.031
16	.014	.016	.016	.017
17	.019	.019	.021	.020
18	.024	.024	.024	.025
19	.029	.027	.028	.028
20	.032	.030	.031	.030
21	.041	.042	.038	.039
22	.034	.036	.036	.035
23	.021	.022	.024	.022
24	.029	.029	.030	.029
25	.016	.017	.017	.016
26	.020	.021	.020	.022

7.58 Baseball Bats A hardwoods manufacturing plant has a production line designed to produce baseball bats weighing 32 ounces. During a period of time when the production process was known to be in statistical control, the average bat weight was found to be 31.7 ounces. The observed data were gathered from 50 samples, each consisting of 5 measurements. The standard deviation of all samples was found to be $s = .2064$ ounces. Construct an $\bar{x}$-chart to monitor the 32-ounce bat production process.

7.59 More Baseball Bats Refer to Exercise 7.58 and suppose that during a day when the state of the 32-ounce bat production process was unknown, the following measurements were obtained at hourly intervals.

Hour	$\bar{x}$	Hour	$\bar{x}$
1	31.6	4	33.1
2	32.5	5	31.6
3	33.4	6	31.8

Each measurement represents a statistic computed from a sample of five bat weights selected from the production process during a certain hour. Use the control chart constructed in Exercise 7.58 to monitor the process.

CHAPTER REVIEW

Key Concepts and Formulas

I. Sampling Plans and Experimental Designs

1. Simple random sampling
   a. Each possible sample of size n is equally likely to occur.
   b. Use a computer or a table of random numbers.
   c. Problems are nonresponse, undercoverage, and wording bias.
2. Other sampling plans involving randomization
   a. Stratified random sampling
   b. Cluster sampling
   c. Systematic 1-in-k sampling
3. Nonrandom sampling
   a. Convenience sampling
   b. Judgment sampling
   c. Quota sampling

II. Statistics and Sampling Distributions

1. Sampling distributions describe the possible values of a statistic and how often they occur in repeated sampling.
2. Sampling distributions can be derived mathematically, approximated empirically, or found using statistical theorems.
3. The Central Limit Theorem states that sums and averages of measurements from a nonnormal population with finite mean μ and standard deviation σ have approximately normal distributions for large samples of size n.

III. Sampling Distribution of the Sample Mean

1. When samples of size n are randomly drawn from a normal population with mean μ and variance σ^2, the sample mean $\bar{x}$ has a normal distribution with mean μ and standard deviation $\sigma/\sqrt{n}$.

2. When samples of size n are randomly drawn from a nonnormal population with mean μ and variance σ^2, the Central Limit Theorem ensures that the sample mean $\bar{x}$ will have an approximately normal distribution with mean μ and standard deviation $\sigma/\sqrt{n}$ when n is large ($n \geq 30$).

3. Probabilities involving the sample mean can be calculated by standardizing the value of $\bar{x}$ using z:

$$z = \frac{\bar{x} - \mu}{\sigma/\sqrt{n}}$$

IV. Sampling Distribution of the Sample Proportion

1. When samples of size n are drawn from a binomial population with parameter p, the sample proportion $\hat{p}$ will have an approximately normal distribution with mean p and standard deviation $\sqrt{pq/n}$ as long as $np > 5$ and $nq > 5$.

2. Probabilities involving the sample proportion can be calculated by standardizing the value $\hat{p}$ using z:

$$z = \frac{\hat{p} - p}{\sqrt{\dfrac{pq}{n}}}$$

V. Statistical Process Control

1. To monitor a quantitative process, use an $\bar{x}$ chart. Select k samples of size n and calculate the overall mean $\bar{\bar{x}}$ and the standard deviation s of all nk measurements. Create upper and lower control limits as

$$\bar{\bar{x}} \pm 3\frac{s}{\sqrt{n}}$$

If a sample mean exceeds these limits, the process is out of control.

2. To monitor a *binomial* process, use a p chart. Select k samples of size n and calculate the average of the sample proportions as

$$\bar{p} = \frac{\Sigma \hat{p}_i}{k}$$

Create upper and lower control limits as

$$\bar{p} \pm 3\sqrt{\frac{\bar{p}(1 - \bar{p})}{n}}$$

If a sample proportion exceeds these limits, the process is out of control.

 TECHNOLOGY TODAY

The Central Limit Theorem at Work—*Microsoft Excel*

Microsoft Excel can be used to explore the way the Central Limit Theorem works in practice. Remember that, according to the Central Limit Theorem, if random samples of size n are drawn from a nonnormal population with mean μ and standard deviation σ, then when n is large, the sampling distribution of the sample mean $\bar{x}$ will be approximately normal with the same mean μ and with standard error $\sigma/\sqrt{n}$. Let's try sampling from a nonnormal population using *Excel*.

In a new spreadsheet, generate 100 samples of size $n = 30$ from a continuous uniform distribution (Example 6.1) over the interval (0, 10). Label column A as "Sample" and enter the numbers 1 to 100 in that column. Then select **Data ▶ Data Analysis ▶ Random Number Generation,** to obtain the Dialog box in Figure 7.15(a). Type **30** for the number of variables and 100 for the number of random numbers. In the drop-down "Distribution" list, choose uniform, with parameters between **0** and **10**. We will leave the first row of our spreadsheet empty, starting the "Output Range" at cell B2. Press **OK** to see the 100 random samples of size $n = 30$. You can look at the distribution of the entire set of data using **Data ▶ Data Analysis ▶ Histogram,** choosing bins 1, 2, ..., 9, 10 and using the procedures described in the "Technology Today" section in Chapter 2. For our data, the distribution, shown in Figure 7.15(b) is not mound-shaped, but is fairly flat, as expected for the uniform distribution.

For the uniform distribution that we have used, the mean and standard deviation are $\mu = 5$ and $\sigma = 2.89$, respectively. Check the descriptive statistics for the 30×100

FIGURE 7.15 ●

(a)

(b)

= 3000 measurements (use the functions **=AVERAGE(B2:AE101)** and **=STDEV (B2:AE101)**), and you will find that the 100 observations have a sample mean and standard deviation *close to* but not exactly equal to $\mu = 5$ and $\sigma = 2.89$, respectively.

Now, generate 100 values of $\bar{x}$ based on samples of size $n = 30$ by creating a column of means for the 100 rows. First, label column AF as "x-bar" and place your cursor in cell AF2. Use **Insert Function ▶ Statistical ▶ Average** (or type **=AVERAGE (B2:AE2)**) to obtain the first average. Then copy the formula into the other 99 cells in column AF. You can now look at the distribution of these 100 sample means using **Data ▶ Data Analysis ▶ Histogram** and choosing bins 1, 1.5, 2, 2.5, ..., 9, 9.5, 10. The distribution for our 100 sample means is shown in Figure 7.16.

FIGURE 7.16 ●

Notice the distinct mound shape of the distribution in Figure 7.16 compared to the original distribution in Figure 7.15(b). Also, if you check the mean and standard deviation for the 100 sample means in column AE, you will find that they are not too different from the theoretical values, $\mu = 5$ and $\sigma/\sqrt{n} = 2.89/\sqrt{30} = .53$. (For our data, the sample mean is 4.98 and the standard deviation is .49.) Since we had only 100 samples, our results are not *exactly* equal to the theoretical values. If we had generated an *infinite* number of samples, we would have gotten an exact match. This is the Central Limit Theorem at work!

The Central Limit Theorem at Work—*MINITAB*

MINITAB provides a perfect tool for exploring the way the Central Limit Theorem works in practice. Remember that, according to the Central Limit Theorem, if random samples of size n are drawn from a nonnormal population with mean μ and standard deviation σ, then when n is large, the sampling distribution of the sample mean $\bar{x}$ will be approximately normal with the same mean μ and with standard error $\sigma/\sqrt{n}$. Let's try sampling from a nonnormal population with the help of *MINITAB*.

In a new *MINITAB* worksheet, generate 100 samples of size $n = 30$ from a non-normal distribution called the exponential distribution. Use **Calc ▶ Random Data ▶ Exponential.** Type **100** for the number of rows of data, and store the results in C1–C30 (see Figure 7.17(a)). Leave the mean at the default of 1.0, the threshold at 0.0, and click **OK.** The data are generated and stored in the worksheet. Use **Graph ▶ Histogram ▶ Simple** to look at the distribution of some of the data—say, C1 (as in Figure 7.17(b)). Notice that the distribution is not mound-shaped; it is highly skewed to the right.

(a)   (b)

FIGURE 7.17

For the exponential distribution that we have used, the mean and standard deviation are $\mu = 1$ and $\sigma = 1$, respectively. Check the descriptive statistics for one of the columns (use **Stat ▶ Basic Statistics ▶ Display Descriptive Statistics**), and you will find that the 100 observations have a sample mean and standard deviation that are both *close to* but not exactly equal to 1. Now, generate 100 values of $\bar{x}$ based on samples of size $n = 30$ by creating a column of means for the 100 rows. Use **Calc ▶ Row Statistics,** and select **Mean.** To average the entries in all 30 columns, select or type **C1–C30** in the Input variables box, and store the results in **C31** (see Figure 7.18(a)). You can now look at the distribution of the sample means using **Graph ▶ Histogram ▶ Simple,** selecting **C31** and clicking **OK.** The distribution of the 100 sample means generated for our example is shown in Figure 7.18(b).

(a)   (b)

FIGURE 7.18

Notice the distinct mound shape of the distribution in Figure 7.18(b) compared to the original distribution in Figure 7.17(b). Also, if you check the descriptive statistics for C31, you will find that the mean and standard deviation of our 100 sample means are not too different from the theoretical values, $\mu = 1$ and $\sigma/\sqrt{n} = 1/\sqrt{30} = .18$. (For our data, the sample mean is 1.0024 and the standard deviation is .1813.) Since we had only 100 samples, our results are not *exactly* equal to the theoretical values. If we had generated an *infinite* number of samples, we would have gotten an exact match. This is the Central Limit Theorem at work!

Supplementary Exercises

7.60 A finite population consists of four elements: 6, 1, 3, 2.

a. How many different samples of size $n = 2$ can be selected from this population if you sample without replacement? (Sampling is said to be *without replacement* if an element cannot be selected twice for the same sample.)

b. List the possible samples of size $n = 2$.

c. Compute the sample mean for each of the samples given in part b.

d. Find the sampling distribution of $\bar{x}$. Use a probability histogram to graph the sampling distribution of $\bar{x}$.

e. If all four population values are equally likely, calculate the value of the population mean μ. Do any of the samples listed in part b produce a value of $\bar{x}$ exactly equal to μ?

7.61 Refer to Exercise 7.60. Find the sampling distribution for $\bar{x}$ if random samples of size $n = 3$ are selected *without replacement*. Graph the sampling distribution of $\bar{x}$.

7.62 Suppose a random sample of $n = 5$ observations is selected from a population that is normally distributed, with mean equal to 1 and standard deviation equal to .36.

a. Give the mean and standard deviation of the sampling distribution of $\bar{x}$.

b. Find the probability $\bar{x}$ that exceeds 1.3.

c. Find the probability that the sample mean $\bar{x}$ is less than .5.

d. Find the probability that the sample mean deviates from the population mean $\mu = 1$ by more than .4.

7.63 Batteries A certain type of automobile battery is known to last an average of 1110 days with a standard deviation of 80 days. If 400 of these batteries are selected, find the following probabilities for the average length of life of the selected batteries:

a. The average is between 1100 and 1110.

b. The average is greater than 1120.

c. The average is less than 900.

7.64 Lead Pipes Studies indicate that drinking water supplied by some old lead-lined city piping systems may contain harmful levels of lead. An important study of the Boston water supply system showed that the distribution of lead content readings for individual water specimens had a mean and standard deviation of approximately .033 milligrams per liter (mg/l) and .10 mg/l, respectively.[15]

a. Explain why you believe this distribution is or is not normally distributed.

b. Because the researchers were concerned about the shape of the distribution in part a, they calculated the average daily lead levels at 40 different locations on each of 23 randomly selected days. What can you say about the shape of the distribution of the average daily lead levels from which the sample of 23 days was taken?

c. What are the mean and standard deviation of the distribution of average lead levels in part b?

7.65 Biomass Studies[16] indicate that the earth's vegetative mass, or biomass for tropical woodlands, thought to be about 35 kilograms per square meter (kg/m^2), may in fact be too high and that tropical biomass values vary regionally—from about 5 to 55 kg/m^2. Suppose you measure the tropical biomass in 400 randomly selected square-meter plots.

a. Approximate σ, the standard deviation of the biomass measurements.

b. What is the probability that your sample average is within two units of the true average tropical biomass?

c. If your sample average is $\bar{x} = 31.75$, what would you conclude about the overestimation that concerns the scientists?

7.66 Hard Hats The safety requirements for hard hats worn by construction workers and others, established by the American National Standards Institute (ANSI), specify that each of three hats pass the following test. A hat is mounted on an aluminum head form. An 8-pound steel ball is dropped on the hat from a height of 5 feet, and the resulting force is measured at the bottom of the head form. The force exerted on the head form by each of the three hats must be less than 1000 pounds, and the average of the three must be less than 850 pounds. (The relationship between this test and actual human head damage is unknown.) Suppose the exerted force is normally distributed, and hence a sample mean of three force measurements is normally distributed. If a random sample of three hats is selected from a shipment with a mean equal to 900 and $\sigma = 100$, what is the probability that the sample mean will satisfy the ANSI standard?

7.67 Imagery and Memory A research psychologist is planning an experiment to determine whether the use of imagery—picturing a word in your mind—

affects people's ability to memorize. He wants to use two groups of subjects: a group that memorizes a set of 20 words using the imagery technique, and a control group that does not use imagery.

a. Use a randomization technique to divide a group of 20 subjects into two groups of equal size.

b. How can the researcher randomly select the group of 20 subjects?

c. Suppose the researcher offers to pay subjects $50 each to participate in the experiment and uses the first 20 students who apply. Would this group behave as if it were a simple random sample of size $n = 20$?

7.68 Same-Sex Marriage The results of a *CBS News Poll* concerning views on same-sex marriage and gay rights given in the table that follows show that there is no consensus on this issue among Americans.[17]

CBS News Poll. August 20–24, 2010. N = 1,082 adults nationwide. MoE ± 3.

"Which comes closest to your view? Gay couples should be allowed to legally marry. OR, Gay couples should be allowed to form civil unions but not legally marry. OR, There should be no legal recognition of a gay couple's relationship."

	Legal Marriage (%)	Civil Unions (%)	No Legal Recognition (%)	Unsure (%)
All adults	40	30	25	5
Republicans	25	34	37	4
Democrats	46	27	20	7
Independents	44	29	21	6

---

a. Is this an observational study or a planned experiment?

b. Is there the possibility of problems in responses arising because of the somewhat sensitive nature of the subject? What kinds of biases might occur?

7.69 Sprouting Radishes A biology experiment was designed to determine whether sprouting radish seeds inhibit the germination of lettuce seeds.[18] Three 10-centimeter petri dishes were used. The first contained 26 lettuce seeds, the second contained 26 radish seeds, and the third contained 13 lettuce seeds and 13 radish seeds.

a. Assume that the experimenter had a package of 50 radish seeds and another of 50 lettuce seeds.

Devise a plan for randomly assigning the radish and lettuce seeds to the three treatment groups.

b. What assumptions must the experimenter make about the packages of 50 seeds in order to assure randomness in the experiment?

7.70 9/11 A study of about $n = 1000$ individuals in the United States during September 21–22, 2001, revealed that 43% of the respondents indicated that they were less willing to fly following the events of September 11, 2001.[19]

a. Is this an observational study or a designed experiment?

b. What problems might or could have occurred because of the sensitive nature of the subject? What kinds of biases might have occurred?

7.71 Telephone Service Suppose a telephone company executive wishes to select a random sample of $n = 20$ out of 7000 customers for a survey of customer attitudes concerning service. If the customers are numbered for identification purposes, indicate the customers whom you will include in your sample. Use the random number table and explain how you selected your sample.

7.72 Rh Positive The proportion of individuals with an Rh-positive blood type is 85%. You have a random sample of $n = 500$ individuals.

a. What are the mean and standard deviation of $\hat{p}$, the sample proportion with Rh-positive blood type?

b. Is the distribution of $\hat{p}$ approximately normal? Justify your answer.

c. What is the probability that the sample proportion $\hat{p}$ exceeds 82%?

d. What is the probability that the sample proportion lies between 83% and 88%?

e. 99% of the time, the sample proportion would lie between what two limits?

7.73 What survey design is used in each of these situations?

a. A random sample of $n = 50$ city blocks is selected, and a census is done for each single-family dwelling on each block.

b. The highway patrol stops every 10th vehicle on a given city artery between 9:00 A.M. and 3:00 P.M. to perform a routine traffic safety check.

c. One hundred households in each of four city wards are surveyed concerning a pending city tax relief referendum.

d. Every 10th tree in a managed slash pine plantation is checked for pine needle borer infestation.

e. A random sample of $n = 1000$ taxpayers from the city of San Bernardino is selected by the Internal Revenue Service and their tax returns are audited.

7.74 Elevator Loads The maximum load (with a generous safety factor) for the elevator in an office building is 2000 pounds. The relative frequency distribution of the weights of all men and women using the elevator is mound-shaped (slightly skewed to the heavy weights), with mean μ equal to 150 pounds and standard deviation σ equal to 35 pounds. What is the largest number of people you can allow on the elevator if you want their total weight to exceed the maximum weight with a small probability (say, near .01)? (HINT: Use the alternative statement of the Central Limit Theorem for Σx_i given in Section 7.4.)

7.75 Wiring Packages The number of wiring packages that can be assembled by a company's employees has a normal distribution, with a mean equal to 16.4 per hour and a standard deviation of 1.3 per hour.

a. What are the mean and standard deviation of the number x of packages produced per worker in an 8-hour day?

b. Do you expect the probability distribution for x to be mound-shaped and approximately normal? Explain.

c. What is the probability that a worker will produce at least 135 packages per 8-hour day?

7.76 Wiring Packages, continued Refer to Exercise 7.75. Suppose the company employs 10 assemblers of wiring packages.

a. Find the mean and standard deviation of the company's daily (8-hour day) production of wiring packages.

b. What is the probability that the company's daily production is less than 1280 wiring packages per day?

7.77 Defective Lightbulbs The table lists the number of defective 60-watt lightbulbs found in samples of 100 bulbs selected over 25 days from a manufacturing process. Assume that during this time the manufacturing process was not producing an excessively large fraction of defectives.

Day	1	2	3	4	5	6	7	8	9	10
Defectives	4	2	5	8	3	4	4	5	6	1

Day	11	12	13	14	15	16	17	18	19	20
Defectives	2	4	3	4	0	2	3	1	4	0

Day	21	22	23	24	25
Defectives	2	2	3	5	3

a. Construct a p chart to monitor the manufacturing process, and plot the data.

b. How large must the fraction of defective items be in a sample selected from the manufacturing process before the process is assumed to be out of control?

c. During a given day, a sample of 100 items is selected from the manufacturing process and 15 defective bulbs are found. If a decision is made to shut down the manufacturing process in an attempt to locate the source of the implied controllable variation, explain how this decision might lead to erroneous conclusions.

7.78 Lightbulbs, continued A hardware store chain purchases large shipments of lightbulbs from the manufacturer described in Exercise 7.77 and specifies that each shipment must contain no more than 4% defectives. When the manufacturing process is in control, what is the probability that the hardware store's specifications are met?

7.79 Lightbulbs, again Refer to Exercise 7.77. During a given week the number of defective bulbs in each of five samples of 100 were found to be 2, 4, 9, 7, and 11. Is there reason to believe that the production process has been producing an excessive proportion of defectives at any time during the week?

7.80 Canned Tomatoes During long production runs of canned tomatoes, the average weights (in ounces) of samples of five cans of standard-grade tomatoes in pureed form were taken at 30 control points during an 11-day period. These results are shown in the table.[20] When the machine is performing normally, the average weight per can is 21 ounces with a standard deviation of 1.20 ounces.

a. Compute the upper and lower control limits and the centerline for the $\bar{x}$ chart.

b. Plot the sample data on the $\bar{x}$ chart and determine whether the performance of the machine is in control.

Sample Number	Average Weight	Sample Number	Average Weight
1	23.1	16	21.4
2	21.3	17	20.4
3	22.0	18	22.8
4	21.4	19	21.1
5	21.8	20	20.7
6	20.6	21	21.6
7	20.1	22	22.4
8	21.4	23	21.3
9	21.5	24	21.1
10	20.2	25	20.1
11	20.3	26	21.2
12	20.1	27	19.9
13	21.7	28	21.1
14	21.0	29	21.6
15	21.6	30	21.3

Source: Adapted from J. Hackl, *Journal of Quality Technology,* April 1991. Used with permission.

7.81 Pepsi or Coke? The battle for consumer preference continues between Pepsi and Coke. How can you make your preferences known? There is a web page where you can vote for one of these colas if you click on the link that says PAY CASH for your opinion. Explain why the respondents do not represent a random sample of the opinions of purchasers or drinkers of these drinks. Explain the types of distortions that could creep into an Internet opinion poll.

7.82 Strawberries An experimenter wants to find an appropriate temperature at which to store fresh strawberries to minimize the loss of ascorbic acid. There are 20 storage containers, each with controllable temperature, in which strawberries can be stored. If two storage temperatures are to be used, how would the experimenter assign the 20 containers to one of the two storage temperatures?

7.83 Filling Soda Cans A bottler of soft drinks packages cans in six-packs. Suppose that the fill per can has an approximate normal distribution with a mean of 12 fluid ounces and a standard deviation of 0.2 fluid ounces.

a. What is the distribution of the total fill for a case of 24 cans?

b. What is the probability that the total fill for a case is less than 286 fluid ounces?

c. If a six-pack of soda can be considered a random sample of size $n = 6$ from the population, what is the probability that the average fill per can for a six-pack of soda is less than 11.8 fluid ounces?

7.84 Total Packing Weight Packages of food whose average weight is 16 ounces with a standard deviation of 0.6 ounces are shipped in boxes of 24 packages. If the package weights are approximately normally distributed, what is the probability that a box of 24 packages will weigh more than 392 ounces (24.5 pounds)?

7.85 Electronic Components A manufacturing process is designed to produce an electronic component for use in small portable television sets. The components are all of standard size and need not conform to any measurable characteristic, but are sometimes inoperable when emerging from the manufacturing process. Fifteen samples were selected from the process at times when the process was known to be in statistical control. Fifty components were observed within each sample, and the number of inoperable components was recorded.

6, 7, 3, 5, 6, 8, 4, 5, 7, 3, 1, 6, 5, 4, 5

Construct a p chart to monitor the manufacturing process.

CASE STUDY **Sampling the Roulette at Monte Carlo**

The technique of simulating a process that contains random elements and repeating the process over and over to see how it behaves is called a **Monte Carlo procedure.** It is widely used in business and other fields to investigate the properties of an operation that is subject to random effects, such as weather, human behavior, and so on. For example, you could model the behavior of a manufacturing company's inventory by creating, on paper, daily arrivals, and departures of manufactured products from the company's warehouse. Each day a random number of items produced by the company would be received into inventory. Similarly, each day a random number of orders of varying random sizes would be shipped. Based on the input and output of items, you could calculate the inventory—that is, the number of items on hand at the end of each day. The values of the random variables, the number of items produced, the number of orders, and the number of items per order needed for each day's simulation would be obtained from theoretical distributions of observations that closely model the corresponding distributions of the variables that have been observed over time in the manufacturing operation. By repeating the simulation of the supply, the shipping, and the calculation of daily inventory for a large number of days (a sampling of what might really happen), you can observe the behavior of the plant's daily inventory. The Monte Carlo procedure is particularly valuable because it enables the manufacturer to see how the daily inventory would behave when certain changes are made in the supply pattern or in some other aspect of the operation that could be controlled.

In an article entitled "The Road to Monte Carlo," Daniel Seligman comments on the Monte Carlo method, noting that although the technique is widely used in business schools to study capital budgeting, inventory planning, and cash flow management, no one seems to have used the procedure to study how well we might do if we were to gamble at Monte Carlo.[21]

To follow up on this thought, Seligman programmed his personal computer to simulate the game of roulette. Roulette involves a wheel with its rim divided into 38 pockets. Thirty-six of the pockets are numbered 1 to 36 and are alternately colored red and black. The two remaining pockets are colored green and are marked 0 and 00. To play the game, you bet a certain amount of money on one or more pockets. The wheel is spun and turns until it stops. A ball falls into a slot on the wheel to indicate the winning number. If you have money on that number, you win a specified amount.

For example, if you were to play the number 20, the payoff is 35 to 1. If the wheel does not stop at that number, you lose your bet. Seligman decided to see how his nightly gains (or losses) would fare if he were to bet $5 on each turn of the wheel and repeat the process 200 times each night. He did this 365 times, thereby simulating the outcomes of 365 nights at the casino. Not surprisingly, the mean "gain" per $1000 evening for the 365 nights was a *loss* of $55, the average of the winnings retained by the gambling house. The surprise, according to Seligman, was the extreme variability of the nightly "winnings." Seven times out of the 365 evenings, the fictitious gambler lost the $1000 stake, and only once did he win a maximum of $1160. On 141 nights, the loss exceeded $250.

1. To evaluate the results of Seligman's Monte Carlo experiment, first find the probability distribution of the gain x on a single $5 bet.

2. Find the expected value and variance of the gain x from part 1.

3. Find the expected value and variance for the evening's gain, the sum of the gains or losses for the 200 bets of $5 each.

4. Use the results of part 2 to evaluate the probability of 7 out of 365 evenings resulting in a loss of the total $1000 stake.

5. Use the results of part 3 to evaluate the probability that the largest evening's winnings were as great as $1160.

Large-Sample Estimation

© Associated Press

8

? NEED TO KNOW...

How to Estimate a Population Mean or Proportion
How to Choose the Sample Size

How Reliable Is That Poll?

Do the national polls conducted by the Gallup and Harris organizations, the news media, and others provide accurate estimates of the percentages of people in the United States who have a variety of eating habits? The case study at the end of this chapter examines the reliability of a poll conducted by *CBS News* using the theory of large-sample estimation.

8.1 WHERE WE'VE BEEN

The first seven chapters of this book have given you the building blocks you will need to understand statistical inference and how it can be applied in practical situations. The first three chapters were concerned with using descriptive statistics, both graphical and numerical, to describe and interpret sets of measurements. In the next three chapters, you learned about probability and probability distributions—the basic tools used to describe *populations* of measurements. The binomial and the normal distributions were emphasized as important for practical applications.

The seventh chapter provided the link between probability and statistical inference. Many statistics are either sums or averages calculated from sample measurements. The Central Limit Theorem states that, even if the sampled populations are not normal, the sampling distributions of these *statistics* will be approximately normal when the sample size n is large. These statistics are the tools you use for *inferential statistics*—making inferences about a population using information contained in a sample.

8.2 WHERE WE'RE GOING— STATISTICAL INFERENCE

Inference—specifically, decision making and prediction—is centuries old and plays a very important role in most peoples' lives. Here are some applications:

- The government needs to predict short- and long-term interest rates.
- A broker wants to forecast the behavior of the stock market.
- A metallurgist wants to decide whether a new type of steel is more resistant to high temperatures than the current type.
- A consumer wants to estimate the selling price of her house before putting it on the market.

 NEED A TIP?

Parameter ⇔ Population
Statistic ⇔ Sample

There are many ways to make these decisions or predictions, some subjective and some more objective in nature. How good will your predictions or decisions be? Although you may feel that your own built-in decision-making ability is quite good, experience suggests that this may not be the case. It is the job of the mathematical statistician to provide methods of statistical inference making that are better and more reliable than just subjective guesses.

Statistical inference is concerned with making decisions or predictions about **parameters**—the numerical descriptive measures that characterize a population. Three parameters you encountered in earlier chapters are the population mean μ, the population standard deviation σ, and the binomial proportion p. In statistical inference, a practical problem is restated in the framework of a population with a specific parameter of interest. For example, the metallurgist could measure the *average* coefficients of expansion for both types of steel and then compare their values.

Methods for making inferences about population parameters fall into one of two categories:

- **Estimation:** Estimating or predicting the value of the parameter
- **Hypothesis testing:** Making a decision about the value of a parameter based on some preconceived idea about what its value might be

EXAMPLE (8.1) The circuits in computers and other electronics equipment consist of one or more printed circuit boards (PCB), and computers are often repaired by simply replacing one or more defective PCBs. In an attempt to find the proper setting of a plating process applied to one side of a PCB, a production supervisor might *estimate* the average thickness of copper plating on PCBs using samples from several days of operation. Since he has no knowledge of the average thickness μ before observing the production process, his is an *estimation* problem.

EXAMPLE (8.2) The supervisor in Example 8.1 is told by the plant owner that the thickness of the copper plating must not be less than .001 inch in order for the process to be in control. To decide whether or not the process is in control, the supervisor might formulate a test. He could *hypothesize* that the process is in control—that is, assume that the average thickness of the copper plating is .001 or greater—and use samples from several days of operation to decide whether or not his hypothesis is correct. The supervisor's decision-making approach is called a *test of hypothesis.*

Which method of inference should be used? That is, should the parameter be estimated, or should you test a hypothesis concerning its value? The answer is dictated by the practical question posed and is often determined by personal preference. Since both estimation and tests of hypotheses are used frequently in scientific literature, we include both methods in this and the next chapter.

A statistical problem, which involves planning, analysis, and inference making, is incomplete without a measure of the **goodness of the inference.** That is, how accurate or reliable is the method you have used? If a stockbroker predicts that the price of a stock will be $80 next Monday, will you be willing to take action to buy or sell your stock without knowing how reliable her prediction is? Will the prediction be within $1, $2, or $10 of the actual price next Monday? Statistical procedures are important because they provide two types of information:

- Methods for making the inference
- A numerical measure of the goodness or reliability of the inference

TYPES OF ESTIMATORS
(8.3)

To estimate the value of a population parameter, you can use information from the sample in the form of an **estimator.** Estimators are calculated using information from the sample observations, and hence, by definition they are also *statistics.*

Definition An **estimator** is a rule, usually expressed as a formula, that tells us how to calculate an estimate based on information in the sample.

Estimators are used in two different ways:

- **Point estimation:** Based on sample data, a single number is calculated to estimate the population parameter. The rule or formula that describes this calculation is called the **point estimator,** and the resulting number is called a **point estimate.**

- **Interval estimation:** Based on sample data, two numbers are calculated to form an interval within which the parameter is expected to lie. The rule or formula that describes this calculation is called the **interval estimator,** and the resulting pair of numbers is called an **interval estimate** or **confidence interval.**

EXAMPLE (8.3)

A veterinarian wants to estimate the average weight gain per month of 4-month-old golden retriever pups that have been placed on a lamb and rice diet. The *population* consists of the weight gains per month of all 4-month-old golden retriever pups that are given this particular diet. The veterinarian wants to estimate the unknown parameter μ, the average monthly weight gain for this *hypothetical* population.

One possible *estimator* based on sample data is the sample mean, $\bar{x} = \Sigma x_i/n$. It could be used in the form of a single number or *point estimate*—for instance, 3.8 pounds—or you could use an *interval estimate* and estimate that the average weight gain will be between 2.7 and 4.9 pounds.

Both point and interval estimation procedures use information provided by the sampling distribution of the specific estimator you have chosen to use. We will begin by discussing *point estimation* and its use in estimating population means and proportions.

(8.4)
POINT ESTIMATION

In a practical situation, there may be several statistics that could be used as point estimators for a population parameter. To decide which of several choices is best, you need to know how the estimator behaves in repeated sampling, described by its *sampling distribution.*

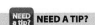
NEED A TIP?

Parameter = Target's bull's-eye

Estimator = Bullet or arrow

By way of analogy, think of firing a revolver at a target. The parameter of interest is the bull's-eye, at which you are firing bullets. Each bullet represents a single sample estimate, fired by the revolver, which represents the estimator.

Suppose your friend fires a single shot and hits the bull's-eye. Can you conclude that he is an excellent shot? Would you stand next to the target while he fires a second shot? Probably not, because you have no measure of how well he performs in repeated trials. Does he always hit the bull's-eye, or is he consistently too high or too low? Do his shots cluster closely around the target, or do they consistently miss the target by a wide margin? Figure 8.1 shows several target configurations. Which target would you pick as belonging to the best shot?

FIGURE 8.1
Which marksman is best?

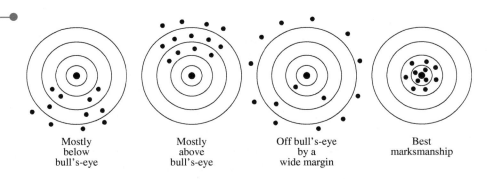

Mostly below bull's-eye   Mostly above bull's-eye   Off bull's-eye by a wide margin   Best marksmanship

Sampling distributions provide information that can be used to select the **best estimator.** What characteristics would be valuable? First, the **sampling distribution of the point estimator should be centered over the true value of the parameter to be estimated.** That is, the estimator should not constantly underestimate or overestimate the parameter of interest. Such an estimator is said to be **unbiased.**

Definition   An estimator of a parameter is said to be **unbiased** if the mean of its distribution is equal to the true value of the parameter. Otherwise, the estimator is said to be **biased.**

The sampling distributions for an unbiased estimator and a biased estimator are shown in Figure 8.2. The sampling distribution for the biased estimator is shifted to the right of the true value of the parameter. This biased estimator is more likely than an unbiased one to overestimate the value of the parameter.

FIGURE 8.2

Distributions for biased and unbiased estimators

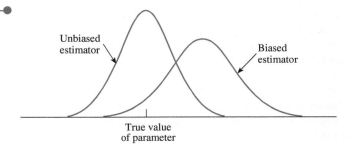

The second desirable characteristic of an estimator is that **the spread (as measured by the variance) of the sampling distribution should be as small as possible.** This ensures that, with a high probability, an individual estimate will fall close to the true value of the parameter. The sampling distributions for two unbiased estimators, one with a small variance[†] and the other with a larger variance, are shown in Figure 8.3.

FIGURE 8.3

Comparison of estimator variability

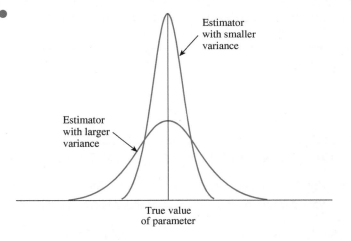

[†]Statisticians usually use the term *variance of an estimator* when in fact they mean the variance of the sampling distribution of the estimator. This contractive expression is used almost universally.

Naturally, you would prefer the estimator with the smaller variance because the estimates tend to lie closer to the true value of the parameter than in the distribution with the larger variance.

In real-life sampling situations, you may know that the sampling distribution of an estimator centers about the parameter that you are attempting to estimate, but all you have is the estimate computed from the *n* measurements contained in the sample. How far from the true value of the parameter will your estimate lie? How close is the marksman's bullet to the bull's-eye? The distance between the estimate and the true value of the parameter is called the **error of estimation.**

Definition   The distance between an estimate and the estimated parameter is called the **error of estimation.**

In this chapter, you may assume that the sample sizes are always large and, therefore, that the *unbiased* estimators you will study have sampling distributions that can be approximated by a normal distribution (because of the Central Limit Theorem). Remember that, for any point estimator with a normal distribution, the Empirical Rule states that approximately 95% of all the point estimates will lie within two (or more exactly, 1.96) standard deviations of the mean of that distribution.

For *unbiased* estimators, this implies that the difference between the point estimator and the true value of the parameter will be less than 1.96 standard deviations or 1.96 standard errors (SE). This quantity, called the 95% **margin of error** (or simply the **"margin of error"**), provides a practical upper bound for the error of estimation (see Figure 8.4). It is possible that the error of estimation will exceed this margin of error, but that is very unlikely.

FIGURE 8.4

Sampling distribution of an unbiased estimator

NEED A TIP?

95% Margin of error = 1.96 × Standard error

POINT ESTIMATION OF A POPULATION PARAMETER

- Point estimator: a statistic calculated using sample measurements
- 95% Margin of error: 1.96 × Standard error of the estimator

The sampling distributions for two unbiased point estimators were discussed in Chapter 7. It can be shown that both of these point estimators have the *minimum variability* of all unbiased estimators and are thus the *best estimators* you can find in each situation.

The variability of the estimator is measured using its standard error. However, you might have noticed that the standard error usually depends on unknown parameters such as σ or p. These parameters must be estimated using sample statistics such as s and $\hat{p}$. Although not exactly correct, experimenters generally refer to the estimated standard error as *the standard error*.

NEED TO KNOW...

How to Estimate a Population Mean or Proportion

- To estimate the population mean μ for a quantitative population, the point estimator $\bar{x}$ is *unbiased* with standard error estimated as

$$\text{SE} = \frac{s}{\sqrt{n}}^{\dagger}$$

The 95% margin of error when $n \geq 30$ is estimated as

$$\pm 1.96\left(\frac{s}{\sqrt{n}}\right)$$

- To estimate the population proportion p for a binomial population, the point estimator $\hat{p} = x/n$ is *unbiased,* with standard error estimated as

$$\text{SE} = \sqrt{\frac{\hat{p}\hat{q}}{n}}$$

The 95% margin of error is estimated as

$$\pm 1.96\sqrt{\frac{\hat{p}\hat{q}}{n}}$$

Assumptions: $n\hat{p} > 5$ and $n\hat{q} > 5$.

EXAMPLE 8.4

An environmentalist is conducting a study of the polar bear, a species found in and around the Arctic Ocean. Their range is limited by the availability of sea ice, which they use as a platform to hunt seals, the mainstay of their diet. The destruction of its habitat on the Arctic ice, which has been attributed to global warming, threatens the bear's survival as a species; it may become extinct within the century.[1] A random sample of $n = 50$ polar bears produced an average weight of 980 pounds with a standard deviation of 105 pounds. Use this information to estimate the average weight of all Arctic polar bears.

Solution The random variable measured is weight, a quantitative random variable best described by its mean μ. The point estimate of μ, the average weight of all Arctic polar bears, is $\bar{x} = 980$ pounds. The margin of error is estimated as

$$1.96 \text{ SE} = 1.96\left(\frac{s}{\sqrt{n}}\right) = 1.96\left(\frac{105}{\sqrt{50}}\right) = 29.10 \approx 29 \text{ pounds}$$

[†]When you sample from a normal distribution, the statistic $(\bar{x} - \mu)/(s/\sqrt{n})$ has a t distribution, which will be discussed in Chapter 10. When the sample is *large,* this statistic is approximately normally distributed whether the sampled population is normal or nonnormal.

You can be fairly confident that the sample estimate of 980 pounds is within ±29 pounds of the population mean.

In reporting research results, investigators often attach either the sample standard deviation s (sometimes called SD) or the standard error $s/\sqrt{n}$ (usually called SE or SEM) to the estimates of population means. You should always look for an explanation somewhere in the text of the report that tells you whether the investigator is reporting $\bar{x} \pm$ SD or $\bar{x} \pm$ SE. In addition, the sample means and standard deviations or standard errors are often presented as "error bars" using the graphical format shown in Figure 8.5.

FIGURE 8.5 ●

Plot of treatment means and their standard errors

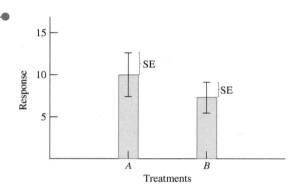

EXAMPLE 8.5

In addition to the average weight of the Arctic polar bear, the environmentalist from Example 8.4 is also interested in the opinions of adults on the subject of global warming. In particular, he wants to estimate the proportion of adults who think that global warming is a very serious problem. In a random sample of $n = 100$ adults, 73% of the sample indicated that global warming is a very serious problem. Estimate the true population proportion of adults who believe that global warming is a very serious problem, and find the margin of error for the estimate.

Solution The parameter of interest is now p, the proportion of individuals in the population who believe that global warming is a very serious problem. The best estimator of p is the sample proportion $\hat{p}$, which for this sample is $\hat{p} = .73$. In order to find the margin of error, you can approximate the value of p with its estimate $\hat{p} = .73$:

$$1.96 \text{ SE} = 1.96 \sqrt{\frac{\hat{p}\hat{q}}{n}} = 1.96 \sqrt{\frac{.73(.27)}{100}} = .09$$

With this margin of error, you can be fairly confident that the estimate of .73 is within ±.09 of the true value of p. Hence, you can conclude that the true value of p could be as small as .64 or as large as .82. This margin of error is quite large when compared to the estimate itself and reflects the fact that large samples are required to achieve a small margin of error when estimating p.

TABLE 8.1		Some Calculated Values of $\sqrt{pq}$				
	p	pq	$\sqrt{pq}$	p	pq	$\sqrt{pq}$
	.1	.09	.30	.6	.24	.49
	.2	.16	.40	.7	.21	.46
	.3	.21	.46	.8	.16	.40
	.4	.24	.49	.9	.09	.30
	.5	.25	.50			

Table 8.1 shows how the numerator of the standard error of $\hat{p}$ changes for various values of p. Notice that, for most values of p—especially when p is between .3 and .7—there is very little change in $\sqrt{pq}$, the numerator of SE, reaching its maximum value when $p = .5$. This means that the margin of error using the estimator $\hat{p}$ will also be a maximum when $p = .5$. Some pollsters routinely use the maximum margin of error—often called the **sampling error**—when estimating p, in which case they calculate

$$1.96 \text{ SE} = 1.96 \sqrt{\frac{.5(.5)}{n}} \quad \text{or sometimes} \quad 2 \text{ SE} = 2\sqrt{\frac{.5(.5)}{n}}$$

Gallup, Harris, and Roper polls generally use sample sizes of approximately 1000, so their margin of error is

$$1.96 \sqrt{\frac{.5(.5)}{1000}} = .031 \quad \text{or approximately 3\%}$$

In this case, the estimate is said to be within ± 3 percentage points of the true population proportion.

8.4 EXERCISES

BASIC TECHNIQUES

8.1 Explain what is meant by "margin of error" in point estimation.

8.2 What are two characteristics of the best point estimator for a population parameter?

8.3 Calculate the margin of error in estimating a population mean μ for these values:

a. $n = 30, \sigma^2 = .2$ **b.** $n = 30, \sigma^2 = .9$
c. $n = 30, \sigma^2 = 1.5$

8.4 Refer to Exercise 8.3. What effect does a larger population variance have on the margin of error?

8.5 Calculate the margin of error in estimating a population mean μ for these values:

a. $n = 50, s^2 = 4$ **b.** $n = 500, s^2 = 4$
c. $n = 5000, s^2 = 4$

8.6 Refer to Exercise 8.5. What effect does an increased sample size have on the margin of error?

8.7 Calculate the margin of error in estimating a binomial proportion for each of the following values of n. Use $p = .5$ to calculate the standard error of the estimator.

a. $n = 30$ **b.** $n = 100$
c. $n = 400$ **d.** $n = 1000$

8.8 Refer to Exercise 8.7. What effect does increasing the sample size have on the margin of error?

8.9 Calculate the margin of error in estimating a binomial proportion p using samples of size $n = 100$ and the following values for p:

a. $p = .1$ **b.** $p = .3$ **c.** $p = .5$
d. $p = .7$ **e.** $p = .9$
f. Which of the values of p produces the largest margin of error?

8.10 Suppose you are writing a questionnaire for a sample survey involving $n = 100$ individuals. The questionnaire will generate estimates for several different binomial proportions. If you want to report a single margin of error for the survey, which margin of error from Exercise 8.9 is the correct one to use?

8.11 A random sample of $n = 900$ observations from a binomial population produced $x = 655$ successes. Estimate the binomial proportion p and calculate the margin of error.

8.12 A random sample of $n = 50$ observations from a quantitative population produced $\bar{x} = 56.4$ and $s^2 = 2.6$. Give the best point estimate for the population mean μ, and calculate the margin of error.

8.13 A random sample of $n = 500$ observations from a binomial population produced $x = 450$ successes. Estimate the binomial proportion p and calculate the margin of error.

8.14 A random sample of $n = 75$ observations from a quantitative population produced $\bar{x} = 29.7$ and $s^2 = 10.8$. Give the best point estimate for the population mean μ and calculate the margin of error.

APPLICATIONS

8.15 The San Andreas Fault One of the most famous large fractures (cracks) in the earth's crust is the San Andreas fault in California. A geologist attempting to study the movement of the earth's crust at a particular location found many fractures in the local rock structure. In an attempt to determine the mean angle of the breaks, she sampled $n = 50$ fractures and found the sample mean and standard deviation to be $39.8°$ and $17.2°$, respectively. Estimate the mean angular direction of the fractures and find the margin of error for your estimate.

8.16 Biomass Estimates of the earth's biomass, the total amount of vegetation held by the earth's forests, are important in determining the amount of unabsorbed carbon dioxide that is expected to remain in the earth's atmosphere.[2] Suppose a sample of 75 one-square-meter plots, randomly chosen in North America's boreal (northern) forests, produced a mean biomass of 4.2 kilograms per square meter (kg/m^2), with a standard deviation of 1.5 kg/m^2. Estimate the average biomass for the boreal forests of North America and find the margin of error for your estimate.

Source: Reprinted with permission from *Science News*, the weekly newsmagazine of *Science*, copyright 1989 by Science Services, Inc.

8.17 Consumer Confidence An increase in the rate of consumer savings is frequently tied to a lack of confidence in the economy and is said to be an indicator of a recessional tendency in the economy. A random sampling of $n = 200$ savings accounts in a local community showed a mean increase in savings account values of 7.2% over the past 12 months, with a standard deviation of 5.6%. Estimate the mean percent increase in savings account values over the past 12 months for depositors in the community. Find the margin of error for your estimate.

8.18 Multimedia Kids Do our children spend as much time enjoying the outdoors and playing with family and friends as previous generations did? Or are our children spending more and more time glued to the television, computer, and other multimedia equipment? A random sample of 250 children between the ages of 8 and 18 showed that 170 children had a TV in their bedroom and that 120 of them had a video game player in their bedroom.

a. Estimate the proportion of all 8- to 18-year-olds who have a TV in their bedroom, and calculate the margin of error for your estimate.

b. Estimate the proportion of all 8- to 18-year-olds who have a video game player in their bedroom, and calculate the margin of error for your estimate.

8.19 Illegal Immigration In a recent poll that included questions about illegal immigration into the United States, and the federal and state responses to the problem, 75% of the $n = 1004$ adults surveyed felt that the United States is not doing enough to keep illegal immigrants from coming into this country.[3]

a. What is a point estimate for the proportion of adults who feel that the United States is not doing enough to keep illegal immigrants from coming into this country? Calculate the margin of error.

b. The poll reports a margin of error of $\pm 3.5\%$. How should the reported margin of error be calculated so that it can be applied to all of the questions in the survey? Is the reported margin of error correct?

8.20 Hotel Costs Even within a particular chain of hotels, lodging during the summer months can vary substantially depending on the type of room and the amenities offered.[4] Suppose that we randomly select 50 billing statements from each of the computer databases of the Marriott, Westin, and the Doubletree hotel chains, and record the nightly room rates.

	Marriott	**Westin**	**Doubletree**
Sample Average ($)	150	165	125
Sample Standard Deviation	17.2	22.5	12.8

a. Describe the sampled population(s).

b. Find a point estimate for the average room rate for the Marriott hotel chain. Calculate the margin of error.

c. Find a point estimate for the average room rate for the Westin hotel chain. Calculate the margin of error.

d. Find a point estimate for the average room rate for the Doubletree hotel chain. Calculate the margin of error.

e. Display the results of parts b, c, and d graphically, using the form shown in Figure 8.5. Use this display to compare the average room rates for the three hotel chains.

8.21 "900" Numbers Radio and television stations often air controversial issues during broadcast time and ask viewers to indicate their agreement or disagreement with a given stand on the issue. A poll is conducted by asking those viewers who *agree* to call a certain 900 telephone number and those who *disagree* to call a second 900 telephone number. All respondents pay a fee for their calls.

a. Does this polling technique result in a random sample?

b. What can be said about the validity of the results of such a survey? Do you need to worry about a margin of error in this case?

8.22 Men On Mars? Do you think that the United States should pursue a program to send humans to Mars? An opinion poll conducted by the Associated Press indicated that 49% of the 1034 adults surveyed think that we should pursue such a program.[5]

a. Estimate the true proportion of Americans who think that the United States should pursue a program to send humans to Mars. Calculate the margin of error.

b. The question posed in part a was only one of many questions concerning our space program that were asked in the opinion poll. If the Associated Press wanted to report one sampling error that would be valid for the entire poll, what value should they report?

8.23 Hungry Rats In an experiment to assess the strength of the hunger drive in rats, 30 previously trained animals were deprived of food for 24 hours. At the end of the 24-hour period, each animal was put into a cage where food was dispensed if the animal pressed a lever. The length of time the animal continued pressing the bar (although receiving no food) was recorded for each animal. If the data yielded a sample mean of 19.3 minutes with a standard deviation of 5.2 minutes, estimate the true mean time and calculate the margin of error.

8.5 INTERVAL ESTIMATION

An *interval estimator* is a rule for calculating two numbers—say, a and b—to create an interval that you are fairly certain contains the parameter of interest. The concept of "fairly certain" means "with high probability." We measure this probability using the **confidence coefficient,** designated by $1 - \alpha$.

Definition The probability that a confidence interval will contain the estimated parameter is called the **confidence coefficient.**

For example, experimenters often construct 95% confidence intervals. This means that the confidence coefficient, or the probability that the interval will contain the estimated parameter, is .95. You can increase or decrease your amount of certainty by changing the confidence coefficient. Some values typically used by experimenters are .90, .95, .98, and .99.

Consider an analogy—this time, throwing a lariat at a fence post. The fence post represents the parameter that you wish to estimate, and the loop formed by the lariat represents the confidence interval. Each time you throw your lariat, you hope to rope the fence post; however, sometimes your lariat misses. In the same way, each time you draw a sample and construct a confidence interval for a parameter, you hope to include the parameter in your interval, but, just like the lariat, sometimes you miss. Your "success rate"—the proportion of intervals that "rope the post" in repeated sampling—is the confidence coefficient.

Constructing a Confidence Interval

When the sampling distribution of a point estimator is approximately normal, an interval estimator or **confidence interval** can be constructed using the following reasoning. For simplicity, assume that the confidence coefficient is .95 and refer to Figure 8.6.

FIGURE 8.6
Parameter ±1.96 SE

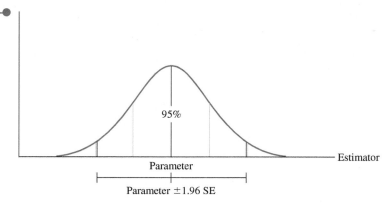

- We know that, of all possible values of the estimator that we might select, 95% of them will be in the interval

    Parameter ±1.96 SE

  shown in Figure 8.6.

- Since the value of the parameter is unknown, consider constructing the interval

    Estimator ±1.96 SE

  which has the same width as the first interval, but has a variable center.

- How often will this interval work properly and enclose the parameter of interest? Refer to Figure 8.7.

FIGURE 8.7
Some 95% confidence intervals

The first two intervals work properly—the parameter (marked with a light gray line) is contained within both intervals. The third interval does not work, since it fails to enclose the parameter. This happened because the value of the estimator at the center of the interval was too far away from the parameter. Fortunately, values of the estimator only fall this far away 5% of the time—our procedure will work properly 95% of the time!

You may want to change the *confidence coefficient* from $(1 - \alpha) = .95$ to another confidence level $(1 - \alpha)$. To accomplish this, you need to change the value $z = 1.96$, which locates an area .95 in the center of the standard normal curve, to a value of z that locates the area $(1 - \alpha)$ in the center of the curve, as shown in Figure 8.8.

Since the total area under the curve is 1, the remaining area in the two tails is α, and each tail contains area $\alpha/2$. The value of z that has "tail area" $\alpha/2$ to its right is called $z_{\alpha/2}$, and the area between $-z_{\alpha/2}$ and $z_{\alpha/2}$ is the confidence coefficient $(1 - \alpha)$. Values of $z_{\alpha/2}$ that are typically used by experimenters will become familiar to you as you begin to construct confidence intervals for different practical situations. Some of these values are given in Table 8.2.

FIGURE 8.8

Location of $z_{\alpha/2}$

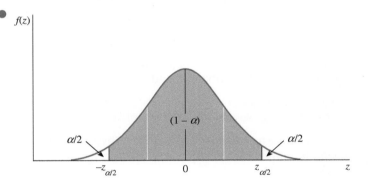

A $(1 - \alpha)100\%$ LARGE-SAMPLE CONFIDENCE INTERVAL

(Point estimator) $\pm z_{\alpha/2} \times$ (Standard error of the estimator)

where $z_{\alpha/2}$ is the z-value with an area $\alpha/2$ in the right tail of a standard normal distribution. This formula generates two values; the **lower confidence limit (LCL)** and the **upper confidence limit (UCL).**

TABLE 8.2 — **Values of z Commonly Used for Confidence Intervals**

Confidence Coefficient, $(1 - \alpha)$	α	$\alpha/2$	$z_{\alpha/2}$
.90	.10	.05	1.645
.95	.05	.025	1.96
.98	.02	.01	2.33
.99	.01	.005	2.58

Large-Sample Confidence Interval for a Population Mean μ

Practical problems very often lead to the estimation of μ, the mean of a population of quantitative measurements. Here are some examples:

- The average achievement of college students at a particular university
- The average strength of a new type of steel
- The average number of deaths per age category
- The average demand for a new cosmetics product

When the sample size n is large, the sample mean $\bar{x}$ is the best point estimator for the population mean μ. Since its sampling distribution is approximately normal, it can be used to construct a confidence interval according to the general approach given earlier.

A $(1 - \alpha)100\%$ LARGE-SAMPLE CONFIDENCE INTERVAL FOR A POPULATION MEAN μ

$$\bar{x} \pm z_{\alpha/2}\frac{\sigma}{\sqrt{n}}$$

where $z_{\alpha/2}$ is the z-value corresponding to an area $\alpha/2$ in the upper tail of a standard normal z distribution, and

n = Sample size
σ = Standard deviation of the sampled population

If σ is unknown, it can be approximated by the sample standard deviation s when the sample size is large ($n \geq 30$) and the approximate confidence interval is

$$\bar{x} \pm z_{\alpha/2}\frac{s}{\sqrt{n}}$$

Another way to find the large-sample confidence interval for a population mean μ is to begin with the statistic

$$z = \frac{\bar{x} - \mu}{\sigma/\sqrt{n}}$$

which has a standard normal distribution. If you write $z_{\alpha/2}$ as the value of z with area $\alpha/2$ to its right, then you can write

$$P\left(-z_{\alpha/2} < \frac{\bar{x} - \mu}{\sigma/\sqrt{n}} < z_{\alpha/2}\right) = 1 - \alpha$$

You can rewrite this inequality as

$$-z_{\alpha/2}\frac{\sigma}{\sqrt{n}} < \bar{x} - \mu < z_{\alpha/2}\frac{\sigma}{\sqrt{n}}$$

$$-\bar{x} - z_{\alpha/2}\frac{\sigma}{\sqrt{n}} < -\mu < -\bar{x} + z_{\alpha/2}\frac{\sigma}{\sqrt{n}}$$

so that

$$P\left(\bar{x} - z_{\alpha/2}\frac{\sigma}{\sqrt{n}} < \mu < \bar{x} + z_{\alpha/2}\frac{\sigma}{\sqrt{n}}\right) = 1 - \alpha$$

Both $\bar{x} - z_{\alpha/2}(\sigma/\sqrt{n})$ and $\bar{x} + z_{\alpha/2}(\sigma/\sqrt{n})$, the lower and upper confidence limits, are actually random quantities that depend on the sample mean $\bar{x}$. Therefore, in repeated sampling, the random interval, $\bar{x} \pm z_{\alpha/2}(\sigma/\sqrt{n})$, will contain the population mean μ with probability $(1 - \alpha)$.

EXAMPLE 8.6

A dietician selected a random sample of $n = 50$ male adults and found that their average daily intake of dairy products was $\bar{x} = 756$ grams per day with a standard deviation of $s = 35$ grams per day. Use this sample information to construct a 95% confidence interval for the mean daily intake of dairy products for men.

Solution Since the sample size of $n = 50$ is large, the distribution of the sample mean $\bar{x}$ is approximately normally distributed with mean μ and standard error estimated by $s/\sqrt{n}$. The approximate 95% confidence interval is

$$\bar{x} \pm 1.96\left(\frac{s}{\sqrt{n}}\right)$$

$$756 \pm 1.96\left(\frac{35}{\sqrt{50}}\right)$$

$$756 \pm 9.70$$

Hence, the 95% confidence interval for μ is from 746.30 to 765.70 grams per day.

Interpreting the Confidence Interval

What does it mean to say you are "95% confident" that the true value of the population mean μ is within a given interval? If you were to construct 20 such intervals, each using different sample information, your intervals might look like those shown in Figure 8.9(a). Of the 20 intervals, you might expect that 95% of them, or 19 out of 20, will perform as planned and contain μ within their upper and lower bounds. If you constructed 100 such intervals (Figure 8.9(b)), you would expect about 95 of them

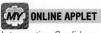

NEED A TIP?

A 95% confidence interval tells you that, if you were to construct many of these intervals (all of which would have slightly different endpoints), 95% of them would enclose the population mean.

FIGURE 8.9

Interpreting Confidence Intervals

MY ONLINE APPLET

Interpreting Confidence Intervals

(a) 20 intervals

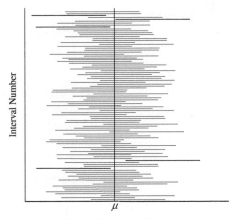

(b) 100 intervals

to perform as planned. Remember that you cannot be absolutely sure that any one particular interval contains the mean μ. You will never know whether your particular interval is one of the 19 that "worked," or whether it is the one interval that "missed." Your confidence in the estimated interval follows from the fact that when repeated intervals are calculated, 95% of these intervals will contain μ.

A good confidence interval has two desirable characteristics:

- It is as narrow as possible. The narrower the interval, the more exactly you have located the estimated parameter.
- It has a large confidence coefficient, near 1. The larger the confidence coefficient, the more likely it is that the interval will contain the estimated parameter.

EXAMPLE 8.7

Construct a 99% confidence interval for the mean daily intake of dairy products for adult men in Example 8.6.

Solution   To change the confidence level to .99, you must find the appropriate value of the standard normal z that puts area $(1 - \alpha) = .99$ in the center of the curve. This value, with tail area $\alpha/2 = .005$ to its right, is found from Table 8.2 to be $z = 2.58$ (see Figure 8.10). The 99% confidence interval is then

$$\bar{x} \pm 2.58\left(\frac{s}{\sqrt{n}}\right)$$

$$756 \pm 2.58(4.95)$$

$$756 \pm 12.77$$

or 743.23 to 768.77 grams per day. This confidence interval is *wider* than the 95% confidence interval in Example 8.6.

FIGURE 8.10

Standard normal values for a 99% confidence interval

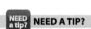

Right Tail Area	z-Value
.05	1.645
.025	1.96
.01	2.33
.005	2.58

The increased width is necessary to increase the confidence, just as you might want a wider loop on your lariat to ensure roping the fence post! The only way to *increase the confidence* without increasing the width of the interval is to *increase the sample size, n.*

The standard error of $\bar{x}$,

$$SE = \frac{\sigma}{\sqrt{n}}$$

measures the variability or spread of the values of $\bar{x}$. The more variable the population data, measured by σ, the more variable will be $\bar{x}$, and the standard error will be larger. On the other hand, if you increase the sample size n, more information is available for estimating μ. The estimates should fall closer to μ and the standard error will be smaller.

The confidence intervals of Examples 8.6 and 8.7 are approximate because you substituted s as an approximation for σ. That is, instead of the confidence coefficient being .95, the value specified in the example, the true value of the coefficient may be .92, .94, or .97. But this discrepancy is of little concern from a practical point of view; as far as your "confidence" is concerned, there is little difference among these confidence coefficients. Most interval estimators used in statistics yield approximate confidence intervals because the assumptions upon which they are based are not satisfied exactly. Having made this point, we will not continue to refer to confidence intervals as "approximate." It is of little practical concern as long as the actual confidence coefficient is near the value specified.

Large-Sample Confidence Interval for a Population Proportion p

Many research experiments or sample surveys have as their objective the estimation of the proportion of people or objects in a large group that possess a certain characteristic. Here are some examples:

- The proportion of sales that can be expected in a large number of customer contacts
- The proportion of seeds that germinate
- The proportion of "likely" voters who plan to vote for a particular political candidate

Each is a practical example of the binomial experiment, and the parameter to be estimated is the binomial proportion p.

When the sample size is large, the sample proportion,

$$\hat{p} = \frac{x}{n} = \frac{\text{Total number of successes}}{\text{Total number of trials}}$$

is the best point estimator for the population proportion p. Since its sampling distribution is approximately normal, with mean p and standard error $\text{SE} = \sqrt{pq/n}$, $\hat{p}$ can be used to construct a confidence interval according to the general approach given in this section.

Right Tail Area	z-Value
.05	1.645
.025	1.96
.01	2.33
.005	2.58

A $(1 - \alpha)100\%$ LARGE-SAMPLE CONFIDENCE INTERVAL FOR A POPULATION PROPORTION p

$$\hat{p} \pm z_{\alpha/2}\sqrt{\frac{pq}{n}}$$

where $z_{\alpha/2}$ is the z-value corresponding to an area $\alpha/2$ in the right tail of a standard normal z distribution. Since p and q are unknown, they are estimated using the best point estimators: $\hat{p}$ and $\hat{q}$. The sample size is considered large when the normal approximation to the binomial distribution is adequate—namely, when $n\hat{p} > 5$ and $n\hat{q} > 5$.

EXAMPLE 8.8

A random sample of 985 "likely" voters—those who are likely to vote in the upcoming election—were polled during a phone-athon conducted by the Republican Party. Of those surveyed, 592 indicated that they intended to vote for the Republican candidate in the upcoming election. Construct a 90% confidence interval for p, the proportion of likely voters in the population who intend to vote for the Republican candidate. Based on this information, can you conclude that the candidate will win the election?

Solution The point estimate for p is

$$\hat{p} = \frac{x}{n} = \frac{592}{985} = .601$$

and the estimated standard error is

$$\sqrt{\frac{\hat{p}\hat{q}}{n}} = \sqrt{\frac{(.601)(399)}{985}} = .016$$

The z-value for a 90% confidence interval is the value that has area $\alpha/2 = .05$ in the upper tail of the z distribution, or $z_{.05} = 1.645$ from Table 8.2. The 90% confidence interval for p is thus

$$\hat{p} \pm 1.645 \sqrt{\frac{\hat{p}\hat{q}}{n}}$$

$$.601 \pm .026$$

or $.575 < p < .627$. You estimate that the percentage of likely voters who intend to vote for the Republican candidate is between 57.5% and 62.7%. Will the candidate win the election? Assuming that she needs more than 50% of the vote to win, and since both the upper and lower confidence limits exceed this minimum value, you can say with 90% confidence that the candidate will win.

There are some problems, however, with this type of sample survey. What if the voters who consider themselves "likely to vote" do not actually go to the polls? What if a voter changes his or her mind between now and election day? What if a surveyed voter does not respond truthfully when questioned by the campaign worker? The 90% confidence interval you have constructed gives you 90% confidence only if you have selected a *random sample from the population of interest.* You can no longer be assured of "90% confidence" if your sample is biased, or if the population of voter responses changes before the day of the election!

You may have noticed that the point estimator with its 95% margin of error looks very similar to a 95% confidence interval for the same parameter. This close relationship exists for most of the parameters estimated in this book, but it is not true in general. Sometimes the best point estimator for a parameter *does not* fall in the middle of the best confidence interval; the best confidence interval may not even be a function of the best point estimator. Although this is a theoretical distinction, you should remember that there is a difference between point and interval estimation, and that the choice between the two depends on the preference of the experimenter.

8.5 **EXERCISES**

BASIC TECHNIQUES

8.24 Find and interpret a 95% confidence interval for a population mean μ for these values:

a. $n = 36, \bar{x} = 13.1, s^2 = 3.42$

b. $n = 64, \bar{x} = 2.73, s^2 = .1047$

8.25 Find a 90% confidence interval for a population mean μ for these values:

a. $n = 125, \bar{x} = .84, s^2 = .086$

b. $n = 50, \bar{x} = 21.9, s^2 = 3.44$

c. Interpret the intervals found in parts a and b.

8.26 Find a $(1 - \alpha)100\%$ confidence interval for a population mean μ for these values:

a. $\alpha = .01, n = 38, \bar{x} = 34, s^2 = 12$

b. $\alpha = .10, n = 65, \bar{x} = 1049, s^2 = 51$

c. $\alpha = .05, n = 89, \bar{x} = 66.3, s^2 = 2.48$

8.27 A random sample of $n = 300$ observations from a binomial population produced $x = 263$ successes. Find a 90% confidence interval for p and interpret the interval.

8.28 Suppose the number of successes observed in $n = 500$ trials of a binomial experiment is 27. Find a 95% confidence interval for p. Why is the confidence interval narrower than the confidence interval in Exercise 8.27?

8.29 A random sample of n measurements is selected from a population with unknown mean μ and known standard deviation $\sigma = 10$. Calculate the width of a 95% confidence interval for μ for these values of n:

a. $n = 100$ **b.** $n = 200$ **c.** $n = 400$

8.30 Compare the confidence intervals in Exercise 8.29. What effect does each of these actions have on the width of a confidence interval?

a. Double the sample size

b. Quadruple the sample size

8.31 Refer to Exercise 8.30.

a. Calculate the width of a 90% confidence interval for μ when $n = 100$.

b. Calculate the width of a 99% confidence interval for μ when $n = 100$.

c. Compare the widths of 90%, 95%, and 99% confidence intervals for μ. What effect does increasing the confidence coefficient have on the width of the confidence interval?

APPLICATIONS

8.32 A Chemistry Experiment In an electrolysis experiment, a class measured the amount of copper precipitated from a saturated solution of copper sulfate over a 30-minute period. The $n = 30$ students calculated a sample mean and standard deviation equal to .145 and .0051 mole, respectively. Find a 90% confidence interval for the mean amount of copper precipitated from the solution over a 30-minute period.

8.33 Acid Rain Acid rain, caused by the reaction of certain air pollutants with rainwater, is a growing problem in the United States. Pure rain falling through clean air registers a pH value of 5.7 (pH is a measure of acidity: 0 is acid; 14 is alkaline). Suppose water samples from 40 rainfalls are analyzed for pH, and $\bar{x}$ and s are equal to 3.7 and .5, respectively. Find a 99% confidence interval for the mean pH in rainfall and interpret the interval. What assumption must be made for the confidence interval to be valid?

8.34 Working Women In an *Advertising Age* white paper concerning the changing role of women as "breadwinners" in the American family, it was reported that according to their survey with JWT, working men reported doing 54 minutes of household chores a day, while working women reported tackling 72 minutes daily. But when examined more closely, Millennial men reported doing just as many household chores as the average working women, 72 minutes, compared to an average of 54 minutes among both Boomer men and Xer men.[6] The information that follows is adapted from these data and is based on random samples of 1136 men and 795 women.

	Mean	Standard Deviation	n
All Women	72	10.4	795
All Men	54	12.7	1136
Millennial	72	9.2	345
Boomers	54	13.9	475
Xers	54	10.5	316

a. Construct a 95% confidence interval for the average time all men spend doing household chores.

b. Construct a 95% confidence interval for the average time women spend doing household chores.

8.35 Hamburger Meat The meat department of a local supermarket chain packages ground beef using meat trays of two sizes: one designed to hold approximately 1 pound of meat, and one that holds approximately 3 pounds. A random sample of 35 packages in the smaller meat trays produced weight measurements with an average of 1.01 pounds and a standard deviation of .18 pound.

a. Construct a 99% confidence interval for the average weight of all packages sold in the smaller meat trays by this supermarket chain.

b. What does the phrase "99% confident" mean?

c. Suppose that the quality control department of this supermarket chain intends that the amount of ground beef in the smaller trays should be 1 pound on average. Should the confidence interval in part a concern the quality control department? Explain.

8.36 Same-Sex Marriage The results of a *CBS News Poll* concerning views on same-sex marriage and gay rights given in Exercise 7.68 showed that of $n = 1082$ adults, 40% favored legal marriage, 30% favored civil unions, and 25% believed there should be no legal recognition.[7] The poll reported a margin of error of plus or minus 3%.

a. Construct a 90% confidence interval for the proportion of adults who favor the "legal marriage" position.

b. Construct a 90% confidence interval for the proportion of adults who favor the "civil unions" position.

c. How did the researchers calculate the margin of error for this survey? Confirm that their margin of error is correct.

8.37 SUVs A sample survey is designed to estimate the proportion of sports utility vehicles being driven in the state of California. A random sample of 500 registrations are selected from a Department of Motor Vehicles database, and 68 are classified as sports utility vehicles.

a. Use a 95% confidence interval to estimate the proportion of sports utility vehicles in California.

b. How can you estimate the proportion of sports utility vehicles in California with a higher degree of accuracy? (HINT: There are two answers.)

8.38 Who Killed the Electric Car? New car models with names such as the "Volt" and the "Leaf" are being hyped by automakers, as they scramble to produce electric cars that are affordable for most Americans. Still in the trial stage, BMW's *Mini E* can be leased for about $600 per month, and is claimed to be able to travel between 100 and 120 miles per battery charge.[8] Suppose that $n = 60$ field trials are conducted and that the average time between charges is 112.5 miles with a standard deviation of 4.6 miles.

a. Construct a 95% confidence interval for μ, the average time between battery charges for BMW's *Mini E*.

b. Does the confidence interval in part a confirm the claim of 100 to 120 miles per battery charge? Why or why not?

8.39 What's Normal? What *is* normal, when it comes to people's body temperatures? A random sample of 130 human body temperatures, provided by Allen Shoemaker[9] in the *Journal of Statistical Education,* had a mean of 98.25° and a standard deviation of 0.73°.

a. Construct a 99% confidence interval for the average body temperature of healthy people.

b. Does the confidence interval constructed in part a contain the value 98.6°, the usual average temperature cited by physicians and others? If not, what conclusions can you draw?

8.40 Gonna' Vote? How likely are you to vote in the next national election? In a survey by *Pew Research*,[10] fully 77% of the registered Republican voters are *absolutely* going to vote this year while only 65% of Democrats are *absolutely* going to vote in the next election. The sample consisted of 469 registered Republicans, 490 registered Democrats, and 480 registered Independents.

a. Construct a 98% confidence interval for the proportion of registered Republicans who say they are *absolutely* going to vote in the next election. If a Republican senator predicts that at least 85% of registered Republicans will *absolutely* vote in the next election, is this figure realistic?

b. Construct a 99% confidence interval for the proportion of registered Democrats who say they are *absolutely* going to vote in the next election.

ESTIMATING THE DIFFERENCE BETWEEN TWO POPULATION MEANS

8.6

A problem equally as important as the estimation of a single population mean μ for a quantitative population is the comparison of two population means. You may want to make comparisons like these:

- The average scores on the Medical College Admission Test (MCAT) for students whose major was biochemistry and those whose major was biology
- The average yields in a chemical plant using raw materials furnished by two different suppliers
- The average stem diameters of plants grown on two different types of nutrients

For each of these examples, there are two populations: the first with mean and variance μ_1 and σ_1^2, and the second with mean and variance μ_2 and σ_2^2. A random sample of n_1 measurements is drawn from population 1, and a second random sample of size n_2 is independently drawn from population 2. Finally, the estimates of the population parameters are calculated from the sample data using the estimators $\bar{x}_1$, s_1^2, $\bar{x}_2$, and s_2^2 as shown in Table 8.3.

TABLE 8.3 ● **Samples From Two Quantitative Populations**

	Population 1	Population 2
Mean	μ_1	μ_2
Variance	σ_1^2	σ_2^2

	Sample 1	Sample 2
Mean	$\bar{x}_1$	$\bar{x}_2$
Variance	s_1^2	s_2^2
Sample Size	n_1	n_2

Intuitively, the difference between two sample means would provide the maximum information about the actual difference between two population means, and this is in fact the case. The best point estimator of the difference $(\mu_1 - \mu_2)$ between the population means is $(\bar{x}_1 - \bar{x}_2)$. The sampling distribution of this estimator is not difficult to derive, but we state it here without proof.

PROPERTIES OF THE SAMPLING DISTRIBUTION OF $(\bar{x}_1 - \bar{x}_2)$, THE DIFFERENCE BETWEEN TWO SAMPLE MEANS

When independent random samples of n_1 and n_2 observations have been selected from populations with means μ_1 and μ_2 and variances σ_1^2 and σ_2^2, respectively, the sampling distribution of the difference $(\bar{x}_1 - \bar{x}_2)$ has the following properties:

1. The mean of $(\bar{x}_1 - \bar{x}_2)$ is

$$\mu_1 - \mu_2$$

and the standard error is

$$SE = \sqrt{\frac{\sigma_1^2}{n_1} + \frac{\sigma_2^2}{n_2}}$$

which can be estimated as

$$SE = \sqrt{\frac{s_1^2}{n_1} + \frac{s_2^2}{n_2}} \text{ when the sample sizes are large.}$$

2. **If the sampled populations are normally distributed,** then the sampling distribution of $(\bar{x}_1 - \bar{x}_2)$ is **exactly** normally distributed, regardless of the sample size.

3. **If the sampled populations are not normally distributed,** then the sampling distribution of $(\bar{x}_1 - \bar{x}_2)$ is **approximately** normally distributed when n_1 and n_2 are both 30 or more, due to the Central Limit Theorem.

Since $(\mu_1 - \mu_2)$ is the mean of the sampling distribution, it follows that $(\bar{x}_1 - \bar{x}_2)$ is an unbiased estimator of $(\mu_1 - \mu_2)$ with an approximately normal distribution when n_1 and n_2 are large. That is, the statistic

$$z = \frac{(\bar{x}_1 - \bar{x}_2) - (\mu_1 - \mu_2)}{\sqrt{\frac{s_1^2}{n_1} + \frac{s_2^2}{n_2}}}$$

has an approximately standard normal z distribution, and the general procedures of Section 8.5 can be used to construct point and interval estimates. Although the choice between point and interval estimation depends on your personal preference, most experimenters choose to construct confidence intervals for two-sample problems. The appropriate formulas for both methods are given next.

LARGE-SAMPLE POINT ESTIMATION OF $(\mu_1 - \mu_2)$

Point estimator: $(\bar{x}_1 - \bar{x}_2)$

95% Margin of error: ± 1.96 SE $= \pm 1.96 \sqrt{\frac{s_1^2}{n_1} + \frac{s_2^2}{n_2}}$

A $(1 - \alpha)100\%$ LARGE-SAMPLE CONFIDENCE INTERVAL FOR $(\mu_1 - \mu_2)$

$$(\bar{x}_1 - \bar{x}_2) \pm z_{\alpha/2} \sqrt{\frac{s_1^2}{n_1} + \frac{s_2^2}{n_2}}$$

NEED A TIP?

Right Tail Area	z-Value
.05	1.645
.025	1.96
.01	2.33
.005	2.58

EXAMPLE 8.9

The wearing qualities of two types of automobile tires were compared by road-testing samples of $n_1 = n_2 = 100$ tires for each type and recording the number of miles until wearout, defined as a specific amount of tire wear. The test results are given in Table 8.4. Estimate $(\mu_1 - \mu_2)$, the difference in mean miles to wearout, using a 99% confidence interval. Is there a difference in the average wearing quality for the two types of tires?

TABLE 8.4   ● **Sample Data Summary for Two Types of Tires**

Tire 1	Tire 2
$\bar{x}_1 = 26{,}400$ miles	$\bar{x}_2 = 25{,}100$ miles
$s_1^2 = 1{,}440{,}000$	$s_2^2 = 1{,}960{,}000$

Solution   The point estimate of $(\mu_1 - \mu_2)$ is

$$(\bar{x}_1 - \bar{x}_2) = 26{,}400 - 25{,}100 = 1300 \text{ miles}$$

and the standard error of $(\bar{x}_1 - \bar{x}_2)$ is estimated as

$$\text{SE} = \sqrt{\frac{s_1^2}{n_1} + \frac{s_2^2}{n_2}} = \sqrt{\frac{1{,}440{,}000}{100} + \frac{1{,}960{,}000}{100}} = 184.4 \text{ miles}$$

The 99% confidence interval is calculated as

$$(\bar{x}_1 - \bar{x}_2) \pm 2.58 \sqrt{\frac{s_1^2}{n_1} + \frac{s_2^2}{n_2}}$$

$$1300 \pm 2.58(184.4)$$

$$1300 \pm 475.8$$

or $824.2 < (\mu_1 - \mu_2) < 1775.8$. The difference in the average miles to wearout for the two types of tires is estimated to lie between LCL = 824.2 and UCL = 1775.8 miles.

Based on this confidence interval, can you conclude that there is a difference in the average miles to wearout for the two types of tires? If there were no difference in the two population means, then μ_1 and μ_2 would be equal and $(\mu_1 - \mu_2) = 0$. If you look at the confidence interval you constructed, you will see that 0 is not one of the possible values for $(\mu_1 - \mu_2)$. Therefore, it is not likely that the means are the same; you can conclude that there is a difference in the average miles to wearout for the two types of tires. The confidence interval has allowed you to *make a decision* about the equality of the two population means.

 NEED A TIP?

If 0 is not in the interval, you *can* conclude that there is a difference in the population means.

EXAMPLE (**8.10**)   The scientist in Example 8.6 wondered whether there was a difference in the average daily intakes of dairy products between men and women. He took a sample of $n_1 = 50$ adult men and $n_2 = 50$ adult women and recorded their daily intakes of dairy products in grams per day. A summary of his sample results is listed in Table 8.5. Construct a 95% confidence interval for the difference in the average daily intakes of dairy products for men and women. Can you conclude that there is a difference in the average daily intakes for men and women?

TABLE 8.5   ● **Sample Values for Daily Intakes of Dairy Products**

	Men	Women
Sample Size	50	50
Sample Mean	756	762
Sample Standard Deviation	35	30

Solution The confidence interval is constructed using a value of z with tail area $\alpha/2 = .025$ to its right; that is, $z_{.025} = 1.96$. Using the sample standard deviations to approximate the unknown population standard deviations, the 95% confidence interval is

$$(\bar{x}_1 - \bar{x}_2) \pm 1.96\sqrt{\frac{s_1^2}{n_1} + \frac{s_2^2}{n_2}}$$

$$(756 - 762) \pm 1.96\sqrt{\frac{35^2}{50} + \frac{30^2}{50}}$$

$$-6 \pm 12.78$$

or $-18.78 < (\mu_1 - \mu_2) < 6.78$. Look at the possible values for $(\mu_1 - \mu_2)$ in the confidence interval. It is possible that the difference $(\mu_1 - \mu_2)$ could be negative (indicating that the average for women exceeds the average for men), it could be positive (indicating that men have the higher average), or it could be 0 (indicating no difference between the averages). Based on this information, you *should not be willing to conclude* that there is a difference in the average daily intakes of dairy products for men and women.

Examples 8.9 and 8.10 deserve further comment with regard to using sample estimates in place of unknown parameters. The sampling distribution of

$$\frac{(\bar{x}_1 - \bar{x}_2) - (\mu_1 - \mu_2)}{\sqrt{\frac{\sigma_1^2}{n_1} + \frac{\sigma_2^2}{n_2}}}$$

has a standard normal distribution for all sample sizes when both sampled populations are normal and an *approximate* standard normal distribution when the sampled populations are not normal but the sample sizes are large (≥ 30). When σ_1^2 and σ_2^2 are not known and are estimated by the sample estimates s_1^2 and s_2^2, the resulting statistic will still have an approximate standard normal distribution when the sample sizes are large. The behavior of this statistic when the population variances are unknown and the sample sizes are small will be discussed in Chapter 10.

8.6 EXERCISES

BASIC TECHNIQUES

8.41 Independent random samples were selected from populations 1 and 2. The sample sizes, means, and variances are as follows:

	Population	
	1	2
Sample Size	35	49
Sample Mean	12.7	7.4
Sample Variance	1.38	4.14

a. Find a 95% confidence interval for estimating the difference in the population means $(\mu_1 - \mu_2)$.

b. Based on the confidence interval in part a, can you conclude that there is a difference in the means for the two populations? Explain.

8.42 Independent random samples were selected from populations 1 and 2. The sample sizes, means, and variances are as follows:

	Population	
	1	2
Sample Size	64	64
Sample Mean	2.9	5.1
Sample Variance	0.83	1.67

a. Find a 90% confidence interval for the difference in the population means. What does the phrase "90% confident" mean?

b. Find a 99% confidence interval for the difference in the population means. Can you conclude that there is a difference in the two population means? Explain.

8.43 Independent random samples of size $n_1 = n_2 = 100$ were selected from each of two populations. The mean and standard deviations for the two samples were $\bar{x}_1 = 125.2$, $\bar{x}_2 = 123.7$, $s_1 = 5.6$, and $s_2 = 6.8$.

a. Construct a 99% confidence interval for estimating the difference in the two population means.

b. Does the confidence interval in part a provide sufficient evidence to conclude that there is a difference in the two population means? Explain.

8.44 Independent random samples of size $n_1 = n_2 = 500$ were selected from each of two populations. The mean and standard deviations for the two samples were $\bar{x}_1 = 125.2$, $\bar{x}_2 = 123.7$, $s_1 = 5.6$, and $s_2 = 6.8$.

a. Find a point estimate for the difference in the two population means. Calculate the margin of error.

b. Based on the results in part a, can you conclude that there is a difference in the two population means? Explain.

APPLICATIONS

8.45 Selenium A small amount of the trace element selenium, 50–200 micrograms (μg) per day, is considered essential to good health. Suppose that random samples of $n_1 = n_2 = 30$ adults were selected from two regions of the United States and that a day's intake of selenium was recorded for each person. The mean and standard deviation of the selenium daily intakes for the 30 adults from region 1 were $\bar{x}_1 = 167.1$ and $s_1 = 24.3$ μg, respectively. The corresponding statistics for the 30 adults from region 2 were $\bar{x}_2 = 140.9$ and $s_2 = 17.6$ μg. Find a 95% confidence interval for the difference in the mean selenium intakes for the two regions. Interpret this interval.

8.46 9-1-1 A study was conducted to compare the mean numbers of police emergency calls per 8-hour shift in two districts of a large city. Samples of 100 8-hour shifts were randomly selected from the police records for each of the two regions, and the number of emergency calls was recorded for each shift. The sample statistics are listed here:

	Region	
	1	2
Sample Size	100	100
Sample Mean	2.4	3.1
Sample Variance	1.44	2.64

Find a 90% confidence interval for the difference in the mean numbers of police emergency calls per shift between the two districts of the city. Interpret the interval.

8.47 Teaching Biology An experiment was conducted to compare a teacher-developed curriculum, "Biology: A Community Context" (BACC) that was standards-based, activity-oriented, and inquiry-centered to the traditional presentation using lecture, vocabulary, and memorized facts. The perhaps not-so-startling results when students were tested on biology concepts, published in *The American Biology Teacher*, are shown in the following table.[11]

	Mean	Sample Size	Standard Deviation
Pretest: All BACC Classes	13.38	372	5.59
Pretest: All Traditional	14.06	368	5.45
Posttest: All BACC Classes	18.5	365	8.03
Posttest: All Traditional	16.5	298	6.96

a. Find a 95% confidence interval for the mean score for the posttest for all BACC classes.

b. Find a 95% confidence interval for the mean score for the posttest for all traditional classes.

c. Find a 95% confidence interval for the difference in mean scores for the posttest BACC classes and the posttest traditional classes.

d. Does the confidence interval in c provide evidence that there is a real difference in the posttest BACC and traditional class scores? Explain.

Source: From "Performance Assessment of a Standards-Based High School Biology Curriculum," by W. Leonard, B. Speziale, and J. Pernick in *The American Biology Teacher,* 2001, 63(5), 310–316. Reprinted by permission of National Association of Biology Teachers.

8.48 Are You Dieting? To compare two weight reduction diets A and B, 60 dieters were randomly selected. One group of 30 dieters was placed on diet A and the other 30 on diet B, and their weight losses were recorded over a 30-day period. The means and standard deviations of the weight-loss measurements for the two groups are shown in the table. Find a 95% confidence interval for the difference in mean weight loss for the two diets. Interpret your confidence interval.

Diet A	Diet B
$\bar{x}_A = 21.3$	$\bar{x}_B = 13.4$
$s_A = 2.6$	$s_B = 1.9$

8.49 Starting Salaries As a group, students majoring in the engineering disciplines have the highest salary expectations, followed by those studying the computer science fields, according to results of NACE's *2010 Student Survey*.[12] To compare the starting salaries of college graduates majoring in engineering and computer science, random samples of 50 recent college graduates in each major were selected and the following information obtained:

Major	Mean ($)	SD
Engineering	56,202	2225
Computer science	50,657	2375

a. Find a point estimate for the difference in the average starting salaries of college students majoring in engineering and computer science. What is the margin of error for your estimate?

b. Based upon the results in part a, do you think that there is a significant difference in the average starting salaries for engineers and computer scientists? Explain.

8.50 Biology Skills Refer to Exercise 8.47. In addition to tests involving biology concepts, students were also tested on process skills. The results of pretest and posttest scores, published in *The American Biology Teacher,* are given below.[11]

	Mean	Sample Size	Standard Deviation
Pretest: All BACC Classes	10.52	395	4.79
Pretest: All Traditional	11.97	379	5.39
Posttest: All BACC Classes	14.06	376	5.65
Posttest: All Traditional	12.96	308	5.93

a. Find a 95% confidence interval for the mean score on process skills for the posttest for all BACC classes.

b. Find a 95% confidence interval for the mean score on process skills for the posttest for all traditional classes.

c. Find a 95% confidence interval for the difference in mean scores on process skills for the posttest BACC classes and the posttest traditional classes.

d. Does the confidence interval in c provide evidence that there is a real difference in the mean process skills scores between posttest BACC and traditional class scores? Explain.

Source: From "Peformance Assessment of a Standards-Based High School Biology Curriculum," by W. Leonard, B. Speziale, and J. Pernick in The American Biology Teacher, 2001, 63(5), 310–316. Reprinted by permission of National Association of Biology Teachers.

8.51 Hotel Costs Refer to Exercise 8.20. The means and standard deviations for 50 billing statements from each of the computer databases of each of the three hotel chains are given in the table:[4]

	Marriott	Westin	Doubletree
Sample Average ($)	150	165	125
Sample Standard Deviation	17.2	22.5	12.8

a. Find a 95% confidence interval for the difference in the average room rates for the Marriott and the Westin hotel chains.

b. Find a 99% confidence interval for the difference in the average room rates for the Westin and the Doubletree hotel chains.

c. Do the intervals in parts a and b contain the value $(\mu_1 - \mu_2) = 0$? Why is this of interest to the researcher?

d. Do the data indicate a difference in the average room rates between the Marriott and the Westin chains? Between the Westin and the Doubletree chains?

8.52 Noise and Stress To compare the effect of stress in the form of noise on the ability to perform a simple task, 70 subjects were divided into two groups. The first group of 30 subjects acted as a control, while the second group of 40 were the experimental group. Although each subject performed the task, the experimental group subjects had to perform the task while loud rock music was played. The time to finish the task was recorded for each subject and the following summary was obtained:

	Control	Experimental
n	30	40
$\bar{x}$	15 minutes	23 minutes
s	4 minutes	10 minutes

a. Find a 99% confidence interval for the difference in mean completion times for these two groups.

b. Based on the confidence interval in part a, is there sufficient evidence to indicate a difference in the average time to completion for the two groups? Explain.

8.53 What's Normal, continued Of the 130 people in Exercise 8.39, 65 were female and 65 were male.[9] The means and standard deviation of their temperatures are shown below.

	Men	Women
Sample Mean	98.11	98.39
Standard Deviation	0.70	0.74

Find a 95% confidence interval for the difference in the average body temperatures for males versus females. Based on this interval, can you conclude that there is a difference in the average temperatures for males versus females? Explain.

ESTIMATING THE DIFFERENCE BETWEEN TWO BINOMIAL PROPORTIONS

8.7

A simple extension of the estimation of a binomial proportion p is the estimation of the difference between two binomial proportions. You may wish to make comparisons like these:

- The proportion of defective items manufactured in two production lines
- The proportion of male and female voters who favor an equal rights amendment
- The germination rates of untreated seeds and seeds treated with a fungicide

These comparisons can be made using the difference $(p_1 - p_2)$ between two binomial proportions, p_1 and p_2. Independent random samples consisting of n_1 and n_2 trials are drawn from populations 1 and 2, respectively, and the sample estimates $\hat{p}_1$ and $\hat{p}_2$ are calculated. The unbiased estimator of the difference $(p_1 - p_2)$ is the sample difference $(\hat{p}_1 - \hat{p}_2)$.

PROPERTIES OF THE SAMPLING DISTRIBUTION OF THE DIFFERENCE $(\hat{p}_1 - \hat{p}_2)$ BETWEEN TWO SAMPLE PROPORTIONS

Assume that independent random samples of n_1 and n_2 observations have been selected from binomial populations with parameters p_1 and p_2, respectively. The sampling distribution of the difference between sample proportions

$$(\hat{p}_1 - \hat{p}_2) = \left(\frac{x_1}{n_1} - \frac{x_2}{n_2} \right)$$

has these properties:

1. The mean of $(\hat{p}_1 - \hat{p}_2)$ is

   $$p_1 - p_2$$

   and the standard error is

   $$SE = \sqrt{\frac{p_1 q_1}{n_1} + \frac{p_2 q_2}{n_2}}$$

   which is estimated as

   $$SE = \sqrt{\frac{\hat{p}_1 \hat{q}_1}{n_1} + \frac{\hat{p}_2 \hat{q}_2}{n_2}}$$

2. The sampling distribution of $(\hat{p}_1 - \hat{p}_2)$ can be approximated by a normal distribution when n_1 and n_2 are large, due to the Central Limit Theorem.

Although the range of a single proportion is from 0 to 1, the difference between two proportions ranges from -1 to 1. To use a normal distribution to approximate the distribution of $(\hat{p}_1 - \hat{p}_2)$, both $\hat{p}_1$ and $\hat{p}_2$ should be approximately normal; that is, $n_1 \hat{p}_1 > 5$, $n_1 \hat{q}_1 > 5$, and $n_2 \hat{p}_2 > 5$, $n_2 \hat{q}_2 > 5$.

The appropriate formulas for point and interval estimation are given next.

LARGE-SAMPLE POINT ESTIMATION OF $(p_1 - p_2)$

Point estimator: $(\hat{p}_1 - \hat{p}_2)$

95% Margin of error: $\pm 1.96 \, \text{SE} = \pm 1.96 \sqrt{\dfrac{\hat{p}_1 \hat{q}_1}{n_1} + \dfrac{\hat{p}_2 \hat{q}_2}{n_2}}$

A $(1 - \alpha)100\%$ LARGE-SAMPLE CONFIDENCE INTERVAL FOR $(p_1 - p_2)$

$(\hat{p}_1 - \hat{p}_2) \pm z_{\alpha/2} \sqrt{\dfrac{\hat{p}_1 \hat{q}_1}{n_1} + \dfrac{\hat{p}_2 \hat{q}_2}{n_2}}$

Assumption: n_1 and n_2 must be sufficiently large so that the sampling distribution of $(\hat{p}_1 - \hat{p}_2)$ can be approximated by a normal distribution—namely, if $n_1 \hat{p}_1$, $n_1 \hat{q}_1$, $n_2 \hat{p}_2$, and $n_2 \hat{q}_2$ are all greater than 5.

EXAMPLE **8.11**

A bond proposal for school construction will be submitted to the voters at the next municipal election. A major portion of the money derived from this bond issue will be used to build schools in a rapidly developing section of the city, and the remainder will be used to renovate and update school buildings in the rest of the city. To assess the viability of the bond proposal, a random sample of $n_1 = 50$ residents in the developing section and $n_2 = 100$ residents from the other parts of the city were asked whether they plan to vote for the proposal. The results are tabulated in Table 8.6.

TABLE 8.6 ● **Sample Values for Opinion on Bond Proposal**

	Developing Section	Rest of the City
Sample Size	50	100
Number Favoring Proposal	38	65
Proportion Favoring Proposal	.76	.65

1. Estimate the difference in the true proportions favoring the bond proposal with a 99% confidence interval.

2. If both samples were pooled into one sample of size $n = 150$, with 103 in favor of the proposal, provide a point estimate of the proportion of city residents who will vote for the bond proposal. What is the margin of error?

Solution

1. The best point estimate of the difference $(p_1 - p_2)$ is given by

$$(\hat{p}_1 - \hat{p}_2) = .76 - .65 = .11$$

and the standard error of $(\hat{p}_1 - \hat{p}_2)$ is estimated as

$$\sqrt{\frac{\hat{p}_1 \hat{q}_1}{n_1} + \frac{\hat{p}_2 \hat{q}_2}{n_2}} = \sqrt{\frac{(.76)(.24)}{50} + \frac{(.65)(.35)}{100}} = .0770$$

For a 99% confidence interval, $z_{.005} = 2.58$, and the approximate 99% confidence interval is found as

$$(\hat{p}_1 - \hat{p}_2) \pm z_{.005} \sqrt{\frac{\hat{p}_1 \hat{q}_1}{n_1} + \frac{\hat{p}_2 \hat{q}_2}{n_2}}$$

$$.11 \pm (2.58)(.0770)$$

$$.11 \pm .199$$

or $(-.089, .309)$. Since this interval contains the value $(p_1 - p_2) = 0$, it is possible that $p_1 = p_2$, which implies that there may be no difference in the proportions favoring the bond issue in the two sections of the city.

2. If there is no difference in the two proportions, then the two samples are not really different and might well be combined to obtain an overall estimate of the proportion of the city residents who will vote for the bond issue. If both samples are pooled, then $n = 150$ and

$$\hat{p} = \frac{103}{150} = .69$$

Therefore, the point estimate of the overall value of p is .69, with a margin of error given by

$$\pm 1.96 \sqrt{\frac{(.69)(.31)}{150}} = \pm 1.96(.0378) = \pm.074$$

Notice that $.69 \pm .074$ produces the interval .62 to .76, which includes only proportions greater than .5. Therefore, if voter attitudes do not change adversely prior to the election, the bond proposal should pass by a reasonable majority.

 8.7 **EXERCISES**

BASIC TECHNIQUES

8.54 Independent random samples of $n_1 = 500$ and $n_2 = 500$ observations were selected from binomial populations 1 and 2, and $x_1 = 120$ and $x_2 = 147$ successes were observed.

a. What is the best point estimator for the difference $(p_1 - p_2)$ in the two binomial proportions?

b. Calculate the approximate standard error for the statistic used in part a.

c. What is the margin of error for this point estimate?

8.55 Independent random samples of $n_1 = 800$ and $n_2 = 640$ observations were selected from binomial populations 1 and 2, and $x_1 = 337$ and $x_2 = 374$ successes were observed.

a. Find a 90% confidence interval for the difference $(p_1 - p_2)$ in the two population proportions. Interpret the interval.

b. What assumptions must you make for the confidence interval to be valid? Are these assumptions met?

8.56 Independent random samples of $n_1 = 1265$ and $n_2 = 1688$ observations were selected from binomial populations 1 and 2, and $x_1 = 849$ and $x_2 = 910$ successes were observed.

a. Find a 99% confidence interval for the difference $(p_1 - p_2)$ in the two population proportions. What does "99% confidence" mean?

b. Based on the confidence interval in part a, can you conclude that there is a difference in the two binomial proportions? Explain.

APPLICATIONS

8.57 M&M'S Does Mars, Incorporated use the same proportion of red candies in its plain and peanut varieties? A random sample of 56 plain M&M'S contained 12 red candies, and another random sample of 32 peanut M&M'S contained 8 red candies.

a. Construct a 95% confidence interval for the difference in the proportions of red candies for the plain and peanut varieties.

b. Based on the confidence interval in part a, can you conclude that there is a difference in the proportions of red candies for the plain and peanut varieties? Explain.

8.58 Different Priorities As we approached the midterm elections, in the summer of 2010, Democrats and Republicans were split about our nation's top priorities.[13] A sample of $n = 900$ registered voters were asked the following question: "Which ONE of the following items do you think is most important for the federal government to be working on right now?" The list of options is shown in the table below. Options were rotated to reduce bias, and voters were allowed to indicate "All," "None," or "Unsure."

	All (%)	Democrats (%)	Republicans (%)	Independents (%)
Economy and Jobs	47	55	37	48
Deficit, Spending	15	8	22	16
Terrorism, Security	8	6	10	10
Iraq and Afghanistan	7	9	5	4
Immigration	5	3	7	6
All (vol.)	16	18	16	13
None/ Other (vol.)	1	-	1	2
Unsure	1	-	1	-

Suppose that there were 400 Democrats, 350 Republicans, and 150 Independents in the sample. Use a large-sample estimation procedure to compare the proportions of Republicans and Democrats who mentioned the economy and jobs as the most important item for the federal government to work on. Compare the proportions of Republicans and Independents who mentioned deficit spending as the most important item. Explain your conclusions.

8.59 Baseball Fans The first day of baseball comes in late March, ending in October with the World Series. Does fan support grow as the season goes on? Two CNN/*USA Today*/Gallup polls, one conducted in March and one in November, both involved random samples of 1001 adults aged 18 and older. In the March sample, 45% of the adults claimed to be fans of professional baseball, while 51% of the adults in the November sample claimed to be fans.[14]

a. Construct a 99% confidence interval for the difference in the proportion of adults who claim to be fans in March versus November.

b. Does the data indicate that the proportion of adults who claim to be fans increases in November, around the time of the World Series? Explain.

8.60 When Bargaining Pays Off According to a national representative survey done by *Consumer Reports*, you should always try to negotiate for a better deal when shopping or paying for services.[15] Tips include researching prices at other stores and on the Internet, timing your visit late in the month when sales-

people are trying to meet quotas, and talking to a manager rather than a salesperson. Suppose that random samples of 200 men and 200 women are taken, and that the men were more likely than the women to say they "always or often" bargained (30% compared with 25%).

a. Construct a 95% confidence interval for the difference in the proportion of men and women who say they "always or often" negotiate for a better deal.

b. Do the data indicate that there is a difference in the proportion of men and women who say they "always or often" negotiate for a better deal? Explain.

8.61 Catching a Cold Do well-rounded people get fewer colds? A study on the *Chronicle of Higher Education* was conducted by scientists at Carnegie Mellon University, the University of Pittsburgh, and the University of Virginia. They found that people who have only a few social outlets get more colds than those who are involved in a variety of social activities.[16] Suppose that of the 276 healthy men and women tested, $n_1 = 96$ had only a few social outlets and $n_2 = 105$ were busy with six or more activities. When these people were exposed to a cold virus, the following results were observed:

	Few Social Outlets	Many Social Outlets
Sample Size	96	105
Percent with Colds	62%	35%

a. Construct a 99% confidence interval for the difference in the two population proportions.

b. Does there appear to be a difference in the population proportions for the two groups?

c. You might think that coming into contact with more people would lead to more colds, but the data show the opposite effect. How can you explain this unexpected finding?

8.62 Union, Yes! A sampling of political candidates—200 randomly chosen from the West and 200 from the East—was classified according to whether the candidate received backing by a national labor union and whether the candidate won. In the West, 120 winners had union backing, and in the East, 142 winners were backed by a national union. Find a 95% confidence interval for the difference between the proportions of union-backed winners in the West versus the East. Interpret this interval.

8.63 Birth Order and College Success In a study of the relationship between birth order and college success, an investigator found that 126 in a sample of 180 college graduates were firstborn or only children. In a sample of 100 nongraduates of comparable age and socioeconomic background, the number of firstborn or only children was 54. Estimate the difference between the proportions of firstborn or only children in the two populations from which these samples were drawn. Use a 90% confidence interval and interpret your results.

8.64 Generation Next Born between 1980 and 1990, Generation Next is engaged with technology, and the vast majority is dependent upon it.[17] Suppose that in a survey of 500 female and 500 male students in Generation Next, 345 of the females and 365 of the males reported that they decided to attend college in order to make more money.

a. Construct a 98% confidence interval for the difference in the proportions of female and male students who decided to attend college in order to make more money.

b. What does it mean to say that you are "98% confident"?

c. Based on the confidence interval in part a, can you conclude that there is a difference in the proportions of female and male students who decided to attend college in order to make more money?

8.65 Excedrin or Tylenol? In a study to compare the effects of two pain relievers it was found that of $n_1 = 200$ randomly selected individuals who used the first pain reliever, 93% indicated that it relieved their pain. Of $n_2 = 450$ randomly selected individuals who

used the second pain reliever, 96% indicated that it relieved their pain.

a. Find a 99% confidence interval for the difference in the proportions experiencing relief from pain for these two pain relievers.

b. Based on the confidence interval in part a, is there sufficient evidence to indicate a difference in the proportions experiencing relief for the two pain relievers? Explain.

8.66 Auto Accidents Last year's records of auto accidents occurring on a given section of highway were classified according to whether the resulting damage was $1000 or more and to whether a physical injury resulted from the accident. The data follows:

	Under $1000	$1000 or More
Number of Accidents	32	41
Number Involving Injuries	10	23

a. Estimate the true proportion of accidents involving injuries when the damage was $1000 or more for similar sections of highway and find the margin of error.

b. Estimate the true difference in the proportion of accidents involving injuries for accidents with damage under $1000 and those with damage of $1000 or more. Use a 95% confidence interval.

ONE-SIDED CONFIDENCE BOUNDS

8.8

The confidence intervals discussed in Sections 8.5 to 8.7 are sometimes called **two-sided confidence intervals** because they produce both an upper (UCL) and a lower (LCL) bound for the parameter of interest. Sometimes, however, an experimenter is interested in only one of these limits; that is, he needs only an upper bound (or possibly a lower bound) for the parameter of interest. In this case, you can construct a **one-sided confidence bound** for the parameter of interest, such as μ, p, $\mu_1 - \mu_2$, or $p_1 - p_2$.

When the sampling distribution of a point estimator is approximately normal, an argument similar to the one in Section 8.5 can be used to show that one-sided confidence bounds, constructed using the following equations *when the sample size is large,* will contain the true value of the parameter of interest $(1 - \alpha)100\%$ of the time in repeated sampling.

A $(1 - \alpha)100\%$ LOWER CONFIDENCE BOUND (LCB)

(Point estimator) $- z_\alpha \times$ (Standard error of the estimator)

A $(1 - \alpha)100\%$ UPPER CONFIDENCE BOUND (UCB)

(Point estimator) $+ z_\alpha \times$ (Standard error of the estimator)

The z-value used for a $(1 - \alpha)100\%$ one-sided confidence bound, z_α, locates an area α in a single tail of the normal distribution as shown in Figure 8.11.

FIGURE 8.11 ●

z-value for a one-sided confidence bound

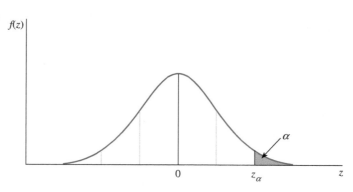

EXAMPLE **8.12**

A corporation plans to issue some short-term notes and is hoping that the interest it will have to pay will not exceed 11.5%. To obtain some information about this problem, the corporation marketed 40 notes, one through each of 40 brokerage firms. The mean and standard deviation for the 40 interest rates were 10.3% and .31%, respectively. Since the corporation is interested in only an upper limit on the interest rates, find a 95% upper confidence bound for the mean interest rate that the corporation will have to pay for the notes.

Solution Since the parameter of interest is μ, the point estimator is $\bar{x}$ with standard error SE $\approx \dfrac{s}{\sqrt{n}}$. The confidence coefficient is .95, so that $\alpha = .05$ and $z_{.05} = 1.645$. Therefore, the 95% upper confidence bound is

$$\text{UCB} = \bar{x} + 1.645\left(\frac{s}{\sqrt{n}}\right) = 10.3 + 1.645\left(\frac{.31}{\sqrt{40}}\right) = 10.3 + .0806 = 10.3806$$

Thus, you can estimate that the mean interest rate that the corporation will have to pay on its notes will be less than 10.3806%. The corporation should not be concerned about its interest rates exceeding 11.5%. How confident are you of this conclusion? Fairly confident, because intervals constructed in this manner contain μ 95% of the time.

8.9 CHOOSING THE SAMPLE SIZE

Designing an experiment is essentially a plan for buying a certain amount of information. Just as the price you pay for a video game varies depending on where and when you buy it, the price of statistical information varies depending on how and where the information is collected. As when you buy any product, you should buy as much statistical information as you can for the minimum possible cost.

The total amount of relevant information in a sample is controlled by two factors:

- The **sampling plan** or **experimental design**: the procedure for collecting the information

- The **sample size** n: the amount of information you collect

You can increase the amount of information you collect by *increasing* the sample size, or perhaps by *changing* the type of sampling plan or experimental design you are using. We will discuss the simplest sampling plan—random sampling from a relatively large population—and focus on ways to choose the sample size n needed to purchase a given amount of information.

A researcher makes little progress in planning an experiment before encountering the problem of sample size. **How many measurements should be included in the sample?** How much information does the researcher want to buy? In order to answer these questions, the researcher must first specify:

- The reliability he wishes to achieve, and
- The accuracy needed for his estimate

In a statistical estimation problem, the accuracy of the estimate is measured by the *margin of error* or the *width of the confidence interval,* both of which have a specified reliability. Since both of these measures are a function of the sample size, specifying the reliability and accuracy allows you to determine the necessary sample size.

For instance, suppose you want to estimate the average daily yield μ of a chemical process and you need the margin of error to be less than 4 tons. This means that, approximately 95% of the time in repeated sampling, the "reliability," the distance between the sample mean $\bar{x}$ and the population mean μ will be less than 1.96 SE. You want this quantity to be less than 4 (the "accuracy"). That is,

$$1.96 \text{ SE} < 4 \quad \text{or} \quad 1.96\left(\frac{\sigma}{\sqrt{n}}\right) < 4$$

Solving for n, you obtain

$$n > \left(\frac{1.96}{4}\right)^2 \sigma^2 \quad \text{or} \quad n > .24\sigma^2$$

If you know σ, the population standard deviation, you can substitute its value into the formula and solve for n. If σ is unknown—which is usually the case—you can use the best approximation available:

- An estimate s obtained from a previous sample
- A range estimate based on knowledge of the largest and smallest possible measurements: $\sigma \approx$ Range/4

For this example, suppose that a prior study of the chemical process produced a sample standard deviation of $s = 21$ tons. Then

$$n > .24\sigma^2 = .24(21)^2 = 105.8$$

Using a sample of size $n = 106$ or larger, you could be reasonably certain (with probability approximately equal to .95) that your estimate of the average yield will be within ±4 tons of the actual average yield.

The solution $n = 106$ is only approximate because you had to use an approximate value for σ to calculate the standard error of the mean. Although this may bother you, it is the best method available for selecting the sample size, and it is certainly better than guessing!

Sometimes researchers request a different reliability, or confidence level other than the 95% confidence specified by the margin of error. In this case, the half-width of the confidence interval provides the accuracy measure for your estimate; that is, the bound B on the error of your estimate is

$$z_{\alpha/2}\left(\frac{\sigma}{\sqrt{n}}\right) < B$$

This method for choosing the sample size can be used for all four estimation procedures presented in this chapter. The general procedure is described next.

NEED TO KNOW...

How to Choose the Sample Size

Determine the parameter to be estimated and the standard error of its point estimator. Then proceed as follows:

1. Choose B, the bound on the error of your estimate, and a confidence coefficient $(1 - \alpha)$.
2. For a one-sample problem, solve this equation for the sample size n:

$$z_{\alpha/2} \times (\text{Standard error of the estimator}) \leq B$$

where $z_{\alpha/2}$ is the value of z having area $\alpha/2$ to its right.
3. For a two-sample problem, set $n_1 = n_2 = n$ and solve the equation in step 2.

[NOTE: For most estimators (all presented in this textbook), the standard error is a function of the sample size n.]

EXAMPLE 8.13   Producers of polyvinyl plastic pipe want to have a supply of pipes sufficient to meet marketing needs. They wish to survey wholesalers who buy polyvinyl pipe in order to estimate the proportion who plan to increase their purchases next year. What sample size is required if they want their estimate to be within .04 of the actual proportion with probability equal to .90?

Solution   For this particular example, the bound B on the error of the estimate is .04. Since the confidence coefficient is $(1 - \alpha) = .90$, α must equal .10 and $\alpha/2$ is .05. The z-value corresponding to an area equal to .05 in the upper tail of the z distribution is $z_{.05} = 1.645$. You then require

$$1.645 \text{ SE} = 1.645\sqrt{\frac{pq}{n}} \leq .04$$

In order to solve this equation for n, you must substitute an approximate value of p into the equation. If you want to be certain that the sample is large enough, you should use $p = .5$ (substituting $p = .5$ will yield the largest possible solution for n because the maximum value of pq occurs when $p = q = .5$). Then

$$1.645\sqrt{\frac{(.5)(.5)}{n}} \leq .04$$

or

$$\sqrt{n} \geq \frac{(1.645)(.5)}{.04} = 20.56$$

$$n \geq (20.56)^2 = 422.7$$

Therefore, the producers must survey at least 423 wholesalers if they want to estimate the proportion p correct to within .04.

EXAMPLE 8.14

A personnel director wishes to compare the effectiveness of two methods of training industrial employees to perform a certain assembly operation. Employees are to be divided into two equal groups: the first receiving training method 1 and the second training method 2. Each will perform the assembly operation, and the length of assembly time will be recorded. It is expected that the assembly times for both groups will have a range of approximately 8 minutes. For the estimate of the difference in mean times to assemble to be correct to within 1 minute with a probability equal to .95, how many workers must be included in each training group?

Solution Since you are estimating the difference between two means, the standard error of the estimate is $\sqrt{\dfrac{\sigma_1^2}{n_1} + \dfrac{\sigma_2^2}{n_2}}$. The bound is $B = 1$ minute, so that you require

$$1.96\sqrt{\frac{\sigma_1^2}{n_1} + \frac{\sigma_2^2}{n_2}} \leq 1$$

Since you want to use two equal groups, you let $n_1 = n_2 = n$ and obtain the equation

$$1.96\sqrt{\frac{\sigma_1^2}{n} + \frac{\sigma_2^2}{n}} \leq 1$$

As noted above, the variability (range) of each method of assembly is approximately the same, so that $\sigma_1^2 = \sigma_2^2 = \sigma^2$. Since the range, equal to 8 minutes, is approximately equal to 4σ, you have

$$4\sigma \approx 8 \quad \text{or} \quad \sigma \approx 2$$

Substituting this value for σ_1 and σ_2 in the earlier equation, you get

$$1.96\sqrt{\frac{(2)^2}{n} + \frac{(2)^2}{n}} \leq 1$$

$$1.96\sqrt{\frac{8}{n}} \leq 1$$

$$\sqrt{n} \geq 1.96\sqrt{8}$$

Solving, you have $n \geq 31$. Thus, each group should contain at least $n = 31$ employees.

Table 8.7 provides a summary of the formulas used to find the sample sizes required for estimation with a given bound on the error of the estimate or confidence interval width W ($W = 2B$). Notice that to estimate p, the sample size formula uses $\sigma^2 = pq$, whereas to estimate ($p_1 - p_2$), the sample size formula uses $\sigma_1^2 = p_1q_1$ and $\sigma_2^2 = p_2q_2$.

TABLE 8.7 ● **Sample Size Formulas**

Parameter	Estimator	Sample Size	Assumptions
μ	$\bar{x}$	$n \geq \dfrac{z_{\alpha/2}^2 \sigma^2}{B^2}$	
$\mu_1 - \mu_2$	$\bar{x}_1 - \bar{x}_2$	$n \geq \dfrac{z_{\alpha/2}^2(\sigma_1^2 + \sigma_2^2)}{B^2}$	$n_1 = n_2 = n$
p	$\hat{p}$	$n \geq \dfrac{z_{\alpha/2}^2 pq}{B^2}$ or $n \geq \dfrac{(.25)z_{\alpha/2}^2}{B^2}$	 $p = .5$
$p_1 - p_2$	$\hat{p}_1 - \hat{p}_2$	$n \geq \dfrac{z_{\alpha/2}^2(p_1 q_1 + p_2 q_2)}{B^2}$ or $n \geq \dfrac{2(.25)z_{\alpha/2}^2}{B^2}$	$n_1 = n_2 = n$ $n_1 = n_2 = n$ and $p_1 = p_2 = .5$

8.9 EXERCISES

BASIC TECHNIQUES

8.67 Find a 90% one-sided upper confidence bound for the population mean μ for these values:

a. $n = 40$, $s^2 = 65$, $\bar{x} = 75$

b. $n = 100$, $s = 2.3$, $\bar{x} = 1.6$

8.68 Find a 99% lower confidence bound for the binomial proportion p when a random sample of $n = 400$ trials produced $x = 196$ successes.

8.69 Independent random samples of size 50 are drawn from two quantitative populations, producing the sample information in the table. Find a 95% upper confidence bound for the difference in the two population means.

	Sample 1	Sample 2
Sample Size	50	50
Sample Mean	12	10
Sample Standard Deviation	5	7

8.70 Suppose you wish to estimate a population mean based on a random sample of n observations, and prior experience suggests that $\sigma = 12.7$. If you wish to estimate μ correct to within 1.6, with probability equal to .95, how many observations should be included in your sample?

8.71 Suppose you wish to estimate a binomial parameter p correct to within .04, with probability equal to .95. If you suspect that p is equal to some value

between .1 and .3 and you want to be certain that your sample is large enough, how large should n be? (HINT: When calculating the standard error, use the value of p in the interval $.1 < p < .3$ that will give the largest sample size.)

8.72 Independent random samples of $n_1 = n_2 = n$ observations are to be selected from each of two populations 1 and 2. If you wish to estimate the difference between the two population means correct to within .17, with probability equal to .90, how large should n_1 and n_2 be? Assume that you know $\sigma_1^2 \approx \sigma_2^2 \approx 27.8$.

8.73 Independent random samples of $n_1 = n_2 = n$ observations are to be selected from each of two binomial populations 1 and 2. If you wish to estimate the difference in the two population proportions correct to within .05, with probability equal to .98, how large should n be? Assume that you have no prior information on the values of p_1 and p_2, but you want to make certain that you have an adequate number of observations in the samples.

APPLICATIONS

8.74 Operating Expenses A random sampling of a company's monthly operating expenses for $n = 36$ months produced a sample mean of $5474 and a standard deviation of $764. Find a 90% upper confidence bound for the company's mean monthly expenses.

8.75 Illegal Immigration Exercise 8.19 discussed a research poll conducted for *ABC News* and the *Washington Post* that included questions about illegal immigration into the United States, and the federal and state responses to the problem.[3] Suppose that you were designing a poll of this type.

a. Explain how you would select your sample. What problems might you encounter in this process?

b. If you wanted to estimate the percentage of the population who agree with a particular statement in your survey questionnaire correct to within 1% with probability .95, approximately how many people would have to be polled?

8.76 Political Corruption A questionnaire is designed to investigate attitudes about political corruption in government. The experimenter would like to survey two different groups—Republicans and Democrats—and compare the responses to various "yes/no" questions for the two groups. The experimenter requires that the sampling error for the difference in the proportion of "yes" responses for the two groups is no more than ±3 percentage points. If the two samples are the same size, how large should the samples be?

8.77 Less Red Meat! As Americans become more conscious of the importance of good nutrition, some researchers believe that we may be eating less red meat. To test this theory, a researcher decides to select hospital nutritional records for subjects surveyed 10 years ago and to compare the average amount of beef consumed per year to the amounts consumed by an equal number of subjects she will interview this year. She knows that the amount of beef consumed annually by Americans ranges from 0 to approximately 104 pounds. How many subjects should the researcher select for each group if she wishes to estimate the difference in the average annual per-capita beef consumption correct to within 5 pounds with 99% confidence?

8.78 Red Meat, continued Refer to Exercise 8.77. The researcher selects two groups of 400 subjects each and collects the following sample information on the annual beef consumption now and 10 years ago:

	Ten Years Ago	This Year
Sample Mean	73	63
Sample Standard Deviation	25	28

a. The researcher would like to show that per-capita beef consumption has decreased in the last 10

years, so she needs to show that the difference in the averages is greater than 0. Find a 99% lower confidence bound for the difference in the average per-capita beef consumptions for the two groups.

b. What conclusions can the researcher draw using the confidence bound from part a?

8.79 Hunting Season A wildlife service wishes to estimate the mean number of days of hunting per hunter for all hunters licensed in the state during a given season. How many hunters must be included in the sample in order to estimate the mean with a bound on the error of estimation equal to 2 hunting days? Assume that data collected in earlier surveys have shown σ to be approximately equal to 10.

8.80 Polluted Rain Suppose you wish to estimate the mean pH of rainfalls in an area that suffers heavy pollution due to the discharge of smoke from a power plant. You know that σ is approximately .5 pH, and you wish your estimate to lie within .1 of μ, with a probability near .95. Approximately how many rainfalls must be included in your sample (one pH reading per rainfall)? Would it be valid to select all of your water specimens from a single rainfall? Explain.

8.81 pH in Rainfall Refer to Exercise 8.80. Suppose you wish to estimate the difference between the mean acidity for rainfalls at two different locations, one in a relatively unpolluted area and the other in an area subject to heavy air pollution. If you wish your estimate to be correct to the nearest .1 pH, with probability near .90, approximately how many rainfalls (pH values) would have to be included in each sample? (Assume that the variance of the pH measurements is approximately .25 at both locations and that the samples will be of equal size.)

8.82 GPAs You want to estimate the difference in grade point averages between two groups of college students accurate to within .2 grade point, with probability approximately equal to .95. If the standard deviation of the grade point measurements is approximately equal to .6, how many students must be included in each group? (Assume that the groups will be of equal size.)

8.83 Selenium, again Refer to the comparison of the daily adult intake of selenium in two different regions of the United States in Exercise 8.45. Suppose you wish to estimate the difference in the mean daily intakes between the two regions correct to within 5 micrograms, with probability equal to .90. If you plan to select an equal number of adults from the two regions (i.e., $n_1 = n_2$), how large should n_1 and n_2 be?

CHAPTER REVIEW

Key Concepts and Formulas

I. Types of Estimators

1. Point estimator: a single number is calculated to estimate the population parameter.

2. Interval estimator: two numbers are calculated to form an interval that, with a certain amount of confidence, contains the parameter.

II. Properties of Good Estimators

1. Unbiased: the average value of the estimator equals the parameter to be estimated.

2. Minimum variance: of all the unbiased estimators, the best estimator has a sampling distribution with the smallest standard error.

3. The margin of error measures the maximum distance between the estimator and the true value of the parameter.

III. Large-Sample Point Estimators

To estimate one of four population parameters when the sample sizes are large, use the following point estimators with the appropriate margins of error.

Parameter	Point Estimator	95% Margin of Error
μ	$\bar{x}$	$\pm 1.96\left(\dfrac{s}{\sqrt{n}}\right)$
p	$\hat{p} = \dfrac{x}{n}$	$\pm 1.96\sqrt{\dfrac{\hat{p}\hat{q}}{n}}$
$\mu_1 - \mu_2$	$\bar{x}_1 - \bar{x}_2$	$\pm 1.96\sqrt{\dfrac{s_1^2}{n_1} + \dfrac{s_2^2}{n_2}}$
$p_1 - p_2$	$(\hat{p}_1 - \hat{p}_2) = \left(\dfrac{x_1}{n_1} - \dfrac{x_2}{n_2}\right)$	$\pm 1.96\sqrt{\dfrac{\hat{p}_1\hat{q}_1}{n_1} + \dfrac{\hat{p}_2\hat{q}_2}{n_2}}$

IV. Large-Sample Interval Estimators

To estimate one of four population parameters when the sample sizes are large, use the following interval estimators.

Parameter	$(1 - \alpha)100\%$ Confidence Interval
μ	$\bar{x} \pm z_{\alpha/2}\left(\dfrac{s}{\sqrt{n}}\right)$
p	$\hat{p} \pm z_{\alpha/2}\sqrt{\dfrac{\hat{p}\hat{q}}{n}}$
$\mu_1 - \mu_2$	$(\bar{x}_1 - \bar{x}_2) \pm z_{\alpha/2}\sqrt{\dfrac{s_1^2}{n_1} + \dfrac{s_2^2}{n_2}}$
$p_1 - p_2$	$(\hat{p}_1 - \hat{p}_2) \pm z_{\alpha/2}\sqrt{\dfrac{\hat{p}_1\hat{q}_1}{n_1} + \dfrac{\hat{p}_2\hat{q}_2}{n_2}}$

1. All values in the interval are possible values for the unknown population parameter.

2. Any values outside the interval are unlikely to be the value of the unknown parameter.

3. To compare two population means or proportions, look for the value 0 in the confidence interval. If 0 is in the interval, it is possible that the two population means or proportions are equal, and you should not declare a difference. If 0 is not in the interval, it is unlikely that the two means or proportions are equal, and you can confidently declare a difference.

V. One-Sided Confidence Bounds

Use either the upper ($+$) or lower ($-$) two-sided bound, with the critical value of z changed from $z_{\alpha/2}$ to z_{α}.

VI. Choosing the Sample Size

1. Determine the size of the margin of error, B, that you are willing to tolerate.

2. Choose the sample size by solving for n or $n = n_1 = n_2$ in the inequality: $z_{\alpha/2} \times \text{SE} \leq B$, where SE is a function of the sample size n.

3. For quantitative populations, estimate the population standard deviation using a previously calculated value of s or the range approximation $\sigma \approx \text{Range}/4$.

4. For binomial populations, use the conservative approach and approximate p using the value $p = .5$.

Supplementary Exercises

8.84 State the Central Limit Theorem. Of what value is the Central Limit Theorem in large-sample statistical estimation?

8.85 A random sample of $n = 64$ observations has a mean $\bar{x} = 29.1$ and a standard deviation $s = 3.9$.

a. Give the point estimate of the population mean μ and find the margin of error for your estimate.

b. Find a 90% confidence interval for μ. What does "90% confident" mean?

c. Find a 90% lower confidence bound for the population mean μ. Why is this bound different from the lower confidence limit in part b?

d. How many observations do you need to estimate μ to within .5, with probability equal to .95?

8.86 Independent random samples of $n_1 = 50$ and $n_2 = 60$ observations were selected from populations 1 and 2, respectively. The sample sizes and computed sample statistics are given in the table:

	Population	
	1	2
Sample Size	50	60
Sample Mean	100.4	96.2
Sample Standard Deviation	0.8	1.3

Find a 90% confidence interval for the difference in population means and interpret the interval.

8.87 Refer to Exercise 8.86. Suppose you wish to estimate $(\mu_1 - \mu_2)$ correct to within .2, with probability equal to .95. If you plan to use equal sample sizes, how large should n_1 and n_2 be?

8.88 A random sample of $n = 500$ observations from a binomial population produced $x = 240$ successes.

a. Find a point estimate for p, and find the margin of error for your estimator.

b. Find a 90% confidence interval for p. Interpret this interval.

8.89 Refer to Exercise 8.88. How large a sample is required if you wish to estimate p correct to within .025, with probability equal to .90?

8.90 Independent random samples of $n_1 = 40$ and $n_2 = 80$ observations were selected from binomial populations 1 and 2, respectively. The number of successes in the two samples were $x_1 = 17$ and $x_2 = 23$. Find a 99% confidence interval for the difference between the two binomial population proportions. Interpret this interval.

8.91 Refer to Exercise 8.90. Suppose you wish to estimate $(p_1 - p_2)$ correct to within .06, with probability equal to .99, and you plan to use equal sample sizes— that is, $n_1 = n_2$. How large should n_1 and n_2 be?

8.92 Ethnic Cuisine Ethnic groups in America buy differing amounts of various food products because of their ethnic cuisine. A researcher interested in market segmentation for Asian and Hispanic households would like to estimate the proportion of households that select certain brands for various products. If the researcher wishes these estimates to be within .03 with probability .95, how many households should she

include in the samples? Assume that the sample sizes are equal.

8.93 Does it Pay to Haggle? In Exercise 8.60, a survey done by *Consumer Reports* indicates that you should always try to negotiate for a better deal when shopping or paying for services.[15] In fact, based on their survey, 37% of the people under age 34 were more likely to "haggle," while only 13% of those 65 and older. Suppose that this survey included 72 people under the age of 34 and 55 people who are 65 or older.

a. What are the values of $\hat{p}_1$ and $\hat{p}_2$ for the two groups in this survey?

b. Find a 95% confidence interval for the difference in the proportion of people who are more likely to "haggle" in the "under 34" versus "65 and older" age groups.

c. What conclusions can you draw regarding the groups compared in part b?

8.94 Smoking and Blood Pressure An experiment was conducted to estimate the effect of smoking on the blood pressure of a group of 35 cigarette smokers, by taking the difference in the blood pressure readings at the beginning of the experiment and again 5 years later. The sample mean increase, measured in millimeters of mercury, was $\bar{x} = 9.7$, and the sample standard deviation was $s = 5.8$. Estimate the mean increase in blood pressure that one would expect for cigarette smokers over the time span indicated by the experiment. Find the margin of error. Describe the population associated with the mean that you have estimated.

8.95 Blood Pressure, continued Using a confidence coefficient equal to .90, place a confidence interval on the mean increase in blood pressure for Exercise 8.94.

8.96 Iodine Concentration Based on repeated measurements of the iodine concentration in a solution, a chemist reports the concentration as 4.614, with an "error margin of .006."

a. How would you interpret the chemist's "error margin"?

b. If the reported concentration is based on a random sample of $n = 30$ measurements, with a sample standard deviation $s = .017$, would you agree that the chemist's "error margin" is .006?

8.97 Heights If it is assumed that the heights of men are normally distributed with a standard deviation of 2.5 inches, how large a sample should be taken to be fairly sure (probability .95) that the sample mean does not differ from the true mean (population mean) by more than .50 in absolute value?

8.98 Chicken Feed An experimenter fed different rations, A and B, to two groups of 100 chicks each. Assume that all factors other than rations are the same for both groups. Of the chicks fed ration A, 13 died, and of the chicks fed ration B, 6 died.

a. Construct a 98% confidence interval for the true difference in mortality rates for the two rations.

b. Can you conclude that there is a difference in the mortality rates for the two rations?

8.99 Antibiotics You want to estimate the mean hourly yield for a process that manufactures an antibiotic. You observe the process for 100 hourly periods chosen at random, with the results $\bar{x} = 34$ ounces per hour and $s = 3$. Estimate the mean hourly yield for the process using a 95% confidence interval.

8.100 Eating Too Much? Partly because of our addiction to fast food, the average American consumes 32.7 pounds of cheese, 14.0 pounds of ice cream, and drinks 48.8 gallons of soda each year, according to the *2010 Statistical Abstract of the United States.*[18] Suppose that we test the accuracy of these reported averages by selecting a random sample of 40 consumers, and recording the following summary statistics:

	Cheese (lbs/yr)	Ice Cream (lbs/yr)	Soda (gal/yr)
Sample Mean	33.1	11.4	49.1
Sample Standard Deviation	3.8	3.2	4.5

Use your knowledge of statistical estimation to estimate the average per-capita annual consumption for these three products. Does this sample cause you to support or to question the accuracy of the reported averages? Explain.

8.101 Fast Food! Even though we know it may not be good for us, many Americans really enjoy their fast food! A survey conducted by *Pew Research Center*[19] graphically illustrated our penchant for eating out, and in particular, eating fast food:

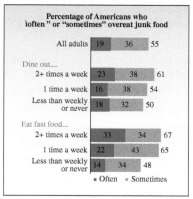

Source: Pew Research Center, pewsocialtrends.org

a. This survey was based on "telephone interviews conducted with a nationally representative sample of 2250 adults, ages 18 years and older, living in continental U.S. telephone households." What problems might arise with this type of sampling?

b. How accurate do you expect the percentages given in the survey to be in estimating the actual population percentages? (HINT: Find the margin of error.)

c. If you want to decrease your margin of error to be ±1%, how large a sample should you take?

8.102 Sunflowers In an article in the *Annals of Botany,* a researcher reported the basal stem diameters of two groups of dicot sunflowers: those that were left to sway freely in the wind and those that were artificially supported.[20] A similar experiment was conducted for monocot maize plants. Although the authors measured other variables in a more complicated experimental design, assume that each group consisted of 64 plants (a total of 128 sunflower and 128 maize plants). The values shown in the table are the sample means plus or minus the standard error.

	Sunflower	Maize
Free-Standing	$35.3 \pm .72$	$16.2 \pm .41$
Supported	$32.1 \pm .72$	$14.6 \pm .40$

Use your knowledge of statistical estimation to compare the free-standing and supported basal diameters for the two plants. Write a paragraph describing your conclusions, making sure to include a measure of the accuracy of your inference.

8.103 Basketball Tickets In a regular NBA season, each team plays 82 games, some at home and some on the road. Can you afford the price of a ticket? The Web site wiki.answers.com indicates that the low prices are around $10 for the high up seats while the court-side seats are around $2000 to $5000 per game and that the average price of a ticket is $75.50 a game.[21] Suppose that we test this claim by selecting a random sample of $n = 50$ ticket purchases from a computer database and find that the average ticket price is $82.50 with a standard deviation of $75.25.

a. Do you think that x, the price of an individual regular season ticket, has a mound-shaped distribution? If not, what shape would you expect?

b. If the distribution of the ticket prices is not normal, you can still use the standard normal distribution to construct a confidence interval for μ, the average price of a ticket. Why?

Fast Food Nights	0	1	2–3	4+
With Kids	47	30	19	4
Without Kids	59	20	16	5

Fast Food Nights	0	1	2–3	4+
Men	46	28	20	6
Women	63	20	15	2

Fifty-three percent of families with kids ate fast food at least once last week, compared with 41% of families without kids. Furthermore, 54% of men ate fast food at least once last week, compared with only 37% of women.

The description of the survey methods that gave rise to this data was stated as follows:

> *"This poll was conducted among a nationwide random sample of 936 adults, interviewed by telephone. The error due to sampling for results based on the entire sample could be plus or minus three percentage points."*

1. Verify the margin of error of ±3 percentage points as stated for the sample of $n = 936$ adults. Suppose that the sample contained an equal number of men and women or 468 men and 468 women. What is the margin of error for men and for women?

2. Do the numbers in the tables indicate the number of people/families in the categories? If not, what do those numbers represent?

3. a. Construct a 95% confidence interval for the proportion of Americans who ate together seven nights a week.

   b. Construct a 95% confidence interval for the difference in the proportion of women and men who personally cook at least 4+ meals per week.

   c. Construct a 95% confidence interval for the proportion of Americans who eat out at restaurants at least once a week.

4. If these questions were asked today, would you expect the responses to be similar to those reported here or would you expect them to differ significantly?

Scott Olson/Getty Images

Large-Sample Tests of Hypotheses

9

GENERAL OBJECTIVE

In this chapter, the concept of a statistical test of hypothesis is formally introduced. The sampling distributions of statistics presented in Chapters 7 and 8 are used to construct large-sample tests concerning the values of population parameters of interest to the experimenter.

CHAPTER INDEX

 NEED TO KNOW...

Rejection Regions, *p*-Values, and Conclusions
How to Calculate β

An Aspirin a Day . . . ?

Will an aspirin a day reduce the risk of heart attack? A very large study of U.S. physicians showed that a single aspirin taken every other day reduced the risk of heart attack in men by one-half. However, 3 days later, a British study reported a completely opposite conclusion. How could this be? The case study at the end of this chapter looks at how the studies were conducted, and you will analyze the data using large-sample techniques.

TESTING HYPOTHESES ABOUT POPULATION PARAMETERS

9.1

In practical situations, statistical inference can involve either estimating a population parameter or making decisions about the value of the parameter. For example, if a pharmaceutical company is fermenting a vat of antibiotic, samples from the vat can be used to *estimate* the mean potency μ for all of the antibiotic in the vat. In contrast, suppose that the company is not concerned about the exact mean potency of the antibiotic, but is concerned only that it meet the minimum government potency standards. Then the company can use samples from the vat to decide between these two possibilities:

- The mean potency μ does not exceed the minimum allowable potency.
- The mean potency μ exceeds the minimum allowable potency.

The pharmaceutical company's problem illustrates a **statistical test of hypothesis.**

The reasoning used in a statistical test of hypothesis is similar to the process in a court trial. In trying a person for theft, the court must decide between innocence and guilt. As the trial begins, the accused person is assumed to be *innocent*. The prosecution collects and presents all available evidence in an attempt to contradict the innocent hypothesis and hence obtain a conviction.

If there is enough evidence against innocence, the court will reject the innocence hypothesis and declare the defendant *guilty*. If the prosecution does not present enough evidence to prove the defendant guilty, the court will find him *not guilty*. Notice that this does not prove that the defendant is innocent, but merely that there was not enough evidence to conclude that the defendant was guilty.

We use this same type of reasoning to explain the basic concepts of hypothesis testing. These concepts are used to test the four population parameters discussed in Chapter 8: a single population mean or proportion (μ or p) and the difference between two population means or proportions ($\mu_1 - \mu_2$ or $p_1 - p_2$). When the sample sizes are large, the point estimators for each of these four parameters have normal sampling distributions, so that all four large-sample statistical tests follow the same general pattern.

A STATISTICAL TEST OF HYPOTHESIS

9.2

A statistical test of hypothesis consists of five parts:

1. The null hypothesis, denoted by H_0
2. The alternative hypothesis, denoted by H_a
3. The test statistic and its p-value
4. The rejection region
5. The conclusion

When you specify these five elements, you define a particular test; changing one or more of the parts creates a new test. Let's look at each part of the statistical test of hypothesis in more detail.

Definition   The two competing hypotheses are the **alternative hypothesis H_a,** generally the hypothesis that the researcher wishes to support, and the **null hypothesis H_0,** a contradiction of the alternative hypothesis.

As you will soon see, it is easier to show support for the alternative hypothesis by proving that the null hypothesis is false. Hence, the statistical researcher always begins by assuming that the null hypothesis H_0 is true. The researcher then uses the sample data to decide whether the evidence favors H_a rather than H_0 and draws one of these two **conclusions:**

- Reject H_0 and conclude that H_a is true.
- Accept (do not reject) H_0 as true.

EXAMPLE 9.1

You wish to show that the average hourly wage of carpenters in the state of California is different from \$19, which is the national average. This is the alternative hypothesis, written as

2

$$H_a : \mu \neq 19$$

The null hypothesis is

1

$$H_0 : \mu = 19$$

You would like to reject the null hypothesis, thus concluding that the California mean is not equal to \$19.

EXAMPLE 9.2

A milling process currently produces an average of 3% defectives. You are interested in showing that a simple adjustment on a machine will decrease p, the proportion of defectives produced in the milling process. Thus, the alternative hypothesis is

2

$$H_a : p < .03$$

and the null hypothesis is

1

$$H_0 : p = .03$$

If you can reject H_0, you can conclude that the adjusted process produces fewer than 3% defectives.

There is a difference in the forms of the alternative hypotheses given in Examples 9.1 and 9.2. In Example 9.1, no directional difference is suggested for the value of μ; that is, μ might be either larger or smaller than \$19 if H_a is true. This type of test is called a **two-tailed test of hypothesis.** In Example 9.2, however, you are specifically interested in detecting a directional difference in the value of p; that is, if H_a is true, the value of p is less than .03. This type of test is called a **one-tailed test of hypothesis.**

The decision to reject or accept the null hypothesis is based on information contained in a sample drawn from the population of interest. This information takes these forms:

NEED A TIP?

Two-tailed ⇔ Look for a ≠ sign in H_a.

One-tailed ⇔ Look for a > or < sign in H_a.

3

- **Test statistic:** a single number calculated from the sample data
- **p-value:** a probability calculated using the test statistic

Either or both of these measures act as decision makers for the researcher in deciding whether to reject or accept H_0.

EXAMPLE (9.3)

For the test of hypothesis in Example 9.1, the average hourly wage $\bar{x}$ for a random sample of 100 California carpenters might provide a good *test statistic* for testing

$$H_0 : \mu = 19 \quad \text{versus} \quad H_a : \mu \neq 19$$

If the null hypothesis H_0 is true, then the sample mean should not be too far from the population mean $\mu = 19$. Suppose that this sample produces a sample mean $\bar{x} = 20$ with standard deviation $s = 2$. Is this sample evidence likely or unlikely to occur, if in fact H_0 is true? You can use two measures to find out. Since the sample size is large, the sampling distribution of $\bar{x}$ is approximately normal with mean $\mu = 19$ and standard error $\sigma/\sqrt{n}$, estimated as

$$SE = \frac{s}{\sqrt{n}} = \frac{2}{\sqrt{100}} = .2$$

• The **test statistic** $\bar{x} = 20$ lies

3

$$z = \frac{\bar{x} - \mu}{\sigma/\sqrt{n}} \approx \frac{20 - 19}{.2} = 5$$

standard deviations from the population mean μ.

• The **p-value** is the probability of observing a test statistic as extreme as or more extreme than the observed value, if in fact H_0 is true. For this example, we define "extreme" as far below or far above what we would have expected. That is,

$$p\text{-value} = P(z > 5) + P(z < -5) \approx 0$$

The *large value of the test statistic* and the *small p-value* mean that you have observed a very unlikely event, if indeed H_0 is true and $\mu = 19$.

4

How do you decide whether to reject or accept H_0? The entire set of values that the test statistic may assume is divided into two sets, or regions. One set, consisting of values that support the alternative hypothesis and lead to rejecting H_0, is called the **rejection region.** The other, consisting of values that support the null hypothesis, is called the **acceptance region.**

For example, in Example 9.1, you would be inclined to believe that California's average hourly wage was different from $19 if the sample mean is either much less than $19 or much greater than $19. The two-tailed rejection region consists of very small and very large values of $\bar{x}$, as shown in Figure 9.1. In Example 9.2, since you want to prove that the percentage of defectives has *decreased,* you would be inclined to reject H_0 for values of $\hat{p}$ that are much smaller than .03. Only *small* values of $\hat{p}$ belong in the left-tailed rejection region shown in Figure 9.2. When the rejection region is in the left tail of the distribution, the test is called a **left-tailed test.** A test with its rejection region in the right tail is called a **right-tailed test.**

FIGURE 9.1

Rejection and acceptance regions for Example 9.1

Rejection region	Acceptance region	Rejection region

Critical value ⎯ $19 ⎯ Critical value $\bar{x}$

FIGURE 9.2 ●

Rejection and acceptance
regions for Example 9.2

If the test statistic falls into the rejection region, then the null hypothesis is rejected.
If the test statistic falls into the acceptance region, then either the null hypothesis is
accepted or the test is judged to be inconclusive. We will clarify the different types
of conclusions that are appropriate as we consider several practical examples of
hypothesis tests.

Finally, how do you decide on the **critical values** that separate the acceptance and
rejection regions? That is, how do you decide how much statistical evidence you need
before you can reject H_0? This depends on the amount of confidence that you, the researcher, want to attach to the test conclusions and the **significance level α,** the risk
you are willing to take of making an incorrect decision.

Definition   A **Type I error** for a statistical test is the error of rejecting the null
hypothesis when it is true. The **level of significance (significance level)** for a statistical test of hypothesis is

$$\alpha = P(\text{Type I error}) = P(\text{falsely rejecting } H_0) = P(\text{rejecting } H_0 \text{ when it is true})$$

This value α represents the *maximum tolerable risk* of incorrectly rejecting H_0.
Once this significance level is fixed, the rejection region can be set to allow the researcher to reject H_0 with a fixed degree of confidence in the decision.

In the next section, we will show you how to use a test of hypothesis to test the
value of a population mean μ. As we continue, we will clarify some of the computational details and add some additional concepts to complete your understanding of
hypothesis testing.

A LARGE-SAMPLE TEST ABOUT A POPULATION MEAN

Consider a random sample of n measurements drawn from a population that has mean
μ and standard deviation σ. You want to test a hypothesis of the form[†]

1    $H_0 : \mu = \mu_0$

where μ_0 is some hypothesized value for μ, versus a one-tailed alternative hypothesis:

2    $H_a : \mu > \mu_0$

The null hypothesis will
always have an "equals"
sign attached.

The subscript zero indicates the value of the parameter specified by H_0. Notice that
H_0 provides an exact value for the parameter to be tested, whereas H_a gives a range
of possible values for μ.

[†]Note that if the test rejects the null hypothesis $\mu = \mu_0$ in favor of the alternative hypothesis $\mu > \mu_0$, then it will
certainly reject a null hypothesis that includes $\mu < \mu_0$, since this is even more contradictory to the alternative hypothesis. For this reason, in this text we state the null hypothesis for a one-tailed test as $\mu = \mu_0$ rather than $\mu \leq \mu_0$.

The Essentials of the Test

The sample mean $\bar{x}$ is the best estimate of the actual value of μ, which is presently in question. What values of $\bar{x}$ would lead you to believe that H_0 is false and μ is, in fact, greater than the hypothesized value? The values of $\bar{x}$ that are extremely *large* would imply that μ is larger than hypothesized. Hence, you should reject H_0 if $\bar{x}$ is too large.

The next problem is to define what is meant by "too large." Values of $\bar{x}$ that lie too many standard deviations to the right of the mean are not very likely to occur. Those values have very little area to their right. Hence, you can define "too large" as being too many standard deviations away from μ_0. But what is "too many"? This question can be answered using the *significance level* α, the probability of rejecting H_0 when H_0 is *true*.

Remember that the standard error of $\bar{x}$ is estimated as

$$\text{SE} = \frac{s}{\sqrt{n}}$$

Since the sampling distribution of the sample mean $\bar{x}$ is approximately normal when **n is large,** the number of standard deviations that $\bar{x}$ lies from μ_0 can be measured using the **standardized test statistic,**

3

$$z = \frac{\bar{x} - \mu_0}{s/\sqrt{n}}$$

which has an approximate standard normal distribution when H_0 is true and $\mu = \mu_0$. The significance level α is equal to the area under the normal curve lying above the rejection region. Thus, if you want $\alpha = .01$, you will reject H_0 when $\bar{x}$ is more than 2.33 standard deviations to the right of μ_0. Equivalently, you will reject H_0 if the standardized test statistic z is greater than 2.33 (see Figure 9.3).

4

FIGURE 9.3

The rejection region for a right-tailed test with $\alpha = .01$

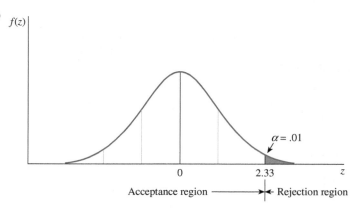

EXAMPLE **9.4**

The average weekly earnings for female social workers is $670. Do men in the same positions have average weekly earnings that are higher than those for women? A random sample of $n = 40$ male social workers showed $\bar{x} = \$725$ and $s = \$102$. Test the appropriate hypothesis using $\alpha = .01$.

NEED A TIP?

For one-tailed tests, look for directional words like "greater," "less than," "higher," "lower," etc.

Solution   You would like to show that the average weekly earnings for men are higher than $670, the women's average. Hence, if μ is the average weekly earnings for male social workers, you can set out the formal test of hypothesis in steps:

1–2

Null and alternative hypotheses:

$$H_0: \mu = 670 \quad \text{versus} \quad H_a: \mu > 670$$

3

Test statistic: Using the sample information, with s as an estimate of the population standard deviation, calculate

$$z \approx \frac{\bar{x} - 670}{s/\sqrt{n}} = \frac{725 - 670}{102/\sqrt{40}} = 3.41$$

4

Rejection region: For this one-tailed test, values of $\bar{x}$ much larger than 670 would lead you to reject H_0; or, equivalently, values of the *standardized test statistic z* in the right tail of the standard normal distribution. To control the risk of making an incorrect decision as $\alpha = .01$, you must set the **critical value** separating the rejection and acceptance regions so that the area in the right tail is exactly $\alpha = .01$. This value is found in Table 3 of Appendix I to be $z = 2.33$, as shown in Figure 9.3. The null hypothesis will be rejected if the observed value of the test statistic, z, is greater than 2.33.

5

Conclusion: Compare the observed value of the test statistic, $z = 3.41$, with the critical value necessary for rejection, $z = 2.33$. Since the observed value of the test statistic falls in the rejection region, you can reject H_0 and conclude that the average weekly earnings for male social workers are higher than the average for female social workers. The probability that you have made an incorrect decision is $\alpha = .01$.

If you wish to detect departures either greater or less than μ_0, then the alternative hypothesis is *two-tailed*, written as

 NEED A TIP?

If the test is two-tailed, you will not see any directional words. The experimenter is only looking for a "difference" from the hypothesized value.

$$H_a : \mu \neq \mu_0$$

which implies either $\mu > \mu_0$ or $\mu < \mu_0$. Values of $\bar{x}$ that are either "too large" or "too small" in terms of their distance from μ_0 are placed in the rejection region. If you choose $\alpha = .01$, the area in the rejection region is equally divided between the two tails of the normal distribution, as shown in Figure 9.4. Using the standardized test statistic z, you can reject H_0 if $z > 2.58$ or $z < -2.58$. For different values of α, the critical values of z that separate the rejection and acceptance regions will change accordingly.

FIGURE 9.4

The rejection region for a two-tailed test with $\alpha = .01$

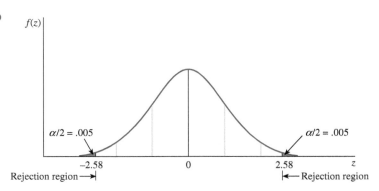

EXAMPLE 9.5

The daily yield for a local chemical plant has averaged 880 tons for the last several years. The quality control manager would like to know whether this average has changed in recent months. She randomly selects 50 days from the computer database and computes the average and standard deviation of the $n = 50$ yields as $\bar{x} = 871$ tons and $s = 21$ tons, respectively. Test the appropriate hypothesis using $\alpha = .05$.

Solution

1-2 **Null and alternative hypotheses:**

$$H_0: \mu = 880 \quad \text{versus} \quad H_a: \mu \neq 880$$

3 **Test statistic:** The point estimate for μ is $\bar{x}$. Therefore, the test statistic is

$$z \approx \frac{\bar{x} - \mu_0}{s/\sqrt{n}} = \frac{871 - 880}{21/\sqrt{50}} = -3.03$$

4 **Rejection region:** For this two-tailed test, you use values of z in both the right and left tails of the standard normal distribution. Using $\alpha = .05$, the **critical values** separating the rejection and acceptance regions cut off areas of $\alpha/2 = .025$ in the right and left tails. These values are $z = \pm 1.96$ and the null hypothesis will be rejected if $z > 1.96$ or $z < -1.96$.

5 **Conclusion:** Since $z = -3.03$, the calculated value of z, falls in the rejection region, the manager can reject the null hypothesis that $\mu = 880$ tons and conclude that it has changed. The probability of rejecting H_0 when H_0 is true is $\alpha = .05$, a fairly small probability. Hence, she is reasonably confident that the decision is correct.

LARGE-SAMPLE STATISTICAL TEST FOR μ

1. Null hypothesis: $H_0 : \mu = \mu_0$
2. Alternative hypothesis:

One-Tailed Test	Two-Tailed Test
$H_a : \mu > \mu_0$	$H_a : \mu \neq \mu_0$
(or, $H_a : \mu < \mu_0$)	

3. Test statistic: $z = \dfrac{\bar{x} - \mu_0}{\sigma/\sqrt{n}}$ estimated as $z = \dfrac{\bar{x} - \mu_0}{s/\sqrt{n}}$

4. Rejection region: Reject H_0 when

One-Tailed Test	Two-Tailed Test
$z > z_\alpha$	$z > z_{\alpha/2}$ or $z < -z_{\alpha/2}$
(or $z < -z_\alpha$ when the alternative hypothesis is $H_a : \mu < \mu_0$)	

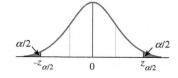

Assumptions: The n observations in the sample are randomly selected from the population and n is large—say, $n \geq 30$.

Calculating the p-Value

In the previous examples, the decision to reject or accept H_0 was made by comparing the calculated value of the test statistic with a critical value of z based on the significance level α of the test. However, different significance levels may lead to different conclusions. For example, if in a right-tailed test, the test statistic is $z = 2.03$, you can reject H_0 at the 5% level of significance because the test statistic exceeds $z = 1.645$. However, you cannot reject H_0 at the 1% level of significance, because the test statistic is less than $z = 2.33$ (see Figure 9.5). To avoid any ambiguity in their conclusions, some experimenters prefer to use a variable level of significance called the **p-value** for the test.

Definition　The **p-value** or observed significance level of a statistical test is the smallest value of α for which H_0 can be rejected. It is the *actual risk* of committing a Type I error, if H_0 is rejected based on the observed value of the test statistic. The p-value measures the strength of the evidence against H_0.

In the right-tailed test with observed test statistic $z = 2.03$, the smallest critical value you can use and still reject H_0 is $z = 2.03$. For this critical value, the risk of an incorrect decision is

$$P(z \geq 2.03) = 1 - .9788 = .0212$$

This probability is the *p-value* for the test. *Notice that it is actually the area to the right of the calculated value of the test statistic.*

FIGURE 9.5

Variable rejection regions

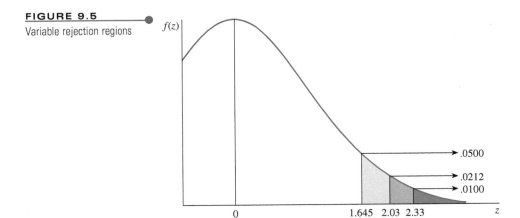

A *small p-value* indicates that the observed value of the test statistic lies far away from the hypothesized value of μ. This presents strong evidence that H_0 is false and should be rejected. *Large p-values* indicate that the observed test statistic is not far from the hypothesized mean and does not support rejection of H_0. How small does the p-value need to be before H_0 can be rejected?

NEED a tip? NEED A TIP?

p-Value = Tail area (one or two tails) "beyond" the observed value of the test statistic

Definition If the p-value is less than or equal to a preassigned significance level α, then the null hypothesis can be rejected, and you can report that the results are **statistically significant** at level α.

In the previous instance, if you choose $\alpha = .05$ as your significance level, H_0 can be rejected because the p-value is less than .05. However, if you choose $\alpha = .01$ as your significance level, the p-value (.0212) is not small enough to allow rejection of H_0. The results are significant at the 5% level, but not at the 1% level. You might see these results reported in professional journals as *significant* $(p < .05)$.[†]

EXAMPLE 9.6

Refer to Example 9.5. The quality control manager wants to know whether the daily yield at a local chemical plant—which has averaged 880 tons for the last several years—has changed in recent months. A random sample of 50 days gives an average yield of 871 tons with a standard deviation of 21 tons. Calculate the p-value for this two-tailed test of hypothesis. Use the p-value to draw conclusions regarding the statistical test.

Solution The rejection region for this two-tailed test of hypothesis is found in both tails of the normal probability distribution. Since the observed value of the test statistic is $z = -3.03$, the smallest rejection region that you can use and still reject H_0 is $|z| > 3.03$. For this rejection region, the value of α is the p-value:

$$p\text{-value} = P(z > 3.03) + P(z < -3.03) = (1 - .9988) + .0012 = .0024$$

Notice that the two-tailed p-value is actually twice the tail area corresponding to the calculated value of the test statistic. If this p-value $= .0024$ is less than or equal to the preassigned level of significance α, H_0 can be rejected. For this test, you can reject H_0 at either the 1% or the 5% level of significance.

If you are reading a research report, how small should the p-value be before you decide to reject H_0? Many researchers use a "sliding scale" to classify their results.

- If the p-value is less than .01, H_0 is rejected. The results are **highly significant.**
- If the p-value is between .01 and .05, H_0 is rejected. The results are **statistically significant.**
- If the p-value is between .05 and .10, H_0 is usually not rejected. The results are only **tending toward statistical significance.**
- If the p-value is greater than .10, H_0 is not rejected. The results are **not statistically significant.**

EXAMPLE 9.7

Standards set by government agencies indicate that Americans should not exceed an average daily sodium intake of 3300 milligrams (mg). To find out whether Americans are exceeding this limit, a sample of 100 Americans is selected, and the mean and standard deviation of daily sodium intake are found to be 3400 mg and 1100 mg, respectively. Use $\alpha = .05$ to conduct a test of hypothesis.

[†]In reporting statistical significance, many researchers write $(p < .05)$ or $(P < .05)$ to mean that the p-value of the test was smaller than .05, making the results significant at the 5% level. The symbol p or P in the expression has no connection with our notation for probability or with the binomial parameter p.

Solution   The hypotheses to be tested are

$$H_0 : \mu = 3300 \quad \text{versus} \quad H_a : \mu > 3300$$

and the test statistic is

$$z \approx \frac{\bar{x} - \mu_0}{s/\sqrt{n}} = \frac{3400 - 3300}{1100/\sqrt{100}} = .91$$

The two approaches developed in this section yield the same conclusions.

NEED A TIP?

Small p-value $\Leftrightarrow$ Large z-value

Small p-value $\Rightarrow$ Reject H_0

How small? p-value $\le \alpha$

- **The critical value approach:** Since the significance level is $\alpha = .05$ and the test is one-tailed, the rejection region is determined by a critical value with tail area equal to $\alpha = .05$; that is, H_0 can be rejected if $z > 1.645$. Since $z = .91$ is not greater than the critical value, H_0 is not rejected (see Figure 9.6).
- **The p-value approach:** Calculate the p-value, the probability that z is greater than or equal to $z = .91$:

$$p\text{-value} = P(z > .91) = 1 - .8186 = .1814$$

The null hypothesis can be rejected only if the *p-value is less than or equal to the specified 5% significance level*. Therefore, H_0 is not rejected and the results are *not statistically significant* (see Figure 9.6). There is not enough evidence to indicate that the average daily sodium intake exceeds 3300 mg.

FIGURE 9.6

Rejection region and p-value for Example 9.7

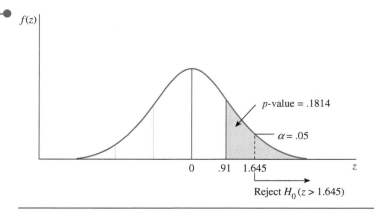

Notice that these two approaches are actually the same, as shown in Figure 9.6. As soon as the calculated value of the test statistic z becomes *larger than* the critical value, z_α, the p-value becomes *smaller than* the significance level α. You can use the most convenient of the two methods; the conclusions you reach will always be the same! The p-value approach does have two advantages, however:

MY **ONLINE APPLET**

Large-Sample Test of a Population Mean

- Statistical output from computer software packages usually reports the p-value of the test.
- Based on the p-value, your test results can be evaluated using any significance level you wish to use. Many researchers report the smallest possible significance level for which their results are *statistically significant.*

Sometimes it is easy to confuse the significance level α with the p-value (or observed significance level). They are both probabilities calculated as areas in the tails of the sampling distribution of the test statistic. However, the significance level α is preset by the experimenter before collecting the data. The p-value is linked directly to the data and actually describes how likely or unlikely the sample results are, assuming that H_0 is true. *The smaller the p-value, the more unlikely it is that H_0 is true!*

NEED TO KNOW...

Rejection Regions, *p*-Values, and Conclusions

The significance level, α, lets you set the risk that you are willing to take of making an incorrect decision in a test of hypothesis.

- To set a rejection region, choose a **critical value** of z so that the area in the tail(s) of the z-distribution is (are) either α for a one-tailed test or $\alpha/2$ for a two-tailed test. Use the right tail for an upper-tailed test and the left tail for a lower-tailed test. Reject H_0 when the test statistic exceeds the critical value and falls in the rejection region.

- To find a **p-value,** find the area in the tail "beyond" the test statistic. If the test is one-tailed, this is the p-value. If the test is two-tailed, this is only half the p-value and must be doubled. Reject H_0 when the p-value is less than α.

Two Types of Errors

You might wonder why, when H_0 was not rejected in the previous example, we did not say that H_0 was definitely true and $\mu = 3300$. This is because, if we choose to *accept H_0*, we must have a measure of the probability of error associated with this decision.

Since there are two choices in a statistical test, there are also two types of errors that can be made. In the courtroom trial, a defendant could be judged not guilty when he's really guilty, or vice versa—the same is true in a statistical test. In fact, the null hypothesis may be either true or false, regardless of the decision the experimenter makes. These two possibilities, along with the two decisions that can be made by the researcher, are shown in Table 9.1.

TABLE 9.1 ● **Decision Table**

Decision	Null Hypothesis	
	True	False
Reject H_0	Type I error	Correct decision
Accept H_0	Correct decision	Type II error

In addition to the Type I error with probability α defined earlier in this section, it is possible to commit a second error, called a **Type II error,** which has probability β.

Definition   A **Type I error** for a statistical test happens if you reject the null hypothesis when it is true. The probability of making a Type I error is denoted by the symbol α.

A **Type II error** for a statistical test happens if you accept the null hypothesis when it is false and some alternative hypothesis is true. The probability of making a Type II error is denoted by the symbol β.

Notice that the probability of a Type I error is exactly the same as the **level of significance** α and is therefore controlled by the researcher. When H_0 is rejected, you have an accurate measure of the reliability of your inference; the probability of an incorrect decision is α. However, the probability β of a Type II error is not always controlled by the experimenter. In fact, when H_0 is false and H_a is true, you may not be able to specify an exact value for μ, but only a range of values. This makes it difficult, if not impossible, to calculate β. Without a measure of reliability, it is not wise to conclude that H_0 is true. Rather than risk an incorrect decision, you should withhold judgment, concluding that you *do not have enough evidence to reject H_0*. Instead of *accepting H_0*, you should *"not reject"* or *"fail to reject"* H_0.

Keep in mind that *"accepting" a particular hypothesis means deciding in its favor.* Regardless of the outcome of a test, you are never *certain* that the hypothesis you "accept" is true. *There is always a risk of being wrong (measured by α or β).* Consequently, you never "accept" H_0 if β is unknown or its value is unacceptable to you. When this situation occurs, you should withhold judgment and collect more data.

The Power of a Statistical Test

The goodness of a statistical test is measured by the size of the two error rates: α, the probability of rejecting H_0 when it is true; and β, the probability of accepting H_0 when H_0 is false and H_a is true. A "good" test is one for which both of these error rates are small. The experimenter begins by selecting α, the probability of a Type I error. If he or she also decides to control the value of β, the probability of accepting H_0 when H_a is true, then an appropriate sample size is chosen.

Another way of evaluating a test is to look at the complement of a Type II error—that is, rejecting H_0 when H_a is true—which has probability

$$1 - \beta = P(\text{reject } H_0 \text{ when } H_a \text{ is true})$$

The quantity $(1 - \beta)$ is called the **power** of the test because it measures the probability of taking the action that we wish to have occur—that is, rejecting the null hypothesis when it is false and H_a is true.

Definition The **power of a statistical test,** given as

$$1 - \beta = P(\text{reject } H_0 \text{ when } H_a \text{ is true})$$

measures the ability of the test to perform as required.

A graph of $(1 - \beta)$, the probability of rejecting H_0 when in fact H_0 is false, as a function of the true value of the parameter of interest is called the **power curve** for the statistical test. Ideally, you would like α to be small and the *power* $(1 - \beta)$ to be large.

EXAMPLE 9.8 Refer to Example 9.5. Calculate β and the power of the test $(1 - \beta)$ when μ is actually equal to 870 tons.

Solution In Example 9.5, you assumed that H_0 was true and that $\mu = 880$. The rejection region with $\alpha = .05$ (using the right-hand curve in Figure 9.7) was set as

$$z = \frac{\bar{x} - \mu_0}{s/\sqrt{n}} > 1.96 \quad \text{or} \quad z = \frac{\bar{x} - \mu_0}{s/\sqrt{n}} < -1.96.$$

This implies that the acceptance region is

$$-1.96 < \frac{\bar{x} - 880}{21/\sqrt{50}} < 1.96 \quad \text{or} \quad 874.18 < \bar{x} < 885.82$$

shown along the horizontal axis in Figure 9.7. When H_0 is false and $\mu = 870$, the sampling distribution of $\bar{x}$ is actually represented by the left-hand curve in Figure 9.7, a normal distribution with $\mu = 870$ and SE $= 21/\sqrt{50} = 2.97$. Then β, the probability of accepting H_0 when $\mu = 870$, is the area under the left-hand normal curve located between 874.18 and 885.82 (see Figure 9.7). Calculating the z-values corresponding to 874.18 and 885.82, you get

FIGURE 9.7

Calculating β in Example 9.8

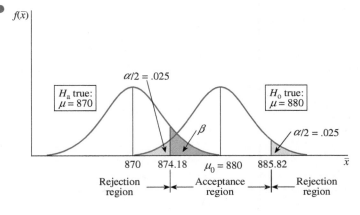

$$z_1 \approx \frac{\bar{x} - \mu}{s/\sqrt{n}} = \frac{874.18 - 870}{21/\sqrt{50}} = 1.41$$

$$z_2 \approx \frac{\bar{x} - \mu}{s/\sqrt{n}} = \frac{885.82 - 870}{21/\sqrt{50}} = 5.33$$

Then

$$\beta = P(\text{accept } H_0 \text{ when } \mu = 870) = P(874.18 < \bar{x} < 885.82 \text{ when } \mu = 870)$$
$$= P(1.41 < z < 5.33)$$

 ONLINE APPLET

Power of a z-Test

You can see from Figure 9.7 that the area under the normal curve with $\mu = 870$ above $\bar{x} = 885.82$ (or $z = 5.33$) is negligible. Therefore,

$$\beta = P(z > 1.41)$$

From Table 3 in Appendix I you can find

$$\beta = 1 - .9207 = .0793$$

Hence, the power of the test is

$$1 - \beta = 1 - .0793 = .9207$$

The probability of correctly rejecting H_0, given that μ is really equal to 870, is .9207, or approximately 92 chances in 100.

Values of $(1 - \beta)$ can be calculated for various values of μ_a different from $\mu_0 = 880$ to measure the power of the test. For example, if $\mu_a = 885$,

$$\beta = P(874.18 < \bar{x} < 885.82 \text{ when } \mu = 885)$$
$$= P(-3.64 < z < .28)$$
$$= .6103 - 0 = .6103$$

and the power is $(1 - \beta) = .3897$. Table 9.2 shows the power of the test for various values of μ_a, and a power curve is graphed in Figure 9.8. Note that the power of the test increases as the distance between μ_a and μ_0 increases. The result is a U-shaped curve for this two-tailed test.

TABLE 9.2   **Value of $(1 - \beta)$ for Various Values of μ_a for Example 9.8**

μ_a	$(1 - \beta)$	μ_a	$(1 - \beta)$
865	.9990	883	.1726
870	.9207	885	.3897
872	.7673	888	.7673
875	.3897	890	.9207
877	.1726	895	.9990
880	.0500		

FIGURE 9.8

Power curve for
Example 9.8

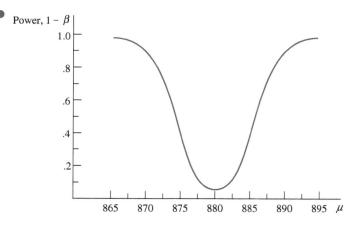

There are many important links among the two error rates, α and β, the power, $(1 - \beta)$, and the sample size, n. Look at the two curves shown in Figure 9.7.

- If α (the sum of the two tail areas in the curve on the right) is increased, the shaded area corresponding to β decreases, and vice versa.

- The only way to decrease β for a fixed α is to "buy" more information—that is, increase the sample size n.

What would happen to the area β as the curve on the left is moved closer to the curve on the right ($\mu = 880$)? With the rejection region in the right curve fixed, the value of β will *increase*. What effect does this have on the power of the test? Look at Figure 9.8.

- As the distance between the true (μ_a) and hypothesized (μ_0) values of the mean increases, the power $(1 - \beta)$ increases. The test is better at detecting *differences* when the distance is *large*.

- The closer the true value (μ_a) gets to the hypothesized value (μ_0), the less power ($1 - \beta$) the test has to detect the difference.
- The only way to increase the power ($1 - \beta$) for a fixed α is to "buy" more information—that is, increase the sample size, n.

The experimenter must decide on the values of α and β—measuring the risks of the possible errors he or she can tolerate. He or she also must decide how much power is needed to detect differences that are practically important in the experiment. Once these decisions are made, the sample size can be chosen by consulting the power curves corresponding to various sample sizes for the chosen test.

NEED TO KNOW...

How to Calculate β

1. Find the critical value or values of $\bar{x}$ used to separate the acceptance and rejection regions.
2. Using one or more values for μ consistent with the alternative hypothesis H_a, calculate the probability that the sample mean $\bar{x}$ falls in the *acceptance region*. This produces the value $\beta = P(\text{accept } H_a \text{ when } \mu = \mu_a)$.
3. Remember that the **power** of the test is ($1 - \beta$).

 EXERCISES

BASIC TECHNIQUES

9.1 Find the appropriate rejection regions for the large-sample test statistic z in these cases:

a. A right-tailed test with $\alpha = .01$

b. A two-tailed test at the 5% significance level

9.2 Refer to Exercise 9.1. Suppose that the observed value of the test statistic was $z = 2.16$. For the rejection regions constructed in parts a and b of Exercise 9.1, draw the appropriate conclusion for the tests. If appropriate, give a measure of the reliability of your conclusion.

9.3 Find the appropriate rejection regions for the large-sample test statistic z in these cases:

a. A left-tailed test at the 1% significance level.

b. A two-tailed test with $\alpha = .01$.

c. Suppose that the observed value of the test statistic was $z = -2.41$. For the rejection regions constructed in parts a and b, draw the appropriate conclusion for the tests. If appropriate, give a measure of the reliability of your conclusion.

9.4 Find the p-value for the following large-sample z tests:

a. A right-tailed test with observed $z = 1.15$

b. A two-tailed test with observed $z = -2.78$

c. A left-tailed test with observed $z = -1.81$

9.5 For the three tests given in Exercise 9.4, use the p-value to determine the significance of the results. Explain what "statistically significant" means in terms of rejecting or accepting H_0 and H_a.

9.6 A random sample of $n = 35$ observations from a quantitative population produced a mean $\bar{x} = 2.4$ and a standard deviation $s = .29$. Suppose your research objective is to show that the population mean μ exceeds 2.3.

a. Give the null and alternative hypotheses for the test.

b. Locate the rejection region for the test using a 5% significance level.

c. Find the standard error of the mean.

d. Before you conduct the test, use your intuition to decide whether the sample mean $\bar{x} = 2.4$ is likely or unlikely, assuming that $\mu = 2.3$. Now conduct the test. Do the data provide sufficient evidence to indicate that $\mu > 2.3$?

9.7 Refer to Exercise 9.6.

a. Calculate the p-value for the test statistic in part d.

b. Use the p-value to draw a conclusion at the 5% significance level.

c. Compare the conclusion in part b with the conclusion reached in part d of Exercise 9.6. Are they the same?

9.8 Refer to Exercise 9.6. You want to test $H_0 : \mu = 2.3$ against $H_a : \mu > 2.3$.

a. Find the critical value of $\bar{x}$ used for rejecting H_0.

b. Calculate $\beta = P(\text{accept } H_0 \text{ when } \mu = 2.4)$.

c. Repeat the calculation of β for $\mu = 2.3, 2.5,$ and 2.6.

d. Use the values of β from parts b and c to graph the power curve for the test.

9.9 A random sample of 100 observations from a quantitative population produced a sample mean of 26.8 and a sample standard deviation of 6.5. Use the p-value approach to determine whether the population mean is different from 28. Explain your conclusions.

APPLICATIONS

9.10 Airline Occupancy Rates Suppose a scheduled airline flight must average at least 60% occupancy in order to be profitable to the airline. An examination of the occupancy rate for 120 10:00 A.M. flights from Atlanta to Dallas showed a mean occupancy per flight of 58% and a standard deviation of 11%.

a. If μ is the mean occupancy per flight and if the company wishes to determine whether or not this scheduled flight is unprofitable, give the alternative and the null hypotheses for the test.

b. Does the alternative hypothesis in part a imply a one- or two-tailed test? Explain.

c. Do the occupancy data for the 120 flights suggest that this scheduled flight is unprofitable? Test using $\alpha = .05$.

9.11 Hamburger Meat Exercise 8.35 involved packages of ground beef in a small tray, intended to hold 1 pound of meat. A random sample of 35 packages in the small tray produced weight measurements with an average of 1.01 pounds and a standard deviation of .18 pound.

a. If you were the quality control manager and wanted to make sure that the average amount of ground

beef was indeed 1 pound, what hypotheses would you test?

b. Find the p-value for the test and use it to perform the test in part a.

c. How would you, as the quality control manager, report the results of your study to a consumer interest group?

9.12 Invasive Species In a study of the pernicious giant hogweed, Jan Pergl[1] and associates compared the density of these plants in two different sites within the Caucasus region of Russia. In its native area, the average density was found to be 5 plants/m². In an invaded area in the Czech Republic, a sample of $n = 50$ plants produced an average density of 11.17 plants/m² with a standard deviation of 3.9 plants/m².

a. Does the invaded area in the Czech Republic have an average density of giant hogweed that is different from $\mu = 5$ at the $\alpha = .05$ level of significance?

b. What is the p-value associated with the test in part a? Can you reject H_0 at the 5% level of significance using the p-value?

9.13 Potency of an Antibiotic A drug manufacturer claimed that the mean potency of one of its antibiotics was 80%. A random sample of $n = 100$ capsules were tested and produced a sample mean of $\bar{x} = 79.7\%$ with a standard deviation of $s = .8\%$. Do the data present sufficient evidence to refute the manufacturer's claim? Let $\alpha = .05$.

a. State the null hypothesis to be tested.

b. State the alternative hypothesis.

c. Conduct a statistical test of the null hypothesis and state your conclusion.

9.14 Flextime A company was contemplating the installation of a flextime schedule in which a worker schedules his or her work hours or compresses work weeks. The company estimates that it needs a minimum mean of 7 hours per day per assembly worker in order to operate effectively. Each of a random sample of 80 of the company's assemblers was asked to submit a tentative flextime schedule. If the mean number of hours per day for Monday was 6.7 hours and the standard deviation was 2.7 hours, do the data provide sufficient evidence to indicate that the mean number of hours worked per day on Mondays, for all of the company's assemblers, will be less than 7 hours? Test using $\alpha = .05$.

9.15 Does College Pay Off? An article in *Time* describing various aspects of American life indicated that higher educational achievement paid off! College grads work 7.4 hours per day, fewer than those with less than a college education.[2] Suppose that the average work day for a random sample of $n = 100$ individuals who had less than a 4-year college education was calculated to be $\bar{x} = 7.9$ hours with a standard deviation of $s = 1.9$ hours.

a. Use the p-value approach to test the hypothesis that the average number of hours worked by individuals having less than a college degree is greater than individuals having a college degree. At what level can you reject H_0?

b. If you were a college graduate, how would you state your conclusion to put yourself in the best possible light?

c. If you were not a college graduate, how might you state your conclusion?

9.16 What's Normal? What *is* normal, when it comes to people's body temperatures? A random sample of 130 human body temperatures, provided by Allen Shoemaker[3] in the *Journal of Statistical Education,* had a mean of 98.25° and a standard deviation of 0.73°. Does the data indicate that the average body temperature for healthy humans is different from 98.6°, the usual average temperature cited by physicians and others? Test using both methods given in this section.

a. Use the p-value approach with $\alpha = .05$.

b. Use the critical value approach with $\alpha = .05$.

c. Compare the conclusions from parts a and b. Are they the same?

d. The 98.6 standard was derived by a German doctor in 1868, who claimed to have recorded 1 million temperatures in the course of his research.[4] What conclusions can you draw about his research in light of your conclusions in parts a and b?

9.17 Sports and Achilles Tendon Injuries Some sports that involve a significant amount of running, jumping, or hopping put participants at risk for Achilles tendinopathy (AT), an inflammation and thickening of the Achilles tendon. A study in *The American Journal of Sports Medicine* looked at the diameter (in mm) of the affected tendons for patients who participated in these types of sports activities.[5] Suppose that the Achilles tendon diameters in the general population have a mean of 5.97 millimeters (mm). When the diameters of the affected tendon were measured for a random sample of 31 patients, the average diameter was 9.80 with a standard deviation of 1.95 mm. Is there sufficient evidence to indicate that the average diameter of the tendon for patients with AT is greater than 5.97 mm? Test at the 5% level of significance.

A LARGE-SAMPLE TEST OF HYPOTHESIS FOR THE DIFFERENCE BETWEEN TWO POPULATION MEANS

9.4

In many situations, the statistical question to be answered involves a comparison of two population means. For example, the U.S. Postal Service is interested in reducing its gasoline costs by replacing gasoline-powered trucks with electric-powered trucks. To determine whether significant savings in operating costs are achieved by changing to electric-powered trucks, a pilot study should be undertaken using, say, 100 conventional gasoline-powered mail trucks and 100 electric-powered mail trucks operated under similar conditions.

The statistic that summarizes the sample information regarding the difference in population means $(\mu_1 - \mu_2)$ is the difference in sample means $(\bar{x}_1 - \bar{x}_2)$. Therefore, in testing whether the difference in sample means indicates that the true difference in population means differs from a specified value, $(\mu_1 - \mu_2) = D_0$, you can use the standard error of $(\bar{x}_1 - \bar{x}_2)$,

$$\sqrt{\frac{\sigma_1^2}{n_1} + \frac{\sigma_2^2}{n_2}} \quad \text{estimated by} \quad SE = \sqrt{\frac{s_1^2}{n_1} + \frac{s_2^2}{n_2}}$$

in the form of a z-statistic to measure how many standard deviations the difference $(\bar{x}_1 - \bar{x}_2)$ lies from the hypothesized difference D_0. The formal testing procedure is described next.

LARGE-SAMPLE STATISTICAL TEST FOR $(\mu_1 - \mu_2)$

1. **Null hypothesis:** $H_0 : (\mu_1 - \mu_2) = D_0$, where D_0 is some specified difference that you wish to test. For many tests, you will hypothesize that there is no difference between μ_1 and μ_2; that is, $D_0 = 0$.

2. **Alternative hypothesis:**

One-Tailed Test	Two-Tailed Test
$H_a : (\mu_1 - \mu_2) > D_0$	$H_a : (\mu_1 - \mu_2) \neq D_0$
[or $H_a : (\mu_1 - \mu_2) < D_0$]	

3. **Test statistic:** $z \approx \dfrac{(\bar{x}_1 - \bar{x}_2) - D_0}{SE} = \dfrac{(\bar{x}_1 - \bar{x}_2) - D_0}{\sqrt{\dfrac{s_1^2}{n_1} + \dfrac{s_2^2}{n_2}}}$

4. **Rejection region:** Reject H_0 when

   One-Tailed Test

   $z > z_\alpha$
   [or $z < -z_\alpha$ when the alternative hypothesis is $H_a : (\mu_1 - \mu_2) < D_0$]

   or when p-value $< \alpha$

   Two-Tailed Test

   $z > z_{\alpha/2}$ or $z < -z_{\alpha/2}$

Assumptions: The samples are randomly and independently selected from the two populations and $n_1 \geq 30$ and $n_2 \geq 30$.

EXAMPLE **9.9**

To determine whether car ownership affects a student's academic achievement, two random samples of 100 male students were each drawn from the student body. The grade point average for the $n_1 = 100$ nonowners of cars had an average and variance equal to $\bar{x}_1 = 2.70$ and $s_1^2 = .36$, while $\bar{x}_2 = 2.54$ and $s_2^2 = .40$ for the $n_2 = 100$ car owners. Do the data present sufficient evidence to indicate a difference in the mean achievements between car owners and nonowners of cars? Test using $\alpha = .05$.

Solution   To detect a difference, if it exists, between the mean academic achievements for nonowners of cars μ_1 and car owners μ_2, you will test the null hypothesis that there is no difference between the means against the alternative hypothesis that $(\mu_1 - \mu_2) \neq 0$; that is,

$$H_0 : (\mu_1 - \mu_2) = D_0 = 0 \quad \text{versus} \quad H_a : (\mu_1 - \mu_2) \neq 0$$

Substituting into the formula for the test statistic, you get

$$z \approx \frac{(\bar{x}_1 - \bar{x}_2) - D_0}{\sqrt{\dfrac{s_1^2}{n_1} + \dfrac{s_2^2}{n_2}}} = \frac{2.70 - 2.54}{\sqrt{\dfrac{.36}{100} + \dfrac{.40}{100}}} = 1.84$$

- **The critical value approach:** Using a two-tailed test with significance level $\alpha = .05$, you place $\alpha/2 = .025$ in each tail of the z distribution and reject H_0 if $z > 1.96$ or $z < -1.96$. Since $z = 1.84$ does not exceed 1.96 and is not less than -1.96, H_0 cannot be rejected (see Figure 9.9). That is, there is insufficient evidence to declare a difference in the average academic achievements for the two groups. Remember that you should not be willing to *accept* H_0—declare the two means to be the same—until β is evaluated for some meaningful values of $(\mu_1 - \mu_2)$.

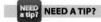
FIGURE 9.9

Rejection region and p-value for Example 9.9

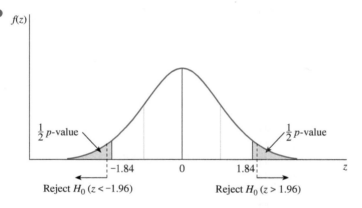

$\frac{1}{2}p$-value $\frac{1}{2}p$-value

-1.84 0 1.84 z

Reject H_0 ($z < -1.96$)   Reject H_0 ($z > 1.96$)

- **The p-value approach:** Calculate the p-value, the probability that z is greater than $z = 1.84$ plus the probability that z is less than $z = -1.84$, as shown in Figure 9.9:

$$p\text{-value} = P(z > 1.84) + P(z < -1.84) = (1 - .9671) + .0329 = .0658$$

The p-value lies between .10 and .05, so you can reject H_0 at the .10 level but not at the .05 level of significance. Since the p-value of .0658 exceeds the specified significance level $\alpha = .05$, H_0 cannot be rejected. Again, you should not be willing to *accept* H_0 until β is evaluated for some meaningful values of $(\mu_1 - \mu_2)$.

Hypothesis Testing and Confidence Intervals

Whether you use the critical value or the p-value approach for testing hypotheses about $(\mu_1 - \mu_2)$, you will always reach the same conclusion because the calculated value of the test statistic and the critical value are related *exactly* in the same way that the p-value and the significance level α are related. You might remember that the confidence intervals constructed in Chapter 8 could also be used to answer questions about the difference between two population means. In fact, for a two-tailed test, the $(1 - \alpha)100\%$ confidence interval for the parameter of interest can be used to test its value, just as you did informally in Chapter 8.

The value of α indicated by the confidence coefficient in the confidence interval is equivalent to the significance level α in the statistical test. For a one-tailed test, the equivalent confidence interval approach would use the one-sided confidence bounds in Section 8.8 with confidence coefficient α. In addition, by using the confidence interval approach, you gain a range of possible values for the parameter of interest, regardless of the outcome of the test of hypothesis.

- If the confidence interval you construct *contains* the value of the parameter specified by H_0, then that value is one of the likely or possible values of the parameter and H_0 should not be rejected.
- If the hypothesized value *lies outside* of the confidence limits, the null hypothesis is rejected at the α level of significance.

EXAMPLE 9.10 — Construct a 95% confidence interval for the difference in average academic achievements between car owners and nonowners. Using the confidence interval, can you conclude that there is a difference in the population means for the two groups of students?

Solution Refer to Section 8.6. For the difference in two population means, the confidence interval is approximated as

$$(\bar{x}_1 - \bar{x}_2) \pm 1.96 \sqrt{\frac{s_1^2}{n_1} + \frac{s_2^2}{n_2}}$$

$$(2.70 - 2.54) \pm 1.96 \sqrt{\frac{.36}{100} + \frac{.40}{100}}$$

$$.16 \pm .17$$

or $-.01 < (\mu_1 - \mu_2) < .33$. This interval gives you a range of possible values for the difference in the population means.

Since the hypothesized difference, $(\mu_1 - \mu_2) = 0$, is contained in the confidence interval, you should not reject H_0. Look at the signs of the possible values in the confidence interval. You cannot tell from the interval whether the difference in the means is negative $(-)$, positive $(+)$, or zero (0)—the latter of the three would indicate that the two means are the same. Hence, you can really reach no conclusion in terms of the question posed. There is not enough evidence to indicate that there is a difference in the average achievements for car owners versus nonowners. The conclusion is the same one reached in Example 9.9.

 9.4 **EXERCISES**

BASIC TECHNIQUES

9.18 Independent random samples of 80 measurements were drawn from two quantitative populations, 1 and 2. Here is a summary of the sample data:

	Sample 1	Sample 2
Sample Size	80	80
Sample Mean	11.6	9.7
Sample Variance	27.9	38.4

a. If your research objective is to show that μ_1 is larger than μ_2, state the alternative and the null hypotheses that you would choose for a statistical test.

b. Is the test in part a one- or two-tailed?

c. Calculate the test statistic that you would use for the test in part a. Based on your knowledge of the standard normal distribution, is this a likely or unlikely observation, assuming that H_0 is true and the two population means are the same?

d. *p-value approach:* Find the *p*-value for the test. Test for a significant difference in the population means at the 1% significance level.

e. *Critical value approach:* Find the rejection region when $\alpha = .01$. Do the data provide sufficient evidence to indicate a difference in the population means?

9.19 Independent random samples of 36 and 45 observations are drawn from two quantitative populations, 1 and 2, respectively. The sample data summary is shown here:

	Sample 1	Sample 2
Sample Size	36	45
Sample Mean	1.24	1.31
Sample Variance	.0560	.0540

Do the data present sufficient evidence to indicate that the mean for population 1 is smaller than the mean for population 2? Use one of the two methods of testing presented in this section, and explain your conclusions.

9.20 Suppose you wish to detect a difference between μ_1 and μ_2 (either $\mu_1 > \mu_2$ or $\mu_1 < \mu_2$) and, instead of running a two-tailed test using $\alpha = .05$, you use the following test procedure. You wait until you have collected the sample data and have calculated $\bar{x}_1$ and $\bar{x}_2$. If $\bar{x}_1$ is larger than $\bar{x}_2$, you choose the alternative hypothesis $H_a : \mu_1 > \mu_2$ and run a one-tailed test placing $\alpha_1 = .05$ in the upper tail of the z distribution. If, on the other hand, $\bar{x}_2$ is larger than $\bar{x}_1$, you reverse the procedure and run a one-tailed test, placing $\alpha_2 = .05$ in the lower tail of the z distribution. If you use this procedure and if μ_1 actually equals μ_2, what is the probability α that you will conclude that μ_1 is not equal to μ_2 (i.e., what is the probability α that you will incorrectly reject H_0 when H_0 is true)? This exercise demonstrates why statistical tests should be formulated *prior* to observing the data.

APPLICATIONS

9.21 Cure for the Common Cold? An experiment was planned to compare the mean time (in days) required to recover from a common cold for persons given a daily dose of 4 milligrams (mg) of vitamin C versus those who were not given a vitamin supplement. Suppose that 35 adults were randomly selected for each

treatment category and that the mean recovery times and standard deviations for the two groups were as follows:

	No Vitamin Supplement	4 mg Vitamin C
Sample Size	35	35
Sample Mean	6.9	5.8
Sample Standard Deviation	2.9	1.2

a. If your research objective is to show that the use of vitamin C reduces the mean time required to recover from a common cold and its complications, give the null and alternative hypotheses for the test. Is this a one- or a two-tailed test?

b. Conduct the statistical test of the null hypothesis in part a and state your conclusion. Test using $\alpha = .05$.

9.22 Healthy Eating As Americans become more conscious about the importance of good nutrition, researchers theorize that the consumption of red meat may have decreased over the last 10 years. A researcher selects hospital nutrition records for 400 subjects surveyed 10 years ago and compares the average amount of beef consumed per year to amounts consumed by an equal number of subjects interviewed this year. The data are given in the table.

	Ten Years Ago	This Year
Sample Mean	73	63
Sample Standard Deviation	25	28

a. Do the data present sufficient evidence to indicate that per-capita beef consumption has decreased in the last 10 years? Test at the 1% level of significance.

b. Find a 99% lower confidence bound for the difference in the average per-capita beef consumptions for the two groups. (This calculation was done as part of Exercise 8.78.) Does your confidence bound confirm your conclusions in part a? Explain. What additional information does the confidence bound give you?

9.23 Lead Levels in Drinking Water Analyses of drinking water samples for 100 homes in each of two different sections of a city gave the following means and standard deviations of lead levels (in parts per million):

	Section 1	Section 2
Sample Size	100	100
Mean	34.1	36.0
Standard Deviation	5.9	6.0

a. Calculate the test statistic and its *p*-value to test for a difference in the two population means. Use the

p-value to evaluate the statistical significance of the results at the 5% level.

b. Use a 95% confidence interval to estimate the difference in the mean lead levels for the two sections of the city.

c. Suppose that the city environmental engineers will be concerned only if they detect a difference of more than 5 parts per million in the two sections of the city. Based on your confidence interval in part b, is the statistical significance in part a of *practical significance* to the city engineers? Explain.

9.24 Starting Salaries, again As a group, students majoring in the engineering disciplines have the highest salary expectations, followed by those studying the computer science fields, according to results of NACE's *2010 Student Survey*.[6] To compare the starting salaries of college graduates majoring in engineering and computer science (see Exercise 8.49), random samples of 50 recent college graduates in each major were selected and the following information obtained.

Major	Mean ($)	SD
Engineering	56,202	2225
Computer Science	50,657	2375

a. Do the data provide sufficient evidence to indicate a difference in average starting salaries for college graduates who majored in engineering and computer science? Test using $\alpha = .05$.

b. Compare your conclusions in part a with the results of part b in Exercise 8.49. Are they the same? Explain.

9.25 Hotel Costs In Exercise 8.20, we explored the average cost of lodging at three different hotel chains.[7] We randomly select 50 billing statements from the computer databases of the Marriott, Westin, and Doubletree hotel chains, and record the nightly room rates. A portion of the sample data is shown in the table.

	Marriott	Westin
Sample Average ($)	150	165
Sample Standard Deviation	17.2	22.5

a. Before looking at the data, would you have any preconceived idea about the direction of the difference between the average room rates for these two hotels? If not, what null and alternative hypotheses should you test?

b. Use the *critical value* approach to determine if there is a significant difference in the average room rates

for the Marriott and the Westin hotel chains. Use $\alpha = .01$.

c. Find the p-value for this test. Does this p-value confirm the results of part b?

9.26 Hotel Costs II Refer to Exercise 9.25. The table below shows the sample data collected to compare the average room rates at the Westin and Doubletree hotel chains.[7]

	Westin	Doubletree
Sample Average ($)	165	125
Sample Standard Deviation	22.5	12.8

a. Do the data provide sufficient evidence to indicate a difference in the average room rates for the Westin and the Doubletree hotel chains? Use $\alpha = .05$.

b. Construct a 95% confidence interval for the difference in the average room rates for the two chains. Does this interval confirm your conclusions in part a?

9.27 Cheaper Airfares Looking for a great airfare? *Consumer Reports*[8] has several hints about how to minimize your costs, which include being flexible about travel dates and times, checking multiple websites, and knowing when to book your flight. One suggestion involved checking fares at "secondary" airports—airports that might be slightly farther from your home, but where fares are lower. For example, the average of all domestic ticket prices at Los Angeles International Airport (LAX) was quoted as $349 compared to an average price of $287 at nearby Ontario International Airport (ONT). Suppose that these estimates were based on random samples of 1000 domestic tickets at each airport and that the standard deviation of the prices at both airports was $200.

a. Is there sufficient evidence to indicate that the mean ticket prices differ for these two airports at the $\alpha = .05$ level of significance? Use the large-sample z-test. What is the p-value of this test?

b. Construct a 95% confidence interval for $(\mu_1 - \mu_2)$. Does this interval confirm your conclusions in part a?

9.28 Noise and Stress In Exercise 8.52, you compared the effect of stress in the form of noise on the ability to perform a simple task. Seventy subjects were divided into two groups; the first group of 30 subjects acted as a control, while the second group of 40 was the experimental group. Although each subject performed the task, the experimental group subjects had to perform the task while loud rock music was played.

The time to finish the task was recorded for each subject and the following summary was obtained:

	Control	Experimental
n	30	40
$\bar{x}$	15 minutes	23 minutes
s	4 minutes	10 minutes

a. Is there sufficient evidence to indicate that the average time to complete the task was longer for the experimental "rock music" group? Test at the 1% level of significance.

b. Construct a 99% one-sided upper bound for the difference (control − experimental) in average times for the two groups. Does this interval confirm your conclusions in part a?

9.29 What's Normal II Of the 130 people in Exercise 9.16, 65 were female and 65 were male.[3] The means and standard deviations of their temperatures are shown below.

	Men	Women
Sample Mean	98.11	98.39
Standard Deviation	0.70	0.74

a. Use the p-value approach to test for a significant difference in the average temperatures for males versus females.

b. Are the results significant at the 5% level? At the 1% level?

A LARGE-SAMPLE TEST OF HYPOTHESIS FOR A BINOMIAL PROPORTION

9.5

When a random sample of n identical trials is drawn from a binomial population, the sample proportion $\hat{p}$ has an approximately normal distribution when n is large, with mean p and standard error

$$SE = \sqrt{\frac{pq}{n}}$$

When you test a hypothesis about p, the proportion in the population possessing a certain attribute, the test follows the same general form as the large-sample tests in Sections 9.3 and 9.4. To test a hypothesis of the form

$$H_0 : p = p_0$$

versus a one- or two-tailed alternative

$$H_a : p > p_0 \quad \text{or} \quad H_a : p < p_0 \quad \text{or} \quad H_a : p \neq p_0$$

the test statistic is constructed using $\hat{p}$, the best estimator of the true population proportion p. The sample proportion $\hat{p}$ is standardized, using the hypothesized mean and standard error, to form a test statistic z, which has a standard normal distribution if H_0 is true. This large-sample test is summarized next.

LARGE-SAMPLE STATISTICAL TEST FOR p

1. Null hypothesis: $H_0 : p = p_0$
2. Alternative hypothesis:

One-Tailed Test	Two-Tailed Test
$H_a : p > p_0$ (or, $H_a : p < p_0$)	$H_a : p \neq p_0$

3. Test statistic: $z = \dfrac{\hat{p} - p_0}{\text{SE}} = \dfrac{\hat{p} - p_0}{\sqrt{\dfrac{p_0 q_0}{n}}}$ with $\hat{p} = \dfrac{x}{n}$

where x is the number of successes in n binomial trials.[†]

4. Rejection region: Reject H_0 when

One-Tailed Test	**Two-Tailed Test**
$z > z_\alpha$	$z > z_{\alpha/2}$ or $z < -z_{\alpha/2}$

(or $z < -z_\alpha$ when the alternative hypothesis is $H_a : p < p_0$)

or when p-value $< \alpha$

 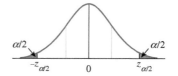

Assumption: The sampling satisfies the assumptions of a binomial experiment (see Section 5.2), and n is large enough so that the sampling distribution of $\hat{p}$ can be approximated by a normal distribution ($np_0 > 5$ and $nq_0 > 5$).

EXAMPLE 9.11

Regardless of age, about 20% of American adults participate in fitness activities at least twice a week. However, these fitness activities change as the people get older, and occasionally participants become nonparticipants as they age. In a local survey of $n = 100$ adults over 40 years old, a total of 15 people indicated that they participated in a fitness activity at least twice a week. Do these data indicate that the participation rate for adults over 40 years of age is significantly less than the 20% figure? Calculate the p-value and use it to draw the appropriate conclusions.

Solution Assuming that the sampling procedure satisfies the requirements of a binomial experiment, you can answer the question posed using a one-tailed test of hypothesis:

$$H_0 : p = .2 \quad \text{versus} \quad H_a : p < .2$$

Begin by assuming that H_0 is true—that is, the true value of p is $p_0 = .2$. Then $\hat{p} = x/n$ will have an approximate normal distribution with mean p_0 and standard error $\sqrt{p_0 q_0/n}$. (NOTE: This is different from the estimation procedure in which the unknown standard error is estimated by $\sqrt{\hat{p}\hat{q}/n}$.) The observed value of $\hat{p}$ is $15/100 = .15$ and the test statistic is

 NEED A TIP?

p-Value $\leq \alpha \Leftrightarrow$ Reject H_0
p-Value $> \alpha \Leftrightarrow$ Do not reject H_0

$$z = \frac{\hat{p} - p_0}{\sqrt{\dfrac{p_0 q_0}{n}}} = \frac{.15 - .20}{\sqrt{\dfrac{(.20)(.80)}{100}}} = -1.25$$

[†] An equivalent test statistic can be found by multiplying the numerator and denominator of z by n to obtain

$$z = \frac{x - np_0}{\sqrt{np_0 q_0}}$$

The p-value associated with this test is found as the area under the standard normal curve to the left of $z = -1.25$ as shown in Figure 9.10. Therefore,

$$p\text{-value} = P(z < -1.25) = .1056$$

FIGURE 9.10

p-value for Example 9.11

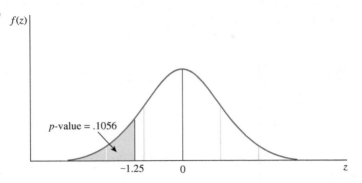

p-value = .1056

$-1.25 \quad 0$

If you use the guidelines for evaluating p-values, then .1056 is greater than .10, and you would not reject H_0. There is insufficient evidence to conclude that the percentage of adults over age 40 who participate in fitness activities twice a week is less than 20%.

Statistical Significance and Practical Importance

It is important to understand the difference between results that are "significant" and results that are practically "important." In statistical language, the word *significant* does not necessarily mean "important," but only that the results could not have occurred by chance. For example, suppose that in Example 9.11, the researcher had used $n = 400$ adults in her experiment and had observed the same sample proportion. The test statistic is now

$$z = \frac{\hat{p} - p_0}{\sqrt{\dfrac{p_0 q_0}{n}}} = \frac{.15 - .20}{\sqrt{\dfrac{(.20)(.80)}{400}}} = -2.50$$

with

$$p\text{-value} = P(z < -2.50) = .0062$$

Now the results are *highly significant: H_0* is rejected, and there is sufficient evidence to indicate that the percentage of adults over age 40 who participate in physical fitness activities is less than 20%. However, is this drop in activity really *important*? Suppose that physicians would be concerned only about a drop in physical activity of more than 10%. If there had been a drop of more than 10% in physical activity, this would imply that the true value of p was less than .10. What is the largest possible value of p? Using a 95% upper one-sided confidence bound, you have

$$\hat{p} + 1.645\sqrt{\frac{\hat{p}\hat{q}}{n}}$$

$$.15 + 1.645\sqrt{\frac{(.15)(.85)}{400}}$$

$$.15 + .029$$

or $p < .179$. The physical activity for adults aged 40 and older has dropped from 20%, but you cannot say that it has dropped below 10%. So, the results, although *statistically significant,* are not *practically important.*

In this book, you will learn how to determine whether results are statistically significant. When you use these procedures in a practical situation, however, you must also make sure the results are practically important.

EXERCISES

BASIC TECHNIQUES

9.30 A random sample of $n = 1000$ observations from a binomial population produced $x = 279$.

a. If your research hypothesis is that p is less than .3, what should you choose for your alternative hypothesis? Your null hypothesis?

b. What is the critical value that determines the rejection region for your test with $\alpha = .05$?

c. Do the data provide sufficient evidence to indicate that p is less than .3? Use a 5% significance level.

9.31 A random sample of $n = 1400$ observations from a binomial population produced $x = 529$.

a. If your research hypothesis is that p differs from .4, what hypotheses should you test?

b. Calculate the test statistic and its p-value. Use the p-value to evaluate the statistical significance of the results at the 1% level.

c. Do the data provide sufficient evidence to indicate that p is different from .4?

9.32 A random sample of 120 observations was selected from a binomial population, and 72 successes were observed. Do the data provide sufficient evidence to indicate that p is greater than .5? Use one of the two methods of testing presented in this section, and explain your conclusions.

APPLICATIONS

9.33 Childhood Obesity According to a survey in *PARADE* magazine, almost half of parents say their children's weight is fine.[9] Only 9% of parents describe their children as overweight. However, the American Obesity Association says the number of overweight children and teens is at least 15%. Suppose that you sample $n = 750$ parents and the number who describe their children as overweight is $x = 68$.

a. How would you test the hypothesis that the proportion of parents who describe their children as overweight is less than the actual proportion reported by the American Obesity Association?

b. What conclusion are you able to draw from these data at the $\alpha = .05$ level of significance?

c. What is the p-value associated with this test?

9.34 Plant Genetics A peony plant with red petals was crossed with another plant having streaky petals. A geneticist states that 75% of the offspring resulting from this cross will have red flowers. To test this claim, 100 seeds from this cross were collected and germinated, and 58 plants had red petals.

a. What hypothesis should you use to test the geneticist's claim?

b. Calculate the test statistic and its p-value. Use the p-value to evaluate the statistical significance of the results at the 1% level.

9.35 Early Detection of Breast Cancer Of those women who are diagnosed to have early-stage breast cancer, one-third eventually die of the disease. Suppose a community public health department instituted a screening program to provide for the early detection of breast cancer and to increase the survival rate p of those diagnosed to have the disease. A random sample of 200 women was selected from among those who were periodically screened by the program and who were diagnosed to have the disease. Let x represent the number of those in the sample who survive the disease.

a. If you wish to determine whether the community screening program has been effective, state the alternative hypothesis that should be tested.

b. State the null hypothesis.

c. If 164 women in the sample of 200 survive the disease, can you conclude that the community screening program was effective? Test using $\alpha = .05$ and explain the practical conclusions from your test.

d. Find the p-value for the test and interpret it.

9.36 Sweet Potato Whitefly Suppose that 10% of the fields in a given agricultural area are infested with the sweet potato whitefly. One hundred fields in this area are randomly selected, and 25 are found to be infested with whitefly.

a. Assuming that the experiment satisfies the conditions of the binomial experiment, do the data indicate that the proportion of infested fields is greater than expected? Use the p-value approach, and test using a 5% significance level.

b. If the proportion of infested fields is found to be significantly greater than .10, why is this of practical significance to the agronomist? What practical conclusions might she draw from the results?

9.37 Taste Testing In a head-to-head taste test of store-brand foods versus national brands, *Consumer Reports* found that it was hard to find a taste difference in the two.[10] If the national brand is indeed better than the store brand, it should be judged as better more than 50% of the time.

a. State the null and alternative hypothesis to be tested. Is this a one- or a two-tailed test?

b. Suppose that, of the 35 food categories used for the taste test, the national brand was found to be better than the store brand in eight of the taste comparisons. Use this information to test the hypothesis in part a. Use $\alpha = .01$. What practical conclusions can you draw from the results?

9.38 Taste Testing, continued In Exercise 9.37, we tried to prove that the national brand tasted better than the store brand.[10] Perhaps, however, the store brand has a better taste than the national brand! If this is true, then the store brand should be judged as better more than 50% of the time.

a. State the null and alternative hypothesis to be tested. Is this a one- or a two-tailed test?

b. Suppose that, of the 35 food categories used for the taste test, the store brand was found to be better than the national brand in six of the taste comparisons. Use this information to test the hypothesis in part a. Use $\alpha = .01$.

c. In the other 21 food comparisons in this experiment, the tasters could find no difference in taste between the store and national brands. What practical conclusions can you draw from this fact and from the two hypothesis tests in Exercises 9.37(b) and 9.38(b)?

9.39 A Cure for Insomnia An experimenter has prepared a drug-dose level that he claims will induce sleep for at least 80% of people suffering from insomnia. After examining the dosage we feel that his claims regarding the effectiveness of his dosage are inflated. In an attempt to disprove his claim, we administer his prescribed dosage to 50 insomniacs and observe that 37 of them have had sleep induced by the drug dose. Is there enough evidence to refute his claim at the 5% level of significance?

9.40 Landlines Passe? According to a new nationwide survey from the Pew Research Center's Social & Demographic Trends project, 62% of Americans consider a landline phone a necessity of life, down from 68% last year, and only 43% of Americans consider a TV a necessity.[11] Both questions were on the second of two forms used in the survey and involved $n = 1483$ individuals contacted by either a landline phone or a cell phone. Use a test of hypothesis to determine whether the $\hat{p} = .62$ figure is a significant drop from last year's value of $p = .68$ with $\alpha = .01$. What conclusions can you draw from this analysis?

9.41 Man's Best Friend The Humane Society reports that there are approximately 77.5 million dogs owned in the United States and that approximately 40% of all U.S. households own at least one dog.[12] In a random sample of 300 households, 114 households said that they owned at least one dog. Does this data provide sufficient evidence to indicate that the proportion of households with at least one dog is different from that reported by the Humane Society? Test using $\alpha = .05$.

A LARGE-SAMPLE TEST OF HYPOTHESIS FOR THE DIFFERENCE BETWEEN TWO BINOMIAL PROPORTIONS

9.6

When random and independent samples are selected from two *binomial* populations, the focus of the experiment may be the difference $(p_1 - p_2)$ in the proportions of individuals or items possessing a specified characteristic in the two populations. In this situation, you can use the difference in the sample proportions $(\hat{p}_1 - \hat{p}_2)$ along with its standard error,

$$SE = \sqrt{\frac{p_1 q_1}{n_1} + \frac{p_2 q_2}{n_2}}$$

in the form of a z-statistic to test for a significant difference in the two population proportions. The null hypothesis to be tested is usually of the form

$$H_0 : p_1 = p_2 \quad \text{or} \quad H_0 : (p_1 - p_2) = 0$$

NEED **a tip? NEED A TIP?**

Remember: Each trial results in one of two outcomes (S or F).

versus either a one- or two-tailed alternative hypothesis. The formal test of hypothesis is summarized in the next display. In estimating the standard error for the z-statistic, you should use the fact that when H_0 is true, the two population proportions are equal to some common value—say, p. To obtain the best estimate of this common value, the sample data are "pooled" and the estimate of p is

$$\hat{p} = \frac{\text{Total number of successes}}{\text{Total number of trials}} = \frac{x_1 + x_2}{n_1 + n_2}$$

Remember that, in order for the difference in the sample proportions to have an approximately normal distribution, the sample sizes must be large and the proportions should not be too close to 0 or 1.

LARGE-SAMPLE STATISTICAL TEST FOR ($p_1 - p_2$)

1. Null hypothesis: $H_0 : (p_1 - p_2) = 0$ or equivalently $H_0 : p_1 = p_2$
2. Alternative hypothesis:

One-Tailed Test	**Two-Tailed Test**
$H_a : (p_1 - p_2) > 0$	$H_a : (p_1 - p_2) \neq 0$
[or $H_a : (p_1 - p_2) < 0$]	

3. Test statistic: $z = \dfrac{(\hat{p}_1 - \hat{p}_2) - 0}{\text{SE}} = \dfrac{\hat{p}_1 - \hat{p}_2}{\sqrt{\dfrac{p_1 q_1}{n_1} + \dfrac{p_2 q_2}{n_2}}} = \dfrac{\hat{p}_1 - \hat{p}_2}{\sqrt{\dfrac{pq}{n_1} + \dfrac{pq}{n_2}}}$

where $\hat{p}_1 = x_1/n_1$ and $\hat{p}_2 = x_2/n_2$. Since the common value of $p_1 = p_2 = p$ (used in the standard error) is unknown, it is estimated by

$$\hat{p} = \frac{x_1 + x_2}{n_1 + n_2}$$

and the test statistic is

$$z = \frac{(\hat{p}_1 - \hat{p}_2) - 0}{\sqrt{\dfrac{\hat{p}\hat{q}}{n_1} + \dfrac{\hat{p}\hat{q}}{n_2}}} \quad \text{or} \quad z = \frac{\hat{p}_1 - \hat{p}_2}{\sqrt{\hat{p}\hat{q}\left(\dfrac{1}{n_1} + \dfrac{1}{n_2}\right)}}$$

4. Rejection region: Reject H_0 when

One-Tailed Test	**Two-Tailed Test**
$z > z_\alpha$	$z > z_{\alpha/2} \quad \text{or} \quad z < -z_{\alpha/2}$
[or $z < -z_\alpha$ when the alternative hypothesis is $H_a : (p_1 - p_2) < 0$]	

or when p-value $< \alpha$

Assumptions: Samples are selected in a random and independent manner from two binomial populations, and n_1 and n_2 are large enough so that the sampling distribution of $(\hat{p}_1 - \hat{p}_2)$ can be approximated by a normal distribution. That is, $n_1\hat{p}_1$, $n_1\hat{q}_1$, $n_2\hat{p}_2$, and $n_2\hat{q}_2$ should all be greater than 5.

EXAMPLE 9.12

The records of a hospital show that 52 men in a sample of 1000 men versus 23 women in a sample of 1000 women were admitted because of heart disease. Do these data present sufficient evidence to indicate a higher rate of heart disease among men admitted to the hospital? Use $\alpha = .05$.

Solution Assume that the number of patients admitted for heart disease has an approximate binomial probability distribution for both men and women with parameters p_1 and p_2, respectively. Then, since you wish to determine whether $p_1 > p_2$, you will test the null hypothesis $p_1 = p_2$—that is, $H_0 : (p_1 - p_2) = 0$—against the alternative hypothesis $H_a : p_1 > p_2$ or, equivalently, $H_a : (p_1 - p_2) > 0$. To conduct this test, use the z-test statistic and approximate the standard error using the pooled estimate of p. Since H_a implies a one-tailed test, you can reject H_0 only for large values of z. Thus, for $\alpha = .05$, you can reject H_0 if $z > 1.645$ (see Figure 9.11).

The pooled estimate of p required for the standard error is

$$\hat{p} = \frac{x_1 + x_2}{n_1 + n_2} = \frac{52 + 23}{1000 + 1000} = .0375$$

FIGURE 9.11

Location of the rejection region in Example 9.12

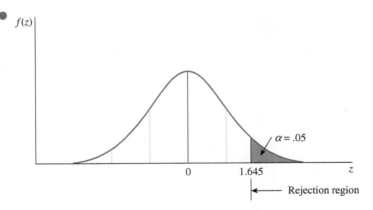

and the test statistic is

$$z = \frac{\hat{p}_1 - \hat{p}_2}{\sqrt{\hat{p}\hat{q}\left(\dfrac{1}{n_1} + \dfrac{1}{n_2}\right)}} = \frac{.052 - .023}{\sqrt{(.0375)(.9625)\left(\dfrac{1}{1000} + \dfrac{1}{1000}\right)}} = 3.41$$

Since the computed value of z falls in the rejection region, you can reject the hypothesis that $p_1 = p_2$. The data present sufficient evidence to indicate that the percentage of men entering the hospital because of heart disease is higher than that of women. (NOTE: This does not imply that the *incidence* of heart disease is higher in men. Perhaps fewer women enter the hospital when afflicted with the disease!)

How *much higher* is the proportion of men than women entering the hospital with heart disease? A 95% lower one-sided confidence bound will help you find the lowest likely value for the difference.

$$(\hat{p}_1 - \hat{p}_2) - 1.645\sqrt{\frac{\hat{p}_1\hat{q}_1}{n_1} + \frac{\hat{p}_2\hat{q}_2}{n_2}}$$

$$(.052 - .023) - 1.645\sqrt{\frac{.052(.948)}{1000} + \frac{.023(.977)}{1000}}$$

$$.029 - .014$$

or $(p_1 - p_2) > .015$. The proportion of men is roughly 1.5% higher than women. Is this of *practical importance*? This is a question for the researcher to answer.

In some situations, you may need to test for a difference D_0 (other than 0) between two binomial proportions. If this is the case, the test statistic is modified for testing $H_0 : (p_1 - p_2) = D_0$, and a pooled estimate for a common p is no longer used in the standard error. The modified test statistic is

$$z = \frac{(\hat{p}_1 - \hat{p}_2) - D_0}{\sqrt{\frac{\hat{p}_1\hat{q}_1}{n_1} + \frac{\hat{p}_2\hat{q}_2}{n_2}}}$$

Although this test statistic is not used often, the procedure is no different from other large-sample tests you have already mastered!

9.6 EXERCISES

BASIC TECHNIQUES

9.42 Independent random samples of $n_1 = 140$ and $n_2 = 140$ observations were randomly selected from binomial populations 1 and 2, respectively. Sample 1 had 74 successes, and sample 2 had 81 successes.

a. Suppose you have no preconceived idea as to which parameter, p_1 or p_2, is the larger, but you want only to detect a difference between the two parameters if one exists. What should you choose as the alternative hypothesis for a statistical test? The null hypothesis?

b. Calculate the standard error of the difference in the two sample proportions, $(\hat{p}_1 - \hat{p}_2)$. Make sure to use the pooled estimate for the common value of p.

c. Calculate the test statistic that you would use for the test in part a. Based on your knowledge of the standard normal distribution, is this a likely or

unlikely observation, assuming that H_0 is true and the two population proportions are the same?

d. *p-value approach:* Find the *p*-value for the test. Test for a significant difference in the population proportions at the 1% significance level.

e. *Critical value approach:* Find the rejection region when $\alpha = .01$. Do the data provide sufficient evidence to indicate a difference in the population proportions?

9.43 Refer to Exercise 9.42. Suppose, for practical reasons, you know that p_1 cannot be larger than p_2.

a. Given this knowledge, what should you choose as the alternative hypothesis for your statistical test? The null hypothesis?

b. Does your alternative hypothesis in part a imply a one- or two-tailed test? Explain.

c. Conduct the test and state your conclusions. Test using $\alpha = .05$.

9.44 Independent random samples of 280 and 350 observations were selected from binomial populations 1 and 2, respectively. Sample 1 had 132 successes, and sample 2 had 178 successes. Do the data present sufficient evidence to indicate that the proportion of successes in population 1 is smaller than the proportion in population 2? Use one of the two methods of testing presented in this section, and explain your conclusions.

APPLICATIONS

9.45 Treatment versus Control An experiment was conducted to test the effect of a new drug on a viral infection. After the infection was induced in 100 mice, the mice were randomly split into two groups of 50. The first group, the *control group,* received no treatment for the infection, and the second group received the drug. After a 30-day period, the proportions of survivors, $\hat{p}_1$ and $\hat{p}_2$, in the two groups were found to be .36 and .60, respectively.

a. Is there sufficient evidence to indicate that the drug is effective in treating the viral infection? Use $\alpha = .05$.

b. Use a 95% confidence interval to estimate the actual difference in the survival rates for the treated versus the control groups.

9.46 Tai Chi and Fibromyalgia A new study (Exercise 7.12) indicates that tai chi, an ancient Chinese practice of exercise and meditation, may relieve symptoms of chronic painful fibromyalgia. The study assigned 66 fibromyalgia patients to take either a 12-week tai chi class ($n_1 = 33$), or attend a wellness education class ($n_2 = 33$).[13] The results of the study are shown in the following table:

	Tai Chi	Wellness Education
Number Who Felt Better at End of Course	26	13

a. Is there a significant difference in the proportion of all fibromyalgia patients who would admit to feeling better after taking the tai chi class compared to the wellness education class? Use $\alpha = .01$.

b. Find the *p*-value for the test in part a. How would you describe the significance or nonsignificance of the test?

9.47 M&M'S In Exercise 8.57, you investigated whether Mars, Inc., uses the same proportion of red M&M'S in its plain and peanut varieties. Random samples of plain and peanut M&M'S provide the following sample data for the experiment:

	Plain	Peanut
Sample Size	56	32
Number of Red M&M'S	12	8

Use a test of hypothesis to determine whether there is a significant difference in the proportions of red candies for the two types of M&M'S. Let $\alpha = .05$ and compare your results with those of Exercise 8.57.

9.48 Hormone Therapy and Alzheimer's Disease A 4-year experiment involving 4532 women, reported in *The Press Enterprise,* was conducted at 39 medical centers to study the benefits and risks of hormone replacement therapy (HRT). Half of the women took placebos and half took Prempro, a widely prescribed type of hormone replacement therapy. There were 40 cases of dementia in the hormone group and 21 in the placebo group.[14] Is there sufficient evidence to indicate that the risk of dementia is higher for patients using Prempro? Test at the 1% level of significance.

9.49 HRT, continued Refer to Exercise 9.48. Calculate a 99% lower one-sided confidence bound for the difference in the risk of dementia for women using hormone replacement therapy versus those who do not. Would this difference be of *practical importance* to a woman considering HRT? Explain.

9.50 Clopidogrel and Aspirin A large study was conducted to test the effectiveness of clopidogrel in combination with aspirin in warding off heart attacks and strokes.[15] The trial involved more than 15,500 people 45 years of age or older from 32 countries, including the United States, who had been diagnosed with cardiovascular disease or had multiple risk factors. The subjects were randomly assigned to one of two groups. After 2 years, there was no difference in the risk of heart attack, stroke, or dying from heart disease between those who took clopidogrel and low-dose aspirin daily and those who took low-dose aspirin plus a dummy pill. The 2-drug combination actually increased the risk of dying (5.4% versus 3.8%) or dying specifically from cardiovascular disease (3.9% versus 2.2%).

a. The subjects were randomly assigned to one of the two groups. Explain how you could use the random number table to make these assignments.

b. No sample sizes were given in the article: however, let us assume that the sample sizes for each group were $n_1 = 7720$ and $n_2 = 7780$. Determine whether the risk of dying was significantly different for the two groups.

c. What do the results of the study mean in terms of *practical significance*?

9.51 Baby's Sleeping Position Does a baby's sleeping position affect the development of motor skills? In one study, published in the *Archives of Pediatric Adolescent Medicine,* 343 full-term infants were examined at their 4-month checkups for various developmental milestones, such as rolling over, grasping a rattle, reaching for an object, and so on.[16] The baby's predominant sleep position—either prone (on the stomach) or supine (on the back) or

side—was determined by a telephone interview with the parent. The sample results for 320 of the 343 infants for whom information was received are shown here:

	Prone	Supine or Side
Number of Infants	121	199
Number That Roll Over	93	119

The researcher reported that infants who slept in the side or supine position were less likely to roll over at the 4-month checkup than infants who slept primarily in the prone position ($P < .001$). Use a large-sample test of hypothesis to confirm or refute the researcher's conclusion.

9.7 SOME COMMENTS ON TESTING HYPOTHESES

A statistical test of hypothesis is a fairly clear-cut procedure that enables an experimenter to either reject or accept the null hypothesis H_0, with measured risks α and β. The experimenter can control the risk of falsely rejecting H_0 by selecting an appropriate value of α. On the other hand, the value of β depends on the sample size and the values of the parameter under test that are of practical importance to the experimenter. When this information is not available, an experimenter may decide to select an affordable sample size, in the hope that the sample will contain sufficient information to reject the null hypothesis. The chance that this decision is in error is given by α, whose value has been set in advance. If the sample does not provide sufficient evidence to reject H_0, the experimenter may wish to state the results of the test as "The data do not support the rejection of H_0" rather than accepting H_0 without knowing the chance of error β.

Some experimenters prefer to use the observed p-value of the test to evaluate the strength of the sample information in deciding to reject H_0. These values can usually be generated by computer and are often used in reports of statistical results:

- If the p-value is greater than .05, the results are reported as NS—not significant at the 5% level.

- If the p-value lies between .05 and .01, the results are reported as $P < .05$—significant at the 5% level.

- If the p-value lies between .01 and .001, the results are reported as $P < .01$—"highly significant" or significant at the 1% level.

- If the p-value is less than .001, the results are reported as $P < .001$—"very highly significant" or significant at the .1% level.

Still other researchers prefer to construct a confidence interval for a parameter and perform a test informally. If the value of the parameter specified by H_0 is included within the upper and lower limits of the confidence interval, then "H_0 is not rejected." If the value of the parameter specified by H_0 is not contained within the interval, then "H_0 is rejected." These results will agree with a two-tailed test; one-sided confidence bounds are used for one-tailed alternatives.

Finally, consider the choice between a one- and two-tailed test. In general, experimenters wish to know whether a treatment causes what could be a beneficial increase in a parameter or what might be a harmful decrease in a parameter. Therefore, most tests are two-tailed unless a one-tailed test is strongly dictated by practical considerations. For example, assume you will sustain a large financial loss if the mean μ is greater than μ_0 but not if it is less. You will then want to detect values larger than μ_0 with a high probability and thereby use a right-tailed test. In the same vein, if pollution levels higher than μ_0 cause critical health risks, then you will certainly wish to detect levels higher than μ_0 with a right-tailed test of hypothesis. In any case, the choice of a one- or two-tailed test should be dictated by the practical consequences that result from a decision to reject or not reject H_0 in favor of the alternative.

CHAPTER REVIEW

Key Concepts and Formulas

I. Parts of a Statistical Test

1. **Null hypothesis:** a contradiction of the alternative hypothesis

2. **Alternative hypothesis:** the hypothesis the researcher wants to support

3. **Test statistic** and its *p*-value: sample evidence calculated from the sample data

4. **Rejection region—critical values** and **significance levels:** values that lead to rejection and nonrejection of the null hypothesis

5. **Conclusion:** Reject or do not reject the null hypothesis, stating the practical significance of your conclusion

II. Errors and Statistical Significance

1. The **significance level** α is the probability of rejecting H_0 when it is in fact true.

2. The ***p*-value** is the probability of observing a test statistic as extreme as or more extreme than the one observed; also, the smallest value of α for which H_0 can be rejected.

3. When the ***p*-value** is less than the **significance level** α, the null hypothesis is rejected. This happens when the **test statistic** exceeds the **critical value.**

4. A **Type II error,** β, is the probability of accepting H_0 when it is in fact false. The **power of the test** is $(1 - \beta)$, the probability of rejecting H_0 when it is false.

III. Large-Sample Test Statistics Using the *z* Distribution

To test one of the four population parameters when the sample sizes are large, use the following test statistics:

Parameter	Test Statistic
μ	$z = \dfrac{\bar{x} - \mu_0}{s/\sqrt{n}}$
p	$z = \dfrac{\hat{p} - p_0}{\sqrt{\dfrac{p_0 q_0}{n}}}$
$\mu_1 - \mu_2$	$z = \dfrac{(\bar{x}_1 - \bar{x}_2) - D_0}{\sqrt{\dfrac{s_1^2}{n_1} + \dfrac{s_2^2}{n_2}}}$
$p_1 - p_2$	$z = \dfrac{\hat{p}_1 - \hat{p}_2}{\sqrt{\hat{p}\hat{q}\left(\dfrac{1}{n_1} + \dfrac{1}{n_2}\right)}}$ or $z = \dfrac{(\hat{p}_1 - \hat{p}_2) - D_0}{\sqrt{\dfrac{\hat{p}_1 \hat{q}_1}{n_1} + \dfrac{\hat{p}_2 \hat{q}_2}{n_2}}}$

Supplementary Exercises

Starred (*) exercises are optional.

9.52 a. Define α and β for a statistical test of hypothesis.

b. For a fixed sample size n, if the value of α is decreased, what is the effect on β?

c. In order to decrease both α and β for a particular alternative value of μ, how must the sample size change?

9.53 What is the p-value for a test of hypothesis? How is it calculated for a large-sample test?

9.54 What conditions must be met so that the z test can be used to test a hypothesis concerning a population mean μ?

9.55 Define the power of a statistical test. As the alternative value of μ gets farther from μ_0, how is the power affected?

9.56 Acidity in Rainfall Refer to Exercise 8.33 and the collection of water samples to estimate the mean acidity (in pH) of rainfalls. As noted, the pH for pure rain falling through clean air is approximately 5.7. The sample of $n = 40$ rainfalls produced pH readings with $\bar{x} = 3.7$ and $s = .5$. Do the data provide sufficient evidence to indicate that the mean pH for rainfalls is more acidic ($H_a : \mu < 5.7$ pH) than pure rainwater? Test using $\alpha = .05$. Note that this inference is appropriate only for the area in which the rainwater specimens were collected.

9.57 Washing Machine Colors A manufacturer of automatic washers provides a particular model in one of three colors. Of the first 1000 washers sold, it is noted that 400 were of the first color. Can you conclude that more than one-third of all customers have a preference for the first color?

a. Find the p-value for the test.

b. If you plan to conduct your test using $\alpha = .05$, what will be your test conclusions?

9.58 Generation Next Born between 1980 and 1990, Generation Next was the topic of Exercise 8.64.[17] In a survey of 500 female and 500 male students in Generation Next, 345 of the females and 365 of the males reported that they decided to attend college in order to make more money.

a. Is there a significant difference in the population proportions of female and male students who decided to attend college in order to make more money? Use $\alpha = .01$.

b. Can you think of any reason why a statistically significant difference in these population proportions might be of *practical importance*? To whom might this difference be important?

9.59 Bass Fishing The pH factor is a measure of the acidity or alkalinity of water. A reading of 7.0 is neutral; values in excess of 7.0 indicate alkalinity; those below 7.0 imply acidity. Loren Hill states that the best chance of catching bass occurs when the pH of the water is in the range 7.5 to 7.9.[18] Suppose you suspect that acid rain is lowering the pH of your favorite fishing spot and you wish to determine whether the pH is less than 7.5.

a. State the alternative and null hypotheses that you would choose for a statistical test.

b. Does the alternative hypothesis in part a imply a one- or a two-tailed test? Explain.

c. Suppose that a random sample of 30 water specimens gave pH readings with $\bar{x} = 7.3$ and $s = .2$. Just glancing at the data, do you think that the difference $\bar{x} - 7.5 = -.2$ is large enough to indicate that the mean pH of the water samples is less than 7.5? (Do *not* conduct the test.)

d. Now conduct a statistical test of the hypotheses in part a and state your conclusions. Test using $\alpha = .05$. Compare your statistically based decision with your intuitive decision in part c.

9.60 Boomers, Xers, and Millennial Men In Exercise 8.34, an *Advertising Age* white paper reported the number of minutes spent doing household chores for women when compared to men. However, there may be a difference among men, depending on the generation to which they belong.[19] The information that follows is adapted from these data and is based on random samples of 1136 men and 795 women.

	Mean	Standard Deviation	n
All Women	72	10.4	795
All Men	54	12.7	1136
Millennial	72	9.2	345
Boomers	54	13.9	475
Xers	54	10.5	316

a. Is there a significant difference in the average number of minutes spent doing household chores for men and women? Use $\alpha = .01$.

b. Is there a significant difference in the average number of minutes spent doing household chores for Millennial men and men classified as Boomers? Use $\alpha = .01$.

c. Is there a significant difference in the average number of minutes spent doing household chores for men classified as Xers and men classified as Boomers? Use $\alpha = .01$.

d. Write a conclusion which explains the practical conclusions which can be drawn from parts a, b, and c.

9.61 White-Tailed Deer In an article entitled "A Strategy for Big Bucks," Charles Dickey discusses studies of the habits of white-tailed deer that indicate that they live and feed within very limited ranges—approximately 150 to 205 acres.[20] To determine whether there was a difference between the ranges of deer located in two different geographic areas, 40 deer were caught, tagged, and fitted with small radio transmitters. Several months later, the deer were tracked and identified, and the distance x from the release point was recorded. The mean and standard deviation of the distances from the release point were as follows:

	Location 1	Location 2
Sample Size	40	40
Sample Mean (ft)	2980	3205
Sample Standard Deviation (ft)	1140	963

a. If you have no preconceived reason for believing one population mean is larger than another, what would you choose for your alternative hypothesis? Your null hypothesis?

b. Does your alternative hypothesis in part a imply a one- or a two-tailed test? Explain.

c. Do the data provide sufficient evidence to indicate that the mean distances differ for the two geographic locations? Test using $\alpha = .05$.

9.62 Female Models In a study to assess various effects of using a female model in automobile advertising, 100 men were shown photographs of two automobiles matched for price, color, and size, but of different makes. One of the automobiles was shown with a female model to 50 of the men (group A), and both automobiles were shown without the model to the other 50 men (group B). In group A, the automobile shown with the model was judged as more expensive by 37 men; in group B, the same automobile was judged as the more expensive by 23 men. Do these results indicate that using a female model influences the perceived cost of an automobile? Use a one-tailed test with $\alpha = .05$.

9.63 Bolts Random samples of 200 bolts manufactured by a type A machine and 200 bolts manufactured by a type B machine showed 16 and 8 defective bolts, respectively. Do these data present sufficient evidence to suggest a difference in the performance of the machine types? Use $\alpha = .05$.

9.64 Biomass Exercise 7.65 reported that the biomass for tropical woodlands, thought to be about 35 kilograms per square meter (kg/m^2), may in fact be too high and that tropical biomass values vary regionally—from about 5 to 55 kg/m^2.[21] Suppose you measure the tropical biomass in 400 randomly selected square-meter plots and obtain $\bar{x} = 31.75$ and $s = 10.5$. Do the data present sufficient evidence to indicate that scientists are overestimating the mean biomass for tropical woodlands and that the mean is in fact lower than estimated?

a. State the null and alternative hypotheses to be tested.

b. Locate the rejection region for the test with $\alpha = .01$.

c. Conduct the test and state your conclusions.

9.65 Adolescents and Social Stress In a study to compare ethnic differences in adolescents' social stress, researchers recruited subjects from three middle schools in Houston, Texas.[22] A tabulation of student responses to a question regarding their socioeconomic status (SES) compared with other families in which the students chose one of five responses (much worse off, somewhat worse off, about the same, better off, or much better off) resulted in the tabulation that follows.

	European American	African American	Hispanic American	Asian American
Sample Size	144	66	77	19
About the Same	68	42	48	8

a. Do these data support the hypothesis that the proportion of adolescent African Americans who state that their SES is "about the same" exceeds that for adolescent Hispanic Americans?

b. Find the p-value for the test.

c. If you plan to test using $\alpha = .05$, what is your conclusion?

9.66* Adolescents and Social Stress, continued Refer to Exercise 9.65. Some thought should have been given to designing a test for which β is tolerably low when p_1 exceeds p_2 by an important amount. For example, find a common sample size n for a test with $\alpha = .05$ and $\beta \leq .20$ when in fact p_1 exceeds p_2 by 0.1. (HINT: The maximum value of $p(1 - p) = .25$.)

9.67 Losing Weight In a comparison of the mean 1-month weight losses for women aged 20–30 years, these sample data were obtained for each of two diets:

	Diet I	Diet II
Sample Size n	40	40
Sample Mean $\bar{x}$(lb)	10	8
Sample Variance s^2	4.3	5.7

Do the data provide sufficient evidence to indicate that diet I produces a greater mean weight loss than diet II? Use $\alpha = .05$.

9.68 Increased Yield An agronomist has shown experimentally that a new irrigation/fertilization regimen produces an increase of 2 bushels per quadrat (significant at the 1% level) when compared with the regimen currently in use. The cost of implementing and using the new regimen will not be a factor if the increase in yield exceeds 3 bushels per quadrat. Is statistical significance the same as practical importance in this situation? Explain.

9.69 Breaking Strengths of Cables A test of the breaking strengths of two different types of cables was conducted using samples of $n_1 = n_2 = 100$ pieces of each type of cable.

Cable I	Cable II
$\bar{x}_1 = 1925$	$\bar{x}_2 = 1905$
$s_1 = 40$	$s_2 = 30$

Do the data provide sufficient evidence to indicate a difference between the mean breaking strengths of the two cables? Use $\alpha = .05$.

9.70 Put on the Brakes The braking ability was compared for two 2012 automobile models. Random samples of 64 automobiles were tested for each type. The recorded measurement was the distance (in feet) required to stop when the brakes were applied at 50 miles per hour. These are the computed sample means and variances:

Model I	Model II
$\bar{x}_1 = 118$	$\bar{x}_2 = 109$
$s_1^2 = 102$	$s_2^2 = 87$

Do the data provide sufficient evidence to indicate a difference between the mean stopping distances for the two models?

9.71 Spraying Fruit Trees A fruit grower wants to test a new spray that a manufacturer claims will *reduce* the loss due to insect damage. To test the claim, the grower sprays 200 trees with the new spray and 200 other trees with the standard spray. The following data were recorded:

	New Spray	Standard Spray
Mean Yield per Tree $\bar{x}$ (lb)	240	227
Variance s^2	980	820

a. Do the data provide sufficient evidence to conclude that the mean yield per tree treated with the new spray exceeds that for trees treated with the standard spray? Use $\alpha = .05$.

b. Construct a 95% confidence interval for the difference between the mean yields for the two sprays.

9.72 Actinomycin D A biologist hypothesizes that high concentrations of actinomycin D inhibit RNA synthesis in cells and hence the production of proteins as well. An experiment conducted to test this theory compared the RNA synthesis in cells treated with two concentrations of actinomycin D: .6 and .7 microgram per milliliter. Cells treated with the lower concentration (.6) of actinomycin D showed that 55 out of 70 developed normally, whereas only 23 out of 70 appeared to develop normally for the higher concentration (.7). Do these data provide sufficient evidence to indicate a difference between the rates of normal RNA synthesis for cells exposed to the two different concentrations of actinomycin D?

a. Find the p-value for the test.

b. If you plan to conduct your test using $\alpha = .05$, what will be your test conclusions?

9.73 SAT Scores How do California high school students compare to students nationwide in their college readiness, as measured by their SAT scores? The national average scores for the class of 2010 were 501 on the critical reading portion, 516 on the math portion, and 492 on the writing portion.[23] Suppose that 100 California students from the class of 2010 were randomly selected and their SAT scores recorded in the following table:

	Critical Reading	Math	Writing
Sample Average	499	514	490
Sample Standard Deviation	98	96	92

a. Do the data provide sufficient evidence to indicate that the average critical reading score for all California students in the class of 2010 is different from the national average? Test using $\alpha = .05$.

b. Do the data provide sufficient evidence to indicate that the average math score for all California students in the class of 2010 is different from the national average? Test using $\alpha = .05$.

c. Could you use this data to determine if there is a difference between the average math and critical reading scores for all California students in the class of 2010? Explain your answer.

9.74 A Maze Experiment In a maze running study, a rat is run in a T maze and the result of each run recorded.

A reward in the form of food is always placed at the right exit. If learning is taking place, the rat will choose the right exit more often than the left. If no learning is taking place, the rat should randomly choose either exit. Suppose that the rat is given $n = 100$ runs in the maze and that he chooses the right exit $x = 64$ times. Would you conclude that learning is taking place? Use the p-value approach, and make a decision based on this p-value.

9.75 PCBs Polychlorinated biphenyls (PCBs) have been found to be dangerously high in some game birds found along the marshlands of the southeastern coast of the United States. The Federal Drug Administration (FDA) considers a concentration of PCBs higher than 5 parts per million (ppm) in these game birds to be dangerous for human consumption. A sample of 38 game birds produced an average of 7.2 ppm with a standard deviation of 6.2 ppm. Is there sufficient evidence to indicate that the mean ppm of PCBs in the population of game birds exceeds the FDA's recommended limit of 5 ppm? Use $\alpha = .01$.

9.76* PCBs, continued Refer to Exercise 9.75.
a. Calculate β and $1 - \beta$ if the true mean ppm of PCBs is 6 ppm.
b. Calculate β and $1 - \beta$ if the true mean ppm of PCBs is 7 ppm.
c. Find the power, $1 - \beta$, when $\mu = 8, 9, 10,$ and 12. Use these values to construct a power curve for the test in Exercise 9.75.
d. For what values of μ does this test have power greater than or equal to .90?

9.77 Goin' to the Chapel? If you choose to marry, what type of wedding site will you pick? A *USA Today* snapshot claims that 43% of all brides-to-be choose a house of worship for their wedding site.[24]

Brides-to-be pick their wedding sites
House of worship 43%
Outdoor location 35%
Club/hall/restaurant 17%
Some other place 4%
Not sure 1%

By Michelle Healey and Veronica Salazar, *USA TODAY*
Source: 'BRIDES' Magazine

In a follow-up study, a random sample of 100 brides-to-be found that 46 of those sampled had chosen or would choose a house of worship for their wedding site. Does this sample contradict the reported 43% figure? Test at the $\alpha = .05$ level of significance.

9.78 Heights and Gender It is a well-accepted fact that males are taller on the average than females. But how much taller? The genders of 105 biomedical students (Exercise 1.54) were also recorded and the data are summarized below:

	Males	Females
Sample Size	48	77
Sample Mean	69.58	64.43
Sample Standard Deviation	2.62	2.58

a. Perform a test of hypothesis to either confirm or refute our initial claim that males are taller on the average than females? Use $\alpha = .01$.
b. If the results of part a show that our claim was correct, construct a 99% confidence one-sided lower confidence bound for the average difference in heights between male and female college students. How much taller are males than females?

9.79 English as a Second Language The state of California monitors the progress of elementary-aged students whose native language is not English using the California English Language Development Test.[25] The test results for two school districts in Riverside County for the 2009–2010 school year follow. The proportion given is the proportion of 6th grade students judged as Advanced or Early-Advanced in English proficiency.

District	Riverside Unified	Palm Springs
Number Tested	602	459
Proportion	.54	.31

Does the data provide sufficient evidence to indicate that there is a difference in the proportion of students who are Advanced or Early-Advanced in English proficiency for these two school districts? Test using $\alpha = .01$.

9.80 Breaststroke Swimmers How much training time does it take to become a world-class breaststroke swimmer? A survey published in *The American Journal of Sports Medicine* reported the number of meters per week swum by two groups of swimmers—those who competed only in breaststroke and those who competed in the individual medley (which includes breaststroke). The number of meters per week practicing the breaststroke swim was recorded and the summary statistics are shown below.[26]

	Breaststroke	Individual Medley
Sample Size	130	80
Sample Mean	9017	5853
Sample Standard Deviation	7162	1961

Is there sufficient evidence to indicate a difference in the average number of meters swum by these two groups of swimmers? Test using $\alpha = .01$.

9.81 Breaststroke, continued Refer to Exercise 9.80.

a. Construct a 99% confidence interval for the difference in the average number of meters swum by breaststroke versus individual medley swimmers.

b. How much longer do pure breaststroke swimmers practice that stroke than individual medley swimmers? What is the practical reason for this difference?

CASE STUDY **An Aspirin a Day . . . ?**

On Wednesday, January 27, 1988, the front page of the *New York Times* read, "Heart attack risk found to be cut by taking aspirin: Lifesaving effects seen." A very large study of U.S. physicians showed that a single aspirin tablet taken every other day reduced by one-half the risk of heart attack in men.[27] Three days later, a headline in the *Times* read, "Value of daily aspirin disputed in British study of heart attacks." How could two seemingly similar studies, both involving doctors as participants, reach such opposite conclusions?

The U.S. physicians' health study consisted of two randomized clinical trials in one. The first tested the hypothesis that 325 milligrams (mg) of aspirin taken every other day reduces mortality from cardiovascular disease. The second tested whether 50 mg of β-carotene taken on alternate days decreases the incidence of cancer. From names on an American Medical Association computer tape, 261,248 male physicians between the ages of 40 and 84 were invited to participate in the trial. Of those who responded, 59,285 were willing to participate. After the exclusion of those physicians who had a history of medical disorders, or who were currently taking aspirin or had negative reactions to aspirin, 22,071 physicians were randomized into one of four treatment groups: (1) buffered aspirin and β-carotene, (2) buffered aspirin and a β-carotene placebo, (3) aspirin placebo and β-carotene, and (4) aspirin placebo and β-carotene placebo. Thus, half were assigned to receive aspirin and half to receive β-carotene.

The study was conducted as a double-blind study, in which neither the participants nor the investigators responsible for following the participants knew to which group a participant belonged. The results of the American study concerning myocardial infarctions (the technical name for heart attacks) are given in the following table:

	American Study	
	Aspirin ($n = 11{,}037$)	Placebo ($n = 11{,}034$)
Myocardial Infarction		
Fatal	5	18
Nonfatal	99	171
Total	104	189

The objective of the British study was to determine whether 500 mg of aspirin taken daily would reduce the incidence of and mortality from cardiovascular disease. In 1978 all male physicians in the United Kingdom were invited to participate. After the usual exclusions, 5139 doctors were randomly allocated to take aspirin, unless some problem developed, and one-third were randomly allocated to *avoid* aspirin. Placebo tablets were not used, so the study was not blind! The results of the British study are given here:

	British Study	
	Aspirin ($n = 3429$)	Control ($n = 1710$)
Myocardial Infarction		
Fatal	89 (47.3)	47 (49.6)
Nonfatal	80 (42.5)	41 (43.3)
Total	169 (89.8)	88 (92.9)

To account for unequal sample sizes, the British study reported rates per 10,000 subject-years alive (given in parentheses).

1. Test whether the American study does in fact indicate that the rate of heart attacks for physicians taking 325 mg of aspirin every other day is significantly different from the rate for those on the placebo. Is the American claim justified?

2. Repeat the analysis using the data from the British study in which one group took 500 mg of aspirin every day and the control group took none. Based on their data, is the British claim justified?

3. Can you think of some possible reasons why the results of these two studies, which were alike in some respects, produced such different conclusions?

Inference from Small Samples

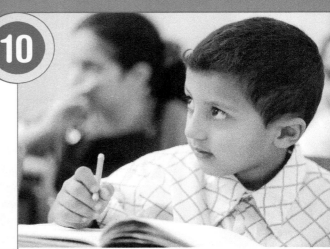

ZouZou/Shutterstock.com

GENERAL OBJECTIVE

The basic concepts of large-sample statistical estimation and hypothesis testing for practical situations involving population means and proportions were introduced in Chapters 8 and 9. Because all of these techniques rely on the Central Limit Theorem to justify the normality of the estimators and test statistics, they apply only when the samples are large. This chapter supplements the large-sample techniques by presenting small-sample tests and confidence intervals for population means and variances. Unlike their large-sample counterparts, these small-sample techniques require the sampled populations to be normal, or approximately so.

CHAPTER INDEX

 NEED TO KNOW...
How to Decide Which Test to Use

School Accountability Study—How is Your School Doing?

Schools are being held responsible for guidelines that have been set forth by states and the federal government in an attempt to quantify students' progress. It has been said that these mandates simply detract from the time that teachers actually teach. In California, one such accountability report is based upon the API, the California state Academic Performance Index. See how the inference methods for small samples can be used to examine this index.

INTRODUCTION

Suppose you need to run an experiment to estimate a population mean or the difference between two means. The process of collecting the data may be very expensive or very time-consuming. If you cannot collect a *large sample,* the estimation and test procedures of Chapters 8 and 9 are of no use to you.

This chapter introduces some equivalent statistical procedures that can be used when the *sample size is small.* The estimation and testing procedures involve these familiar parameters:

- A single population mean, μ
- The difference between two population means, $(\mu_1 - \mu_2)$
- A single population variance, σ^2
- The comparison of two population variances, σ_1^2 and σ_2^2

Small-sample tests and confidence intervals for binomial proportions will be omitted from our discussion.[†]

STUDENT'S *t* DISTRIBUTION

In conducting an experiment to evaluate a new but very costly process for producing synthetic diamonds, you are able to study only six diamonds generated by the process. How can you use these six measurements to make inferences about the average weight μ of diamonds from this process?

In discussing the sampling distribution of $\bar{x}$ in Chapter 7, we made these points:

- When the original sampled population is normal, $\bar{x}$ and $z = (\bar{x} - \mu)/(\sigma/\sqrt{n})$ both have normal distributions, *for any sample size.*
- When the original sampled population is *not* normal, $\bar{x}$, $z = (\bar{x} - \mu)/(\sigma/\sqrt{n})$, and $z \approx (\bar{x} - \mu)/(s/\sqrt{n})$ all have approximately normal distributions, if the sample size is *large.*

 NEED A TIP?

When $n < 30$, the Central Limit Theorem will not guarantee that

$$\frac{\bar{x} - \mu}{s/\sqrt{n}}$$

is approximately normal.

Unfortunately, when the sample size n is small, the statistic $(\bar{x} - \mu)/(s/\sqrt{n})$ *does not* have a normal distribution. Therefore, all the critical values of z that you used in Chapters 8 and 9 are no longer correct. For example, you *cannot say* that $\bar{x}$ will lie within 1.96 standard errors of μ 95% of the time.

This problem is not new; it was studied by statisticians and experimenters in the early 1900s. To find the sampling distribution of this statistic, there are two ways to proceed:

- Use an empirical approach. Draw repeated samples and compute $(\bar{x} - \mu)/(s/\sqrt{n})$ for each sample. The relative frequency distribution that you construct using these values will approximate the shape and location of the sampling distribution.
- Use a mathematical approach to derive the actual density function or curve that describes the sampling distribution.

[†]A small-sample test for the binomial parameter p will be presented in Chapter 15.

This second approach was used by an Englishman named W.S. Gosset in 1908. He derived a complicated formula for the density function of

$$t = \frac{\bar{x} - \mu}{s/\sqrt{n}}$$

for random samples of size n from a normal population, and he published his results under the pen name "Student." Ever since, the statistic has been known as **Student's t.** It has the following characteristics:

- It is mound-shaped and symmetric about $t = 0$, just like z.

- It is more variable than z, with "heavier tails"; that is, the t curve does not approach the horizontal axis as quickly as z does. This is because the t statistic involves two random quantities, $\bar{x}$ and s, whereas the z statistic involves only the sample mean, $\bar{x}$. You can see this phenomenon in Figure 10.1.

- The shape of the t distribution depends on the sample size n. As n increases, the variability of t decreases because the estimate s of σ is based on more and more information. Eventually, when n is infinitely large, the t and z distributions are identical!

FIGURE 10.1

Standard normal z and the t distribution with 5 degrees of freedom

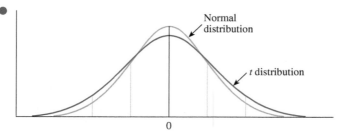

The divisor $(n - 1)$ in the formula for the sample variance s^2 is called the **number of degrees of freedom (df) associated with s^2.** It determines the *shape* of the t distribution. The origin of the term *degrees of freedom* is theoretical and refers to the number of independent squared deviations in s^2 that are available for estimating σ^2. These degrees of freedom may change for different applications and, since they specify the correct t distribution to use, you need to remember to calculate the correct degrees of freedom for each application.

The table of probabilities for the standard normal z distribution is no longer useful in calculating critical values or p-values for the t statistic. Instead, you will use Table 4 in Appendix I, which is partially reproduced in Table 10.1. When you index a particular number of degrees of freedom, the table records t_α, a value of t that has tail area a to its right, as shown in Figure 10.2.

NEED A TIP?

For a one-sample t,
$df = n - 1$.

FIGURE 10.2

Tabulated values of Student's t

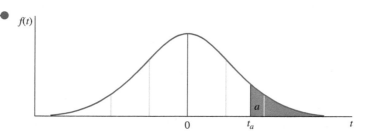

TABLE 10.1 ● **Format of the Student's *t* Table from Table 4 in Appendix I**

df	$t_{.100}$	$t_{.050}$	$t_{.025}$	$t_{.010}$	$t_{.005}$	df
1	3.078	6.314	12.706	31.821	63.657	1
2	1.886	2.920	4.303	6.965	9.925	2
3	1.638	2.353	3.182	4.541	5.841	3
4	1.533	2.132	2.776	3.747	4.604	4
5	1.476	2.015	2.571	3.365	4.032	5
6	1.440	1.943	2.447	3.143	3.707	6
7	1.415	1.895	2.365	2.998	3.499	7
8	1.397	1.860	2.306	2.896	3.355	8
9	1.383	1.833	2.262	2.821	3.250	9
.	.	.	.	.	.	.
.	.	.	.	.	.	.
.	.	.	.	.	.	.
26	1.315	1.706	2.056	2.479	2.779	26
27	1.314	1.703	2.052	2.473	2.771	27
28	1.313	1.701	2.048	2.467	2.763	28
29	1.311	1.699	2.045	2.462	2.756	29
inf.	1.282	1.645	1.960	2.326	2.576	inf.

MY ONLINE APPLET

Student's *t* Probabilities
Comparing *t* and *z*

EXAMPLE **10.1** For a *t* distribution with 5 degrees of freedom, the value of *t* that has area .05 to its right is found in row 5 in the column marked $t_{.050}$. For this particular *t* distribution, the area to the right of $t = 2.015$ is .05; only 5% of all values of the *t* statistic will exceed this value.

EXAMPLE **10.2** Suppose you have a sample of size $n = 10$ from a normal distribution. Find a value of *t* such that only 1% of all values of *t* will be smaller.

Solution The degrees of freedom that specify the correct *t* distribution are $df = n - 1 = 9$, and the necessary *t*-value must be in the lower portion of the distribution, with area .01 to its left, as shown in Figure 10.3. Since the *t* distribution is symmetric about 0, this value is simply the negative of the value on the right-hand side with area .01 to its right, or $-t_{.01} = -2.821$.

FIGURE 10.3 ●

t Distribution for
Example 10.2

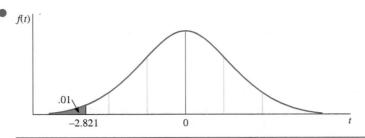

You might wonder why the degrees of freedom (*df*) in Table 10.1 jump from $df = 29$ to $df = inf.$ (infinity). The critical values of *t* for various degrees of freedom between 29 and 300 are given in Figure 10.4. You will notice that the value of *t* for the same right-tail area decreases as the degrees of freedom increase. When the degrees of freedom become infinitely large (*inf.*) the value *t* equals the value of *z* which is given in the last row of Figure 10.4.

FIGURE 10.4

Critical Values of Student's *t* for Degrees of Freedom between *df* = 29 and *df* = *infinity*

	Right-Tail Area		
df	0.05	0.025	0.01
29	1.699	2.045	2.462
49	1.677	2.010	2.405
69	1.667	1.995	2.382
100	1.660	1.984	2.364
200	1.653	1.972	2.345
300	1.650	1.968	2.339
inf.	1.645	1.96	2.326

At the same time, as the degrees of freedom increase, the shape of the *t* distribution becomes less variable until it ultimately looks like (and is) the standard normal distribution. Notice that when the degrees of freedom with *t* are *df* = 300, there is almost no difference. When *df* = 29 and *n* = 30 the critical values of *t* are quite close to their normal counterparts; this may explain why we use the arbitrary dividing line between large and small sample as *n* = 30. Rather than produce a *t*-table with many more critical values, the critical values of *z* are sufficient when *n* reaches 30.

Assumptions behind Student's *t* Distribution

The critical values of *t* allow you to make reliable inferences *only if* you follow all the rules; that is, your sample must meet these requirements specified by the *t* distribution:

NEED A TIP?

Assumptions for one-sample *t*:
• Random sample
• Normal distribution

• The sample must be randomly selected.
• The population from which you are sampling must be normally distributed.

These requirements may seem quite restrictive. How can you possibly know the shape of the probability distribution for the entire population if you have only a sample? If this were a serious problem, however, the *t* statistic could be used in only very limited situations. Fortunately, the shape of the *t* distribution is not affected very much as long as the sampled population has an *approximately mound-shaped* distribution. Statisticians say that the *t* statistic is **robust,** meaning that the distribution of the statistic does not change significantly when the normality assumption is violated.

How can you tell whether your sample is from a normal population? Although there are statistical procedures designed for this purpose, the easiest and quickest way to check for normality is to use the graphical techniques of Chapter 2: Draw a dotplot or construct a stem and leaf plot. As long as your plot tends to "mound up" in the center, you can be fairly safe in using the *t* statistic for making inferences.

The random sampling requirement, on the other hand, is quite critical if you want to produce reliable inferences. If the sample is not random, or if it does not at *least behave as* a random sample, then your sample results may be affected by some unknown factor and your conclusions may be incorrect. When you design an experiment or read about experiments conducted by others, look critically at the way the data have been collected!

SMALL-SAMPLE INFERENCES CONCERNING A POPULATION MEAN

10.3

As with large-sample inference, small-sample inference can involve either **estimation** or **hypothesis testing,** depending on the preference of the experimenter. We explained the basics of these two types of inference in the earlier chapters, and we use them again now, with a different sample statistic, $t = (\bar{x} - \mu)/(s/\sqrt{n})$, and a different sampling distribution, the Student's t, with $(n - 1)$ degrees of freedom.

SMALL-SAMPLE HYPOTHESIS TEST FOR μ

1. Null hypothesis: $H_0 : \mu = \mu_0$
2. Alternative hypothesis:

One-Tailed Test	**Two-Tailed Test**
$H_a : \mu > \mu_0$ (or, $H_a : \mu < \mu_0$)	$H_a : \mu \neq \mu_0$

3. Test statistic: $t = \dfrac{\bar{x} - \mu_0}{s/\sqrt{n}}$

4. Rejection region: Reject H_0 when

One-Tailed Test	**Two-Tailed Test**
$t > t_\alpha$ (or $t < -t_\alpha$ when the alternative hypothesis is $H_a : \mu < \mu_0$)	$t > t_{\alpha/2}$ or $t < -t_{\alpha/2}$

or when p-value $< \alpha$

  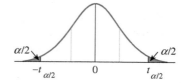

The critical values of t, t_α, and $t_{\alpha/2}$ are based on $(n - 1)$ degrees of freedom. These tabulated values can be found using Table 4 of Appendix I.

Assumption: The sample is randomly selected from a normally distributed population.

SMALL-SAMPLE $(1 - \alpha)100\%$ CONFIDENCE INTERVAL FOR μ

$$\bar{x} \pm t_{\alpha/2}\frac{s}{\sqrt{n}}$$

where $s/\sqrt{n}$ is the estimated standard error of $\bar{x}$, often referred to as the **standard error of the mean.**

EXAMPLE 10.3

A new process for producing synthetic diamonds can be operated at a profitable level only if the average weight of the diamonds is greater than .5 karat. To evaluate the profitability of the process, six diamonds are generated, with recorded weights .46, .61, .52, .48, .57, and .54 karat. Do the six measurements present sufficient evidence to indicate that the average weight of the diamonds produced by the process is in excess of .5 karat?

Solution The population of diamond weights produced by this new process has mean μ, and you can set out the formal test of hypothesis in steps, as you did in Chapter 9:

1–2

Null and alternative hypotheses:

$$H_0: \mu = .5 \quad \text{versus} \quad H_a: \mu > .5$$

3

Test statistic: You can use your calculator to verify that the mean and standard deviation for the six diamond weights are .53 and .0559, respectively. The test statistic is a t statistic, calculated as

$$t = \frac{\bar{x} - \mu_0}{s/\sqrt{n}} = \frac{.53 - .5}{.0559/\sqrt{6}} = 1.32$$

As with the large-sample tests, the test statistic provides evidence for either rejecting or accepting H_0 depending on how far from the center of the t distribution it lies.

4

Rejection region: If you choose a 5% level of significance ($\alpha = .05$), the right-tailed rejection region is found using the critical values of t from Table 4 of Appendix I. With $df = n - 1 = 5$, you can reject H_0 if $t > t_{.05} = 2.015$, as shown in Figure 10.5.

5

Conclusion: Since the calculated value of the test statistic, 1.32, does not fall in the rejection region, you cannot reject H_0. The data do not present sufficient evidence to indicate that the mean diamond weight exceeds .5 karat.

FIGURE 10.5

Rejection region for Example 10.3

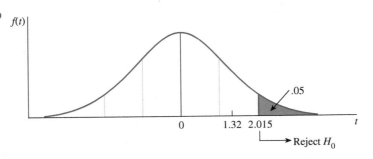

As in Chapter 9, the conclusion to *accept* H_0 would require the difficult calculation of β, the probability of a Type II error. To avoid this problem, we choose to *not reject* H_0. We can then calculate the lower bound for μ using a small-sample lower one-sided confidence bound. This bound is similar to the large-sample one-sided confidence bound, except that the critical z_α is replaced by a critical t_α from Table 4. For this example, a 95% lower one-sided confidence bound for μ is:

$$\bar{x} - t_\alpha \frac{s}{\sqrt{n}}$$

$$.53 - 2.015\frac{.0559}{\sqrt{6}}$$

$$.53 - .046$$

The 95% lower bound for μ is $\mu > .484$. The range of possible values includes mean diamond weights both smaller and greater than .5; this confirms the failure of our test to show that μ exceeds .5.

Remember from Chapter 9 that there are two ways to conduct a test of hypothesis:

- **The critical value approach:** Set up a rejection region based on the critical values of the statistic's sampling distribution. If the test statistic falls in the rejection region, you can reject H_0.
- **The p-value approach:** Calculate the p-value based on the observed value of the test statistic. If the p-value is smaller than the significance level, α, you can reject H_0. If there is no *preset* significance level, use the guidelines in Section 9.3 to judge the statistical significance of your sample results.

We used the first approach in the solution to Example 10.3. We use the second approach to solve Example 10.4.

EXAMPLE 10.4

Labels on 1-gallon cans of paint usually indicate the drying time and the area that can be covered in one coat. Most brands of paint indicate that, in one coat, a gallon will cover between 250 and 500 square feet, depending on the texture of the surface to be painted. One manufacturer, however, claims that a gallon of its paint will cover 400 square feet of surface area. To test this claim, a random sample of ten 1-gallon cans of white paint were used to paint 10 identical areas using the same kind of equipment. The actual areas (in square feet) covered by these 10 gallons of paint are given here:

| 310 | 311 | 412 | 368 | 447 |
| 376 | 303 | 410 | 365 | 350 |

Do the data present sufficient evidence to indicate that the average coverage differs from 400 square feet? Find the p-value for the test, and use it to evaluate the statistical significance of the results.

NEED A TIP?

Remember from Chapter 2 how to calculate $\bar{x}$ and s using the data entry method on your calculator.

Solution To test the claim, the hypotheses to be tested are

$$H_0 : \mu = 400 \quad \text{versus} \quad H_a : \mu \neq 400$$

The sample mean and standard deviation for the recorded data are

$$\bar{x} = 365.2 \qquad s = 48.417$$

and the test statistic is

$$t = \frac{\bar{x} - \mu_0}{s/\sqrt{n}} = \frac{365.2 - 400}{48.417/\sqrt{10}} = -2.27$$

The p-value for this test is the probability of observing a value of the t statistic as contradictory to the null hypothesis as the one observed for this set of data—namely, $t = -2.27$. Since this is a two-tailed test, the p-value is the probability that either $t \leq -2.27$ or $t \geq 2.27$.

Unlike the z-table, the table for t gives the values of t corresponding to upper-tail areas equal to .100, .050, .025, .010, and .005. Consequently, you can only approximate the upper-tail area that corresponds to the probability that $t > 2.27$. Since the t statistic for this test is based on 9 df, we refer to the row corresponding to $df = 9$ in Table 4. The five critical values for various tail areas are shown in Figure 10.6, an enlargement of the tail of the t distribution with 9 degrees of freedom. The value $t = 2.27$ falls between $t_{.025} = 2.262$ and $t_{.010} = 2.821$. Therefore, the right-tail area corresponding to the probability that $t > 2.27$ lies between .01 and .025. Since this area represents only half of the p-value, you can write

$$.01 < \frac{1}{2}(p\text{-value}) < .025 \quad \text{or} \quad .02 < p\text{-value} < .05$$

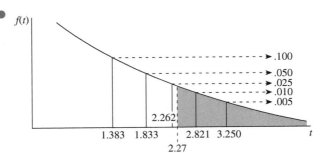

What does this tell you about the significance of the statistical results? For you to reject H_0, the p-value must be less than the specified significance level, α. Hence, you could reject H_0 at the 5% level, but not at the 2% or 1% level. Therefore, the p-value for this test would typically be reported by the experimenter as

$$p\text{-value} < .05 \quad \text{(or sometimes } P < .05\text{)}$$

Small Sample Test of a Population Mean

For this test of hypothesis, H_0 is rejected at the 5% significance level. There is sufficient evidence to indicate that the average coverage differs from 400 square feet.

Within what limits does this average coverage *really* fall? A 95% confidence interval gives the upper and lower limits for μ as

$$\bar{x} \pm t_{\alpha/2}\left(\frac{s}{\sqrt{n}}\right)$$

$$365.2 \pm 2.262\left(\frac{48.417}{\sqrt{10}}\right)$$

$$365.2 \pm 34.63$$

Thus, you can estimate that the average area covered by 1 gallon of this brand of paint lies in the interval 330.6 to 399.8. A more precise interval estimate (a shorter interval) can generally be obtained by increasing the sample size. Notice that the upper limit of this interval is very close to the value of 400 square feet, the coverage claimed on the label. This coincides with the fact that the observed value of $t = -2.27$ is just slightly less than the left-tail critical value of $t_{.025} = -2.262$, making the p-value just slightly less than .05.

Many statistical computing packages contain programs that will implement the Student's t-test or construct a confidence interval for μ. Although *MS Excel* does not have a single command to implement these procedures, you can use the function tool in *Excel* to find the test statistic, the p-value, and/or the upper and lower confidence limits yourself. *MINITAB*, however, calculates and reports all of these values with one set of commands, allowing you to quickly and accurately draw conclusions about the statistical significance of the results. The results of the *MINITAB* one-sample t-test and confidence interval procedures are given in Figure 10.7. Besides the observed value of $t = -2.27$ and the confidence interval (330.6, 399.8), the output gives the sample mean, the sample standard deviation, the standard error of the mean (SE Mean $= s/\sqrt{n}$), and the exact p-value of the test (P = .049). This is consistent with the range for the p-value that we found using Table 4 in Appendix I:

$$.02 < p\text{-value} < .05$$

You will find instructions for generating this *MINITAB* output in the "Technology Today" section at the end of this chapter.

FIGURE 10.7

MINITAB output for Example 10.4

```
One-Sample T: Area
Test of mu = 400  vs  not = 400

Variable      N      Mean       StDev      SE Mean
Area         10     365.2       48.4        15.3

Variable      95% CI                   T         P
Area        (330.6, 399.8)          -2.27      0.049
```

You can see the value of using the computer output to evaluate statistical results:

- The exact p-value eliminates the need for tables and critical values.
- All of the numerical calculations are done for you.

The most important job—which is left for the experimenter—is to *interpret* the results in terms of their practical significance!

10.3 EXERCISES

BASIC TECHNIQUES

10.1 Find the following t-values in Table 4 of Appendix I:

a. $t_{.05}$ for 5 *df*   **b.** $t_{.025}$ for 8 *df*

c. $t_{.10}$ for 18 *df*   **d.** $t_{.025}$ for 30 *df*

10.2 Find the critical value(s) of t that specify the rejection region in these situations:

a. A two-tailed test with $\alpha = .01$ and 12 *df*

b. A right-tailed test with $\alpha = .05$ and 16 *df*

c. A two-tailed test with $\alpha = .05$ and 25 *df*

d. A left-tailed test with $\alpha = .01$ and 7 *df*

10.3 Use Table 4 in Appendix I to approximate the p-value for the t statistic in each situation:

a. A two-tailed test with $t = 2.43$ and 12 *df*

b. A right-tailed test with $t = 3.21$ and 16 *df*

c. A two-tailed test with $t = -1.19$ and 25 *df*

d. A left-tailed test with $t = -8.77$ and 7 *df*

 10.4 Test Scores The test scores on a 100-point test were recorded for 20 students:

71	93	91	86	75
73	86	82	76	57
84	89	67	62	72
77	68	65	75	84

a. Can you reasonably assume that these test scores have been selected from a normal population? Use a stem and leaf plot to justify your answer.

b. Calculate the mean and standard deviation of the scores.

c. If these students can be considered a random sample from the population of all students, find a 95% confidence interval for the average test score in the population.

10.5 The following $n = 10$ observations are a sample from a normal population:

7.4	7.1	6.5	7.5	7.6	6.3	6.9	7.7	6.5	7.0

a. Find the mean and standard deviation of these data.

b. Find a 99% upper one-sided confidence bound for the population mean μ.

c. Test $H_0 : \mu = 7.5$ versus $H_a : \mu < 7.5$. Use $\alpha = .01$.

d. Do the results of part b support your conclusion in part c?

APPLICATIONS

Data set **10.6 Tuna Fish** Is there a difference in the
EX1006 prices of tuna, depending on the method of packaging? *Consumer Reports* gives the estimated average price for a 6-ounce can or a 7.06-ounce pouch of tuna, based on prices paid nationally in supermarkets.[1] These prices are recorded for a variety of different brands of tuna.

Light Tuna in Water		White Tuna in Oil	White Tuna in Water	Light Tuna in Oil	
.99	.53	1.27	1.49	2.56	.62
1.92	1.41	1.22	1.29	1.92	.66
1.23	1.12	1.19	1.27	1.30	62
.85	.63	1.22	1.35	1.79	.65
.65	.67		1.29	1.23	.60
.69	.60		1.00		.67
.60	.66		1.27		
			1.28		

Source: Case Study "Pricing of Tuna" Copyright 2001 by Consumers Union of U.S., Inc., Yonkers, NY 10703-1057, a nonprofit organization. Reprinted with permission from the June 2001 issue of *Consumer Reports®* for educational purposes only. No commercial use or reproduction permitted. www.ConsumerReports.org®.

Assume that the tuna brands included in this survey represent a random sample of all tuna brands available in the United States.

a. Find a 95% confidence interval for the average price for light tuna in water. Interpret this interval. That is, what does the "95%" refer to?

b. Find a 95% confidence interval for the average price for white tuna in oil. How does the width of this interval compare to the width of the interval in part a? Can you explain why?

c. Find 95% confidence intervals for the other two samples (white tuna in water and light tuna in oil). Plot the four treatment means and their standard errors in a two-dimensional plot similar to Figure 8.5. What kind of broad comparisons can you make about the four treatments? (We will discuss the procedure for comparing more than two population means in Chapter 11.)

10.7 Dissolved O_2 Content Industrial wastes and sewage dumped into our rivers and streams absorb oxygen and thereby reduce the amount of dissolved oxygen available for fish and other forms of aquatic life. One state agency requires a minimum of 5 parts per million (ppm) of dissolved oxygen in order for the oxygen content to be sufficient to support aquatic life. Six water specimens taken from a river at a specific location during the low-water season (July) gave readings of 4.9, 5.1, 4.9, 5.0, 5.0, and 4.7 ppm of dissolved oxygen. Do the data provide sufficient evidence to indicate that the dissolved oxygen content is less than 5 ppm? Test using $\alpha = .05$.

10.8 Lobsters In a study of the infestation of the *Thenus orientalis* lobster by two types of barnacles, *Octolasmis tridens* and *O. lowei,* the carapace lengths (in millimeters) of 10 randomly selected lobsters caught in the seas near Singapore are measured:[2]

78	66	65	63	60	60	58	56	52	50

Find a 95% confidence interval for the mean carapace length of the *T. orientalis* lobsters.

Data set **10.9 Smoking and Lung Capacity** In a
EX1009 study of the effect of cigarette smoking on the carbon monoxide diffusing capacity (DL) of the lung, researchers found that current smokers had DL readings significantly lower than those of either exsmokers or nonsmokers. The carbon monoxide diffusing capacities for a random sample of $n = 20$ current smokers are listed here:

103.768	88.602	73.003	123.086	91.052
92.295	61.675	90.677	84.023	76.014
100.615	88.017	71.210	82.115	89.222
102.754	108.579	73.154	106.755	90.479

a. Do these data indicate that the mean DL reading for current smokers is significantly lower than 100 DL, the average for nonsmokers? Use $\alpha = .01$.

b. Find a 99% upper one-sided confidence bound for the mean DL reading for current smokers. Does this bound confirm your conclusions in part a?

10.10 Ben Roethlisberger #7 The number of passes completed by Ben Roethlisberger, quarterback for the Pittsburgh Steelers, was recorded for each of the 12 regular season games in which he played during the fall of 2010 (www.ESPN.com):[3]

```
16   19   17   17   30   18
20   22   21   23   22   15
```

a. A stem and leaf plot of the $n = 12$ observations is shown below:

Stem-and-Leaf Display: Roethlisberger
```
Stem-and-leaf of Roethlisberger   N = 12
Leaf Unit = 1.0
  1    1   5
  4    1   677
  6    1   89
  6    2   01
  4    2   223
  1    2
  1    2
  1    2
  1    3   0
```

Based on this plot, is it reasonable to assume that the underlying population is approximately normal, as required for the one-sample t-test? Explain.

b. Calculate the mean and standard deviation for Ben Roethlisberger's per game pass completions.

c. Construct a 95% confidence interval to estimate the average pass completions per game for Ben Roethlisberger.

10.11 Purifying Organic Compound Organic chemists often purify organic compounds by a method known as fractional crystallization. One chemist prepared ten 4.85-g quantities of aniline and purified it to acetanilide. The following dry yields were recorded:

```
3.85   3.80   3.88   3.85   3.90
3.36   3.62   4.01   3.72   3.83
```

Estimate the mean grams of acetanilide that can be recovered from an initial amount of 4.85 g of aniline. Use a 95% confidence interval.

10.12 Organic Compounds, continued Refer to Exercise 10.11. Approximately how many 4.85-g specimens of aniline are required if you wish to estimate the mean number of grams of acetanilide correct to within .06 g with probability equal to .95?

10.13 Bulimia In a study to determine which factors predict who will benefit from treatment for bulimia nervosa, an article in the *British Journal of Clinical*

Psychology indicates that self-esteem was one of these important predictors.[4] The table gives the mean and standard deviation of self-esteem scores prior to treatment, at posttreatment, and during a follow-up:

	Pretreatment	Posttreatment	Follow-up
Sample Mean $\bar{x}$	20.3	26.6	27.7
Standard Deviation s	5.0	7.4	8.2
Sample Size n	21	21	20

a. Use a test of hypothesis to determine whether there is sufficient evidence to conclude that the true pretreatment mean is less than 25.

b. Construct a 95% confidence interval for the true posttreatment mean.

c. In Section 10.4, we will introduce small-sample techniques for making inferences about the difference between two population means. Without the formality of a statistical test, what are you willing to conclude about the differences among the three sampled population means represented by the results in the table?

10.14 RBC Counts Here are the red blood cell counts (in 10^6 cells per microliter) of a healthy person measured on each of 15 days:

```
5.4   5.2   5.0   5.2   5.5
5.3   5.4   5.2   5.1   5.3
5.3   4.9   5.4   5.2   5.2
```

Find a 95% confidence interval estimate of μ, the true mean red blood cell count for this person during the period of testing.

10.15 Hamburger Meat These data are the weights (in pounds) of 27 packages of ground beef in a supermarket meat display:

```
1.08   .99    .97    1.18   1.41   1.28   .83
1.06   1.14   1.38   .75    .96    1.08   .87
.89    .89    .96    1.12   1.12   .93    1.24
.89    .98    1.14   .92    1.18   1.17
```

a. Interpret the accompanying *MINITAB* printouts for the one-sample test and estimation procedures.

MINITAB output for Exercise 10.15

One-Sample T: Weight
```
Test of mu = 1 vs not = 1

Variable      N         Mean     StDev   SE Mean
Weight        27       1.0522   0.1657    0.0319

Variable          95% CI              T        P
Weight     (0.9867, 1.1178)        1.64    0.113
```

b. Verify the calculated values of t and the upper and lower confidence limits.

10.16 Cholesterol The serum cholesterol levels of 50 subjects randomly selected from the L.A. Heart Data, data from an epidemiological heart disease study on Los Angeles County employees,[5] follow.

148	304	300	240	368	139	203	249	265	229
303	315	174	209	253	169	170	254	212	255
262	284	275	229	261	239	254	222	273	299
278	227	220	260	221	247	178	204	250	256
305	225	306	184	242	282	311	271	276	248

a. Construct a histogram for the data. Are the data approximately mound-shaped?

b. Use a *t*-distribution to construct a 95% confidence interval for the average serum cholesterol levels for L.A. County employees.

10.17 Cholesterol, continued Refer to Exercise 10.16. Since $n > 30$, use the methods of Chapter 8 to create a large-sample 95% confidence interval for the average serum cholesterol level for L.A. County employees. Compare the two intervals. (HINT: The two intervals should be quite similar. This is the reason we choose to approximate the sample distribution of $\dfrac{\bar{x} - \mu}{s/\sqrt{n}}$ with a z-distribution when $n > 30$.)

SMALL-SAMPLE INFERENCES FOR THE DIFFERENCE BETWEEN TWO POPULATION MEANS: INDEPENDENT RANDOM SAMPLES

10.4

The physical setting for the problem considered in this section is the same as the one in Section 8.6, except that the sample sizes are no longer large. Independent random samples of n_1 and n_2 measurements are drawn from two populations, with means and variances μ_1, σ_1^2, μ_2, and σ_2^2, and your objective is to make inferences about $(\mu_1 - \mu_2)$, the difference between the two population means.

When the sample sizes are small, you can no longer rely on the Central Limit Theorem to ensure that the sample means will be normal. If the original populations *are normal*, however, then the sampling distribution of the difference in the sample means, $(\bar{x}_1 - \bar{x}_2)$, will be normal (even for small samples) with mean $(\mu_1 - \mu_2)$ and standard error

$$\sqrt{\frac{\sigma_1^2}{n_1} + \frac{\sigma_2^2}{n_2}}$$

NEED A TIP?

Assumptions for the two-sample (independent) *t*-test:

- Random independent samples
- Normal distributions
- $\sigma_1 = \sigma_2$

In Chapters 7 and 8, you used the sample variances, s_1^2 and s_2^2, to calculate an *estimate* of the standard error, which was then used to form a large-sample confidence interval or a test of hypothesis based on the large-sample z statistic:

$$z \approx \frac{(\bar{x}_1 - \bar{x}_2) - (\mu_1 - \mu_2)}{\sqrt{\dfrac{s_1^2}{n_1} + \dfrac{s_2^2}{n_2}}}$$

Unfortunately, when the sample sizes are small, this statistic does not have an approximately normal distribution—nor does it have a Student's *t* distribution. In order to form a statistic with a sampling distribution that can be derived theoretically, you must make one more assumption.

Suppose that the variability of the measurements in the two normal populations is the same and can be measured by a common variance σ^2. That is, *both populations have exactly the same shape,* and $\sigma_1^2 = \sigma_2^2 = \sigma^2$. Then the standard error of the difference in the two sample means is

$$\sqrt{\frac{\sigma_1^2}{n_1} + \frac{\sigma_2^2}{n_2}} = \sqrt{\sigma^2\left(\frac{1}{n_1} + \frac{1}{n_2}\right)}$$

It can be proven mathematically that, if you use the appropriate sample estimate s^2 for the population variance σ^2, then the resulting test statistic,

$$t = \frac{(\bar{x}_1 - \bar{x}_2) - (\mu_1 - \mu_2)}{\sqrt{s^2\left(\frac{1}{n_1} + \frac{1}{n_2}\right)}}$$

has a *Student's t distribution.* The only remaining problem is to find the sample estimate s^2 and the appropriate number of *degrees of freedom* for the t statistic.

Remember that the population variance σ^2 describes the shape of the normal distributions from which your samples come, so that either s_1^2 or s_2^2 would give you an estimate of σ^2. But why use just one when information is provided by both? A better procedure is to combine the information in both sample variances using a *weighted average,* in which the weights are determined by the relative amount of information (the number of measurements) in each sample. For example, if the first sample contained twice as many measurements as the second, you might consider giving the first sample variance twice as much weight. To achieve this result, use this formula:

$$s^2 = \frac{(n_1 - 1)s_1^2 + (n_2 - 1)s_2^2}{n_1 + n_2 - 2}$$

Remember from Section 10.3 that the degrees of freedom for the one-sample t statistic are $(n - 1)$, the denominator of the sample estimate s^2. Since s_1^2 has $(n_1 - 1)$ *df* and s_2^2 has $(n_2 - 1)$ *df,* the total number of degrees of freedom is the sum

$$(n_1 - 1) + (n_2 - 1) = n_1 + n_2 - 2$$

shown in the denominator of the formula for s^2.

CALCULATION OF s^2

- If you have a scientific calculator, calculate each of the two sample standard deviations s_1 and s_2 separately, using the data entry procedure for your particular calculator. These values are squared and used in this formula:

$$s^2 = \frac{(n_1 - 1)s_1^2 + (n_2 - 1)s_2^2}{n_1 + n_2 - 2}$$

NEED A TIP?

For the two-sample
(independent) t-test,
$df = n_1 + n_2 - 2$

It can be shown that s^2 is an unbiased estimator of the common population variance σ^2. If s^2 is used to estimate σ^2 and if the samples have been randomly and independently drawn from normal populations with a common variance, then the statistic

$$t = \frac{(\bar{x}_1 - \bar{x}_2) - (\mu_1 - \mu_2)}{\sqrt{s^2\left(\frac{1}{n_1} + \frac{1}{n_2}\right)}}$$

has a Student's t distribution with $(n_1 + n_2 - 2)$ degrees of freedom. The small-sample estimation and test procedures for the difference between two means are given next.

TEST OF HYPOTHESIS CONCERNING THE DIFFERENCE BETWEEN TWO MEANS: INDEPENDENT RANDOM SAMPLES

1. Null hypothesis: $H_0 : (\mu_1 - \mu_2) = D_0$, where D_0 is some specified difference that you wish to test. For many tests, you will hypothesize that there is no difference between μ_1 and μ_2; that is, $D_0 = 0$.

2. Alternative hypothesis:

One-Tailed Test	**Two-Tailed Test**
$H_a : (\mu_1 - \mu_2) > D_0$	$H_a : (\mu_1 - \mu_2) \neq D_0$
[or $H_a : (\mu_1 - \mu_2) < D_0$]	

3. Test statistic: $t = \dfrac{(\bar{x}_1 - \bar{x}_2) - D_0}{\sqrt{s^2\left(\dfrac{1}{n_1} + \dfrac{1}{n_2}\right)}}$ where

$$s^2 = \frac{(n_1 - 1)s_1^2 + (n_2 - 1)s_2^2}{n_1 + n_2 - 2}$$

4. Rejection region: Reject H_0 when

One-Tailed Test	**Two-Tailed Test**
$t > t_\alpha$	$t > t_{\alpha/2}$ or $t < -t_{\alpha/2}$
[or $t < -t_\alpha$ when the alternative hypothesis is $H_a : (\mu_1 - \mu_2) < D_0$]	

   or when p-value $< \alpha$

The critical values of t, t_α, and $t_{\alpha/2}$ are based on $(n_1 + n_2 - 2)$ df. The tabulated values can be found using Table 4 of Appendix I.

Assumptions: The samples are randomly and independently selected from normally distributed populations. The variances of the populations σ_1^2 and σ_2^2 are equal.

SMALL-SAMPLE $(1 - \alpha)100\%$ CONFIDENCE INTERVAL FOR $(\mu_1 - \mu_2)$ BASED ON INDEPENDENT RANDOM SAMPLES

$$(\bar{x}_1 - \bar{x}_2) \pm t_{\alpha/2}\sqrt{s^2\left(\frac{1}{n_1} + \frac{1}{n_2}\right)}$$

where s^2 is the pooled estimate of σ^2.

EXAMPLE 10.5 — A course can be taken for credit either by attending lecture sessions at fixed times and days, or by doing online sessions that can be done at the student's own pace and at those times the student chooses. The course coordinator wants to determine if these two ways of taking the course resulted in a significant difference in achievement as measured by the final exam for the course. Table 10.2 gives the scores on an examination with 45 possible points for one group of $n_1 = 9$ students who took the course online, and a second group of $n_2 = 9$ students who took the course with conventional lectures. Do these data present sufficient evidence to indicate that the average grade for students who take the course online is significantly higher than for those who attend a conventional class?

TABLE 10.2 ● **Test Scores for Online and Classroom Presentations**

Online	Classroom
32	35
37	31
35	29
28	25
41	34
44	40
35	27
31	32
34	31

Solution Let μ_1 and μ_2 be the mean scores for the online group and the classroom group, respectively. Then, since you seek evidence to support the theory that $\mu_1 > \mu_2$, you can test the null hypothesis

$$H_0 : \mu_1 = \mu_2 \qquad [\text{or } H_0 : (\mu_1 - \mu_2) = 0]$$

versus the alternative hypothesis

$$H_a : \mu_1 > \mu_2 \qquad [\text{or } H_a : (\mu_1 - \mu_2) > 0]$$

To conduct the t-test for these two independent samples, you must assume that the sampled populations are both normal and have the same variance σ^2. Is this reasonable? Stem and leaf plots of the data in Figure 10.8 show at least a "mounding" pattern, so that the assumption of normality is not unreasonable.

FIGURE 10.8 ●
Stem and leaf plots for
Example 10.5

Online		Classroom	
2	8	2	579
3	124	3	1124
3	557	3	5
4	14	4	0

Furthermore, the standard deviations of the two samples, calculated as

NEED A TIP?

Stem and leaf plots can help you decide if the normality assumption is reasonable.

$$s_1 = 4.9441 \quad \text{and} \quad s_2 = 4.4752$$

are not different enough for us to doubt that the two distributions may have the same shape. If you make these two assumptions and calculate (using full accuracy) the pooled estimate of the common variance as

$$s^2 = \frac{(n_1 - 1)s_1^2 + (n_2 - 1)s_2^2}{n_1 + n_2 - 2} = \frac{8(4.9441)^2 + 8(4.4752)^2}{9 + 9 - 2} = 22.2361$$

you can then calculate the test statistic,

$$t = \frac{\bar{x}_1 - \bar{x}_2}{\sqrt{s^2\left(\dfrac{1}{n_1} + \dfrac{1}{n_2}\right)}} = \frac{35.22 - 31.56}{\sqrt{22.2361\left(\dfrac{1}{9} + \dfrac{1}{9}\right)}} = 1.65$$

NEED A TIP?

If you are using a calculator, don't round off until the final step!

The alternative hypothesis $H_a : \mu_1 > \mu_2$ or, equivalently, $H_a : (\mu_1 - \mu_2) > 0$ implies that you should use a one-tailed test in the upper tail of the t distribution with $(n_1 + n_2 - 2) = 16$ degrees of freedom. You can find the appropriate critical value for a rejection region with $\alpha = .05$ in Table 4 of Appendix I, and H_0 will be rejected if $t > 1.746$. Comparing the observed value of the test statistic $t = 1.65$ with the critical value $t_{.05} = 1.746$, you cannot reject the null hypothesis (see Figure 10.9). There is insufficient evidence to indicate that the average online course grade is higher than the average conventional course grade at the 5% level of significance.

FIGURE 10.9

Rejection region for Example 10.5

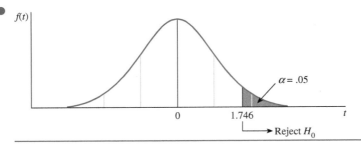

EXAMPLE 10.6

Find the p-value that would be reported for the statistical test in Example 10.5.

Solution The observed value of t for this one-tailed test is $t = 1.65$. Therefore,

$$p\text{-value} = P(t > 1.65)$$

ONLINE APPLET

Two-Sample t-Test: Independent Samples

for a t statistic with 16 degrees of freedom. Remember that you cannot obtain this probability directly from Table 4 in Appendix I; you can only *bound* the p-value using the critical values in the table. Since the observed value, $t = 1.65$, lies between $t_{.100} = 1.337$ and $t_{.050} = 1.746$, the tail area to the right of 1.65 is between .05 and .10. The p-value for this test would be reported as

$$.05 < p\text{-value} < .10$$

Because the p-value is greater than .05, most researchers would report the results as *not significant.*

EXAMPLE 10.7

Use a lower 95% confidence bound to estimate the difference $(\mu_1 - \mu_2)$ in Example 10.5. Does the lower confidence bound indicate that the online average is significantly higher than the classroom average?

Solution The lower confidence bound takes a familiar form—the point estimator $(\bar{x}_1 - \bar{x}_2)$ minus an amount equal to t_α times the standard error of the estimator. Substituting into the formula, you can calculate the 95% lower confidence bound:

$$(\bar{x}_1 - \bar{x}_2) - t_\alpha\sqrt{s^2\left(\dfrac{1}{n_1} + \dfrac{1}{n_2}\right)}$$

$$(35.22 - 31.56) - 1.746 \sqrt{22.2361\left(\frac{1}{9} + \frac{1}{9}\right)}$$

$$3.66 - 3.88$$

or $(\mu_1 - \mu_2) > -.22$. Since the value $(\mu_1 - \mu_2) = 0$ is included in the confidence interval, it is possible that the two means are equal. There is insufficient evidence to indicate that the online average is higher than the classroom average.

NEED A TIP?

Larger s^2/Smaller $s^2 < 3$

⇔

Variance assumption is reasonable

The two-sample procedure that uses a pooled estimate of the common variance σ^2 relies on four important assumptions:

- The samples must be *randomly selected.* Samples not randomly selected may introduce bias into the experiment and thus alter the significance levels you are reporting.
- The samples must be *independent.* If not, this is not the appropriate statistical procedure. We discuss another procedure for dependent samples in Section 10.5.
- The populations from which you sample must be *normal.* However, moderate departures from normality do not seriously affect the distribution of the test statistic, especially if the sample sizes are nearly the same.
- The population *variances should be equal* or nearly equal to ensure that the procedures are valid.

If the population variances are far from equal, there is an alternative procedure for estimation and testing that has an *approximate t* distribution in repeated sampling. As a rule of thumb, you should use this procedure if the ratio of the two sample variances,

$$\frac{\text{Larger } s^2}{\text{Smaller } s^2} > 3$$

Since the population variances are not equal, the pooled estimator s^2 is no longer appropriate, and each population variance must be estimated by its corresponding sample variance. The resulting test statistic is

$$\frac{(\bar{x}_1 - \bar{x}_2) - D_0}{\sqrt{\dfrac{s_1^2}{n_1} + \dfrac{s_2^2}{n_2}}}$$

When the sample sizes are *small,* critical values for this statistic are found using degrees of freedom approximated by the formula

$$df \approx \frac{\left(\dfrac{s_1^2}{n_1} + \dfrac{s_2^2}{n_2}\right)^2}{\dfrac{(s_1^2/n_1)^2}{(n_1 - 1)} + \dfrac{(s_2^2/n_2)^2}{(n_2 - 1)}}$$

The degrees of freedom are taken to be the integer part of this result.

Computer packages such as *MINITAB* and *MS Excel* can be used to implement this procedure, sometimes called *Satterthwaite's approximation,* as well as the *pooled method* described earlier. In fact, some experimenters choose to analyze their data using *both* methods. As long as both analyses lead to the same conclusions, you need not concern yourself with the equality or inequality of variances.

The *MINITAB* and *Excel* outputs resulting from the pooled method of analysis for the data of Example 10.5 are shown in Figure 10.10(a) and (b). Notice that the ratio of the two sample variances, $(4.94/4.48)^2 = 1.22$, is less than 3, which makes the pooled method appropriate. The calculated value of $t = 1.65$ and the exact p-value $= .059$ with 16 degrees of freedom are shown in both of the outputs. The exact p-value makes it quite easy for you to determine the significance or nonsignificance of the sample results. You will find instructions for generating this output in the section "Technology Today" at the end of this chapter.

FIGURE 10.10(a)

MINITAB output for Example 10.5

Two-Sample T-Test and CI: Online, Classroom

```
Two-sample T for Online vs Classroom
              N       Mean     StDev   SE Mean
Online        9      35.22      4.94       1.6
Classroom     9      31.56      4.48       1.5

Difference = mu (Online) - mu (Classroom)
Estimate for difference: 3.67
95% lower bound for difference: -0.21
T-Test of difference = 0 (vs >): T-Value = 1.65  P-Value = 0.059  DF = 16
Both use Pooled StDev = 4.7155
```

FIGURE 10.10(b)

Excel output for Example 10.5

D	E	F
t-Test: Two-Sample Assuming Equal Variances		
	Online	Classroom
Mean	35.222	31.556
Variance	24.444	20.028
Observations	9	9
Pooled Variance	22.236	
Hypothesized Mean Difference	0	
df	16	
t Stat	1.649	
P(T<=t) one-tail	0.059	
t Critical one-tail	1.746	
P(T<=t) two-tail	0.119	
t Critical two-tail	2.120	

If there is reason to believe that the normality assumptions have been violated, you can test for a shift in location of two population distributions using the nonparametric Wilcoxon rank sum test of Chapter 15. This test procedure, which requires fewer assumptions concerning the nature of the population probability distributions, is almost as sensitive in detecting a difference in population means when the conditions necessary for the t-test are satisfied. It may be more sensitive when the normality assumption is not satisfied.

10.4 EXERCISES

BASIC TECHNIQUES

10.18 Give the number of degrees of freedom for s^2, the pooled estimator of σ^2, in these cases:

a. $n_1 = 16, n_2 = 8$

b. $n_1 = 10, n_2 = 12$

c. $n_1 = 15, n_2 = 3$

10.19 Calculate s^2, the pooled estimator for σ^2, in these cases:

a. $n_1 = 10, n_2 = 4, s_1^2 = 3.4, s_2^2 = 4.9$

b. $n_1 = 12, n_2 = 21, s_1^2 = 18, s_2^2 = 23$

10.20 Two independent random samples of sizes $n_1 = 4$ and $n_2 = 5$ are selected from each of two normal populations:

Population 1	12	3	8	5	
Population 2	14	7	7	9	6

a. Calculate s^2, the pooled estimator of σ^2.

b. Find a 90% confidence interval for $(\mu_1 - \mu_2)$, the difference between the two population means.

c. Test $H_0 : (\mu_1 - \mu_2) = 0$ against $H_a : (\mu_1 - \mu_2) < 0$ for $\alpha = .05$. State your conclusions.

10.21 Independent random samples of $n_1 = 16$ and $n_2 = 13$ observations were selected from two normal populations with equal variances:

	Population	
	1	2
Sample Size	16	13
Sample Mean	34.6	32.2
Sample Variance	4.8	5.9

a. Suppose you wish to detect a difference between the population means. State the null and alternative hypotheses for the test.

b. Find the rejection region for the test in part a for $\alpha = .01$.

c. Find the value of the test statistic.

d. Find the approximate p-value for the test.

e. Conduct the test and state your conclusions.

10.22 Refer to Exercise 10.21. Find a 99% confidence interval for $(\mu_1 - \mu_2)$.

10.23 The *MINITAB* printout shows a test for the difference in two population means.

MINITAB output for Exercise 10.23

```
Two-Sample T-Test and CI: Sample 1, Sample 2
Two-sample T for Sample 1 vs Sample 2
            N      Mean     StDev    SE Mean
Sample 1    6     29.00     4.00        1.6
Sample 2    7     28.86     4.67        1.8

Difference = mu (Sample 1) - mu (Sample 2)
Estimate for difference: 0.14
95% CI for difference: (-5.2, 5.5)
T-Test of difference = 0 (vs not =):
T-Value = 0.06  P-Value = 0.95  DF = 11
Both use Pooled StDev = 4.38
```

a. Do the two sample standard deviations indicate that the assumption of a common population variance is reasonable?

b. What is the observed value of the test statistic? What is the p-value associated with this test?

c. What is the pooled estimate s^2 of the population variance?

d. Use the answers to part b to draw conclusions about the difference in the two population means.

e. Find the 95% confidence interval for the difference in the population means. Does this interval confirm your conclusions in part d?

10.24 The *MS Excel* printout shows a test for the difference in two population means.

MS Excel output for Exercise 10.24

D	E	F
t-Test: Two-Sample Assuming Equal Variances		
	Sample 1	Sample 2
Mean	28.667	28.286
Variance	5.067	2.238
Observations	6	7
Pooled Variance	3.524	
Hypothesized Mean Difference	0	
df	11	
t Stat	0.365	
P(T<=t) one-tail	0.361	
t Critical one-tail	1.796	
P(T<=t) two-tail	0.722	
t Critical two-tail	2.201	

a. Do the two sample variances indicate that the assumption of a common population variance is reasonable?

b. What is the observed value of the test statistic? If this is a two-tailed test, what is the p-value associated with the test?

c. What is the pooled estimate s^2 of the population variance?

d. Use the answers to part b to draw conclusions about the difference in the two population means.

e. Use the information in the printout to construct a 95% confidence interval for the difference in the population means. Does this interval confirm your conclusions in part d?

APPLICATIONS

10.25 Healthy Teeth Jan Lindhe conducted a study on the effect of an oral antiplaque rinse on plaque buildup on teeth.[6] Fourteen people whose teeth were thoroughly cleaned and polished were randomly assigned to two groups of seven subjects each. Both groups were assigned to use oral rinses (no brushing) for a 2-week period. Group 1 used a rinse that contained an antiplaque agent. Group 2, the control group, received a similar rinse except that, unknown to the subjects, the rinse contained no antiplaque agent. A plaque index x, a measure of plaque buildup, was recorded at 14 days with means and standard deviations for the two groups shown in the table.

	Control Group	Antiplaque Group
Sample Size	7	7
Mean	1.26	.78
Standard Deviation	.32	.32

a. State the null and alternative hypotheses that should be used to test the effectiveness of the antiplaque oral rinse.

b. Do the data provide sufficient evidence to indicate that the oral antiplaque rinse is effective? Test using $\alpha = .05$.

c. Find the approximate p-value for the test.

Data set **10.26 Tuna, again** In Exercise 10.6 we
EX1026 presented data on the estimated average price for a 6-ounce can or a 7.06-ounce pouch of tuna, based on prices paid nationally in supermarkets. A portion of the data is reproduced in the table below. Use the *MINITAB* printout to answer the questions.

Light Tuna in Water		Light Tuna in Oil	
.99	.53	2.56	.62
1.92	1.41	1.92	.66
1.23	1.12	1.30	.62
.85	.63	1.79	.65
.65	.67	1.23	.60
.69	.60		.67
.60	.66		

MINITAB output for Exercise 10.26

Two-Sample T-Test and CI: Water, Oil

```
Two-sample T for Water vs Oil
          N     Mean    StDev   SE Mean
Water    14    0.896    0.400      0.11
Oil      11    1.147    0.679      0.20

Difference = mu (Water) - mu (Oil)
Estimate for difference: -0.251
95% CI for difference: (-0.700, 0.198)
T-Test of difference = 0 (vs not =):
T-Value = -1.16  P-Value = 0.260  DF = 23
Both use Pooled StDev = 0.5389
```

a. Do the data in the table present sufficient evidence to indicate a difference in the average prices of light tuna in water versus oil? Test using $\alpha = .05$.

b. What is the p-value for the test?

c. The *MINITAB* analysis uses the pooled estimate of σ^2. Is the assumption of equal variances reasonable? Why or why not?

10.27 Runners and Cyclists Chronic anterior compartment syndrome is a condition characterized by exercise-induced pain in the lower leg. Swelling and impaired nerve and muscle function also accompany this pain, which is relieved by rest. Susan Beckham and colleagues conducted an experiment involving 10 healthy runners and 10 healthy cyclists to determine whether there are significant differences in pressure measurements within the anterior muscle compartment for runners and cyclists.[7] The data summary—compartment pressure in millimeters of mercury (Hg)—is as follows:

	Runners		Cyclists	
Condition	Mean	Standard Deviation	Mean	Standard Deviation
Rest	14.5	3.92	11.1	3.98
80% maximal O_2 consumption	12.2	3.49	11.5	4.95
Maximal O_2 consumption	19.1	16.9	12.2	4.47

a. Test for a significant difference in the average compartment pressure between runners and cyclists under the resting condition. Use $\alpha = .05$.

b. Construct a 95% confidence interval estimate of the difference in means for runners and cyclists under the condition of exercising at 80% of maximal oxygen consumption.

c. To test for a significant difference in the average compartment pressures at maximal oxygen consumption, should you use the pooled or unpooled t-test? Explain.

10.28 Disinfectants An experiment published in *The American Biology Teacher* studied the efficacy of using 95% ethanol or 20% bleach as a disinfectant in removing bacterial and fungal contamination when culturing plant tissues. The experiment was repeated 15 times with each disinfectant, using eggplant as the plant tissue being cultured.[8] Five cuttings per plant were placed on a petri dish for each disinfectant and stored at 25°C for 4 weeks. The observation reported was the number of uncontaminated eggplant cuttings after the 4-week storage.

Disinfectant	95% Ethanol	20% Bleach
Mean	3.73	4.80
Variance	2.78095	.17143
n	15	15
Pooled variance 1.47619		

a. Are you willing to assume that the underlying variances are equal?

b. Using the information from part a, are you willing to conclude that there is a significant difference in the mean numbers of uncontaminated eggplants for the two disinfectants tested?

Data set
EX1029

10.29 Titanium A geologist collected 20 different ore samples, all of the same weight, and randomly divided them into two groups. The titanium contents of the samples, found using two different methods, are listed in the table:

Method 1					Method 2				
.011	.013	.013	.015	.014	.011	.016	.013	.012	.015
.013	.010	.013	.011	.012	.012	.017	.013	.014	.015

a. Use an appropriate method to test for a significant difference in the average titanium contents using the two different methods.

b. Determine a 95% confidence interval estimate for $(\mu_1 - \mu_2)$. Does your interval estimate substantiate your conclusion in part a? Explain.

Data set
EX1030

10.30 Raisins The numbers of raisins in each of 14 miniboxes (1/2-ounce size) were counted for a generic brand and for Sunmaid® brand raisins:

Generic Brand				Sunmaid			
25	26	25	28	25	29	24	24
26	28	28	27	28	24	28	22
26	27	24	25	25	28	30	27
26	26			28	24		

a. Although counts cannot have a normal distribution, do these data have approximately normal distributions? (HINT: Use a histogram or stem and leaf plot.)

b. Are you willing to assume that the underlying population variances are equal? Why?

c. Use the p-value approach to determine whether there is a significant difference in the mean numbers of raisins per minibox. What are the implications of your conclusion?

10.31 Dissolved O_2 Content, continued Refer to Exercise 10.7, in which we measured the dissolved oxygen content in river water to determine whether a stream had sufficient oxygen to support aquatic life. A pollution control inspector suspected that a river community was releasing amounts of semitreated sewage into a river. To check his theory, he drew five randomly selected specimens of river water at a location above the town, and another five below. The dissolved oxygen readings (in parts per million) are as follows:

Above Town	4.8	5.2	5.0	4.9	5.1
Below Town	5.0	4.7	4.9	4.8	4.9

a. Do the data provide sufficient evidence to indicate that the mean oxygen content below the town is less than the mean oxygen content above? Test using $\alpha = .05$.

b. Suppose you prefer estimation as a method of inference. Estimate the difference in the mean dissolved oxygen contents for locations above and below the town. Use a 95% confidence interval.

Data set
EX1032

10.32 Freestyle Swimmers In an effort to compare the average swimming times for two swimmers, each swimmer was asked to swim freestyle for a distance of 100 yards at randomly selected times. The swimmers were thoroughly rested between laps and did not race against each other, so that each sample of times was an independent random sample. The times for each of 10 trials are shown for the two swimmers.

Swimmer 1		Swimmer 2	
59.62	59.74	59.81	59.41
59.48	59.43	59.32	59.63
59.65	59.72	59.76	59.50
59.50	59.63	59.64	59.83
60.01	59.68	59.86	59.51

Suppose that swimmer 2 was last year's winner when the two swimmers raced. Does it appear that the average time for swimmer 2 is still faster than the average time for swimmer 1 in the 100-yard freestyle? Find the approximate p-value for the test and interpret the results.

10.33 Freestyle Swimmers, continued Refer to Exercise 10.32. Construct a lower 95% one-sided confidence bound for the difference in the average times for the two swimmers. Does this interval confirm your conclusions in Exercise 10.32?

Data set
EX1034

10.34 Comparing NFL Quarterbacks How does Aaron Rodgers, quarterback for the 2011 Super Bowl Champion Green Bay Packers, compare to Drew Brees, quarterback for the 2010 Super Bowl winners, the New Orleans Saints? The table below shows the number of completed passes for each athlete during the 2010 NFL football season.[9] Use the *Excel* printout to answer the questions that follow.

Aaron Rodgers			Drew Brees		
19	21	7	27	37	25
19	15	25	28	34	29
34	27	19	30	27	35
12	22		33	29	22
27	26		24	23	
18	21		21	24	

E	F	G
t-Test: Two-Sample Assuming Equal Variances		
	Rodgers	Brees
Mean	20.800	28.000
Variance	44.029	23.333
Observations	15.000	16.000
Pooled Variance	33.324	
Hypothesized Mean Difference	0.000	
df	29.000	
t Stat	-3.470	
P(T<=t) one-tail	0.001	
t Critical one-tail	1.699	
P(T<=t) two-tail	0.002	
t Critical two-tail	2.045	

a. The *Excel* analysis uses the pooled estimate of σ^2. Is the assumption of equal variances reasonable? Why or why not?

b. Do the data indicate that there is a difference in the average number of completed passes for the two quarterbacks? Test using $\alpha = .05$.

c. What is the *p*-value for the test?

d. Use the information given in the printout to construct a 95% confidence interval for the difference in the average number of completed passes for the two quarterbacks. Does the confidence interval confirm your conclusion in part b? Explain.

 10.35 An Archeological Find An article in EX1035 *Archaeometry* involved an analysis of 26 samples of Romano-British pottery, found at four different kiln sites in the United Kingdom.[10] The samples were analyzed to determine their chemical composition and the percentage of aluminum oxide in each of 10 samples at two sites is shown below.

Island Thorns	Ashley Rails
18.3	17.7
15.8	18.3
18.0	16.7
18.0	14.8
20.8	19.1

Does the data provide sufficient information to indicate that there is a difference in the average percentage of aluminum oxide at the two sites? Test at the 5% level of significance.

SMALL-SAMPLE INFERENCES FOR THE DIFFERENCE BETWEEN TWO MEANS: A PAIRED-DIFFERENCE TEST

10.5

To compare the wearing qualities of two types of automobile tires, A and B, a tire of type A and one of type B are randomly assigned and mounted on the rear wheels of each of five automobiles. The automobiles are then operated for a specified number of miles, and the amount of wear is recorded for each tire. These measurements appear in Table 10.3. Do the data present sufficient evidence to indicate a difference in the average wear for the two tire types?

TABLE 10.3 ● **Average Wear for Two Types of Tires**

Automobile	Tire A	Tire B
1	10.6	10.2
2	9.8	9.4
3	12.3	11.8
4	9.7	9.1
5	8.8	8.3
	$\bar{x}_1 = 10.24$	$\bar{x}_2 = 9.76$
	$s_1 = 1.316$	$s_2 = 1.328$

Table 10.3 shows a difference of $(\bar{x}_1 - \bar{x}_2) = (10.24 - 9.76) = .48$ between the two sample means, while the standard deviations of both samples are approximately 1.3. Given the variability of the data and the small number of measurements, this is a rather

small difference, and you would probably not suspect a difference in the average wear for the two types of tires. Let's check your suspicions using the methods of Section 10.4.

Look at the *MINITAB* analysis in Figure 10.11. The two-sample *pooled* t-test is used for testing the difference in the means based on two independent random samples. The calculated value of t used to test the null hypothesis $H_0 : \mu_1 = \mu_2$ is $t = .57$ with *p*-value $= .582$, a value that is not nearly small enough to indicate a significant difference in the two population means. The corresponding 95% confidence interval, given as

$$-1.448 < (\mu_1 - \mu_2) < 2.408$$

is quite wide and also does not indicate a significant difference in the population means.

FIGURE 10.11

MINITAB output using t-test for independent samples for the tire data

Two-Sample T-Test and CI: Tire A, Tire B

```
Two-sample T for Tire A vs Tire B
              N       Mean      StDev     SE Mean
Tire A   5       10.24     1.32       0.59
Tire B   5        9.76     1.33       0.59

Difference = mu (Tire A) - mu (Tire B)
Estimate for difference: 0.480
95% CI for difference: (-1.448, 2.408)
T-Test of difference = 0 (vs not =): T-Value = 0.57  P-Value = 0.582  DF = 8
Both use Pooled StDev = 1.3221
```

Take a second look at the data and you will notice that the wear measurement for type A is greater than the corresponding value for type B for *each* of the five automobiles. Wouldn't this be unlikely, if there's really no difference between the two tire types?

Consider a simple intuitive test, based on the binomial distribution of Chapter 5. If there is no difference in the mean tire wear for the two types of tires, then it is just as likely as not that tire A shows more wear than tire B. The five automobiles then correspond to five binomial trials with $p = P$(tire A shows more wear than tire B) $= .5$.

Is the observed value of $x = 5$ positive differences shown in Table 10.4 unusual? The probability of observing $x = 5$ or the equally unlikely value $x = 0$ can be found in Table 1 in Appendix I to be $2(.031) = .062$, which is quite small compared to the likelihood of the more powerful *t*-test, which had a *p*-value of .58. Isn't it peculiar that the *t*-test, which uses more information (the actual sample measurements) than the binomial test, fails to supply sufficient information for rejecting the null hypothesis?

There is an explanation for this inconsistency. The *t*-test described in Section 10.4 is *not* the proper statistical test to be used for our example. The statistical test procedure of Section 10.4 requires that the two samples be *independent and random*. Certainly, the independence requirement is violated by the manner in which the experiment was conducted.

The (pair of) measurements, an A and a B tire, for a particular automobile are definitely related. A glance at the data shows that the readings have approximately the same magnitude for a particular automobile but vary markedly from one automobile to another. This, of course, is exactly what you might expect. Tire wear is largely determined by driver habits, the balance of the wheels, and the road surface. Since each automobile has a different driver, you would expect a large amount of variability in the data from one automobile to another.

In designing the tire wear experiment, the experimenter realized that the measurements would vary greatly from automobile to automobile. If the tires (five of type A and five of type B) were randomly assigned to the 10 wheels, resulting in *independent* random samples, this variability would result in a large standard error and make it difficult to detect a difference in the means. Instead, he chose to "pair" the measurements, comparing the wear for type A and type B tires on each of the five automobiles. This experimental design, sometimes called a **paired-difference** or **matched pairs** design, allows us to eliminate the car-to-car variability by looking at only the five difference measurements shown in Table 10.4. These five differences form a single random sample of size $n = 5$.

TABLE 10.4 ● **Differences in Tire Wear, Using the Data of Table 10.3**

Automobile	A	B	$d = A - B$
1	10.6	10.2	.4
2	9.8	9.4	.4
3	12.3	11.8	.5
4	9.7	9.1	.6
5	8.8	8.3	.5
			$\bar{d} = .48$

Notice that in Table 10.4 the sample mean of the differences, $d = A - B$, is calculated as

$$\bar{d} = \frac{\Sigma d_i}{n} = .48$$

and is exactly the same as the difference of the sample means: $(\bar{x}_1 - \bar{x}_2) = (10.24 - 9.76) = .48$. It should not surprise you that this can be proven to be true in general, and also that the same relationship holds for the population means. That is, the average of the population differences is

$$\mu_d = (\mu_1 - \mu_2)$$

Because of this fact, you can use the sample differences to test for a significant difference in the two population means, $(\mu_1 - \mu_2) = \mu_d$. The test is a single-sample t-test of the difference measurements to test the null hypothesis

$$H_0 : \mu_d = 0 \qquad [\text{or } H_0 : (\mu_1 - \mu_2) = 0]$$

versus the alternative hypothesis

$$H_a : \mu_d \neq 0 \qquad [\text{or } H_a : (\mu_1 - \mu_2) \neq 0]$$

The test procedures take the same form as the procedures used in Section 10.3 and are described next.

PAIRED-DIFFERENCE TEST OF HYPOTHESIS FOR $(\mu_1 - \mu_2) = \mu_d$: DEPENDENT SAMPLES

1. Null hypothesis: $H_0 : \mu_d = 0$
2. Alternative hypothesis:

One-Tailed Test	**Two-Tailed Test**
$H_a : \mu_d > 0$	$H_a : \mu_d \neq 0$
(or $H_a : \mu_d < 0$)	

3. Test statistic: $t = \dfrac{\bar{d} - 0}{s_d / \sqrt{n}} = \dfrac{\bar{d}}{s_d / \sqrt{n}}$

where n = Number of paired differences
$\bar{d}$ = Mean of the sample differences
s_d = Standard deviation of the sample differences

$$= \sqrt{\frac{\Sigma(d_i - \bar{d})^2}{n - 1}} = \sqrt{\frac{\Sigma d_i^2 - \dfrac{(\Sigma d_i)^2}{n}}{n - 1}}$$

4. Rejection region: Reject H_0 when

One-Tailed Test	**Two-Tailed Test**
$t > t_\alpha$	$t > t_{\alpha/2}$ or $t < -t_{\alpha/2}$
(or $t < -t_\alpha$ when the alternative hypothesis is $H_a : \mu_d < 0$)	

or when p-value $< \alpha$

The critical values of t, t_α, and $t_{\alpha/2}$ are based on $(n - 1)$ df. These tabulated values can be found using Table 4 in Appendix I.

(1 − α)100% SMALL-SAMPLE CONFIDENCE INTERVAL FOR $(\mu_1 - \mu_2) = \mu_d$, BASED ON A PAIRED-DIFFERENCE EXPERIMENT

$$\bar{d} \pm t_{\alpha/2}\left(\frac{s_d}{\sqrt{n}}\right)$$

Assumptions: The experiment is designed as a paired-difference test so that the n differences represent a random sample from a normal population.

EXAMPLE 10.8

Do the data in Table 10.3 provide sufficient evidence to indicate a difference in the mean wear for tire types A and B? Test using $\alpha = .05$.

Solution   You can verify using your calculator that the average and standard deviation of the five difference measurements are

$$\bar{d} = .48 \quad \text{and} \quad s_d = \sqrt{\frac{1.18 - \dfrac{(2.4)^2}{5}}{4}} = .0837$$

Then

$$H_0 : \mu_d = 0 \quad \text{and} \quad H_a : \mu_d \neq 0$$

and

$$t = \frac{\bar{d} - 0}{s_d/\sqrt{n}} = \frac{.48}{.0837/\sqrt{5}} = 12.8$$

The critical value of t for a two-tailed statistical test, $\alpha = .05$ and 4 df, is 2.776. Certainly, the observed value of $t = 12.8$ is extremely large and highly significant. Hence, you can conclude that there is a difference in the mean wear for tire types A and B.

EXAMPLE 10.9

Find a 95% confidence interval for $(\mu_1 - \mu_2) = \mu_d$ using the data in Table 10.3.

Solution A 95% confidence interval for the difference between the mean levels of wear is

$$\bar{d} \pm t_{\alpha/2}\left(\frac{s_d}{\sqrt{n}}\right)$$

$$.48 \pm 2.776\left(\frac{.0837}{\sqrt{5}}\right)$$

$$.48 \pm .10$$

NEED A TIP?

Confidence intervals are always interpreted in the same way! In repeated sampling, intervals constructed in this way enclose the true value of the parameter $100(1 - \alpha)$% of the time.

or $.38 < (\mu_1 - \mu_2) < .58$. How does the width of this interval compare with the width of an interval you might have constructed *if* you had designed the experiment in an unpaired manner? It probably would have been of the same magnitude as the interval calculated in Figure 10.11, where the observed data were *incorrectly* analyzed using the unpaired analysis. This interval, $-1.448 < (\mu_1 - \mu_2) < 2.408$, is much wider than the paired interval, which indicates that the paired difference design increased the accuracy of our estimate, and we have gained valuable information by using this design.

The *paired-difference test* or *matched pairs design* used in the tire wear experiment is a simple example of an experimental design called a **randomized block design.** When there is a great deal of variability among the experimental units, even before any experimental procedures are implemented, the effect of this variability can be minimized by **blocking**—that is, comparing the different procedures within groups of relatively similar experimental units called **blocks.** In this way, the "noise" caused by the large variability does not mask the true differences between the procedures. We will discuss randomized block designs in more detail in Chapter 11.

It is important for you to remember that the *pairing* or *blocking* occurs when the experiment is planned, and not after the data are collected. An experimenter may choose to use pairs of identical twins to compare two learning methods. A physician may record a patient's blood pressure before and after a particular medication is given. Once you have used a paired design for an experiment, you no longer have the option of using the unpaired analysis of Section 10.4. The independence assumption has been purposely violated, and your only choice is to use the paired analysis described here!

NEED A TIP?

Paired difference test:
$df = n - 1$

Although pairing was very beneficial in the tire wear experiment, this may not always be the case. In the paired analysis, the degrees of freedom for the t-test are cut in half—from $(n + n - 2) = 2(n - 1)$ to $(n - 1)$. This reduction *increases* the critical value of t for rejecting H_0 and also increases the width of the confidence interval for the difference in the two means. If pairing is not effective, this increase is not offset by a *decrease* in the variability, and you may in fact lose rather than gain information by pairing. This, of course, did not happen in the tire experiment—the large reduction in the standard error more than compensated for the loss in degrees of freedom.

Except for notation, the paired-difference analysis is the same as the single-sample analysis presented in Section 10.3. However, both *MINITAB* and *MS Excel* provide a single procedure (**Paired *t*** in *MINITAB* and ***t*-Test: Paired Two Sample for Means** in *MS Excel*) to analyze the differences. The *MINITAB* output, shown in Figure 10.12, shows the *p*-value for the paired analysis, .000, indicating a *highly significant* difference in the means. You will find instructions for generating both the *MINITAB* and *MS Excel* outputs in the "Technology Today" section at the end of this chapter.

FIGURE 10.12 ●

MINITAB output for paired-difference analysis of tire wear data

Paired T-Test and CI: Tire A, Tire B

Paired T for Tire A - Tire B

	N	Mean	StDev	SE Mean
Tire A	5	10.240	1.316	0.589
Tire B	5	9.760	1.328	0.594
Difference	5	0.4800	0.0837	0.0374

95% CI for mean difference: (0.3761, 0.5839)
T-Test of mean difference = 0 (vs not = 0): T-Value = 12.83 P-Value = 0.000

10.5 EXERCISES

BASIC TECHNIQUES

10.36 A paired-difference experiment was conducted using $n = 10$ pairs of observations.

a. Test the null hypothesis $H_0 : (\mu_1 - \mu_2) = 0$ against $H_a : (\mu_1 - \mu_2) \neq 0$ for $\alpha = .05$, $\bar{d} = .3$, and $s_d^2 = .16$. Give the approximate *p*-value for the test.

b. Find a 95% confidence interval for $(\mu_1 - \mu_2)$.

c. How many pairs of observations do you need if you want to estimate $(\mu_1 - \mu_2)$ correct to within .1 with probability equal to .95?

10.37 A paired-difference experiment consists of $n = 18$ pairs, $\bar{d} = 5.7$, and $s_d^2 = 256$. Suppose you wish to detect $\mu_d > 0$.

a. Give the null and alternative hypotheses for the test.

b. Conduct the test and state your conclusions.

10.38 A paired-difference experiment was conducted to compare the means of two populations:

	Pairs				
Population	1	2	3	4	5
1	1.3	1.6	1.1	1.4	1.7
2	1.2	1.5	1.1	1.2	1.8

a. Do the data provide sufficient evidence to indicate that μ_1 differs from μ_2? Test using $\alpha = .05$.

b. Find the approximate *p*-value for the test and interpret its value.

c. Find a 95% confidence interval for $(\mu_1 - \mu_2)$. Compare your interpretation of the confidence interval with your test results in part a.

d. What assumptions must you make for your inferences to be valid?

APPLICATIONS

 10.39 Auto Insurance In Exercise 2.4, we EX1039 presented the annual 2010 premium for a male, licensed for 6–8 years, who drives a Honda Accord 12,600 to 15,000 miles per year and has no violations or accidents.[11]

City	GEICO ($)	21st Century ($)
Long Beach	2780	2352
Pomona	2411	2462
San Bernardino	2261	2284
Moreno Valley	2263	2520

Source: www.insurance.ca.gov

a. Why would you expect these pairs of observations to be dependent?

b. Do the data provide sufficient evidence to indicate that there is a difference in the average annual premiums between GEICO and 21st Century insurance? Test using $\alpha = .01$.

c. Find the approximate *p*-value for the test and interpret its value.

d. Find a 99% confidence interval for the difference in the average annual premiums for GEICO and 21st Century insurance.

e. Can we use the information in the table to make valid comparisons between GEICO and 21st Century insurance throughout the United States? Why or why not?

10.40 Runners and Cyclists II Refer to Exercise 10.27. In addition to the compartment pressures, the level of creatine phosphokinase (CPK) in blood samples, a measure of muscle damage, was determined for each of 10 runners and 10 cyclists before and after exercise.[7] The data summary—CPK values in units/liter—is as follows:

Condition	Runners Mean	Runners Standard Deviation	Cyclists Mean	Cyclists Standard Deviation
Before Exercise	255.63	115.48	173.8	60.69
After Exercise	284.75	132.64	177.1	64.53
Difference	29.13	21.01	3.3	6.85

a. Test for a significant difference in mean CPK values for runners and cyclists before exercise under the assumption that $\sigma_1^2 \neq \sigma_2^2$; use $\alpha = .05$. Find a 95% confidence interval estimate for the corresponding difference in means.

b. Test for a significant difference in mean CPK values for runners and cyclists after exercise under the assumption that $\sigma_1^2 \neq \sigma_2^2$; use $\alpha = .05$. Find a 95% confidence interval estimate for the corresponding difference in means.

c. Test for a significant difference in mean CPK values for runners before and after exercise.

d. Find a 95% confidence interval estimate for the difference in mean CPK values for cyclists before and after exercise. Does your estimate indicate that there is no significant difference in mean CPK levels for cyclists before and after exercise?

Data set EX1041 **10.41 America's Market Basket** An advertisement for a popular supermarket chain claims that it has had consistently lower prices than one of its competitors. As part of a survey conducted by an independent price-checking company, the average weekly total, based on the prices of approximately 95 items, is given for this chain and for its competitor recorded during four consecutive weeks in a particular month.

Week	Advertiser ($)	Competitor ($)
1	254.26	256.03
2	240.62	255.65
3	231.90	255.12
4	234.13	261.18

a. Is there a significant difference in the average prices for these two different supermarket chains?

b. What is the approximate *p*-value for the test conducted in part a?

c. Construct a 99% confidence interval for the difference in the average prices for the two supermarket chains. Interpret this interval.

Data set EX1042 **10.42 No Left Turn** An experiment was conducted to compare the mean reaction times to two types of traffic signs: prohibitive (No Left Turn) and permissive (Left Turn Only). Ten drivers were included in the experiment. Each driver was presented with 40 traffic signs, 20 prohibitive and 20 permissive, in random order. The mean time to reaction (in milliseconds) was recorded for each driver and is shown here.

Driver	Prohibitive	Permissive
1	824	702
2	866	725
3	841	744
4	770	663
5	829	792
6	764	708
7	857	747
8	831	685
9	846	742
10	759	610

MS Excel printout for Exercise 10.42

D	E	F
t-Test: Paired Two Sample for Means		
	Prohibitive	Permissive
Mean	818.7	711.8
Variance	1573.344444	2596.4
Observations	10	10
Pearson Correlation	0.693852753	
Hypothesized Mean Difference	0	
df	9	
t Stat	9.14983257	
P(T<=t) one-tail	3.72891E-06	
t Critical one-tail	1.833112933	
P(T<=t) two-tail	7.45782E-06	
t Critical two-tail	2.262157163	

a. Explain why this is a paired-difference experiment and give reasons why the pairing should be useful in increasing information on the difference between the mean reaction times to prohibitive and permissive traffic signs.

b. Use the *Excel* printout to determine whether there is a significant difference in mean reaction times to prohibitive and permissive traffic signs. Use the *p*-value approach.

10.43 Healthy Teeth II Exercise 10.25 describes a dental experiment conducted to investigate the effectiveness of an oral rinse used to inhibit the growth of plaque on teeth. Subjects were divided into two groups: One group used a rinse with an antiplaque ingredient, and the control group used a rinse containing inactive ingredients. Suppose that the plaque growth on each person's teeth was measured after using the rinse after 4 hours and then again after 8 hours. If you wish to estimate the difference in plaque growth from 4 to 8 hours, should you use a confidence interval based on a paired or an unpaired analysis? Explain.

10.44 Ground or Air? The earth's temperature can be measured using either ground-based sensors or infrared-sensing devices mounted in aircraft or space satellites. Ground-based sensing is very accurate but tedious, while infrared-sensing appears to introduce a bias into the temperature readings—that is, the average temperature reading may not be equal to the average obtained by ground-based sensing. To determine the bias, readings were obtained at five different locations using both ground- and air-based temperature sensors. The readings (in degrees Celsius) are listed here:

Location	Ground	Air
1	46.9	47.3
2	45.4	48.1
3	36.3	37.9
4	31.0	32.7
5	24.7	26.2

a. Do the data present sufficient evidence to indicate a bias in the air-based temperature readings? Explain.

b. Estimate the difference in mean temperatures between ground- and air-based sensors using a 95% confidence interval.

c. How many paired observations are required to estimate the difference between mean temperatures for ground- versus air-based sensors correct to within .2°C, with probability approximately equal to .95?

10.45 Red Dye To test the comparative brightness of two red dyes, nine samples of cloth were taken from a production line and each sample was divided into two pieces. One of the two pieces in each sample was randomly chosen and red dye 1 applied; red dye 2 was applied to the remaining piece. The following data represent a "brightness score" for each piece. Is there sufficient evidence to indicate a difference in mean brightness scores for the two dyes? Use $\alpha = .05$.

Sample	1	2	3	4	5	6	7	8	9
Dye 1	10	12	9	8	15	12	9	10	15
Dye 2	8	11	10	6	12	13	9	8	13

10.46 Tax Assessors In response to a complaint that a particular tax assessor (A) was biased, an experiment was conducted to compare the assessor named in the complaint with another tax assessor (B) from the same office. Eight properties were selected, and each was assessed by both assessors. The assessments (in thousands of dollars) are shown in the table.

Property	Assessor 1	Assessor 2
1	276.3	275.1
2	288.4	286.8
3	280.2	277.3
4	294.7	290.6
5	268.7	269.1
6	282.8	281.0
7	276.1	275.3
8	279.0	279.1

Use the *MINITAB* printout to answer the questions that follow.

MINITAB output for Exercise 10.46

Paired T-Test and CI: Assessor A, Assessor B
```
Paired T for Assessor A - Assessor B
                 N     Mean    StDev   SE Mean
Assessor A   8   280.78     7.99      2.83
Assessor B   8   279.29     6.85      2.42
Difference   8    1.487     1.491     0.527

95% lower bound for mean difference: 0.489
T-Test of mean difference = 0 (vs > 0):
T-Value = 2.82  P-value = 0.013
```

a. Do the data provide sufficient evidence to indicate that assessor A tends to give higher assessments than assessor B?

b. Estimate the difference in mean assessments for the two assessors.

c. What assumptions must you make in order for the inferences in parts a and b to be valid?

d. Suppose that assessor A had been compared with a more stable standard—say, the average $\bar{x}$ of the assessments given by four assessors selected from the tax office. Thus, each property would be assessed by A and also by each of the four other assessors and $(x_A - \bar{x})$ would be calculated. If the test in part a is valid, can you use the paired-difference t-test to test the hypothesis that the bias, the mean difference between A's assessments and the mean of the assessments of the four assessors, is equal to 0? Explain.

10.47 Memory Experiments A psychology class performed an experiment to compare whether a recall score in which instructions to form images of 25 words were given is better than an initial recall score for which no imagery instructions were given. Twenty students participated in the experiment with the following results:

Student	With Imagery	Without Imagery	Student	With Imagery	Without Imagery
1	20	5	11	17	8
2	24	9	12	20	16
3	20	5	13	20	10
4	18	9	14	16	12
5	22	6	15	24	7
6	19	11	16	22	9
7	20	8	17	25	21
8	19	11	18	21	14
9	17	7	19	19	12
10	21	9	20	23	13

Does it appear that the average recall score is higher when imagery is used?

10.48 Music in the Workplace Before contracting to have stereo music piped into each of his suites of offices, an executive had his office manager randomly select seven offices in which to have the system installed. The average time (in minutes) spent outside these offices per excursion among the employees involved was recorded before and after the music system was installed with the following results.

Office Number	1	2	3	4	5	6	7
No Music	8	9	5	6	5	10	7
Music	5	6	7	5	6	7	8

Would you suggest that the executive proceed with the installation? Conduct an appropriate test of hypothesis. Find the approximate p-value and interpret your results.

INFERENCES CONCERNING A POPULATION VARIANCE

10.6

You have seen in the preceding sections that an estimate of the population variance σ^2 is usually needed before you can make inferences about population means. Sometimes, however, the population variance σ^2 is the primary objective in an experimental investigation. It may be *more* important to the experimenter than the population mean! Consider these examples:

- Scientific measuring instruments must provide unbiased readings with a very small error of measurement. An aircraft altimeter that measures the correct altitude on the *average* is fairly useless if the measurements are in error by as much as 1000 feet above or below the correct altitude.

- Machined parts in a manufacturing process must be produced with minimum variability in order to reduce out-of-size and hence defective parts.

- Aptitude tests must be designed so that scores *will* exhibit a reasonable amount of variability. For example, an 800-point test is not very discriminatory if all students score between 601 and 605.

In previous chapters, you have used

$$s^2 = \frac{\Sigma(x_i - \bar{x})^2}{n - 1}$$

as an unbiased estimator of the population variance σ^2. This means that, in repeated sampling, the average of all your sample estimates will equal the target parameter, σ^2. But how close or far from the target is your estimator s^2 likely to be? To answer this question, we use the sampling distribution of s^2, which describes its behavior in repeated sampling.

Consider the distribution of s^2 based on repeated *random* sampling from a *normal* distribution with a specified mean and variance. We can show theoretically that the distribution begins at $s^2 = 0$ (since the variance cannot be negative) with a mean equal to σ^2. Its shape is *nonsymmetric* and changes with each different sample size and each different value of σ^2. Finding critical values for the sampling distribution of s^2 would be quite difficult and would require separate tables for each population variance. Fortunately, we can simplify the problem by *standardizing,* as we did with the z distribution.

Definition The standardized statistic

$$\chi^2 = \frac{(n-1)s^2}{\sigma^2}$$

is called a **chi-square variable** and has a sampling distribution called the **chi-square probability distribution,** with $n-1$ degrees of freedom.

The equation of the density function for this statistic is quite complicated to look at, but it traces the curve shown in Figure 10.13.

FIGURE 10.13

A chi-square distribution

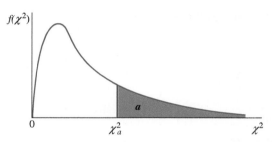

Certain critical values of the chi-square statistic, which are used for making inferences about the population variance, have been tabulated by statisticians and appear in Table 5 of Appendix I. Since the shape of the distribution varies with the sample size n or, more precisely, the degrees of freedom, $n-1$, associated with s^2, Table 5, partially reproduced in Table 10.5, is constructed in exactly the same way as the t table, with the degrees of freedom in the first and last columns. The symbol χ_a^2 indicates that the tabulated χ^2-value has an area a to its right (see Figure 10.13).

TABLE 10.5

Format of the Chi-Square Table from Table 5 in Appendix I

df	$\chi^2_{.995}$	$\cdots$	$\chi^2_{.950}$	$\chi^2_{.900}$	$\chi^2_{.100}$	$\chi^2_{.050}$	$\cdots$	$\chi^2_{.005}$	df
1	.0000393		.0039321	.0157908	2.70554	3.84146		7.87944	1
2	.0100251		.102587	.210720	4.60517	5.99147		10.5966	2
3	.0717212		.351846	.584375	6.25139	7.81473		12.8381	3
4	.206990		.710721	1.063623	7.77944	9.48773		14.8602	4
5	.411740		1.145476	1.610310	9.23635	11.0705		16.7496	5
6	.0675727		1.63539	2.204130	10.6446	12.5916		18.5476	6
.			.	.	.	.		.	.
.			.	.	.	.		.	.
15	4.60094		7.26094	8.54675	22.3072	24.9958		32.8013	15
16	5.14224		7.96164	9.31223	23.5418	26.2962		34.2672	16
17	5.69724		8.67176	10.0852	24.7690	27.5871		35.7185	17
18	6.26481		9.39046	10.8649	25.9894	28.8693		37.1564	18
19	6.84398		10.1170	11.6509	27.2036	30.1435		38.5822	19
.			.	.	.	.		.	.
.			.	.	.	.		.	.
.			.	.	.	.		.	.

NEED A TIP?

Testing one variance: $df = n-1$

ONLINE APPLET

Chi-Square Probabilities

You can see in Table 10.5 that, because the distribution is nonsymmetric and starts at 0, both upper and lower tail areas must be tabulated for the chi-square statistic. For example, the value $\chi^2_{.95}$ is the value that has 95% of the area under the curve to its right and 5% of the area to its left. This value cuts off an area equal to .05 in the lower tail of the chi-square distribution.

EXAMPLE 10.10

Check your ability to use Table 5 in Appendix I by verifying the following statements:

1. The probability that χ^2, based on $n = 16$ measurements ($df = 15$), exceeds 24.9958 is .05.
2. For a sample of $n = 6$ measurements, 95% of the area under the χ^2 distribution lies to the right of 1.145476.

These values are shaded in Table 10.5.

The statistical test of a null hypothesis concerning a population variance

$$H_0 : \sigma^2 = \sigma_0^2$$

uses the test statistic

$$\chi^2 = \frac{(n - 1)s^2}{\sigma_0^2}$$

Notice that when H_0 is true, s^2/σ_0^2 should be near 1, so χ^2 should be close to $(n - 1)$, the degrees of freedom. If σ^2 is really greater than the hypothesized value σ_0^2, the test statistic will tend to be larger than $(n - 1)$ and will probably fall toward the upper tail of the distribution. If $\sigma^2 < \sigma_0^2$, the test statistic will tend to be smaller than $(n - 1)$ and will probably fall toward the lower tail of the chi-square distribution. As in other testing situations, you may use either a one- or a two-tailed statistical test, depending on the alternative hypothesis. This test of hypothesis and the $(1 - \alpha)100\%$ confidence interval for σ^2 are both based on the chi-square distribution and are described next.

TEST OF HYPOTHESIS CONCERNING A POPULATION VARIANCE

1. Null hypothesis: $H_0 : \sigma^2 = \sigma_0^2$
2. Alternative hypothesis:

One-Tailed Test	**Two-Tailed Test**
$H_a : \sigma^2 > \sigma_0^2$ (or $H_a : \sigma^2 < \sigma_0^2$)	$H_a : \sigma^2 \neq \sigma_0^2$

3. Test statistic: $\chi^2 = \dfrac{(n - 1)s^2}{\sigma_0^2}$

4. Rejection region: Reject H_0 when

One-Tailed Test

$\chi^2 > \chi^2_\alpha$
(or $\chi^2 < \chi^2_{(1-\alpha)}$ when the alternative
hypothesis is $H_a : \sigma^2 < \sigma^2_0$), where
χ^2_α and $\chi^2_{(1-\alpha)}$ are, respectively, the
upper- and lower-tail values of χ^2
that place α in the tail areas

or when p-value $< \alpha$

Two-Tailed Test

$\chi^2 > \chi^2_{\alpha/2}$ or $\chi^2 < \chi^2_{(1-\alpha/2)}$,
where $\chi^2_{\alpha/2}$ and $\chi^2_{(1-\alpha/2)}$ are,
respectively, the upper- and
lower-tail values of χ^2 that
place $\alpha/2$ in the tail areas

The critical values of χ^2 are based on $(n - 1)$ df. These tabulated values can be found using Table 5 of Appendix I.

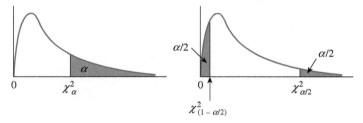

$(1 - \alpha)100\%$ CONFIDENCE INTERVAL FOR σ^2

$$\frac{(n - 1)s^2}{\chi^2_{\alpha/2}} < \sigma^2 < \frac{(n - 1)s^2}{\chi^2_{(1-\alpha/2)}}$$

where $\chi^2_{\alpha/2}$ and $\chi^2_{(1-\alpha/2)}$ are the upper and lower χ^2-values, which locate one-half of α in each tail of the chi-square distribution.
Assumption: The sample is randomly selected from a normal population.

EXAMPLE 10.11

A cement manufacturer claims that concrete prepared from his product has a relatively stable compressive strength and that the strength measured in kilograms per square centimeter (kg/cm^2) lies within a range of 40 kg/cm^2. A sample of $n = 10$ measurements produced a mean and variance equal to, respectively,

$$\bar{x} = 312 \quad \text{and} \quad s^2 = 195$$

Do these data present sufficient evidence to reject the manufacturer's claim?

Solution   In Section 2.5, you learned that the range of a set of measurements should be approximately four standard deviations. The manufacturer's claim that the range of the strength measurements is within 40 kg/cm^2 must mean that the standard deviation of the measurements is roughly 10 kg/cm^2 or less. To test his claim, the appropriate hypotheses are

$$H_0 : \sigma^2 = 10^2 = 100 \quad \text{versus} \quad H_a : \sigma^2 > 100$$

If the sample variance is much larger than the hypothesized value of 100, then the test statistic

$$\chi^2 = \frac{(n - 1)s^2}{\sigma^2_0} = \frac{1755}{100} = 17.55$$

will be unusually large, favoring rejection of H_0 and acceptance of H_a. There are two ways to use the test statistic to make a decision for this test.

- **The critical value approach:** The appropriate test requires a one-tailed rejection region in the right tail of the χ^2 distribution. The critical value for $\alpha = .05$ and $(n - 1) = 9$ df is $\chi^2_{.05} = 16.9190$ from Table 5 in Appendix I. Figure 10.14 shows the rejection region; you can reject H_0 if the test statistic exceeds 16.9190. Since the observed value of the test statistic is $\chi^2 = 17.55$, you can conclude that the null hypothesis is false and that the range of concrete strength measurements exceeds the manufacturer's claim.

FIGURE 10.14

Rejection region and p-value (shaded) for Example 10.11

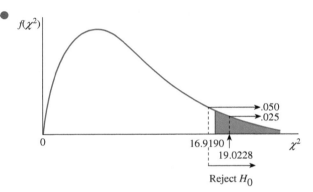

- **The p-value approach:** The p-value for a statistical test is the smallest value of α for which H_0 can be rejected. It is calculated, as in other one-tailed tests, as the area in the tail of the χ^2 distribution to the right of the observed value, $\chi^2 = 17.55$. Although computer packages allow you to calculate this area exactly, Table 5 in Appendix I allows you only to bound the p-value. Since the value 17.55 lies between $\chi^2_{.050} = 16.9190$ and $\chi^2_{.025} = 19.0228$, the p-value lies between .025 and .05. Most researchers would reject H_0 and report these results as significant at the 5% level, or $P < .05$. Again, you can reject H_0 and conclude that the range of measurements exceeds the manufacturer's claim.

EXAMPLE (10.12)

An experimenter is convinced that her measuring instrument had a variability measured by standard deviation $\sigma = 2$. During an experiment, she recorded the measurements 4.1, 5.2, and 10.2. Do these data confirm or disprove her assertion? Test the appropriate hypothesis, and construct a 90% confidence interval to estimate the true value of the population variance.

Solution   Since there is no preset level of significance, you should choose to use the p-value approach in testing these hypotheses:

$$H_0 : \sigma^2 = 4 \quad \text{versus} \quad H_a : \sigma^2 \neq 4$$

Use your scientific calculator to verify that the sample variance is $s^2 = 10.57$ and the test statistic is

$$\chi^2 = \frac{(n - 1)s^2}{\sigma_0^2} = \frac{2(10.57)}{4} = 5.285$$

Since this is a two-tailed test, the rejection region is divided into two parts, half in each tail of the χ^2 distribution. If you approximate the area to the right of the observed test statistic, $\chi^2 = 5.285$, you will have only *half* of the p-value for the test. Since an equally unlikely value of χ^2 might occur in the lower tail of the distribution, with equal probability, you must *double* the upper area to obtain the p-value. With 2 *df*, the observed value, 5.29, falls between $\chi^2_{.10}$ and $\chi^2_{.05}$ so that

$$.05 < \frac{1}{2}(p\text{-value}) < .10 \quad \text{or} \quad .10 < p\text{-value} < .20$$

Since the p-value is greater than .10, the results are not statistically significant. There is insufficient evidence to reject the null hypothesis $H_0 : \sigma^2 = 4$.

The corresponding 90% confidence interval is

$$\frac{(n-1)s^2}{\chi^2_{\alpha/2}} < \sigma^2 < \frac{(n-1)s^2}{\chi^2_{(1-\alpha/2)}}$$

The values of $\chi^2_{(1-\alpha/2)}$ and $\chi^2_{\alpha/2}$ are

$$\chi^2_{(1-\alpha/2)} = \chi^2_{.95} = .102587$$

$$\chi^2_{\alpha/2} = \chi^2_{.05} = 5.99147$$

Substituting these values into the formula for the interval estimate, you get

$$\frac{2(10.57)}{5.99147} < \sigma^2 < \frac{2(10.57)}{.102587} \quad \text{or} \quad 3.53 < \sigma^2 < 206.07$$

Thus, you can estimate the population variance to fall into the interval 3.53 to 206.07. This very wide confidence interval indicates how little information on the population variance is obtained from a sample of only three measurements. Consequently, it is not surprising that there is insufficient evidence to reject the null hypothesis $\sigma^2 = 4$. To obtain more information on σ^2, the experimenter needs to increase the sample size.

Although *MS Excel* does not have a single command to implement these procedures, you can use the function tool in *Excel* to find the test statistic, the p-value and/or the upper and lower confidence limits yourself. If you use *MINITAB*, the command **Stat ▶ Basic Statistics ▶ 1 Variance** allows you to enter either raw data or summary statistics to perform the chi-square test for a single variance, and calculate a confidence interval. The pertinent part of the *MINITAB 16* printout for Example 10.12 is shown in Figure 10.15.

FIGURE 10.15

MINITAB output for Example 10.12

```
Test and CI for One Variance: Measurements
Null hypothesis          Sigma-squared = 4
Alternative hypothesis   Sigma-squared not = 4

Statistics
Variable        N    StDev    Variance
Measurements    3    3.25      10.6

90% Confidence Intervals
CI for          CI for
Variable        Method          StDev        Variance
Measurements    Chi-Square    (1.88, 14.36)  (3.5, 206.1)

Test
Variable        Method        Statistic    DF   P-Value
Measurements    Chi-Square       5.28        2    0.142
```

10.6 EXERCISES

BASIC TECHNIQUES

10.49 A random sample of $n = 25$ observations from a normal population produced a sample variance equal to 21.4. Do these data provide sufficient evidence to indicate that $\sigma^2 > 15$? Test using $\alpha = .05$.

10.50 A random sample of $n = 15$ observations was selected from a normal population. The sample mean and variance were $\bar{x} = 3.91$ and $s^2 = .3214$. Find a 90% confidence interval for the population variance σ^2.

10.51 A random sample of size $n = 7$ from a normal population produced these measurements: 1.4, 3.6, 1.7, 2.0, 3.3, 2.8, 2.9.

a. Calculate the sample variance, s^2.

b. Construct a 95% confidence interval for the population variance, σ^2.

c. Test $H_0 : \sigma^2 = .8$ versus $H_a : \sigma^2 \neq .8$ using $\alpha = .05$. State your conclusions.

d. What is the approximate p-value for the test in part c?

APPLICATIONS

10.52 Instrument Precision A precision instrument is guaranteed to read accurately to within 2 units. A sample of four instrument readings on the same object yielded the measurements 353, 351, 351, and 355.

a. Test the null hypothesis that $\sigma = .7$ against the alternative $\sigma > .7$. Use $\alpha = .05$.

b. Find a 90% confidence interval for the population variance.

10.53 Drug Potency To properly treat patients, drugs prescribed by physicians must not only have a mean potency value as specified on the drug's container, but also the variation in potency values must be small. Otherwise, pharmacists would be distributing drug prescriptions that could be harmfully potent or have a low potency and be ineffective. A drug manufacturer claims that his drug has a potency of $5 \pm .1$ milligram per cubic centimeter (mg/cc). A random sample of four containers gave potency readings equal to 4.94, 5.09, 5.03, and 4.90 mg/cc.

a. Do the data present sufficient evidence to indicate that the mean potency differs from 5 mg/cc?

b. Do the data present sufficient evidence to indicate that the variation in potency differs from the error limits specified by the manufacturer? (HINT: It is sometimes difficult to determine exactly what is meant by limits on potency as specified by a manufacturer. Since he implies that the potency values will fall into the interval $5 \pm .1$ mg/cc with very high probability—the implication is almost *always*—let us assume that the range .2; or 4.9 to 5.1, represents 6σ, as suggested by the Empirical Rule).

10.54 Drug Potency, continued Refer to Exercise 10.53. Testing of 60 additional randomly selected containers of the drug gave a sample mean and variance equal to 5.04 and .0063 (for the total of $n = 64$ containers). Using a 95% confidence interval, estimate the variance of the manufacturer's potency measurements.

10.55 Hard Hats A manufacturer of hard safety hats for construction workers is concerned about the mean and the variation of the forces helmets transmit to wearers when subjected to a standard external force. The manufacturer desires the mean force transmitted by helmets to be 800 pounds (or less), well under the legal 1000-pound limit, and σ to be less than 40. A random sample of $n = 40$ helmets was tested, and the sample mean and variance were found to be equal to 825 pounds and 2350 pounds^2, respectively.

a. If $\mu = 800$ and $\sigma = 40$, is it likely that any helmet, subjected to the standard external force, will transmit a force to a wearer in excess of 1000 pounds? Explain.

b. Do the data provide sufficient evidence to indicate that when the helmets are subjected to the standard external force, the mean force transmitted by the helmets exceeds 800 pounds?

10.56 Hard Hats, continued Refer to Exercise 10.55. Do the data provide sufficient evidence to indicate that σ exceeds 40?

10.57 Light Bulbs A manufacturer of industrial light bulbs likes its bulbs to have a mean length of life that is acceptable to its customers and a variation in length of life that is relatively small. A sample of 20 bulbs tested produced the following lengths of life (in hours):

2100	2302	1951	2067	2415	1883	2101	2146	2278	2019
1924	2183	2077	2392	2286	2501	1946	2161	2253	1827

The manufacturer wishes to control the variability in length of life so that σ is less than 150 hours. Do the data provide sufficient evidence to indicate that the manufacturer is achieving this goal? Test using $\alpha = .01$.

COMPARING TWO POPULATION VARIANCES

10.7

Just as a single population variance is sometimes important to an experimenter, you might also need to compare two population variances. You might need to compare the precision of one measuring device with that of another, the stability of one manufacturing process with that of another, or even the variability in the grading procedure of one college professor with that of another.

One way to compare two population variances, σ_1^2 and σ_2^2, is to use the ratio of the sample variances, s_1^2/s_2^2. If s_1^2/s_2^2 is nearly equal to 1, you will find little evidence to indicate that σ_1^2 and σ_2^2 are unequal. On the other hand, a very large or very small value for s_1^2/s_2^2 provides evidence of a difference in the population variances.

How large or small must s_1^2/s_2^2 be for sufficient evidence to exist to reject the following null hypothesis?

$$H_0 : \sigma_1^2 = \sigma_2^2$$

The answer to this question may be found by studying the distribution of s_1^2/s_2^2 in repeated sampling.

When independent random samples are drawn from two *normal* populations with *equal variances*—that is, $\sigma_1^2 = \sigma_2^2$—then s_1^2/s_2^2 has a probability distribution in repeated sampling that is known to statisticians as an **F distribution.** The equation of the density function for this statistic is quite complicated to look at, but it traces the curve shown in Figure 10.16.

FIGURE 10.16

An *F* distribution with $df_1 = 10$ and $df_2 = 10$

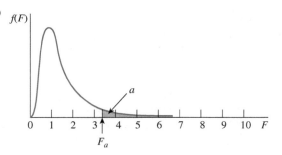

ASSUMPTIONS FOR s_1^2/s_2^2 TO HAVE AN *F* DISTRIBUTION

- Random and independent samples are drawn from each of two normal populations.
- The variability of the measurements in the two populations is the same and can be measured by a common variance, σ^2; that is, $\sigma_1^2 = \sigma_2^2 = \sigma^2$.

ONLINE APPLET

F Probabilities

It is not important for you to know the complex equation of the density function for *F*. For your purposes, you need only to use the well-tabulated critical values of *F* given in Table 6 in Appendix I.

Like the χ^2 distribution, the shape of the *F* distribution is nonsymmetric and depends on the number of degrees of freedom associated with s_1^2 and s_2^2, represented as

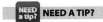

NEED A TIP?

Testing two variances:
$df_1 = n_1 - 1$ and
$df_2 = n_2 - 1$

$df_1 = (n_1 - 1)$ and $df_2 = (n_2 - 1)$, respectively. This complicates the tabulation of critical values of the F distribution because a table is needed for each different combination of df_1, df_2, and a.

In Table 6 in Appendix I, critical values of F for right-tailed areas corresponding to $a = .100, .050, .025, .010,$ and $.005$ are tabulated for various combinations of df_1 numerator degrees of freedom and df_2 denominator degrees of freedom. A portion of Table 6 is reproduced in Table 10.6. The numerator degrees of freedom df_1 are listed across the top margin, and the denominator degrees of freedom df_2 are listed along the side margin. The values of a are listed in the second column. For a fixed combination of df_1 and df_2, the appropriate critical values of F are found in the line indexed by the value of a required.

EXAMPLE **10.13**

Check your ability to use Table 6 in Appendix I by verifying the following statements:

1. The value of F with area .05 to its right for $df_1 = 6$ and $df_2 = 9$ is 3.37.
2. The value of F with area .05 to its right for $df_1 = 5$ and $df_2 = 10$ is 3.33.
3. The value of F with area .01 to its right for $df_1 = 6$ and $df_2 = 9$ is 5.80.

These values are shaded in Table 10.6.

TABLE 10.6 ● **Format of the F Table from Table 6 in Appendix I**

		df_1					
df_2	a	1	2	3	4	5	6
1	.100	39.86	49.50	53.59	55.83	57.24	58.20
	.050	161.4	199.5	215.7	224.6	230.2	234.0
	.025	647.8	799.5	864.2	899.6	921.8	937.1
	.010	4052	4999.5	5403	5625	5764	5859
	.005	16211	20000	21615	22500	23056	23437
2	.100	8.53	9.00	9.16	9.24	9.29	9.33
	.050	18.51	19.00	19.16	19.25	19.30	19.33
	.025	38.51	39.00	39.17	39.25	39.30	39.33
	.010	98.50	99.00	99.17	99.25	99.30	99.33
	.005	198.5	199.0	199.2	199.2	199.3	199.3
3	.100	5.54	5.46	5.39	5.34	5.31	5.28
	.050	10.13	9.55	9.28	9.12	9.01	8.94
	.025	17.44	16.04	15.44	15.10	14.88	14.73
	.010	34.12	30.82	29.46	28.71	28.24	27.91
	.005	55.55	49.80	47.47	46.19	45.39	44.84
⋮	⋮	⋮		⋮			⋮
9	.100	3.36	3.01	2.81	2.69	2.61	2.55
	.050	5.12	4.26	3.86	3.63	3.48	3.37
	.025	7.21	5.71	5.08	4.72	4.48	4.32
	.010	10.56	8.02	6.99	6.42	6.06	5.80
	.005	13.61	10.11	8.72	7.96	7.47	7.13
10	.100	3.29	2.92	2.73	2.61	2.52	2.46
	.050	4.96	4.10	3.71	3.48	3.33	3.22
	.025	6.94	5.46	4.83	4.47	4.24	4.07
	.010	10.04	7.56	6.55	5.99	5.64	5.39
	.005	12.83	9.43	8.08	7.34	6.87	6.54

The statistical test of the null hypothesis

$$H_0 : \sigma_1^2 = \sigma_2^2$$

uses the test statistic

$$F = \frac{s_1^2}{s_2^2}$$

When the alternative hypothesis implies a one-tailed test—that is,

$$H_a : \sigma_1^2 > \sigma_2^2$$

you can find the right-tailed critical value for rejecting H_0 directly from Table 6 in Appendix I. However, when the alternative hypothesis requires a two-tailed test—that is,

$$H_0 : \sigma_1^2 \neq \sigma_2^2$$

the rejection region is divided between the upper and lower tails of the F distribution. These left-tailed critical values are *not given* in Table 6 for the following reason: You are free to decide which of the two populations you want to call "Population 1." If you always choose to call the population with the *larger* sample variance "Population 1," then the observed value of your test statistic will always be in the right tail of the F distribution. Even though half of the rejection region, the area $\alpha/2$ to its left, will be in the lower tail of the distribution, you will never need to use it! Remember these points, though, for a two-tailed test:

- The area in the right tail of the rejection region is only $\alpha/2$.
- The area to the right of the observed test statistic is only (p-value)/2.

The formal procedures for a test of hypothesis and a $(1 - \alpha)100\%$ confidence interval for two population variances are shown next.

TEST OF HYPOTHESIS CONCERNING THE EQUALITY OF TWO POPULATION VARIANCES

1. Null hypothesis: $H_0 : \sigma_1^2 = \sigma_2^2$
2. Alternative hypothesis:

One-Tailed Test	**Two-Tailed Test**
$H_a : \sigma_1^2 > \sigma_2^2$ (or $H_a : \sigma_1^2 < \sigma_2^2$)	$H_a : \sigma_1^2 \neq \sigma_2^2$

3. Test statistic:

One-Tailed Test	**Two-Tailed Test**
$F = \dfrac{s_1^2}{s_2^2}$	$F = \dfrac{s_1^2}{s_2^2}$

where s_1^2 is the larger sample variance.

4. Rejection region: Reject H_0 when

One-Tailed Test	**Two-Tailed Test**
$F > F_\alpha$	$F > F_{\alpha/2}$

or when p-value $< \alpha$

(continued)

TEST OF HYPOTHESIS CONCERNING THE EQUALITY OF TWO POPULATION VARIANCES

(continued)

The critical values of F_α and $F_{\alpha/2}$ are based on $df_1 = (n_1 - 1)$ and $df_2 = (n_2 - 1)$. These tabulated values, for $a = .100, .050, .025, .010,$ and $.005$, can be found using Table 6 in Appendix I.

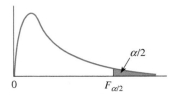

Assumptions: The samples are randomly and independently selected from normally distributed populations.

CONFIDENCE INTERVAL FOR σ_1^2/σ_2^2

$$\left(\frac{s_1^2}{s_2^2}\right)\frac{1}{F_{df_1,df_2}} < \frac{\sigma_1^2}{\sigma_2^2} < \left(\frac{s_1^2}{s_2^2}\right)F_{df_2,df_1}$$

where $df_1 = (n_1 - 1)$ and $df_2 = (n_2 - 1)$. F_{df_1,df_2} is the tabulated critical value of F corresponding to df_1 and df_2 degrees of freedom in the numerator and denominator of F, respectively, with area $\alpha/2$ to its right.

Assumptions: The samples are randomly and independently selected from normally distributed populations.

EXAMPLE 10.14

An experimenter is concerned that the variability of responses using two different experimental procedures may not be the same. Before conducting his research, he conducts a prestudy with random samples of 10 and 8 responses and gets $s_1^2 = 7.14$ and $s_2^2 = 3.21$, respectively. Do the sample variances present sufficient evidence to indicate that the population variances are unequal?

Solution Assume that the populations have probability distributions that are reasonably mound-shaped and hence satisfy, for all practical purposes, the assumption that the populations are normal. You wish to test these hypotheses:

$$H_0 : \sigma_1^2 = \sigma_2^2 \quad \text{versus} \quad H_a : \sigma_1^2 \neq \sigma_2^2$$

Using Table 6 in Appendix I for $\alpha/2 = .025$, you can reject H_0 when $F > 4.82$ with $\alpha = .05$. The calculated value of the test statistic is

$$F = \frac{s_1^2}{s_2^2} = \frac{7.14}{3.21} = 2.22$$

Because the test statistic does not fall into the rejection region, you cannot reject $H_0 : \sigma_1^2 = \sigma_2^2$. Thus, there is insufficient evidence to indicate a difference in the population variances.

EXAMPLE (10.15) Refer to Example 10.14 and find a 90% confidence interval for σ_1^2/σ_2^2.

Solution The 90% confidence interval for σ_1^2/σ_2^2 is

$$\left(\frac{s_1^2}{s_2^2}\right)\frac{1}{F_{df_1,df_2}} < \frac{\sigma_1^2}{\sigma_2^2} < \left(\frac{s_1^2}{s_2^2}\right)F_{df_2,df_1}$$

where

$$s_1^2 = 7.14 \qquad\qquad s_2^2 = 3.21$$
$$df_1 = (n_1 - 1) = 9 \qquad df_2 = (n_2 - 1) = 7$$
$$F_{9,7} = 3.68 \qquad\qquad F_{7,9} = 3.29$$

Substituting these values into the formula for the confidence interval, you get

$$\left(\frac{7.14}{3.21}\right)\frac{1}{3.68} < \frac{\sigma_1^2}{\sigma_2^2} < \left(\frac{7.14}{3.21}\right)3.29 \quad \text{or} \quad .60 < \frac{\sigma_1^2}{\sigma_2^2} < 7.32$$

The calculated interval estimate .60 to 7.32 includes 1.0, the value hypothesized in H_0. This indicates that it is quite possible that $\sigma_1^2 = \sigma_2^2$ and therefore agrees with the test conclusions. Do not reject $H_0 : \sigma_1^2 = \sigma_2^2$.

The *Excel* function called **F.TEST (FTEST** in *Excel 2007* and earlier versions) can be used to perform the *F*-test for the equality of variances when you have entered the raw data into the spreadsheet. The *MINITAB* command **Stat ▶ Basic Statistics ▶ 2 Variances** is a little more flexible, since it allows you to enter either raw data or summary statistics to perform the *F*-test. In addition, *MINITAB 16* calculates confidence intervals for the ratio of two variances or two standard deviations (which we have not discussed). The *MINITAB 16* printout for Example 10.14, containing the *F* statistic and its *p*-value, is shown in Figure 10.17.

FIGURE 10.17
MINITAB output for
Example 10.14

Test and CI for Two Variances

```
Null hypothesis          Sigma(1) / Sigma(2) = 1
Alternative hypothesis   Sigma(1) / Sigma(2) not = 1
Significance level       Alpha = 0.05

Statistics
Sample    N    StDev    Variance
1        10    2.672       7.140
2         8    1.792       3.210
Ratio of standard deviations = 1.491
Ratio of variances = 2.224


95% Confidence Intervals
CI for
Distribution    CI for StDev       Variance
of Data             Ratio             Ratio
Normal        (0.679, 3.055)   (0.461, 9.335)
Test
Method            DF1  DF2   Statistic   P-Value
F Test (normal)     9    7        2.22     0.304
```

FIGURE 10.18

Distributions of impurity measurements for two production lines

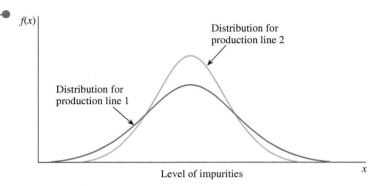

EXAMPLE 10.16

The variability in the amount of impurities present in a batch of chemical used for a particular process depends on the length of time the process is in operation. A manufacturer using two production lines, 1 and 2, has made a slight adjustment to line 2, hoping to reduce the variability as well as the average amount of impurities in the chemical. Samples of $n_1 = 25$ and $n_2 = 25$ measurements from the two batches yield these means and variances:

$$\bar{x}_1 = 3.2 \qquad s_1^2 = 1.04$$
$$\bar{x}_2 = 3.0 \qquad s_2^2 = .51$$

Do the data present sufficient evidence to indicate that the process variability is less for line 2?

Solution The experimenter believes that the average levels of impurities are the same for the two production lines but that her adjustment may have decreased the variability of the levels for line 2, as illustrated in Figure 10.18. This adjustment would be good for the company because it would decrease the probability of producing shipments of the chemical with unacceptably high levels of impurities.

To test for a decrease in variability, the test of hypothesis is

$$H_0 : \sigma_1^2 = \sigma_2^2 \quad \text{versus} \quad H_a : \sigma_1^2 > \sigma_2^2$$

and the observed value of the test statistic is

$$F = \frac{s_1^2}{s_2^2} = \frac{1.04}{.51} = 2.04$$

Using the p-value approach, you can bound the one-tailed p-value using Table 6 in Appendix I with $df_1 = df_2 = (25 - 1) = 24$. The observed value of F falls between $F_{.050} = 1.98$ and $F_{.025} = 2.27$, so that $.025 < p$-value $< .05$. The results are judged significant at the 5% level, and H_0 is rejected. You can conclude that the variability of line 2 is less than that of line 1.

The F-test for the difference in two population variances completes the battery of tests you have learned in this chapter for making inferences about population parameters under these conditions:

- The sample sizes are small.
- The sample or samples are drawn from normal populations.

You will find that the F and χ^2 distributions, as well as the Student's t distribution, are very important in other applications in the chapters that follow. They will be used for different estimators designed to answer different types of inferential questions, but the basic techniques for making inferences remain the same.

In the next section, we review the assumptions required for all of these inference tools, and discuss options that are available when the assumptions do not seem to be reasonably correct.

10.7 EXERCISES

BASIC TECHNIQUES

10.58 Independent random samples from two normal populations produced the variances listed here:

Sample Size	Sample Variance
16	55.7
20	31.4

a. Do the data provide sufficient evidence to indicate that σ_1^2 differs from σ_2^2? Test using $\alpha = .05$.

b. Find the approximate p-value for the test and interpret its value.

10.59 Refer to Exercise 10.58 and find a 95% confidence interval for σ_1^2/σ_2^2.

10.60 Independent random samples from two normal populations produced the given variances:

Sample Size	Sample Variance
13	18.3
13	7.9

a. Do the data provide sufficient evidence to indicate that $\sigma_1^2 > \sigma_2^2$? Test using $\alpha = .05$.

b. Find the approximate p-value for the test and interpret its value.

APPLICATIONS

10.61 SAT Scores The SAT subject tests in chemistry and physics[12] for two groups of 15 students each electing to take these tests are given below.

Chemistry	Physics
$\bar{x} = 644$	$\bar{x} = 658$
$s = 114$	$s = 103$
$n = 15$	$n = 15$

To use the two-sample t-test with a pooled estimate of σ^2, you must assume that the two population variances are equal. Test this assumption using the F-test for equality of variances. What is the approximate p-value for the test?

10.62 Lithium Batteries The stability of measurements on a manufactured product is important in maintaining product quality. A manufacturer of lithium batteries, such as the ones used for digital cameras, suspected that one of the production lines was producing batteries with a wide variation in length of life. To test this theory, he randomly selected $n = 50$ batteries from the suspect line and $n = 50$ from a line that was judged to be "in control." He then measured the length of time (in hours) until depletion to 0.85V with a 5-Ohm load for both samples. The sample means and variances for the two samples were as follows:

Suspect Line	Line "in Control"
$\bar{x}_1 = 9.40$	$\bar{x}_2 = 9.25$
$s_1 = .25$	$s_2 = .12$

a. Do the data provide sufficient evidence to indicate that batteries produced by the "suspect line" have a larger variance in length of life than those produced by the line that is assumed to be in control? Test using $\alpha = .05$.

b. Find the approximate p-value for the test and interpret its value.

c. Construct a 90% confidence interval for the variance ratio.

 10.63 Roethlisberger and Rodgers Quarterbacks not only need to have a good passing percentage, but they need to be consistent. That is, the variability in the number of passes completed per game should be small. The table below gives the number of passes completed for Ben Roethlisberger and Aaron Rodgers, quarterbacks for the Pittsburgh Steelers and Green Bay Packers, respectively, during the 2010 NFL season.[13]

Aaron Rodgers			Ben Roethlisberger	
19	21	7	16	20
19	15	25	19	22
34	27	19	17	21
12	22		17	23
27	26		30	22
18	21		18	15

a. Does the data indicate that there is a difference in the variability in the number of passes completed for the two quarterbacks? Use $\alpha = .01$.

b. If you were going to test for a difference in the two population means, would it be appropriate to use the two-sample t-test that assumes equal variances? Explain.

10.64 Tuna III In Exercise 10.26 and dataset EX1026, you conducted a test to detect a difference in the average prices of light tuna in water versus light tuna in oil.

a. What assumption had to be made concerning the population variances so that the test would be valid?

b. Do the data present sufficient evidence to indicate that the variances violate the assumption in part a? Test using $\alpha = .05$.

10.65 Runners and Cyclists III Refer to Exercise 10.27. Susan Beckham and colleagues conducted an experiment involving 10 healthy runners and 10 healthy cyclists to determine if there are significant differences in pressure measurements within the anterior muscle compartment for runners and cyclists.[7] The data—compartment pressure, in millimeters of mercury (Hg)—are reproduced here:

Condition	Runners Mean	Runners Standard Deviation	Cyclists Mean	Cyclists Standard Deviation
Rest	14.5	3.92	11.1	3.98
80% maximal O$_2$ consumption	12.2	3.49	11.5	4.95
Maximal O$_2$ consumption	19.1	16.9	12.2	4.47

For each of the three variables measured in this experiment, test to see whether there is a significant difference in the variances for runners versus cyclists. Find the approximate p-values for each of these tests. Will a two-sample t-test with a pooled estimate of σ^2 be appropriate for all three of these variables? Explain.

10.66 Impurities A pharmaceutical manufacturer purchases a particular material from two different suppliers. The mean level of impurities in the raw material is approximately the same for both suppliers, but the manufacturer is concerned about the variability of the impurities from shipment to shipment. To compare the variation in percentage impurities for the two suppliers, the manufacturer selects 10 shipments from each of the two suppliers and measures the percentage of impurities in the raw material for each shipment. The sample means and variances are shown in the table.

Supplier A	Supplier B
$\bar{x}_1 = 1.89$	$\bar{x}_2 = 1.85$
$s_1^2 = .273$	$s_2^2 = .094$
$n_1 = 10$	$n_2 = 10$

a. Do the data provide sufficient evidence to indicate a difference in the variability of the shipment impurity levels for the two suppliers? Test using $\alpha = .01$. Based on the results of your test, what recommendation would you make to the pharmaceutical manufacturer?

b. Find a 99% confidence interval for σ_2^2 and interpret your results.

 NEED TO KNOW...

How to Decide Which Test to Use

Are you interested in testing means? If the design involves:

a. One random sample, use the one-sample t statistic.

b. Two independent random samples, are the population variances equal?
  i. If equal, use the two-sample t statistic with pooled s^2.
  ii. If unequal, use the unpooled t with estimated df.

c. Two paired samples with random pairs, use a one-sample t for analyzing differences.

Are you interested in testing variances? If the design involves:

a. One random sample, use the χ^2 test for a single variance.

b. Two independent random samples, use the F-test to compare two variances.

REVISITING THE SMALL-SAMPLE ASSUMPTIONS

All of the tests and estimation procedures discussed in this chapter require that the data satisfy certain conditions in order that the error probabilities (for the tests) and the confidence coefficients (for the confidence intervals) be equal to the values you have specified. For example, if you construct what you believe to be a 95% confidence interval, you want to be certain that, in repeated sampling, 95% (and not 85% or 75% or less) of all such intervals will contain the parameter of interest. These conditions are summarized in these assumptions:

ASSUMPTIONS

1. For all tests and confidence intervals described in this chapter, it is assumed that **samples are randomly selected from normally distributed populations.**

2. When two samples are selected, it is assumed that they are **selected in an independent manner** except in the case of the paired-difference experiment.

3. For tests or confidence intervals concerning the difference between two population means μ_1 and μ_2 based on independent random samples, it is assumed that $\sigma_1^2 = \sigma_2^2$.

In reality, you will never know everything about the sampled population. If you did, there would be no need for sampling or statistics. It is also highly unlikely that a population will *exactly* satisfy the assumptions given in the box. Fortunately, the procedures presented in this chapter give good inferences even when the data exhibit moderate departures from the necessary conditions.

A statistical procedure that is not sensitive to departures from the conditions on which it is based is said to be **robust.** The Student's t-tests are quite robust for moderate departures from normality. Also, as long as the sample sizes are nearly equal, there is not much difference between the pooled and unpooled t statistics for the difference in two population means. However, if the sample sizes are not clearly equal, and if the population variances are unequal, the pooled t statistic provides inaccurate conclusions.

If you are concerned that your data do not satisfy the assumptions, other options are available:

- If you can select relatively large samples, you can use one of the large-sample procedures of Chapters 8 and 9, which do not rely on the normality or equal variance assumptions.

- You may be able to use a *nonparametric test* to answer your inferential questions. These tests have been developed specifically so that few or no distributional assumptions are required for their use. Tests that can be used to compare the locations or variability of two populations are presented in Chapter 15.

CHAPTER REVIEW

Key Concepts and Formulas

I. Experimental Designs for Small Samples

1. **Single random sample:** The sampled population must be normal.

2. **Two independent random samples:** Both sampled populations must be normal.
   a. Populations have a common variance σ^2.
   b. Populations have different variances: σ_1^2 and σ_2^2.

3. **Paired-difference** or **matched pairs design:** The samples are not independent.

II. Statistical Tests of Significance

1. Based on the t, F, and χ^2 distributions

2. Use the same procedure as in Chapter 9

3. **Rejection region—critical values** and **significance levels:** based on the t, F, or χ^2 distributions with the appropriate degrees of freedom

4. **Tests of population parameters:** a single mean, the difference between two means, a single variance, and the ratio of two variances

III. Small-Sample Test Statistics

To test one of the population parameters when the sample sizes are small, use the following test statistics:

Parameter	Test Statistic	Degrees of Freedom
μ	$t = \dfrac{\bar{x} - \mu_0}{s/\sqrt{n}}$	$n - 1$
$\mu_1 - \mu_2$ (equal variances)	$t = \dfrac{(\bar{x}_1 - \bar{x}_2) - (\mu_1 - \mu_2)}{\sqrt{s^2\left(\dfrac{1}{n_1} + \dfrac{1}{n_2}\right)}}$	$n_1 + n_2 - 2$
$\mu_1 - \mu_2$ (unequal variances)	$t \approx \dfrac{(\bar{x}_1 - \bar{x}_2) - (\mu_1 - \mu_2)}{\sqrt{\dfrac{s_1^2}{n_1} + \dfrac{s_2^2}{n_2}}}$	Satterthwaite's approximation
$\mu_1 - \mu_2$ (paired samples)	$t = \dfrac{\bar{d} - \mu_d}{s_d/\sqrt{n}}$	$n - 1$
σ^2	$\chi^2 = \dfrac{(n-1)s^2}{\sigma_0^2}$	$n - 1$
σ_1^2/σ_2^2	$F = s_1^2/s_2^2$	$n_1 - 1$ and $n_2 - 1$

 TECHNOLOGY TODAY

Small-Sample Testing—*Microsoft Excel*

The tests of hypotheses for two population means based on the Student's t distribution and the F-test for the ratio of two variances can be found using the *Microsoft Excel* command **Data ▶ Data Analysis.** Remember that you need to have loaded the *Excel* add-ins called **Analysis ToolPak** (see the instructions in the "Technology Today." section of Chapter 1). You will find three choices for the two-sample t-tests and one F-test in the list of "Analysis Tools." To choose the proper t-tests, you must first decide whether the samples are independent or paired; for the independent samples test, you must decide whether or not the population variances can be assumed equal.

 EXAMPLE 10.17

(**Two-Sample t-Test Assuming Equal Variances**) The test scores on the same algebra test were recorded for nine students randomly selected from a classroom taught by teacher A and eight students randomly selected from a classroom taught by Teacher B. Is there a difference in the average scores for students taught by these two teachers?

Teacher A	65	88	93	95	80	76	79	77	
Teacher B	91	85	70	82	92	68	86	87	75

Enter the data into columns A and B of an *Excel* spreadsheet.

1. Use **Data ▶ Data ▶ Analysis ▶ Descriptive Statistics** or **▶ Statistical ▶ STDEV.S** (**STDEV** in *Excel 2007* and earlier versions) to find the standard deviations for the two samples, $s_1 = 9.913$ and $s_2 = 8.800$. Since the ratio of the two variances is $s_1^2/s_2^2 = 1.27$ (less than 3), you are safe in assuming that the population variances are the same.

2. Select **Data ▶ Data Analysis ▶ t-Test: Two-Sample Assuming Equal Variances** to generate the Dialog box in Figure 10.19(a). Highlight or type the **Variable 1 Range** and **Variable 2 Range** (the data in the first and second columns) into the first two boxes. In the box marked "Hypothesized Mean Difference" type **0** (since we are testing $H_0: \mu_1 - \mu_2 = 0$) and check "Labels" if necessary.

3. The default significance level is $\alpha = .05$ in *Excel*. Change this significance level if necessary. Enter a cell location for the **Output Range** and click **OK**. The output will appear in the selected cell location, and should be adjusted using **Format ▶ AutoFit Column Width** on the **Home** tab in the **Cells** group while it is still highlighted. You can decrease the decimal accuracy if you like, using ⬆.0 .00 on the **Home** tab in the **Number** group.

4. The observed value of the test statistic $t = -.0337$ is found in Figure 10.19(b) in the row labeled "*t Stat*" followed by the one-tailed p-value "*P(T<=t) one-tail*" and the critical value marking the rejection region for a one-tailed test with $\alpha = .05$. The last two rows of output give the p-value and critical t-value for a two-tailed test.

5. For this example, the p-value $= .9736$ indicates that there is no significant difference in the average scores for students taught by the two teachers.

FIGURE 10.19

(a)

t-Test: Two-Sample Assuming Equal Variances

Input		
Variable 1 Range:	$A$1:$A$9	
Variable 2 Range:	$B$1:$B$10	
Hypothesized Mean Difference:	0	
☑ Labels		
Alpha:	0.05	
Output options		
◉ Output Range:	$D$1	
○ New Worksheet Ply:		
○ New Workbook		

OK Cancel Help

(b)

D	E	F
t-Test: Two-Sample Assuming Equal Variances		
	A	*B*
Mean	81.625	81.778
Variance	98.268	77.444
Observations	8	9
Pooled Variance	87.162	
Hypothesized Mean Difference	0	
df	15	
t Stat	-0.0337	
P(T<=t) one-tail	0.4868	
t Critical one-tail	1.7531	
P(T<=t) two-tail	0.9736	
t Critical two-tail	2.1314	

6. In Section 10.7, we presented a formal test of hypothesis for the equality of two variances using the F-test. To implement this test using *Excel*, select **Data ▶ Data Analysis ▶ F-Test: Two-Sample for Variances.** Follow the directions for the Equal Variances t-test, but replace the "Alpha" value with 0.025, and you will generate the output in Figure 10.20.

FIGURE 10.20

F-Test Two-Sample for Variances		
	A	*B*
Mean	81.625	81.78
Variance	98.268	77.44
Observations	8	9
df	7	8
F	1.2689	
P(F<=f) one-tail	0.3701	
F Critical one-tail	4.5286	

Notice that only the one-tailed p-value and critical value are given in the output, which is why we specified the single tail to be 0.025. Hence, for our two-tailed test, $\alpha = 0.05$ and p-value $= .7402$. There is no significant difference in the two variances.

EXAMPLE 10.18

(Two-Sample t-Test Assuming Unequal Variances)

1. Refer to Example 10.17. If the ratio of the two sample variances had been so large that you could not assume equal variances (we use "greater than 3" as a rule of thumb), you should select **Data ▶ Data Analysis ▶ t-Test: Two-Sample Assuming Unequal Variances.**

2. Follow the directions for the Equal Variances t-test, and you will generate similar output. If we use this test on the data from Example 10.17, the following output results (Figure 10.21).

FIGURE 10.21

t-Test: Two-Sample Assuming Unequal Variances		
	A	B
Mean	81.625	81.778
Variance	98.268	77.444
Observations	8	9
Hypothesized Mean Difference	0	
df	14	
t Stat	-0.0334	
P(T<=t) one-tail	0.4869	
t Critical one-tail	1.7613	
P(T<=t) two-tail	0.9738	
t Critical two-tail	2.1448	

3. You will see slight differences in the observed value of the test statistic, the degrees of freedom and the p-values for the test, but the conclusions did not change.

4. NOTE: When calculating the degrees of freedom for *Satterthwaite's Approximation,* the **Data Analysis Tool** in *Excel* rounds to the nearest integer. An alternative *Excel* function for calculating the p-value for this test (f_x ▶ **Statistical ▶ T.TEST**) uses the exact value of df given by Satterthwaite's formula. Because of these different approaches to determining the degrees of freedom, the results of **T.TEST** and the t-Test tool will differ slightly in the Unequal Variances case, and will *also* differ slightly from the *MINITAB* output.

EXAMPLE 10.19

(Paired t-Test) Refer to the tire wear data from Table 10.3.

1. To perform a paired-difference test for these dependent samples, enter the data into the first two columns of an *Excel* spreadsheet and select **Data ▶ Data Analysis ▶ t-Test: Paired Two Sample for Means.**

2. Follow the directions for the Equal Variances t-test, and you will generate similar output. For the data in Table 10.3, you obtain the output in Figure 10.22. Again, you can decrease the decimal accuracy if you like, using ⁎.₀ on the **Home** tab in the **Number** group.

3. Using the observed value of the test statistic ($t = 12.83$) with two-tailed p-value $= .0002$, there is strong evidence to indicate a difference in the two population means.

FIGURE 10.22

D	E	F
t-Test: Paired Two Sample for Means		
	Tire A	Tire B
Mean	10.24	9.76
Variance	1.733	1.763
Observations	5	5
Pearson Correlation	0.998	
Hypothesized Mean Difference	0	
df	4	
t Stat	12.8285	
P(T<=t) one-tail	0.0001	
t Critical one-tail	2.1318	
P(T<=t) two-tail	0.0002	
t Critical two-tail	2.7764	

Small-Sample Testing and Estimation—*MINITAB*

The tests of hypotheses for two population means based on the Student's *t* distribution and the *F*-test for the ratio of two variances can be found using the *MINITAB* command **Stat ▶ Basic Statistics.** You will find choices for **1-Sample t, 2-Sample t, Paired t,** and **2 Variances,** which will perform the tests and estimation procedures of Sections 10.3, 10.4, 10.5, and 10.7. To choose the proper two sample *t*-tests, you must first decide whether the samples are independent or paired; for the independent samples test, you must decide whether or not the population variances can be assumed equal.

EXAMPLE 10.20 **(One Sample *t*-Test)** Refer to Example 10.3, in which the average weight of diamonds using a new process was compared to an average weight of .5 karat.

1. Enter the six recorded weights—.46, .61, .52, .48, .57, .54—in column C1 and name them "Weights." Use **Stat ▶ Basic Statistics ▶ 1-Sample t** to generate the Dialog boxes in Figure 10.23.

FIGURE 10.23

2. To test $H_0: \mu = .5$ versus $H_a: \mu > .5$, use the list on the left to select "Weights" for the box marked "Samples in Columns." Check the box marked "Perform hypothesis test." Then, place your cursor in the box marked "Hypothesized mean:" and enter **.5** as the test value. Finally, use **Options** and the drop-down menu marked "Alternative" to select "greater than." You can change the default confidence coefficient of **.95** if you wish. Click **OK** twice to obtain the output in Figure 10.24.

FIGURE 10.24 ●

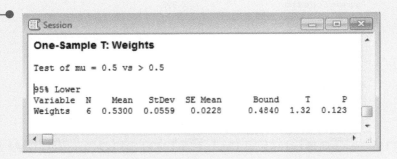

```
Session                                          _  □  ✕

One-Sample T: Weights

Test of mu = 0.5 vs > 0.5

95% Lower
Variable   N    Mean    StDev   SE Mean    Bound     T      P
Weights    6  0.5300   0.0559   0.0228    0.4840   1.32  0.123
```

3. Notice that *MINITAB* produces a one- or a two-sided confidence interval for the single population mean, consistent with the alternative hypothesis you have chosen.

EXAMPLE **(Two Sample *t*-Test)** The test scores on the same algebra test were recorded for nine students randomly selected from a classroom taught by teacher A and eight students randomly selected from a classroom taught by Teacher B. Is there a difference in the average scores for students taught by these two teachers?

Teacher A	65	88	93	95	80	76	79	77	
Teacher B	91	85	70	82	92	68	86	87	75

1. The data can be entered into the worksheet in one of three ways:
   - Enter measurements from both samples into a single column and enter letters (A or B) in a second column to identify the sample from which the measurement comes.
   - Enter the samples in two separate columns.
   - If you do not have the raw data, but rather have summary statistics, *MINITAB 16* will allow you to use these values by selecting "Summarized data" and entering the appropriate values in the boxes.

2. Use the second method and enter the data into two columns of the worksheet. Use **Stat ▶ Basic Statistics ▶ Display Descriptive Statistics** to find the standard deviations for the two samples, $s_1 = 9.91$ and $s_2 = 8.80$. Since the ratio of the two variances is $s_1^2/s_2^2 = 1.27$ (less than 3), you are safe in assuming that the population variances are the same.

3. Select **Stat ▶ Basic Statistics ▶ 2-Sample t** to generate the Dialog box in Figure 10.25(a). Check "samples in different columns," selecting the appropriate columns from the list at the left. Check the "Assume equal variances" box and select the proper alternative in the **Options** box. The two-sample output when you click **OK** twice automatically contains a 95% one- or two-sided confidence interval as well as the test statistic and *p*-value (you can change the confidence coefficient if you wish). The output is shown in Figure 10.25(b).

FIGURE 10.25

(a)

(b)

4. The observed value of the test statistic $t = -.03$ is labeled "*T-Value*" followed by the two-tailed "*P-Value*." For this example, the *p*-value $= .974$ indicates that there is no significant difference in the average scores for students taught by the two teachers.

5. In Section 10.7, we presented a formal test of hypothesis for the equality of two variances using the *F*-test. To implement this test using *MINITAB*, select **Stat ▶ Basic Statistics ▶ 2 Variances.** In the drop-down list, select "Samples in two columns" and enter the appropriate columns from the list on the left. The pertinent portion of the output is shown in Figure 10.26. For our two-tailed test with $\alpha = 0.05$, the test statistic is $F = 1.27$ and the *p*-value $= .740$. There is no significant difference in the two variances.

FIGURE 10.26

EXAMPLE 10.22

(Two-Sample *t*-Test Assuming Unequal Variances)

1. Refer to Example 10.21. If the ratio of the two sample variances had been so large that you could not assume equal variances (we use "greater than 3" as a rule of thumb), you should select **Stat ▶ Basic Statistics ▶ 2-Sample t,** but DO NOT check the box marked "Assume Equal Variances." If we use this test on the data from Example 10.17, the following output results (Figure 10.27).

FIGURE 10.27

2. You will see slight differences in the degrees of freedom, there is no "Pooled StDev" listed, but the conclusions did not change.

3. NOTE: When calculating the degrees of freedom for *Satterthwaite's Approximation*, *MINITAB* uses the integer part of the calculated value, which is different from the procedures used in *MS Excel*. Because of these different approaches to determining the degrees of freedom, the results of the outputs from *MS Excel* and *MINITAB* will differ slightly.

 EXAMPLE 10.23 (**Paired-Difference Test**) Refer to the tire wear data from Table 10.3.

1. To perform a paired-difference test for these dependent samples, enter the data into the first two columns of a *MINITAB* worksheet and select **Stat ▶ Basic Statistics ▶ Paired t.**

2. Follow the directions for the independent samples *t*-test, and you will generate similar output. For the data in Table 10.3, you obtain the output in Figure 10.28.

3. Using the observed value of the test statistic ($t = 12.83$) with two-tailed *p*-value = .000, there is strong evidence to indicate a difference in the two population means.

 FIGURE 10.28

```
Session

Paired T-Test and CI: Tire A, Tire B

             N    Mean   StDev   SE Mean
Tire A       5   10.240  1.316   0.589
Tire B       5    9.760  1.328   0.594
Difference   5    0.4800 0.0837  0.0374

95% CI for mean difference: (0.3761, 0.5839)
T-Test of mean difference = 0 (vs not = 0): T-Value = 12.83   P-Value = 0.000
```

Supplementary Exercises

10.67 What assumptions are made when Student's *t*-test is used to test a hypothesis concerning a population mean?

10.68 What assumptions are made about the populations from which random samples are obtained when the *t* distribution is used in making small-sample inferences concerning the difference in population means?

10.69 Why use paired observations to estimate the difference between two population means rather than estimation based on independent random samples selected from the two populations? Is a paired experiment always preferable? Explain.

10.70 Use Table 4 in Appendix I to find the following critical values:
a. An upper one-tailed rejection region with $\alpha = .05$ and 11 *df*.

b. A two-tailed rejection region with $\alpha = .05$ and 7 *df*.

c. A lower one-tailed rejection region with $\alpha = .01$ and 15 *df*.

10.71 Use Table 4 in Appendix I to bound the following *p*-values:
a. $P(t > 1.2)$ with 5 *df*
b. $P(t > 2) + P(t < -2)$ with 10 *df*
c. $P(t < -3.3)$ with 8 *df*
d. $P(t > 0.6)$ with 12 *df*

10.72 A random sample of $n = 12$ observations from a normal population produced $\bar{x} = 47.1$ and $s^2 = 4.7$. Test the hypothesis $H_0: \mu = 48$ against $H_0: \mu \neq 48$ at the 5% level of significance.

10.73 Impurities II A manufacturer can tolerate a small amount (.05 milligrams per liter (mg/l)) of impurities in a raw material needed for manufacturing its product. Because the laboratory test for the impurities is subject to experimental error, the manufacturer tests each batch 10 times. Assume that the mean value of the experimental error is 0 and hence that the mean value of the 10 test readings is an unbiased estimate of the true amount of the impurities in the batch. For a particular batch of the raw material, the mean of the 10 test readings is .058 mg/l, with a standard deviation of .012 mg/l. Do the data provide sufficient evidence to indicate that the amount of impurities in the batch exceeds .05 mg/l? Find the p-value for the test and interpret its value.

10.74 Red Pine The main stem growth measured for a sample of seventeen 4-year-old red pine trees produced a mean and standard deviation equal to 11.3 and 3.4 inches, respectively. Find a 90% confidence interval for the mean growth of a population of 4-year-old red pine trees subjected to similar environmental conditions.

10.75 Sodium Hydroxide The object of a general chemistry experiment is to determine the amount (in milliliters) of sodium hydroxide (NaOH) solution needed to neutralize 1 gram of a specified acid. This will be an exact amount, but when the experiment is run in the laboratory, variation will occur as the result of experimental error. Three titrations are made using phenolphthalein as an indicator of the neutrality of the solution (pH equals 7 for a neutral solution). The three volumes of NaOH required to attain a pH of 7 in each of the three titrations are as follows: 82.10, 75.75, and 75.44 milliliters. Use a 99% confidence interval to estimate the mean number of milliliters required to neutralize 1 gram of the acid.

10.76 Sodium Chloride Measurements of water intake, obtained from a sample of 17 rats that had been injected with a sodium chloride solution, produced a mean and standard deviation of 31.0 and 6.2 cubic centimeters (cm^3), respectively. Given that the average water intake for noninjected rats observed over a comparable period of time is 22.0 cm^3, do the data indicate that injected rats drink more water than noninjected rats? Test at the 5% level of significance. Find a 90% confidence interval for the mean water intake for injected rats.

10.77 Sea Urchins An experimenter was interested in determining the mean thickness of the cortex of the sea urchin egg. The thickness was measured for

$n = 10$ sea urchin eggs. These measurements were obtained:

4.5	6.1	3.2	3.9	4.7
5.2	2.6	3.7	4.6	4.1

Estimate the mean thickness of the cortex using a 95% confidence interval.

10.78 Fabricating Systems A production plant has two complex fabricating systems, both of which are maintained at 2-week intervals. However, one system is twice as old as the other. The number of finished products fabricated daily by each of the systems is recorded for 30 working days, with the results given in the table. Do these data present sufficient evidence to conclude that the variability in daily production warrants increased maintenance of the older fabricating system? Use the p-value approach.

New System	Old System
$\bar{x}_1 = 246$	$\bar{x}_2 = 240$
$s_1 = 15.6$	$s_2 = 28.2$

Data set **10.79 Fossils** The data in the table are the **EX1079** diameters and heights of 10 fossil specimens of a species of small shellfish, *Rotularia (Annelida) fallax,* that were unearthed in a mapping expedition near the Antarctic Peninsula.[14] The table gives an identification symbol for the fossil specimen, the fossil's diameter and height in millimeters, and the ratio of diameter to height.

Specimen	Diameter	Height	D/H
OSU 36651	185	78	2.37
OSU 36652	194	65	2.98
OSU 36653	173	77	2.25
OSU 36654	200	76	2.63
OSU 36655	179	72	2.49
OSU 36656	213	76	2.80
OSU 36657	134	75	1.79
OSU 36658	191	77	2.48
OSU 36659	177	69	2.57
OSU 36660	199	65	3.06
$\bar{x}$:	184.5	73	2.54
s:	21.5	5	.37

a. Find a 95% confidence interval for the mean diameter of the species.

b. Find a 95% confidence interval for the mean height of the species.

c. Find a 95% confidence interval for the mean ratio of diameter to height.

d. Compare the three intervals constructed in parts a, b, and c. Is the average of the ratios the same as the ratio of the average diameter to average height?

10.80 Fossils, continued Refer to Exercise 10.79 and data set EX1079. Suppose you want to estimate the mean diameter of the fossil specimens correct to within 5 millimeters with probability equal to .95. How many fossils do you have to include in your sample?

Data set **10.81 Alcohol and Reaction Times** To test
EX1081 the effect of alcohol in increasing the reaction time to respond to a given stimulus, the reaction times of seven people were measured. After consuming 3 ounces of 40% alcohol, the reaction time for each of the seven people was measured again. Do the following data indicate that the mean reaction time after consuming alcohol was greater than the mean reaction time before consuming alcohol? Use $\alpha = .05$.

Person	1	2	3	4	5	6	7
Before	4	5	5	4	3	6	2
After	7	8	3	5	4	5	5

Data set **10.82 Cheese, Please** Here are the prices per
EX1082 ounce of $n = 13$ different brands of individually wrapped cheese slices:

29.0	24.1	23.7	19.6	27.5
28.7	28.0	23.8	18.9	23.9
21.6	25.9	27.4		

Construct a 95% confidence interval estimate of the underlying average price per ounce of individually wrapped cheese slices.

10.83 Drug Absorption An experiment was conducted to compare the mean lengths of time required for the bodily absorption of two drugs A and B. Ten people were randomly selected and assigned to receive one of the drugs. The length of time (in minutes) for the drug to reach a specified level in the blood was recorded, and the data summary is given in the table:

Drug A	Drug B
$\bar{x}_1 = 27.2$	$\bar{x}_2 = 33.5$
$s_1^2 = 16.36$	$s_2^2 = 18.92$

a. Do the data provide sufficient evidence to indicate a difference in mean times to absorption for the two drugs? Test using $\alpha = .05$.

b. Find the approximate p-value for the test. Does this value confirm your conclusions?

c. Find a 95% confidence interval for the difference in mean times to absorption. Does the interval confirm your conclusions in part a?

10.84 Drug Absorption, continued Refer to Exercise 10.83. Suppose you wish to estimate the difference in mean times to absorption correct to within 1 minute with probability approximately equal to .95.

a. Approximately how large a sample is required for each drug (assume that the sample sizes are equal)?

b. If conducting the experiment using the sample sizes of part a will require a large amount of time and money, can anything be done to reduce the sample sizes and still achieve the 1-minute margin of error for estimation?

Data set **10.85 Ring-Necked Pheasants** The weights
EX1085 in grams of 10 males and 10 female juvenile ring-necked pheasants are given below.

Males		Females	
1384	1672	1073	1058
1286	1370	1053	1123
1503	1659	1038	1089
1627	1725	1018	1034
1450	1394	1146	1281

a. Use a statistical test to determine if the population variance of the weights of the male birds differs from that of the females.

b. Test whether the average weight of juvenile male ring-necked pheasants exceeds that of the females by more than 300 grams. (HINT: The procedure that you use should take into account the results of the analysis in part a.)

Data set **10.86 Bees** Insects hovering in flight expend
EX1086 enormous amounts of energy for their size and weight. The data shown here were taken from a much larger body of data collected by T.M. Casey and colleagues.[15] They show the wing stroke frequencies (in hertz) for two different species of bees, $n_1 = 4$ *Euglossa mandibularis* Friese and $n_2 = 6$ *Euglossa imperialis* Cockerell.

E. mandibularis Friese	*E. imperialis* Cockerell
235	180
225	169
190	180
188	185
	178
	182

a. Based on the observed ranges, do you think that a difference exists between the two population variances?

b. Use an appropriate test to determine whether a difference exists.

c. Explain why a Student's *t*-test with a pooled estimator s^2 is unsuitable for comparing the mean wing stroke frequencies for the two species of bees.

Data set **10.87 Calcium** The calcium (Ca) content of
EX1087 a powdered mineral substance was analyzed 10 times with the following percent compositions recorded:

.0271	.0282	.0279	.0281	.0268
.0271	.0281	.0269	.0275	.0276

a. Find a 99% confidence interval for the true calcium content of this substance.

b. What does the phrase "99% confident" mean?

c. What assumptions must you make about the sampling procedure so that this confidence interval will be valid? What does this mean to the chemist who is performing the analysis?

10.88 Sun or Shade? Karl Niklas and T.G. Owens examined the differences in a particular plant, *Plantago Major L.,* when grown in full sunlight versus shade conditions.[16] In this study, shaded plants received direct sunlight for less than 2 hours each day, whereas full-sun plants were never shaded. A partial summary of the data based on $n_1 = 16$ full-sun plants and $n_2 = 15$ shade plants is shown here:

	Full Sun		Shade	
	$\bar{x}$	s	$\bar{x}$	s
Leaf Area (cm²)	128.00	43.00	78.70	41.70
Overlap Area (cm²)	46.80	2.21	8.10	1.26
Leaf Number	9.75	2.27	6.93	1.49
Thickness (mm)	.90	.03	.50	.02
Length (cm)	8.70	1.64	8.91	1.23
Width (cm)	5.24	.98	3.41	.61

a. What assumptions are required in order to use the small-sample procedures given in this chapter to compare full-sun versus shade plants? From the summary presented, do you think that any of these assumptions have been violated?

b. Do the data present sufficient evidence to indicate a difference in mean leaf area for full-sun versus shade plants?

c. Do the data present sufficient evidence to indicate a difference in mean overlap area for full-sun versus shade plants?

10.89 Orange Juice A comparison of the precisions of two machines developed for extracting juice from oranges is to be made using the following data:

Machine A	Machine B
$s^2 = 3.1$ ounces²	$s^2 = 1.4$ ounces²
$n = 25$	$n = 25$

a. Is there sufficient evidence to indicate that there is a difference in the precision of the two machines at the 5% level of significance?

b. Find a 95% confidence interval for the ratio of the two population variances. Does this interval confirm your conclusion from part a? Explain.

Data set **10.90 At Home or at Preschool?** Four sets
EX1090 of identical twins (pairs A, B, C, and D) were selected at random from a computer database of identical twins. One child was selected at random from each pair to form an "experimental group." These four children were sent to preschool. The other four children were kept at home as a control group. At the end of the year, the following IQ scores were obtained:

Pair	Experimental Group	Control Group
A	110	111
B	125	120
C	139	128
D	142	135

Does this evidence justify the conclusion that lack of preschool experience has a depressing effect on IQ scores? Use the *p*-value approach.

Data set **10.91 Dieting** Eight obese persons were
EX1091 placed on a diet for 1 month, and their weights, at the beginning and at the end of the month, were recorded:

	Weights	
Subjects	Initial	Final
1	310	263
2	295	251
3	287	249
4	305	259
5	270	233
6	323	267
7	277	242
8	299	265

Estimate the mean weight loss for obese persons when placed on the diet for a 1-month period. Use a 95% confidence interval and interpret your results. What assumptions must you make so that your inference is valid?

10.92 Price Wars Many seniors are ordering their drugs online to take advantage of lower costs for these pharmacies. A random sample of nine online pharmacies was selected and the cost of a 10-mg Buspar (Buspirone) tablet recorded, as given in the following table.[17]

Pharmacy	Brand ($)	Generic ($)
CanadaDrugStop.com	1.33	.79
CanadaDrugCenter	1.33	.79
Big Mountain Drugs	1.16	.74
Blue Sky Drugs	1.17	.75
CanadaDrugsPharmacy	1.33	.79
Canada Drugs Online	1.11	.75
PharmStore.com	1.13	.59
Buy Low Drugs	1.45	.45
Planetdrugsdirect.com	1.14	.59

a. Test the hypothesis of no difference in costs between brand and generic Buspar 10-mg tablets at the $\alpha = .05$ level of significance. (HINT: the observations are paired by the online pharmacies.)

b. Find the estimated savings per tablet by purchasing the generic as opposed to the brand-name tablets with a 95% confidence interval.

10.93 Breathing Patterns Research psychologists measured the baseline breathing patterns—the total ventilation (in liters of air per minute) adjusted for body size—for each of $n = 30$ patients, so that they could estimate the average total ventilation for patients before any experimentation was done. The data, along with some *MINITAB* output, are presented here:

5.23	5.72	5.77	4.99	5.12	4.82
5.54	4.79	5.16	5.84	4.51	5.14
5.92	6.04	5.83	5.32	6.19	5.70
4.72	5.38	5.48	5.37	4.96	5.58
4.67	5.17	6.34	6.58	4.35	5.63

MINITAB output for Exercise 10.93

Stem-and-Leaf Display: Ltrs/min
```
Stem-and-leaf of Ltrs/min  N = 30
Leaf Unit = 0.10
    1    4 3
    2    4 5
    5    4 677
    8    4 899
   12    5 1111
   (4)   5 2333
   14    5 455
   11    5 6777
    7    5 889
    4    6 01
    2    6 3
    1    6 5
```

Descriptive Statistics: Ltrs/min
```
Variable   N   N*   Mean    SE Mean   StDev
Ltrs/min   30   0  5.3953   0.0997   0.5462

Minimum     Q1    Median     Q3    Maximum
 4.3500   4.9825  5.3750   5.7850   6.5800
```

a. What information does the stem and leaf plot give you about the data? Why is this important?

b. Use the output to construct a 99% confidence interval for the average total ventilation for patients.

10.94 Reaction Times A comparison of reaction times (in seconds) for two different stimuli in a psychological word-association experiment produced the following results when applied to a random sample of 16 people:

Stimulus 1	1	3	2	1	2	1	3	2
Stimulus 2	4	2	3	3	1	2	3	3

Do the data present sufficient evidence to indicate a difference in mean reaction times for the two stimuli? Test using $\alpha = .05$.

10.95 Reaction Times II Refer to Exercise 10.94. Suppose that the word-association experiment is conducted using eight people as blocks and making a comparison of reaction times within each person; that is, each person is subjected to both stimuli in a random order. The reaction times (in seconds) for the experiment are as follows:

Person	Stimulus 1	Stimulus 2
1	3	4
2	1	2
3	1	3
4	2	1
5	1	2
6	2	3
7	3	3
8	2	3

Do the data present sufficient evidence to indicate a difference in mean reaction times for the two stimuli? Test using $\alpha = .05$.

10.96 Refer to Exercises 10.94 and 10.95. Calculate a 95% confidence interval for the difference in the two population means for each of these experimental designs. Does it appear that blocking increased the amount of information available in the experiment?

Data set **10.97 Impact Strength** The following data
EX1097 are readings (in foot-pounds) of the impact
strengths of two kinds of packaging material:

A	B
1.25	.89
1.16	1.01
1.33	.97
1.15	.95
1.23	.94
1.20	1.02
1.32	.98
1.28	1.06
1.21	.98

MS Excel output for Exercise 10.97

D	E	F
t-Test: Two-Sample Assuming Equal Variances		
	A	B
Mean	1.2367	0.9778
Variance	0.0042	0.0024
Observations	9	9
Pooled Variance	0.0033	
Hypothesized Mean Difference	0	
df	16	
t Stat	9.5641	
P(T<=t) one-tail	0.0000	
t Critical one-tail	1.7459	
P(T<=t) two-tail	0.0000	
t Critical two-tail	2.1199	

a. Use the *MS Excel* printout to determine whether
there is evidence of a difference in the mean
strengths for the two kinds of material.

b. Are there practical implications to your results?

Data set **10.98 Cake Mixes** An experiment was con-
EX1098 ducted to compare the densities (in ounces per
cubic inch) of cakes prepared from two different cake
mixes. Six cake pans were filled with batter A, and six
were filled with batter B. Expecting a variation in oven
temperature, the experimenter placed a pan filled with
batter A and another with batter B *side by side* at six
different locations in the oven. The six paired observations of densities are as follows:

Location	1	2	3	4	5	6
Batter A	.135	.102	.098	.141	.131	.144
Batter B	.129	.120	.112	.152	.135	.163

a. Do the data present sufficient evidence to indicate
a difference between the average densities of cakes
prepared using the two types of batter?

b. Construct a 95% confidence interval for the difference between the average densities for the two
mixes.

10.99 Under what assumptions can the F distribution
be used in making inferences about the ratio of population variances?

10.100 Got Milk? A dairy is in the market for a
new container-filling machine and is considering two
models, manufactured by company A and company B.
Ruggedness, cost, and convenience are comparable in
the two models, so the deciding factor is the variability
of fills. The model that produces fills with the smaller
variance is preferred. If you obtain samples of fills for
each of the two models, an F-test can be used to test
for the equality of population variances. Which type of
rejection region would be most favored by each of
these individuals?

a. The manager of the dairy—Why?

b. A sales representative for company A—Why?

c. A sales representative for company B—Why?

10.101 Got Milk II Refer to Exercise 10.100.
Wishing to demonstrate that the variability of fills is
less for her model than for her competitor's, a sales
representative for company A acquired a sample of
30 fills from her company's model and a sample of
10 fills from her competitor's model. The sample
variances were $s_A^2 = .027$ and $s_B^2 = .065$, respectively. Does this result provide statistical support at
the .05 level of significance for the sales representative's claim?

10.102 Chemical Purity A chemical manufacturer
claims that the purity of his product never varies
by more than 2%. Five batches were tested and
given purity readings of 98.2, 97.1, 98.9, 97.7, and
97.9%.

a. Do the data provide sufficient evidence to contradict the manufacturer's claim? (HINT: To be generous, let a range of 2% equal 4σ.)

b. Find a 90% confidence interval for σ^2.

10.103 16-Ounce Cans? A cannery prints "weight
16 ounces" on its label. The quality control supervisor
selects nine cans at random and weighs them. She
finds $\bar{x} = 15.7$ and $s = .5$. Do the data present sufficient evidence to indicate that the mean weight is less
than that claimed on the label?

10.104 Reaction Time III A psychologist wishes to
verify that a certain drug increases the reaction time to
a given stimulus. The following reaction times (in

tenths of a second) were recorded before and after injection of the drug for each of four subjects:

	Reaction Time	
Subject	Before	After
1	7	13
2	2	3
3	12	18
4	12	13

Test at the 5% level of significance to determine whether the drug significantly increases reaction time.

10.105 Food Production At a time when energy conservation is so important, some scientists think closer scrutiny should be given to the cost (in energy) of producing various forms of food. Suppose you wish to compare the mean amount of oil required to produce 1 acre of corn versus 1 acre of cauliflower. The readings (in barrels of oil per acre), based on 20-acre plots, seven for each crop, are shown in the table.

Corn	Cauliflower
5.6	15.9
7.1	13.4
4.5	17.6
6.0	16.8
7.9	15.8
4.8	16.3
5.7	17.1

a. Use these data to find a 90% confidence interval for the difference between the mean amounts of oil required to produce these two crops.

b. Based on the interval in part a, is there evidence of a difference in the average amount of oil required to produce these two crops? Explain.

10.106 Alcohol and Altitude The effect of alcohol consumption on the body appears to be much greater at high altitudes than at sea level. To test this theory, a scientist randomly selects 12 subjects and randomly divides them into two groups of six each. One group is put into a chamber that simulates conditions at an altitude of 12,000 feet, and each subject ingests a drink containing 100 cubic centimeters (cc) of alcohol. The second group receives the same drink in a chamber that simulates conditions at sea level. After 2 hours, the amount of alcohol in the blood (grams per 100 cc) for each subject is measured. The data are shown in the table. Do the data provide sufficient evidence to support the theory that average amount of alcohol in the blood after 2 hours is greater at high altitudes?

Sea Level	12,000 Feet
.07	.13
.10	.17
.09	.15
.12	.14
.09	.10
.13	.14

10.107 Stock Risks The closing prices of two common stocks were recorded for a period of 15 days. The means and variances are

$$\bar{x}_1 = 40.33 \quad \bar{x}_2 = 42.54$$
$$s_1^2 = 1.54 \quad s_2^2 = 2.96$$

a. Do these data present sufficient evidence to indicate a difference between the variabilities of the closing prices of the two stocks for the populations associated with the two samples? Give the p-value for the test and interpret its value.

b. Construct a 99% confidence interval for the ratio of the two population variances.

10.108 Auto Design An experiment is conducted to compare two new automobile designs. Twenty people are randomly selected, and each person is asked to rate each design on a scale of 1 (poor) to 10 (excellent). The resulting ratings will be used to test the null hypothesis that the mean level of approval is the same for both designs against the alternative hypothesis that one of the automobile designs is preferred. Do these data satisfy the assumptions required for the Student's t-test of Section 10.4? Explain.

10.109 Safety Programs The data shown here were collected on lost-time accidents (the figures given are mean work-hours lost per month over a period of 1 year) before and after an industrial safety program was put into effect. Data were recorded for six industrial plants. Do the data provide sufficient evidence to indicate whether the safety program was effective in reducing lost-time accidents? Test using $\alpha = .01$.

	Plant Number					
	1	2	3	4	5	6
Before Program	38	64	42	70	58	30
After Program	31	58	43	65	52	29

10.110 Two Different Entrees To compare the demand for two different entrees, the manager of a cafeteria recorded the number of purchases of each entree on seven consecutive days. The data are shown in the table. Do the data provide sufficient evidence to indicate a greater mean demand for one of the entrees? Use the *Excel* printout.

Day	A	B
Monday	420	391
Tuesday	374	343
Wednesday	434	469
Thursday	395	412
Friday	637	538
Saturday	594	521
Sunday	679	625

MS Excel output for Exercise 10.110

E	F	G
t-Test: Paired Two Sample for Means		
	A	B
Mean	504.714	471.286
Variance	16191.238	9495.571
Observations	7	7
Pearson Correlation	0.945	
Hypothesized Mean Difference	0	
df	6	
t Stat	1.862	
P(T<=t) one-tail	0.056	
t Critical one-tail	1.943	
P(T<=t) two-tail	0.112	
t Critical two-tail	2.447	

10.111 Pollution Control The EPA limit on the allowable discharge of suspended solids into rivers and streams is 60 milligrams per liter (mg/l) per day. A study of water samples selected from the discharge at a phosphate mine shows that over a long period, the mean daily discharge of suspended solids is 48 mg/l, but day-to-day discharge readings are variable. State inspectors measured the discharge rates of suspended solids for $n = 20$ days and found $s^2 = 39$ (mg/l)^2. Find a 90% confidence interval for σ^2. Interpret your results.

10.112 Enzymes Two methods were used to EX10112 measure the specific activity (in units of enzyme activity per milligram of protein) of an enzyme. One unit of enzyme activity is the amount that catalyzes the formation of 1 micromole of product per minute under specified conditions. Use an appropriate test or estimation procedure to compare the two methods of measurement. Comment on the validity of any assumptions you need to make.

Method 1	125	137	130	151	142
Method 2	137	143	151	156	149

10.113 Connector Rods A producer of machine parts claimed that the diameters of the connector rods produced by his plant had a variance of at most .03 inch^2. A random sample of 15 connector rods from his plant produced a sample mean and variance of .55 inch and .053 inch^2, respectively.

a. Is there sufficient evidence to reject his claim at the $\alpha = .05$ level of significance?

b. Find a 95% confidence interval for the variance of the rod diameters.

10.114 Sleep and the College Student How much sleep do you get on a typical school night? A group of 10 college students were asked to report the number of hours that they slept on the previous night with the following results:

7, 6, 7.25, 7, 8.5, 5, 8, 7, 6.75, 6

a. Find a 99% confidence interval for the average number of hours that college students sleep.

b. What assumptions are required in order for this confidence interval to be valid?

10.115 Arranging Objects The following EX10115 data are the response times in seconds for $n = 25$ first graders to arrange three objects by size.

5.2	3.8	5.7	3.9	3.7
4.2	4.1	4.3	4.7	4.3
3.1	2.5	3.0	4.4	4.8
3.6	3.9	4.8	5.3	4.2
4.7	3.3	4.2	3.8	5.4

Find a 95% confidence interval for the average response time for first graders to arrange three objects by size. Interpret this interval.

10.116 The NBA Finals Want to attend a EX10116 pro-basketball finals game? The average prices for the NBA rematch of the Boston Celtics and the LA Lakers in 2010 compared to the average ticket prices in 2008 are given in the table that follows.[18]

Game	2008 ($)	2010 ($)
1	593	532
2	684	855
3	727	541
4	907	458
5	769	621
6	753	681
7	533	890

a. If we were to assume that the prices given in the table have been randomly selected, test for a significant difference between the 2008 and 2010 prices. Use $\alpha = .01$.

b. Find a 98% confidence interval for the mean difference, $\mu_d = \mu_{08} - \mu_{10}$. Does this estimate confirm the results of part a?

10.117 Mall Rats An article in *American Demographics* investigated consumer habits at the mall. We tend to spend the most money shopping on the weekends, and, in particular, on Sundays from 4 to 6 P.M. Wednesday morning shoppers spend the least![19] Suppose that a random sample of 20 weekend shoppers and a random sample of 20 weekday shoppers were selected, and the amount spent per trip to the mall was recorded.

	Weekends	Weekdays
Sample Size	20	20
Sample Mean ($)	78	67
Sample Standard Deviation ($)	22	20

a. Is it reasonable to assume that the two population variances are equal? Use the F-test to test this hypothesis with $\alpha = .05$.

b. Based on the results of part a, use the appropriate test to determine whether there is a difference in the average amount spent per trip on weekends versus weekdays. Use $\alpha = .05$.

10.118 Books or iPads? As part of a larger pilot study, students at a Riverside, California middle school, will compare the learning of algebra by students using iPads versus students using the traditional algebra textbook with the same author and publisher.[20] To remove teacher-to-teacher variation, the same teacher will teach both classes, and the iPad and textbook material are both provided by the same author and publisher. Suppose that after 1 month, 10 students were selected from each class and their scores on an algebra advancement test recorded. The summarized data follow.

	iPad	Textbook
Mean	86.4	79.7
Standard Deviation	8.95	10.7
Sample Size	10	10

a. Use the summary data to test for a significant difference in advancement scores for the two groups using $\alpha = .05$.

b. Find a 95% confidence interval for the difference in mean scores for the two groups.

c. In light of parts a and b, what can we say about the efficacy of using an iPad versus a traditional textbook in learning algebra at the middle school level?

CASE STUDY

 API
Riverside
School

School Accountability Study—How Is Your School Doing?

If you are a California resident, the API (Academic Performance Index) may be published by your local paper. The 2009–2010 accountability progress report for Riverside County, California, was published in the *Press-Enterprise* listing the API and Growth (difference in 2010 and 2009 API) for all school districts in the county, together with whether each school has met the federal Adequate Yearly Progress guidelines.[21] The first two of these measures are given for a sample of elementary schools from the two school districts in Riverside, California.

Alvord Schools	API	Growth	Riverside Unified Schools	API	Growth
Stokoe	786	−5	Lake Matthews	891	18
Orrenmaa	760	−2	Beatty	778	8
Collett	768	8	Alcott	827	10
La Granada	745	40	Emerson	789	23
Terrace	731	−6	Harrison	805	28
McAuliffe	815	0	Highgrove	762	6
Valley View	690	−16	Hyatt	798	35
			Jefferson	796	34
			Longfellow	731	−4
			Magnolia	818	28
			Mountain View	769	3
			Victoria	815	7
			Woodcrest	841	8
			Franklin	876	3

1. To study these data statistically, begin by finding the means and standard deviations for the API and the Growth scores for each of the two districts.

2. Test for a significant growth in API scores for both the Alvord and Riverside Unified school districts. Use $\alpha = .05$.

3. Test for a significant difference in API for Alvord and Riverside Unified school districts using $\alpha = .05$.

4. Summarize your results on the progress of these two school districts in the form of a report.

chirapbogdan/Shutterstock.com

The Analysis of Variance

GENERAL OBJECTIVE

The quantity of information contained in a sample is affected by various factors that the experimenter may or may not be able to control. This chapter introduces three different *experimental designs,* two of which are direct extensions of the unpaired and paired designs of Chapter 10. A new technique called the *analysis of variance* is used to determine how the different experimental factors affect the average response.

CHAPTER INDEX

NEED TO KNOW...
How to Determine Whether My Calculations Are Accurate

How to Save Money on Groceries!

Canning or freezing produce that you buy in bulk will almost always save you money compared to buying in supermarkets. You can save more than 75% by canning—and more than 80% by freezing—produce purchased in bulk. The case study at the end of this chapter investigates the costs of purchasing in bulk, canning, and freezing using the analysis of variance procedures presented in this chapter.

THE DESIGN OF AN EXPERIMENT

11.1

The way that a sample is selected is called the *sampling plan* or *experimental design* and determines the amount of information in the sample. Some research involves an **observational study,** in which the researcher does not actually produce the data but only *observes* the characteristics of data that already exist. Most sample surveys, in which information is gathered with a questionnaire, fall into this category. The researcher forms a plan for collecting the data—called the *sampling plan*—and then uses the appropriate statistical procedures to draw conclusions about the population or populations from which the sample comes.

Other research involves **experimentation.** The researcher may deliberately impose one or more experimental conditions on the experimental units in order to determine their effect on the response. Here are some new terms we will use to discuss the design of a statistical experiment.

Definition An **experimental unit** is the object on which a measurement (or measurements) is taken.

A **factor** is an independent variable whose values are controlled and varied by the experimenter.

A **level** is the intensity setting of a factor.

A **treatment** is a specific combination of factor levels.

The **response** is the variable being measured by the experimenter.

EXAMPLE 11.1

A group of people is randomly divided into an experimental and a control group. The control group is given an aptitude test after having eaten a full breakfast. The experimental group is given the same test without having eaten any breakfast. What are the factors, levels, and treatments in this experiment?

Solution The *experimental units* are the people on which the *response* (test score) is measured. The *factor* of interest could be described as "meal" and has two *levels:* "breakfast" and "no breakfast." Since this is the only factor controlled by the experimenter, the two levels—"breakfast" and "no breakfast"—also represent the *treatments* of interest in the experiment.

EXAMPLE 11.2

Suppose that the experimenter in Example 11.1 began by randomly selecting 20 men and 20 women for the experiment. These two groups were then randomly divided into 10 each for the experimental and control groups. What are the factors, levels, and treatments in this experiment?

Solution Now there are two *factors* of interest to the experimenter, and each factor has two *levels:*

- "Gender" at two levels: men and women
- "Meal" at two levels: breakfast and no breakfast

In this more complex experiment, there are four *treatments,* one for each specific combination of factor levels: men without breakfast, men with breakfast, women without breakfast, and women with breakfast.

In this chapter, we will concentrate on experiments that have been designed in three different ways, and we will use a technique called the *analysis of variance* to judge the effects of various factors on the experimental response. Two of these *experimental designs* are extensions of the unpaired and paired designs from Chapter 10.

WHAT IS AN ANALYSIS OF VARIANCE?

11.2

The responses that are generated in an experimental situation always exhibit a certain amount of *variability*. In an **analysis of variance,** you divide the total variation in the response measurements into portions that may be attributed to various *factors* of interest to the experimenter. If the experiment has been properly designed, these portions can then be used to answer questions about the effects of the various factors on the response of interest.

You can better understand the logic underlying an analysis of variance by looking at a simple experiment. Consider two sets of samples randomly selected from populations 1 (◆) and 2 (○), each with identical pairs of means, $\bar{x}_1$ and $\bar{x}_2$. The two sets are shown in Figure 11.1. Is it easier to detect the difference in the two means when you look at set A or set B? You will probably agree that set A shows the difference much more clearly. In set A, the variability of the measurements *within* the groups (◆s and ○s) is much smaller than the variability *between* the two groups. In set B, there is more variability *within* the groups (◆s and ○s), causing the two groups to "mix" together and making it more difficult to see the *identical* difference in the means.

FIGURE 11.1

Two sets of samples with the same means

The comparison you have just done intuitively is formalized by the analysis of variance. Moreover, the analysis of variance can be used not only to compare two means but also to make comparisons of *more than two* population means and to determine the effects of various factors in more complex experimental designs. The analysis of variance relies on statistics with sampling distributions that are modeled by the *F* distribution of Section 10.7.

THE ASSUMPTIONS FOR AN ANALYSIS OF VARIANCE

11.3

The assumptions required for an analysis of variance are similar to those required for the Student's *t* and *F* statistics of Chapter 10. Regardless of the experimental design used to generate the data, you must assume that the observations within each treatment group are **normally distributed** with a **common variance** σ^2. As in Chapter 10, the analysis of variance procedures are fairly **robust** when the sample sizes are equal and when the data are fairly mound-shaped. Violating the assumption of a common variance is more serious, especially when the sample sizes are not nearly equal.

ASSUMPTIONS FOR ANALYSIS OF VARIANCE TEST AND ESTIMATION PROCEDURES

- The observations within each population are normally distributed with a common variance σ^2.
- Assumptions regarding the sampling procedure are specified for each design in the sections that follow.

This chapter describes the analysis of variance for three different experimental designs. The first design is based on independent random sampling from several populations and is an extension of the *unpaired t-test* of Chapter 10. The second is an extension of the *paired-difference* or *matched pairs* design and involves a random assignment of treatments within matched sets of observations. The third is a design that allows you to judge the effect of two experimental factors on the response. The sampling procedures necessary for each design will be restated in their respective sections.

THE COMPLETELY RANDOMIZED DESIGN: A ONE-WAY CLASSIFICATION

11.4

One of the simplest experimental designs is the **completely randomized design,** in which random samples are selected independently from each of k populations. This design involves only one *factor,* the population from which the measurement comes—hence the designation as a **one-way classification.** There are k different *levels* corresponding to the k populations, which are also the *treatments* for this one-way classification. Are the k population means all the same, or is at least one mean different from the others?

Why do you need a new procedure, the *analysis of variance,* to compare the population means when you already have the Student's *t*-test available? In comparing $k = 3$ means, you could test each of three pairs of hypotheses:

$$H_0 : \mu_1 = \mu_2 \qquad H_0 : \mu_1 = \mu_3 \qquad H_0 : \mu_2 = \mu_3$$

to find out where the differences lie. However, you must remember that each test you perform is subject to the possibility of error. To compare $k = 4$ means, you would need six tests, and you would need 10 tests to compare $k = 5$ means. The more tests you perform on a set of measurements, the more likely it is that at least one of your conclusions will be incorrect. The analysis of variance procedure provides one overall test to judge the equality of the k population means. Once you have determined whether there is *actually* a difference in the means, you can use another procedure to find out where the differences lie.

How can you select these k random samples? Sometimes the populations actually exist in fact, and you can use a computerized random number generator or a random number table to randomly select the samples. For example, in a study to compare the average sizes of health insurance claims in four different states, you could use a computer database provided by the health insurance companies to select random samples from the four states. In other situations, the populations may be *hypothetical,* and responses can be generated only after the experimental treatments have been applied.

EXAMPLE **11.3**

A researcher is interested in the effects of five types of insecticides for use in controlling the boll weevil in cotton fields. Explain how to implement a completely randomized design to investigate the effects of the five insecticides on crop yield.

Solution The only way to generate the equivalent of five random samples from the hypothetical populations corresponding to the five insecticides is to use a method called a **randomized assignment.** A fixed number of cotton plants are chosen for treatment, and each is assigned a random number. Suppose that each sample is to have an equal number of measurements. Using a randomization device, you can assign the first n_1 plants chosen to receive insecticide 1, the second n_2 plants to receive insecticide 2, and so on, until all five treatments have been assigned.

Whether by *random selection* or *random assignment,* both of these examples result in a completely randomized design, or one-way classification, for which the analysis of variance is used.

THE ANALYSIS OF VARIANCE FOR A COMPLETELY RANDOMIZED DESIGN

11.5

Suppose you want to compare k population means, $\mu_1, \mu_2, \ldots, \mu_k$, based on independent random samples of size $n_1, n_2, \ldots, n_k$ from normal populations with a common variance σ^2. That is, each of the normal populations has the same shape, but their locations might be different, as shown in Figure 11.2.

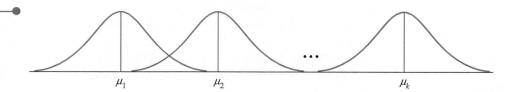

Partitioning the Total Variation in an Experiment

Let x_{ij} be the jth measurement ($j = 1, 2, \ldots, n_i$) in the ith sample. The analysis of variance procedure begins by considering the total variation in the experiment, which is measured by a quantity called the **total sum of squares** (**TSS**):

$$\text{Total SS} = \Sigma(x_{ij} - \bar{x})^2 = \Sigma x_{ij}^2 - \frac{(\Sigma x_{ij})^2}{n}$$

This is the familiar numerator in the formula for the sample variance for the entire set of $n = n_1 + n_2 + \cdots + n_k$ measurements. The second part of the calculational formula is sometimes called the **correction for the mean** (**CM**). If we let G represent the *grand total* of all n observations, then

$$\text{CM} = \frac{(\Sigma x_{ij})^2}{n} = \frac{G^2}{n}$$

This Total SS is partitioned into two components. The first component, called the **sum of squares for treatments** (**SST**), measures the variation among the k sample means:

$$\text{SST} = \Sigma n_i(\bar{x}_i - \bar{x})^2 = \Sigma \frac{T_i^2}{n_i} - \text{CM}$$

where T_i is the total of the observations for treatment i. The second component, called the **sum of squares for error (SSE)**, is used to measure the pooled variation within the k samples:

$$SSE = (n_1 - 1)s_1^2 + (n_2 - 1)s_2^2 + \cdots + (n_k - 1)s_k^2$$

This formula is a direct extension of the numerator in the formula for the pooled estimate of σ^2 from Chapter 10. We can show algebraically that, in the analysis of variance,

$$\text{Total SS} = \text{SST} + \text{SSE}$$

Therefore, you need to calculate only two of the three sums of squares—Total SS, SST, and SSE—and the third can be found by subtraction.

Each of the sources of variation, when divided by its appropriate **degrees of freedom,** provides an estimate of the variation in the experiment. Since Total SS involves n squared observations, its degrees of freedom are $df = (n - 1)$. Similarly, the sum of squares for treatments involves k squared observations, and its degrees of freedom are $df = (k - 1)$. Finally, the sum of squares for error, a direct extension of the pooled estimate in Chapter 10, has

$$df = (n_1 - 1) + (n_2 - 1) + \cdots + (n_k - 1) = n - k$$

Notice that the degrees of freedom for treatments and error are additive—that is,

$$df(\text{total}) = df(\text{treatments}) + df(\text{error})$$

These two sources of variation and their respective degrees of freedom are combined to form the **mean squares** as MS = SS/df. The total variation in the experiment is then displayed in an **analysis of variance** (or **ANOVA**) **table.**

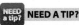 **NEED A TIP?**

The column labeled "SS" satisfies:
Total SS = SST + SSE.

ANOVA TABLE FOR k INDEPENDENT RANDOM SAMPLES: COMPLETELY RANDOMIZED DESIGN

Source	df	SS	MS	F
Treatments	$k - 1$	SST	MST = SST/($k - 1$)	MST/MSE
Error	$n - k$	SSE	MSE = SSE/($n - k$)	
Total	$n - 1$	Total SS		

where

$$\text{Total SS} = \Sigma x_{ij}^2 - \text{CM}$$
$$= (\text{Sum of squares of all } x\text{-values}) - \text{CM}$$

with

$$\text{CM} = \frac{(\Sigma x_{ij})^2}{n} = \frac{G^2}{n}$$

$$\text{SST} = \Sigma \frac{T_i^2}{n_i} - \text{CM} \qquad \text{MST} = \frac{\text{SST}}{k - 1}$$

$$\text{SSE} = \text{Total SS} - \text{SST} \qquad \text{MSE} = \frac{\text{SSE}}{n - k}$$

and

G = Grand total of all n observations
T_i = Total of all observations in sample i
n_i = Number of observations in sample i
$n = n_1 + n_2 + \cdots + n_k$

NEED A TIP?

The column labeled "df" always adds up to $n - 1$.

EXAMPLE 11.4

In an experiment to determine the effect of nutrition on the attention spans of elementary school students, a group of 15 students were randomly assigned to each of three meal plans: no breakfast, light breakfast, and full breakfast. Their attention spans (in minutes) were recorded during a morning reading period and are shown in Table 11.1. Construct the analysis of variance table for this experiment.

TABLE 11.1 ● **Attention Spans of Students After Three Meal Plans**

No Breakfast	Light Breakfast	Full Breakfast
8	14	10
7	16	12
9	12	16
13	17	15
10	11	12
$T_1 = 47$	$T_2 = 70$	$T_3 = 65$

Solution To use the calculational formulas, you need the $k = 3$ treatment totals together with $n_1 = n_2 = n_3 = 5$, $n = 15$, and $\Sigma x_{ij} = 182$. Then

$$CM = \frac{(182)^2}{15} = 2208.2667$$

Total SS $= (8^2 + 7^2 + \cdots + 12^2) - CM = 2338 - 2208.2667 = 129.7333$ with $(n - 1) = (15 - 1) = 14$ degrees of freedom,

$$SST = \frac{47^2 + 70^2 + 65^2}{5} - CM = 2266.8 - 2208.2667 = 58.5333$$

with $(k - 1) = (3 - 1) = 2$ degrees of freedom, and by subtraction,

SSE = Total SS $-$ SST $= 129.7333 - 58.5333 = 71.2$

with $(n - k) = (15 - 3) = 12$ degrees of freedom. These three sources of variation, their degrees of freedom, sums of squares, and mean squares are shown in the shaded area of the ANOVA tables generated by both *MS Excel* and *MINITAB* and given in Figure 11.3. You will find instructions for generating this output in the "Technology Today" section at the end of this chapter.

FIGURE 11.3(a)

MINITAB output for Example 11.4

● **One-way ANOVA: Span versus Meal**

```
Source   DF       SS      MS      F       P
Meal      2     58.53   29.27   4.93   0.027
Error    12     71.20    5.93
Total    14    129.73

S = 2.436   R-Sq = 45.12%   R-Sq(adj) = 35.97%
                            Individual 95% CIs For Mean
                            Based on Pooled StDev
Level   N     Mean   StDev   --+---------+---------+---------+-------
1       5    9.400   2.302   (---------*--------)
2       5   14.000   2.550                    (--------*--------)
3       5   13.000   2.449            (--------*--------)
                            --+---------+---------+---------+-------
                            7.5       10.0      12.5      15.0

Pooled StDev = 2.436
```

FIGURE 11.3(b)

MS Excel output for Example 11.4

Anova: Single Factor

SUMMARY

Groups	Count	Sum	Average	Variance
None	5	47	9.4	5.3
Light	5	70	14	6.5
Full	5	65	13	6

ANOVA

Source of Variation	SS	df	MS	F	P-value	F crit
Between Groups	58.533	2	29.267	4.933	0.027	3.885
Within Groups	71.2	12	5.933			
Total	129.733	14				

The computer outputs give some additional information about the variation in the experiment. The lower section in *MINITAB* and the upper section in *MS Excel* show the means and standard deviations (or variances) for the three meal plans. More important, you can see in the upper section in *MINITAB* and the lower section in *MS Excel* two columns marked "F" and "P" ("*F-*" and "*P*-value" in *Excel*). We can use these values to test a hypothesis concerning the equality of the three treatment means.

Testing the Equality of the Treatment Means

The *mean squares* in the analysis of variance table can be used to test the null hypothesis

$$H_0 : \mu_1 = \mu_2 = \cdots = \mu_k$$

versus the alternative hypothesis

$$H_a : \text{At least one of the means is different from the others}$$

using the following theoretical argument:

- Remember that σ^2 is the common variance for all k populations. The quantity

$$\text{MSE} = \frac{\text{SSE}}{n - k}$$

is a pooled estimate of σ^2, a weighted average of all k sample variances, whether or not H_0 is true.

- If H_0 is true, then the variation in the sample means, measured by MST = [SST/($k - 1$)], also provides an unbiased estimate of σ^2. However, if H_0 is false and the population means are different, then MST—which measures the variation in the sample means—will be unusually *large*, as shown in Figure 11.4.

FIGURE 11.4

Sample means drawn from identical versus different populations

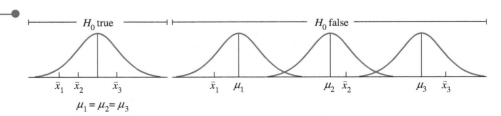

- The test statistic

$$F = \frac{\text{MST}}{\text{MSE}}$$

NEED A TIP?

F-tests for ANOVA tables are **always** upper (right) tailed.

tends to be larger than usual if H_0 is false. Hence, you can reject H_0 for large values of F, using a *right-tailed* statistical test. When H_0 is true, this test statistic has an F distribution with $df_1 = (k - 1)$ and $df_2 = (n - k)$ degrees of freedom, and *right-tailed* critical values of the F distribution (from Table 6 in Appendix I) or computer-generated *p*-values can be used to draw statistical conclusions about the equality of the population means.

F-TEST FOR COMPARING *k* POPULATION MEANS

1. Null hypothesis: $H_0 : \mu_1 = \mu_2 = \cdots = \mu_k$
2. Alternative hypothesis: H_a : One or more pairs of population means differ
3. Test statistic: $F = \text{MST/MSE}$, where F is based on $df_1 = (k - 1)$ and $df_2 = (n - k)$
4. Rejection region: Reject H_0 if $F > F_\alpha$, where F_α lies in the upper tail of the F distribution (with $df_1 = k - 1$ and $df_2 = n - k$) or if the *p*-value $< \alpha$.

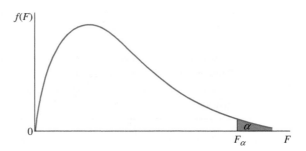

Assumptions:

- The samples are randomly and independently selected from their respective populations.
- The populations are normally distributed with means $\mu_1, \mu_2, \ldots, \mu_k$ and equal variances, $\sigma_1^2 = \sigma_2^2 = \cdots = \sigma_k^2 = \sigma^2$.

EXAMPLE 11.5

Do the data in Example 11.4 provide sufficient evidence to indicate a difference in the average attention spans depending on the type of breakfast eaten by the student?

Solution To test $H_0 : \mu_1 = \mu_2 = \mu_3$ versus the alternative hypothesis that the average attention span is different for at least one of the three treatments, you use the analysis of variance F statistic, calculated as

$$F = \frac{\text{MST}}{\text{MSE}} = \frac{29.2667}{5.9333} = 4.93$$

and shown in the columns marked "F" in Figure 11.3(a) and "F" in Figure 11.3(b). It will not surprise you to know that the value in the column marked "P" in Figure 11.3(a) and "P-value" in Figure 11.3(b) is the exact p-value for this statistical test.

The test statistic MST/MSE calculated above has an F distribution with $df_1 = 2$ and $df_2 = 12$ degrees of freedom. Using the critical value approach with $\alpha = .05$, you can reject H_0 if $F > F_{.05} = 3.89$ from Table 6 in Appendix I (see Figure 11.5). Since the observed value, $F = 4.93$, exceeds the critical value, you reject H_0. There is sufficient evidence to indicate that at least one of the three average attention spans is different from at least one of the others.

FIGURE 11.5

Rejection region for Example 11.5

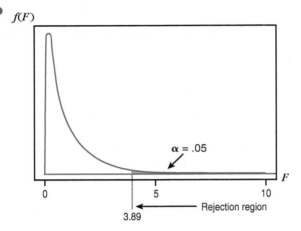

You could have reached this same conclusion using the exact p-value, .027, given in Figure 11.3. Since the p-value is less than $\alpha = .05$, the results are statistically significant at the 5% level. You still conclude that at least one of the three average attention spans is different from at least one of the others.

Estimating Differences in the Treatment Means

The next obvious question you might ask involves the nature of the differences in the population means. Which means are different from the others? How can you estimate the difference, or possibly the individual means for each of the three treatments? In Section 11.6, we will present a procedure that you can use to compare all possible pairs of treatment means simultaneously. However, if you have a special interest in a particular mean or pair of means, you can construct confidence intervals using the small-sample procedures of Chapter 10, based on the Student's t distribution. For a single population mean, μ_i, the confidence interval is

$$\bar{x}_i \pm t_{\alpha/2}\left(\frac{s}{\sqrt{n_i}}\right)$$

where $\bar{x}_i$ is the sample mean for the ith treatment. Similarly, for a comparison of two population means—say, μ_i and μ_j—the confidence interval is

$$(\bar{x}_i - \bar{x}_j) \pm t_{\alpha/2}\sqrt{s^2\left(\frac{1}{n_i} + \frac{1}{n_j}\right)}$$

Before you can use these confidence intervals, however, two questions remain:

- How do you calculate s or s^2, the best estimate of the common variance σ^2?
- How many degrees of freedom are used for the critical value of t?

To answer these questions, remember that in an analysis of variance, the mean square for error, MSE, always provides an unbiased estimator of σ^2 and uses information from the entire set of measurements. Hence, it is the best available estimator of σ^2, regardless of what test or estimation procedure you are using. You should *always* use

$$s^2 = \text{MSE} \qquad \text{with } df = (n - k)$$

to estimate σ^2! You can find the positive square root of this estimator, $s = \sqrt{\text{MSE}}$, on the last line of Figure 11.3(a) labeled "Pooled StDev."

COMPLETELY RANDOMIZED DESIGN: $(1 - \alpha)100\%$ CONFIDENCE INTERVALS FOR A SINGLE TREATMENT MEAN AND THE DIFFERENCE BETWEEN TWO TREATMENT MEANS

NEED A TIP?

Degrees of freedom for confidence intervals are the *df* for **error**.

Single treatment mean:

$$\bar{x}_i \pm t_{\alpha/2}\left(\frac{s}{\sqrt{n_i}}\right)$$

Difference between two treatment means:

$$(\bar{x}_i - \bar{x}_j) \pm t_{\alpha/2}\sqrt{s^2\left(\frac{1}{n_i} + \frac{1}{n_j}\right)}$$

with

$$s = \sqrt{s^2} = \sqrt{\text{MSE}} = \sqrt{\frac{\text{SSE}}{n - k}}$$

where $n = n_1 + n_2 + \cdots + n_k$ and $t_{\alpha/2}$ is based on $(n - k)\ df$.

EXAMPLE 11.6

The researcher in Example 11.4 believes that students who have no breakfast will have significantly shorter attention spans but that there may be no difference between those who eat a light or a full breakfast. Find a 95% confidence interval for the average attention span for students who eat no breakfast, as well as a 95% confidence interval for the difference in the average attention spans for light versus full breakfast eaters.

Solution For $s^2 = \text{MSE} = 5.9333$ so that $s = \sqrt{5.9333} = 2.436$ with $df = (n - k) = 12$, you can calculate the two confidence intervals:

- For no breakfast:

$$\bar{x}_1 \pm t_{\alpha/2}\left(\frac{s}{\sqrt{n_1}}\right)$$

$$9.4 \pm 2.179\left(\frac{2.436}{\sqrt{5}}\right)$$

$$9.4 \pm 2.37$$

or between 7.03 and 11.77 minutes.

- For light versus full breakfast:

$$(\bar{x}_2 - \bar{x}_3) \pm t_{\alpha/2}\sqrt{s^2\left(\frac{1}{n_2} + \frac{1}{n_3}\right)}$$

$$(14 - 13) \pm 2.179\sqrt{5.9333\left(\frac{1}{5} + \frac{1}{5}\right)}$$

$$1 \pm 3.36$$

a difference of between -2.36 and 4.36 minutes.

You can see that the second confidence interval does not indicate a difference in the average attention spans for students who ate light versus full breakfasts, as the researcher suspected. If the researcher, because of prior beliefs, wishes to test the other two possible pairs of means—none versus light breakfast, and none versus full breakfast—the methods given in Section 11.6 should be used for testing all three pairs.

Some computer programs have graphics options that provide a powerful visual description of data and the k treatment means. One such option in the *MINITAB* program is shown in Figure 11.6. The treatment means are indicated by the symbol $\oplus$ and are connected with straight lines. Notice that the "no breakfast" mean appears to be somewhat different from the other two means, as the researcher suspected, although there is a bit of overlap in the box plots. In the next section, we present a formal procedure for testing the significance of the differences between all pairs of treatment means.

FIGURE 11.6

Box plots for Example 11.6

NEED TO KNOW...

How to Determine Whether My Calculations Are Accurate

The following suggestions apply to all the analyses of variance in this chapter:

1. When calculating sums of squares, be certain to carry at least six significant figures before performing subtractions.

2. Remember, sums of squares can never be negative. If you obtain a negative sum of squares, you have made a mistake in arithmetic.

3. Always check your analysis of variance table to make certain that the degrees of freedom sum to the total degrees of freedom $(n - 1)$ and that the sums of squares sum to Total SS.

11.5 EXERCISES

BASIC TECHNIQUES

11.1 Suppose you wish to compare the means of six populations based on independent random samples, each of which contains 10 observations. Insert, in an ANOVA table, the sources of variation and their respective degrees of freedom.

11.2 The values of Total SS and SSE for the experiment in Exercise 11.1 are Total SS = 21.4 and SSE = 16.2.

a. Complete the ANOVA table for Exercise 11.1.

b. How many degrees of freedom are associated with the F statistic for testing $H_0 : \mu_1 = \mu_2 = \cdots = \mu_6$?

c. Give the rejection region for the test in part b for $\alpha = .05$.

d. Do the data provide sufficient evidence to indicate differences among the population means?

e. Estimate the p-value for the test. Does this value confirm your conclusions in part d?

11.3 The sample means corresponding to populations 1 and 2 in Exercise 11.1 are $\bar{x}_1 = 3.07$ and $\bar{x}_2 = 2.52$.

a. Find a 95% confidence interval for μ_1.

b. Find a 95% confidence interval for the difference $(\mu_1 - \mu_2)$.

11.4 Suppose you wish to compare the means of four populations based on independent random samples, each of which contains six observations. Insert, in an ANOVA table, the sources of variation and their respective degrees of freedom.

11.5 The values of Total SS and SST for the experiment in Exercise 11.4 are Total SS = 473.2 and SST = 339.8.

a. Complete the ANOVA table for Exercise 11.4.

b. How many degrees of freedom are associated with the F statistic for testing $H_0 : \mu_1 = \mu_2 = \mu_3 = \mu_4$?

c. Give the rejection region for the test in part b for $\alpha = .05$.

d. Do the data provide sufficient evidence to indicate differences among the population means?

e. Approximate the p-value for the test. Does this confirm your conclusions in part d?

11.6 The sample means corresponding to populations 1 and 2 in Exercise 11.4 are $\bar{x}_1 = 88.0$ and $\bar{x}_2 = 83.9$.

a. Find a 90% confidence interval for μ_1.

b. Find a 90% confidence interval for the difference $(\mu_1 - \mu_2)$.

 11.7 These data are observations collected
EX1107 using a completely randomized design:

Sample 1	Sample 2	Sample 3
3	4	2
2	3	0
4	5	2
3	2	1
2	5	

a. Calculate CM and Total SS.

b. Calculate SST and MST.

c. Calculate SSE and MSE.

d. Construct an ANOVA table for the data.

e. State the null and alternative hypotheses for an analysis of variance F-test.

f. Use the p-value approach to determine whether there is a difference in the three population means.

11.8 Refer to Exercise 11.7 and data set EX1107. Do the data provide sufficient evidence to indicate a difference between μ_2 and μ_3? Test using the t-test of Section 10.4 with $\alpha = .05$.

11.9 Refer to Exercise 11.7 and data set EX1107.

a. Find a 90% confidence interval for μ_1.

b. Find a 90% confidence interval for the difference $(\mu_1 - \mu_3)$.

APPLICATIONS

Data set **11.10 Reducing Hostility** A clinical psychologist wished to compare three methods for EX1110 reducing hostility levels in university students using a certain psychological test (HLT). High scores on this test were taken to indicate great hostility, and 11 students who got high and nearly equal scores were used in the experiment. Five were selected at random from among the 11 students and treated by method A, three were taken at random from the remaining six students and treated by method B, and the other three students were treated by method C. All treatments continued throughout a semester, when the HLT test was given again. The results are shown in the table.

Method	Scores on the HLT Test				
A	73	83	76	68	80
B	54	74	71		
C	79	95	87		

a. Perform an analysis of variance for this experiment.

b. Do the data provide sufficient evidence to indicate a difference in mean student scores after treatment for the three methods?

11.11 Hostility, continued Refer to Exercise 11.10. Let μ_A and μ_B, respectively, denote the mean scores at the end of the semester for the populations of extremely hostile students who were treated throughout that semester by method A and method B.

a. Find a 95% confidence interval for μ_A.

b. Find a 95% confidence interval for μ_B.

c. Find a 95% confidence interval for $(\mu_A - \mu_B)$.

d. Is it correct to claim that the confidence intervals found in parts a, b, and c are jointly valid?

Data set **11.12 Assembling Electronic Equipment** EX1112 An experiment was conducted to compare the effectiveness of three training programs, A, B, and C, in training assemblers of a piece of electronic equipment. Fifteen employees were randomly assigned, five each, to the three programs. After completion of the program, each person was required to assemble four pieces of the equipment, and the average length of time required to complete the assembly was recorded. Several of the employees resigned during the course of the program; the remainder were evaluated, producing the data shown in the accompanying table. Use the *Excel* printout to answer the questions.

Training Program	Average Assembly Time (min)				
A	59	64	57	62	
B	52	58	54		
C	58	65	71	63	64

a. Do the data provide sufficient evidence to indicate a difference in mean assembly times for people trained by the three programs? Give the p-value for the test and interpret its value.

b. Find a 99% confidence interval for the difference in mean assembly times between persons trained by programs A and B.

c. Find a 99% confidence interval for the mean assembly times for persons trained by program A.

d. Do you think the data will satisfy (approximately) the assumption that they have been selected from normal populations? Why?

MS Excel output for Exercise 11.12

Anova: Single Factor

SUMMARY

Groups	Count	Sum	Average	Variance
A	4	242	60.5	9.667
B	3	164	54.667	9.333
C	5	321	64.2	21.7

ANOVA

Source of Variation	SS	df	MS	F	P-value	F crit
Between Groups	170.45	2	85.225	5.704	0.0251	4.256
Within Groups	134.467	9	14.941			
Total	304.917	11				

Data set **11.13 Swampy Sites** An ecological study EX1113 was conducted to compare the rates of growth of vegetation at four swampy undeveloped sites and to determine the cause of any differences that might be observed. Part of the study involved measuring the leaf lengths of a particular plant species on a preselected date in May. Six plants were randomly selected at each of the four sites to be used in the comparison. The data in the table are the mean leaf length per plant (in centimeters) for a random sample of 10 leaves per plant.

The *MINITAB* analysis of variance computer printout for these data is also provided.

Location	Mean Leaf Length (cm)					
1	5.7	6.3	6.1	6.0	5.8	6.2
2	6.2	5.3	5.7	6.0	5.2	5.5
3	5.4	5.0	6.0	5.6	4.9	5.2
4	3.7	3.2	3.9	4.0	3.5	3.6

MINITAB output for Exercise 11.13

```
One-way ANOVA: Length versus Location

Source      DF        SS        MS        F        P
Location     3     19.740     6.580    57.38    0.000
Error       20      2.293     0.115
Total       23     22.033
S = 0.3386    R-Sq = 89.59%    R-Sq(adj) = 88.03%
                                         Individual 95% CIs For Mean
                                       Based on Pooled StDev
Level     N     Mean     StDev    --------+---------+---------+---------+-
1         6    6.0167    0.2317                            (--*---)
2         6    5.6500    0.3937                    (---*--)
3         6    5.3500    0.4087                (---*--)
4         6    3.6500    0.2881    (---*--)
                                  --------+---------+---------+---------+-
Pooled StDev =    0.3386             4.00      4.80      5.60      6.40
```

a. You will recall that the test and estimation procedures for an analysis of variance require that the observations be selected from normally distributed (at least, roughly so) populations. Why might you feel reasonably confident that your data satisfy this assumption?

b. Do the data provide sufficient evidence to indicate a difference in mean leaf length among the four locations? What is the *p*-value for the test?

c. Suppose, prior to seeing the data, you decided to compare the mean leaf lengths of locations 1 and 4. Test the null hypothesis $\mu_1 = \mu_4$ against the alternative $\mu_1 \neq \mu_4$.

d. Refer to part c. Construct a 99% confidence interval for $(\mu_1 - \mu_4)$.

e. Rather than use an analysis of variance *F*-test, it would seem simpler to examine one's data, select the two locations that have the smallest and largest sample mean lengths, and then compare these two means using a Student's *t*-test. If there is evidence to indicate a difference in these means, there is clearly evidence of a difference among the four. (If you were to use this logic, there would be no need for the analysis of variance *F*-test.) Explain why this procedure is invalid.

11.14 Dissolved O₂ Content Water samples were taken at four different locations in a river to determine whether the quantity of dissolved oxygen, a measure of water pollution, varied from one location to another. Locations 1 and 2 were selected above an industrial plant, one near the shore and the other in

midstream; location 3 was adjacent to the industrial water discharge for the plant; and location 4 was slightly downriver in midstream. Five water specimens were randomly selected at each location, but one specimen, corresponding to location 4, was lost in the laboratory. The data and an *MS Excel* analysis of variance computer printout are provided here (the greater the pollution, the lower the dissolved oxygen readings).

Location	Mean Dissolved Oxygen Content				
1	5.9	6.1	6.3	6.1	6.0
2	6.3	6.6	6.4	6.4	6.5
3	4.8	4.3	5.0	4.7	5.1
4	6.0	6.2	6.1	5.8	

MS Excel output for Exercise 11.14

Anova: Single Factor

SUMMARY

Groups	Count	Sum	Average	Variance		
1	5	30.4	6.08	0.022		
2	5	32.2	6.44	0.013		
3	5	23.9	4.78	0.097		
4	4	24.1	6.025	0.0292		

ANOVA

Source of Variation	SS	df	MS	F	P-value	F crit
Between Groups	7.8361	3	2.6120	63.656	9E-09	3.287
Within Groups	0.6155	15	0.0410			
Total	8.4516	18				

a. Do the data provide sufficient evidence to indicate a difference in the mean dissolved oxygen contents for the four locations?

b. Compare the mean dissolved oxygen content in midstream above the plant with the mean content adjacent to the plant (location 2 versus location 3). Use a 95% confidence interval.

11.15 Calcium The calcium content of a powdered mineral substance was analyzed five times by each of three methods, with similar standard deviations:

Method	Percent Calcium				
1	.0279	.0276	.0270	.0275	.0281
2	.0268	.0274	.0267	.0263	.0267
3	.0280	.0279	.0282	.0278	.0283

Use an appropriate test to compare the three methods of measurement. Comment on the validity of any assumptions you need to make.

11.16 Tuna Fish In Exercise 10.6, we reported the estimated average prices for a 6-ounce can or a 7.06-ounce pouch of tuna fish, based on prices paid nationally for a variety of different brands of tuna.[1]

Light Tuna in Water		White Tuna in Oil	White Tuna in Water	Light Tuna in Oil	
.99	.53	1.27	1.49	2.56	.62
1.92	1.41	1.22	1.29	1.92	.66
1.23	1.12	1.19	1.27	1.30	.62
.85	.63	1.22	1.35	1.79	.65
.65	.67		1.29	1.23	.60
.69	.60		1.00		.67
.60	.66		1.27		
			1.28		

Source: From "Pricing of Tuna" Copyright 2001 by Consumers Union of U.S., Inc., Yonkers, NY 10703-1057, a nonprofit organization. Reprinted with permission from the June 2001 issue of *Consumer Reports*® for educational purposes only. No commercial use or reproduction permitted. www.ConsumerReports.org®.

a. Use an analysis of variance for a completely randomized design to determine if there are significant differences in the prices of tuna packaged in these four different ways. Can you reject the hypothesis of no difference in average price for these packages at the $\alpha = .05$ level of significance? At the $\alpha = .01$ level of significance?

b. Find a 95% confidence interval estimate of the difference in price between light tuna in water and light tuna in oil. Does there appear to be a significant difference in the price of these two kinds of packaged tuna?

c. Find a 95% confidence interval estimate of the difference in price between white tuna in water and white tuna in oil. Does there appear to be a significant difference in the price of these two kinds of packaged tuna?

d. What other confidence intervals might be of interest to the researcher who conducted this experiment?

11.17 The Cost of Lumber A national home builder wants to compare the prices per 1000 board feet of standard or better grade green Douglas fir framing lumber. He randomly selects five suppliers in each of the four states where the builder is planning to begin construction. The prices are given in the table.

State			
1 ($)	2 ($)	3 ($)	4 ($)
261	236	250	265
255	240	245	270
258	225	255	258
267	233	248	275
270	240	260	275

a. What type of experimental design has been used?

b. Construct the analysis of variance table for this data.

c. Do the data provide sufficient evidence to indicate that the average price per 1000 board feet of Douglas fir differs among the four states? Test using $\alpha = .05$.

11.18 Good at Math? Twenty third graders were randomly separated into four equal groups, and each group was taught a mathematical concept using a different teaching method. At the end of the teaching period, progress was measured by a unit test. The scores are shown below (one child in group 3 was absent on the day that the test was administered).

Group			
1	2	3	4
112	111	140	101
92	129	121	116
124	102	130	105
89	136	106	126
97	99		119

a. What type of design has been used in this experiment?

b. Construct an ANOVA table for the experiment.

c. Do the data present sufficient evidence to indicate a difference in the average scores for the four teaching methods? Test using $\alpha = .05$.

11.6

RANKING POPULATION MEANS

Many experiments are exploratory in nature. You have no preconceived notions about the results and have not decided (before conducting the experiment) to make specific treatment comparisons. Rather, you want to rank the treatment means, determine which means differ, and identify sets of means for which no evidence of difference exists.

One option might be to order the sample means from the smallest to the largest and then to conduct *t*-tests for adjacent means in the ordering. If two means differ by more than

$$t_{\alpha/2}\sqrt{s^2\left(\frac{1}{n_1} + \frac{1}{n_2}\right)}$$

you conclude that the pair of population means differ. The problem with this procedure is that the probability of making a Type I error—that is, concluding that two means differ when, in fact, they are equal—is α for each test. If you compare a large number of pairs of means, the probability of detecting at least one difference in means, when in fact none exists, is quite large.

A simple way to avoid the high risk of declaring differences when they do not exist is to use the **studentized range,** the difference between the smallest and the largest in a set of k sample means, as the yardstick for determining whether there is a difference in a pair of population means. This method, often called **Tukey's method for paired comparisons,** makes the probability of declaring that a difference exists between at least one pair in a set of k treatment means, when no difference exists, equal to α.

Tukey's method for making paired comparisons is based on the usual analysis of variance assumptions. **In addition, it assumes that the sample means are independent and based on samples of equal size.** The yardstick that determines whether a difference exists between a pair of treatment means is the quantity ω (Greek lowercase omega), which is presented next.

YARDSTICK FOR MAKING PAIRED COMPARISONS

$$\omega = q_\alpha(k, df)\left(\frac{s}{\sqrt{n_t}}\right)$$

where

$\quad k$ = Number of treatments
$\quad s^2$ = MSE = Estimator of the common variance σ^2 and $s = \sqrt{s^2}$
$\quad df$ = Number of degrees of freedom for s^2
$\quad n_t$ = Common sample size—that is, the number of observations in each of the k treatment means
$\quad q_\alpha(k, df)$ = Tabulated value from Tables 11(a) and 11(b) in Appendix I, for α = .05 and .01, respectively, and for various combinations of k and df

Rule: Two population means are judged to differ if the corresponding sample means differ by ω or more.

Tables 11(a) and 11(b) in Appendix I list the values of $q_\alpha(k, df)$ for α = .05 and .01, respectively. To illustrate the use of the tables, refer to the portion of Table 11(a) reproduced in Table 11.2. Suppose you want to make pairwise comparisons of $k = 5$ means with α = .05 for an analysis of variance, where s^2 possesses 9 df. The tabulated value for $k = 5$, $df = 9$, and α = .05, shaded in Table 11.2, is $q_{.05}(5, 9) = 4.76$.

TABLE 11.2

**A Partial Reproduction of Table 11(a) in Appendix I;
Upper 5% Points**

df	2	3	4	5	6	7	8	9	10	11	12
1	17.97	26.98	32.82	37.08	40.41	43.12	45.40	47.36	49.07	50.59	51.96
2	6.08	8.33	9.80	10.88	11.74	12.44	13.03	13.54	13.99	14.39	14.75
3	4.50	5.91	6.82	7.50	8.04	8.48	8.85	9.18	9.46	9.72	9.95
4	3.93	5.04	5.76	6.29	6.71	7.05	7.35	7.60	7.83	8.03	8.21
5	3.64	4.60	5.22	5.67	6.03	6.33	6.58	6.80	6.99	7.17	7.32
6	3.46	4.34	4.90	5.30	5.63	5.90	6.12	6.32	6.49	6.65	6.79
7	3.34	4.16	4.68	5.06	5.36	5.61	5.82	6.00	6.16	6.30	6.43
8	3.26	4.04	4.53	4.89	5.17	5.40	5.60	5.77	5.92	6.05	6.18
9	3.20	3.95	4.41	4.76	5.02	5.24	5.43	5.59	5.74	5.87	5.98
10	3.15	3.88	4.33	4.65	4.91	5.12	5.30	5.46	5.60	5.72	5.83
11	3.11	3.82	4.26	4.57	4.82	5.03	5.20	5.35	5.49	5.61	5.71
12	3.08	3.77	4.20	4.51	4.75	4.95	5.12	5.27	5.39	5.51	5.61

EXAMPLE 11.7

Refer to Example 11.4, in which you compared the average attention spans for students given three different "meal" treatments in the morning: no breakfast, a light breakfast, or a full breakfast. The ANOVA F-test in Example 11.5 indicated a significant difference in the population means. Use Tukey's method for paired comparisons to determine which of the three population means differ from the others.

Solution For this example, there are $k = 3$ treatment means, with $s = \sqrt{\text{MSE}} = 2.436$. Tukey's method can be used, with each of the three samples containing $n_t = 5$ measurements and $(n - k) = 12$ degrees of freedom. Consult Table 11 in Appendix I to find $q_{.05}(k, df) = q_{.05}(3, 12) = 3.77$ and calculate the "yardstick" as

$$\omega = q_{.05}(3, 12)\left(\frac{s}{\sqrt{n_t}}\right) = 3.77\left(\frac{2.436}{\sqrt{5}}\right) = 4.11$$

The three treatment means are arranged in order from the smallest, 9.4, to the largest, 14.0, in Figure 11.7. The next step is to check the difference between every pair of means. The only difference that exceeds $\omega = 4.11$ is the difference between no breakfast and a light breakfast. These two treatments are thus declared significantly different. You cannot declare a difference between the other two pairs of treatments. To indicate this fact visually, Figure 11.7 shows a line under those pairs of means that are not significantly different.

FIGURE 11.7

Ranked means for
Example 11.7

None   Full   Light
9.4   13.0   14.0

The results here may seem confusing. However, it usually helps to think of ranking the means and interpreting nonsignificant differences as our inability to distinctly rank those means underlined by the same line. For this example, the light breakfast definitely ranked higher than no breakfast, but the full breakfast could not be ranked higher than no breakfast, or lower than the light breakfast. The probability that we make at least one error among the three comparisons is at most $\alpha = .05$.

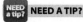
Many computer programs provide an option to perform **paired comparisons,** including Tukey's method, although *MS Excel* does not. The *MINITAB* output in Figure 11.8 shows its form of Tukey's test, which differs slightly from the method we have presented. The three intervals that you see in the printout marked "Lower" and "Upper" represent the difference in the two sample means plus or minus the yardstick ω. If the interval contains the value 0, the two means are judged to be not significantly different. You can see that only means 1 and 2 (none versus light) show a significant difference.

FIGURE 11.8

MINITAB output for Example 11.7

```
Tukey's 95% Simultaneous Confidence Intervals
All Pairwise Comparisons among Levels of Meal
Individual confidence level = 97.94%

Meal = 1 subtracted from:

Meal  Lower   Center  Upper   -----+---------+---------+---------+----
2     0.493   4.600   8.707                    (----------*-----------)
3    -0.507   3.600   7.707                   (----------*-----------)
                              -----+---------+---------+---------+----
                              -3.5      0.0       3.5       7.0

Meal = 2 subtracted from:

Meal  Lower   Center  Upper   -----+---------+---------+---------+----
3    -5.107  -1.000   3.107   (-----------*-----------)
                              -----+---------+---------+---------+----
                              -3.5      0.0       3.5       7.0
```

As you study two more experimental designs in the next sections of this chapter, remember that, once you have found a factor to be significant, you should use Tukey's method or another method of paired comparisons to find out exactly where the differences lie!

11.6 EXERCISES

BASIC TECHNIQUES

11.19 Suppose you wish to use Tukey's method of paired comparisons to rank a set of population means. In addition to the analysis of variance assumptions, what other property must the treatment means satisfy?

11.20 Consult Tables 11(a) and 11(b) in Appendix I and find the values of $q_\alpha(k, df)$ for these cases:

a. $\alpha = .05, k = 5, df = 7$
b. $\alpha = .05, k = 3, df = 10$
c. $\alpha = .01, k = 4, df = 8$
d. $\alpha = .01, k = 7, df = 5$

11.21 If the sample size for each treatment is n_t and if s^2 is based on 12 df, find ω in these cases:

a. $\alpha = .05, k = 4, n_t = 5$
b. $\alpha = .01, k = 6, n_t = 8$

11.22 An independent random sampling design was used to compare the means of six treatments based on

samples of four observations per treatment. The pooled estimator of σ^2 is 9.12, and the sample means follow:

$$\bar{x}_1 = 101.6 \qquad \bar{x}_2 = 98.4 \qquad \bar{x}_3 = 112.3$$
$$\bar{x}_4 = 92.9 \qquad \bar{x}_5 = 104.2 \qquad \bar{x}_6 = 113.8$$

a. Give the value of ω that you would use to make pairwise comparisons of the treatment means for $\alpha = .05$.

b. Rank the treatment means using pairwise comparisons.

APPLICATIONS

11.23 Swamp Sites, again Refer to Exercise 11.13 and data set EX1113. Rank the mean leaf growth for the four locations. Use $\alpha = .01$.

11.24 Calcium Refer to Exercise 11.15 and data set EX1115. The paired comparisons option in *MINITAB* generated the output provided here. What do these results tell you about the differences in the population means? Does this confirm your conclusions in Exercise 11.15?

MINITAB output for Exercise 11.24

```
Tukey's 95% Simultaneous Confidence Intervals
All Pairwise Comparisons among Levels of Method
Individual confidence level = 97.94%

Method = 1 subtracted from:

Method      Lower       Center      Upper
2         -0.0014377  -0.0008400  -0.0002423
3         -0.0001777   0.0004200   0.0010177
Method    --------+---------+---------+---------+
2              (-----*-----)
3                          (-----*-----)
          --------+---------+---------+---------+
             -0.0010    0.0000    0.0010    0.0020

Method = 2 subtracted from:

Method      Lower       Center      Upper
3          0.0006623   0.0012600   0.0018577
Method    --------+---------+---------+---------+
3                           (-----*-----)
          --------+---------+---------+---------+
             -0.0010    0.0000    0.0010    0.0020
```

11.25 Glucose Tolerance Physicians depend on laboratory test results when managing medical problems such as diabetes or epilepsy. In a uniformity test for glucose tolerance, three different laboratories were each sent $n_t = 5$ identical blood samples from a person who had drunk 50 milligrams (mg) of glucose dissolved in water. The laboratory results (in mg/dl) are listed here:

Lab 1	Lab 2	Lab 3
120.1	98.3	103.0
110.7	112.1	108.5
108.9	107.7	101.1
104.2	107.9	110.0
100.4	99.2	105.4

a. Do the data indicate a difference in the average readings for the three laboratories?

b. Use Tukey's method for paired comparisons to rank the three treatment means. Use $\alpha = .05$.

11.26 The Cost of Lumber, continued The analysis of variance F-test in Exercise 11.17 (and data set EX1117) determined that there was indeed a difference in the average cost of lumber for the four states. The following information from Exercise 11.17 is given in the table:

Sample Means	$\bar{x}_1 = 262.2$	MSE	41.25
	$\bar{x}_2 = 234.8$	Error df:	16
	$\bar{x}_3 = 251.6$	n_i:	5
	$\bar{x}_4 = 268.6$	k:	4

Use Tukey's method for paired comparisons to determine which means differ significantly from the others at the $\alpha = .01$ level.

11.27 GRE Scores The quantitative reasoning scores on the Graduate Record Examination $(GRE)^2$ were recorded for students admitted to three different graduate programs at a local university.

Graduate Program					
Life Sciences		Physical Sciences		Social Sciences	
630	660	660	760	440	540
640	660	640	670	330	450
470	480	720	700	670	570
600	650	690	710	570	530
580	710	530	450	590	630

a. Do these data provide sufficient evidence to indicate a difference in the mean GRE scores for applicants admitted to the three programs?

b. Find a 95% confidence interval for the difference in mean GRE scores for Life Sciences and Physical Sciences.

c. If you find a significant difference in the average GRE scores for the three programs, use Tukey's method for paired comparisons to determine which means differ significantly from the others. Use $\alpha = .05$.

THE RANDOMIZED BLOCK DESIGN: A TWO-WAY CLASSIFICATION

11.7

The *completely randomized design* introduced in Section 11.4 is a generalization of the *two independent samples* design presented in Section 10.4. It is meant to be used when the experimental units are quite similar or *homogeneous* in their makeup and when there is only one factor—the *treatment*—that might influence the response.

Any other variation in the response is due to random variation or *experimental error.* Sometimes it is clear to the researcher that the experimental units are *not homogeneous.* Experimental subjects or animals, agricultural fields, days of the week, and other experimental units often add their own variability to the response. Although the researcher is not really interested in this source of variation, but rather in some *treatment* he chooses to apply, he may be able to increase the information by isolating this source of variation using the **randomized block design**—a direct extension of the *matched pairs* or *paired-difference design* in Section 10.5.

In a randomized block design, the experimenter is interested in comparing k treatment means. The design uses *blocks* of k experimental units that are relatively similar, or *homogeneous,* with one unit within each block *randomly* assigned to each treatment. If the randomized block design involves k treatments within each of b blocks, then the total number of observations in the experiment is $n = bk$.

A production supervisor wants to compare the mean times for assembly-line operators to assemble an item using one of three methods: A, B, or C. Expecting variation in assembly times from operator to operator, the supervisor uses a randomized block design to compare the three methods. Five assembly-line operators are selected to serve as blocks, and each is assigned to assemble the item three times, once for each of the three methods. Since the sequence in which the operator uses the three methods may be important (fatigue or increasing dexterity may be factors affecting the response), each operator should be assigned a random sequencing of the three methods. For example, operator 1 might be assigned to perform method C first, followed by A and B. Operator 2 might perform method A first, then C and B, and so on.

To compare four different teaching methods, a group of students might be divided into blocks of size 4, so that the groups are most nearly *matched* according to academic achievement. To compare the average costs for three different cellular phone companies, costs might be compared at each of three usage levels: low, medium, and high. To compare the average yields for three species of fruit trees when a variation in yield is expected because of the field in which the trees are planted, a researcher uses five fields. She divides each field into three *plots* on which the three species of fruit trees are planted.

Matching or *blocking* can take place in many different ways. Comparisons of treatments are often made within blocks of time, within blocks of people, or within similar external environments. The purpose of blocking is to remove or isolate the *block-to-block* variability that might otherwise hide the effect of the treatments. You will find more examples of the use of the randomized block design in the exercises at the end of the next section.

THE ANALYSIS OF VARIANCE FOR A RANDOMIZED BLOCK DESIGN

11.8

The randomized block design identifies two factors: **treatments** and **blocks**—both of which affect the response.

Partitioning the Total Variation in the Experiment

Let x_{ij} be the response when the ith treatment ($i = 1, 2, \ldots, k$) is applied in the jth block ($j = 1, 2, \ldots, b$). The total variation in the $n = bk$ observations is

$$\text{Total SS} = \Sigma(x_{ij} - \overline{x})^2 = \Sigma x_{ij}^2 - \frac{(\Sigma x_{ij})^2}{n}$$

This is partitioned into *three* (rather than two) parts in such a way that

Total SS = SSB + SST + SSE

where

- SSB (sum of squares for blocks) measures the variation among the block means.
- SST (sum of squares for treatments) measures the variation among the treatment means.
- SSE (sum of squares for error) measures the variation of the differences among the treatment observations *within* blocks, which measures the experimental error.

The calculational formulas for the four sums of squares are similar in form to those you used for the completely randomized design in Section 11.5. Although you can simplify your work by using a computer program to calculate these sums of squares, the formulas are given next.

CALCULATING THE SUMS OF SQUARES FOR A RANDOMIZED BLOCK DESIGN, *k* TREATMENTS IN *b* BLOCKS

$$CM = \frac{G^2}{n}$$

where

$$G = \Sigma x_{ij} = \text{Total of all } n = bk \text{ observations}$$

$$\text{Total SS} = \Sigma x_{ij}^2 - CM$$

$$= (\text{Sum of squares of all } x\text{-values}) - CM$$

$$SST = \Sigma \frac{T_i^2}{b} - CM$$

$$SSB = \Sigma \frac{B_j^2}{k} - CM$$

$$SSE = \text{Total SS} - SST - SSB$$

with

T_i = Total of all observations receiving treatment i, $i = 1, 2, \ldots, k$
B_j = Total of all observations in block j, $j = 1, 2, \ldots, b$

NEED A TIP?

Total SS = SST +
SSB + SSE

Each of the three **sources of variation,** when divided by the appropriate **degrees of freedom,** provides an estimate of the variation in the experiment. Since Total SS involves $n = bk$ squared observations, its degrees of freedom are $df = (n - 1)$. Similarly, SST involves k squared totals, and its degrees of freedom are $df = (k - 1)$, while SSB involves b squared totals and has $(b - 1)$ degrees of freedom. Finally, since the degrees of freedom are additive, the remaining degrees of freedom associated with SSE can be shown algebraically to be $df = (b - 1)(k - 1)$.

These three sources of variation and their respective degrees of freedom are combined to form the **mean squares** as MS = SS/*df,* and the total variation in the experiment is then displayed in an **analysis of variance** (or **ANOVA**) **table** as shown here:

ANOVA TABLE FOR A RANDOMIZED BLOCK DESIGN, k TREATMENTS AND b BLOCKS

Source	df	SS	MS	F
Treatments	$k - 1$	SST	MST $= $ SST$/(k - 1)$	MST/MSE
Blocks	$b - 1$	SSB	MSB $= $ SSB$/(b - 1)$	MSB/MSE
Error	$(b - 1)(k - 1)$	SSE	MSE $= $ SSE$/(b - 1)(k - 1)$	
Total	$n - 1 = bk - 1$			

EXAMPLE 11.8

The cell phone industry is involved in a fierce battle for customers, with each company devising its own complex pricing plan to lure customers. Since the cost of a cell phone minute varies drastically depending on the number of minutes per month used by the customer, a consumer watchdog group decided to compare the average costs for four cellular phone companies using three different usage levels as blocks. The monthly costs (in dollars) computed by the cell phone companies for peak-time callers at low (20 minutes per month), middle (150 minutes per month), and high (1000 minutes per month) usage levels are given in Table 11.3. Construct the analysis of variance table for this experiment.

TABLE 11.3 ●

Monthly Phone Costs of Four Companies at Three Usage Levels

			Company		
Usage Level	A	B	C	D	Totals
Low	27	24	31	23	$B_1 = 105$
Middle	68	76	65	67	$B_2 = 276$
High	308	326	312	300	$B_3 = 1246$
Totals	$T_1 = 403$	$T_2 = 426$	$T_3 = 408$	$T_4 = 390$	$G = 1627$

Solution The experiment is designed as a *randomized block design* with $b = 3$ usage levels (blocks) and $k = 4$ companies (treatments), so there are $n = bk = 12$ observations and $G = 1627$. Then

$$CM = \frac{G^2}{n} = \frac{1627^2}{12} = 220,594.0833$$

$$\text{Total SS} = (27^2 + 24^2 + \cdots + 300^2) - CM = 189,798.9167$$

$$SST = \frac{403^2 + \cdots + 390^2}{3} - CM = 222.25$$

$$SSB = \frac{105^2 + 276^2 + 1246^2}{4} - CM = 189,335.1667$$

and by subtraction,

$$SSE = \text{Total SS} - SST - SSB = 241.5$$

These four sources of variation, their degrees of freedom, sums of squares, and mean squares are shown in the shaded areas of the analysis of variance tables, generated by *MINITAB* and *MS Excel* and given in Figures 11.9(a) and 11.9(b). You will find instructions for generating this output in the section "Technology Today" at the end of this chapter.

FIGURE 11.9 (a)

MINITAB output for Example 11.8

Two-way ANOVA: Dollars versus Usage, Company

Source	DF	SS	MS	F	P
Usage	2	189335	94667.6	2351.99	0.000
Company	3	222	74.1	1.84	0.240
Error	6	242	40.3		
Total	11	189799			

S = 6.344   R-Sq = 99.87%   R-Sq(adj) = 99.77%

FIGURE 11.9 (b)

MS Excel output for Example 11.8

Anova: Two-Factor Without Replication

SUMMARY	Count	Sum	Average	Variance
Low	4	105	26.25	12.917
Middle	4	276	69	23.333
High	4	1246	311.5	118.333
A	3	403	134.333	23040.333
B	3	426	142	26068
C	3	408	136	23521
D	3	390	130	22159

ANOVA

Source of Variation	SS	df	MS	F	P-value	F crit
Usage	189335.167	2	94667.583	2351.990	0.000	5.143
Company	222.25	3	74.083	1.841	0.240	4.757
Error	241.5	6	40.25			
Total	189798.917	11				

Notice that both the ANOVA tables show two different F statistics and p-values. It will not surprise you to know that these statistics are used to test hypotheses concerning the equality of both the *treatment* and *block* means.

Testing the Equality of the Treatment and Block Means

The *mean squares* in the analysis of variance table can be used to test the null hypotheses

H_0 : No difference among the k treatment means

or

H_0 : No difference among the b block means

versus the alternative hypothesis

H_a : At least one of the means is different from at least one other

using a theoretical argument similar to the one we used for the completely randomized design.

• Remember that σ^2 is the common variance for the observations in all bk block-treatment combinations. The quantity

$$MSE = \frac{SSE}{(b-1)(k-1)}$$

is an unbiased estimate of σ^2, whether or not H_0 is true.

- The two mean squares, MST and MSB, estimate σ^2 only if H_0 is true and tend to be unusually *large* if H_0 is false and either the treatment or block means are different.
- The test statistics

$$F = \frac{\text{MST}}{\text{MSE}} \quad \text{and} \quad F = \frac{\text{MSB}}{\text{MSE}}$$

are used to test the equality of treatment and block means, respectively. Both statistics tend to be larger than usual if H_0 is false. Hence, you can reject H_0 for large values of F, using *right-tailed* critical values of the F distribution with the appropriate degrees of freedom (see Table 6 in Appendix I) or computer-generated p-values to draw statistical conclusions about the equality of the population means.

TESTS FOR A RANDOMIZED BLOCK DESIGN

For comparing treatment means:

1. Null hypothesis: H_0 : The treatment means are equal
2. Alternative hypothesis: H_a : At least two of the treatment means differ
3. Test statistic: $F = \text{MST/MSE}$, where F is based on $df_1 = (k - 1)$ and $df_2 = (b - 1)(k - 1)$
4. Rejection region: Reject if $F > F_\alpha$, where F_α lies in the upper tail of the F distribution (see the figure), or when the p-value $< \alpha$

For comparing block means:

1. Null hypothesis: H_0 : The block means are equal
2. Alternative hypothesis: H_a : At least two of the block means differ
3. Test statistic: $F = \text{MSB/MSE}$, where F is based on $df_1 = (b - 1)$ and $df_2 = (b - 1)(k - 1)$
4. Rejection region: Reject if $F > F_\alpha$, where F_α lies in the upper tail of the F distribution (see the figure), or when the p-value $< \alpha$

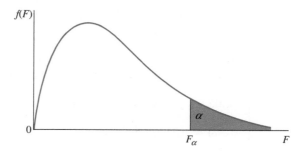

EXAMPLE 11.9

Do the data in Example 11.8 provide sufficient evidence to indicate a difference in the average monthly cell phone cost depending on the company the customer uses?

Solution The cell phone companies represent the *treatments* in this randomized block design, and the differences in their average monthly costs are of primary interest

to the researcher. To test

H_0 : No difference in the average cost among companies

versus the alternative that the average cost is different for at least one of the four companies, you use the analysis of variance F statistic, calculated as

$$F = \frac{\text{MST}}{\text{MSE}} = \frac{74.1}{40.3} = 1.84$$

and shown in the column marked "F" and the row marked "Company" in Figures 11.9(a) and 11.9(b). The exact p-value for this statistical test is also given in Figures 11.9(a) and 11.9(b) as .240, which is too large to allow rejection of H_0. The results do not show a significant difference in the treatment means. That is, there is insufficient evidence to indicate a difference in the average monthly costs for the four companies.

The researcher in Example 11.9 was fairly certain in using a *randomized block design* that there would be a significant difference in the block means—that is, a significant difference in the average monthly costs depending on the usage level. This suspicion is justified by looking at the test of equality of block means. Notice that the observed test statistic is $F = 2351.99$ with p-value $= .000$, showing a highly significant difference, as expected, in the block means.

Identifying Differences in the Treatment and Block Means

Once the overall F-test for equality of the treatment or block means has been performed, what more can you do to identify the nature of any differences you have found? As in Section 11.5, you can use Tukey's method of paired comparisons to determine which pairs of treatment or block means are significantly different from one another. However, if the F-test does not indicate a significant difference in the means, there is no reason to use Tukey's procedure. If you have a special interest in a particular *pair* of treatment or block means, you can estimate the difference using a $(1 - \alpha)100\%$ confidence interval.[†] The formulas for these procedures, shown next, follow a pattern similar to the formulas for the completely randomized design. Remember that MSE always provides an unbiased estimator of σ^2 and uses information from the entire set of measurements. Hence, it is the best available estimator of σ^2, regardless of what test or estimation procedure you are using. You will again use

NEED A TIP?

Degrees of freedom for Tukey's test and for confidence intervals are **error df.**

$$s^2 = \text{MSE} \qquad \text{with } df = (b - 1)(k - 1)$$

to estimate σ^2 in comparing the treatment and block means.

COMPARING TREATMENT AND BLOCK MEANS

Tukey's yardstick for comparing block means:

$$\omega = q_\alpha(b, df)\left(\frac{s}{\sqrt{k}}\right)$$

Tukey's yardstick for comparing treatment means:

$$\omega = q_\alpha(k, df)\left(\frac{s}{\sqrt{b}}\right)$$

[†]You cannot construct a confidence interval for a single mean unless the blocks have been randomly selected from among the population of all blocks. The procedure for constructing intervals for single means is beyond the scope of this book.

$(1 - \alpha)100\%$ confidence interval for the difference in two block means:

$$(\overline{B}_i - \overline{B}_j) \pm t_{\alpha/2}\sqrt{s^2\left(\frac{1}{k} + \frac{1}{k}\right)}$$

where $\overline{B}_i$ is the average of all observations in block i

$(1 - \alpha)100\%$ confidence interval for the difference in two treatment means:

$$(\overline{T}_i - \overline{T}_j) \pm t_{\alpha/2}\sqrt{s^2\left(\frac{1}{b} + \frac{1}{b}\right)}$$

where $\overline{T}_i$ is the average of all observations in treatment i.

Note: The values $q_\alpha(*, df)$ from Table 11 in Appendix I, $t_{\alpha/2}$ from Table 4 in Appendix I, and $s^2 = $ MSE all depend on $df = (b - 1)(k - 1)$ degrees of freedom.

EXAMPLE (11.10)

Identify the nature of any differences you found in the average monthly cell phone costs from Example 11.8.

Solution   Since the F-test did not show any significant differences in the average costs for the four companies, there is no reason to use Tukey's method of paired comparisons. Suppose, however, that you are an executive for company B and your major competitor is company C. Can you claim a significant difference in the two average costs? Using a 95% confidence interval, you can calculate

$$(\overline{T}_2 - \overline{T}_3) \pm t_{.025}\sqrt{\text{MSE}\left(\frac{2}{b}\right)}$$

$$\left(\frac{426}{3} - \frac{408}{3}\right) \pm 2.447\sqrt{40.3\left(\frac{2}{3}\right)}$$

$$6 \pm 12.68$$

NEED A TIP?

You **cannot** form a confidence interval or test an hypothesis about a single treatment mean in a randomized block design!

so the difference between the two average costs is estimated as between $-\$6.68$ and $\$18.68$. Since 0 is contained in the interval, you do not have evidence to indicate a significant difference in your average costs. Sorry!

Some Cautionary Comments on Blocking

Here are some important points to remember:

- A randomized block design should not be used when treatments and blocks both correspond to **experimental** factors of interest to the researcher. In designating one factor as a *block,* you may assume that the effect of the treatment will be the same, regardless of which block you are using. If this is *not* the case, the two factors—blocks and treatments—are said to **interact,** and your analysis could lead to incorrect conclusions regarding the relationship between the treatments and the response. When an *interaction* is suspected between two factors, you should analyze the data as a **factorial experiment,** which is introduced in the next section.

- Remember that blocking may not always be beneficial. When SSB is removed from SSE, the number of degrees of freedom associated with SSE gets smaller. For blocking to be beneficial, the information gained by isolating the block variation must outweigh the loss of degrees of freedom for error. Usually, though, if you suspect that the experimental units are not homogeneous and you can group the units into blocks, it pays to use the *randomized block design!*

- Finally, remember that you cannot construct confidence intervals for individual treatment means unless it is reasonable to assume that the *b* blocks have been randomly selected from a population of blocks. If you construct such an interval, the sample treatment mean will be biased by the positive and negative effects that the blocks have on the response.

11.8 EXERCISES

BASIC TECHNIQUES

11.28 A randomized block design was used to compare the means of three treatments within six blocks. Construct an ANOVA table showing the sources of variation and their respective degrees of freedom.

11.29 Suppose that the analysis of variance calculations for Exercise 11.28 are SST = 11.4, SSB = 17.1, and Total SS = 42.7. Complete the ANOVA table, showing all sums of squares, mean squares, and pertinent *F*-values.

11.30 Do the data of Exercise 11.28 provide sufficient evidence to indicate differences among the treatment means? Test using $\alpha = .05$.

11.31 Refer to Exercise 11.28. Find a 95% confidence interval for the difference between a pair of treatment means A and B if $\bar{x}_A = 21.9$ and $\bar{x}_B = 24.2$.

11.32 Do the data of Exercise 11.28 provide sufficient evidence to indicate that blocking increased the amount of information in the experiment about the treatment means? Justify your answer.

11.33 The data that follow are observations collected from an experiment that compared four treatments, A, B, C, and D, within each of three blocks, using a randomized block design.

Block	A	B	C	D	Total
			Treatment		
1	6	10	8	9	33
2	4	9	5	7	25
3	12	15	14	14	55
Total	22	34	27	30	113

a. Do the data present sufficient evidence to indicate differences among the treatment means? Test using $\alpha = .05$.

b. Do the data present sufficient evidence to indicate differences among the block means? Test using $\alpha = .05$.

c. Rank the four treatment means using Tukey's method of paired comparisons with $\alpha = .01$.

d. Find a 95% confidence interval for the difference in means for treatments A and B.

e. Does it appear that the use of a randomized block design for this experiment was justified? Explain.

 11.34 The data shown here are observations collected from an experiment that compared three treatments, A, B, and C, within each of five blocks, using a randomized block design:

Treatment	1	2	3	4	5	Total
			Block			
A	2.1	2.6	1.9	3.2	2.7	12.5
B	3.4	3.8	3.6	4.1	3.9	18.8
C	3.0	3.6	3.2	3.9	3.9	17.6
Total	8.5	10.0	8.7	11.2	10.5	48.9

MS Excel output for Exercise 11.34

Anova: Two-Factor Without Replication

SUMMARY	Count	Sum	Average	Variance
A	5	12.5	2.5	0.265
B	5	18.8	3.76	0.073
C	5	17.6	3.52	0.167
1	3	8.5	2.833	0.443
2	3	10	3.333	0.413
3	3	8.7	2.9	0.79
4	3	11.2	3.733	0.223
5	3	10.5	3.5	0.48

ANOVA						
Source of Variation	SS	df	MS	F	P-value	F crit
Rows	4.476	2	2.238	79.929	0.000	4.459
Columns	1.796	4	0.449	16.036	0.001	3.838
Error	0.224	8	0.028			
Total	6.496	14				